Pendulums of Time

Pendulums of Time

Volume One: END OF MAN

Fredrick James Terriere

To order additional copies of this book, contact:
Xlibris Corporation
1-888-795-4274
www.Xlibris.com
Orders@Xlibris.com
98156

Contents

I would like to thank all those who helped in the formation of this book, but there are too many to list all of them. A few special people need mention though, Cat, Brett, Ryan, Carrie, Kit, Mom, and Dad. You know what you have given. Xlibris, thanks for your assistance as well.

Author's notes. Many of you will note that the sections in the beginning are quite broad and detailed. We can fully examine the past at our leisure, adding a little here or subtracting there. Many lives have been spent accumulating information. They past is a mountain of data. As we approach the present, there is less time spent, fewer people attacking the information overall. The present is a jumble of rapid intersections of information that we can not reflect on until it is well past us. This is why the sections at the end of the book become less detailed, more vague. As Adam says, the Heisenberg priciple applies to time as well as spatial oddities. The closer we get, the less we know about all that occurs around us. The flow of information is too great. That is what imagination is for.

Prelude:

The Cycles and Scales of Existence

0

000000

Poseidon, Neptune to the Romans, god of the seas, rarely saw fit to calm his domain. Rather, he delighted in the tempestuous, sudden, and violent rages of storm. He pursued with vigor those that demeaned his realm or slighted his powers. Ancient sailors shipped warily, mouthed few obscenities in his direction. His great trident could spear a ship as easily as a man could spear a fish.

He would send, with but a single command, great gyres of water, spouts like tornadoes, or tremendous waves, rogues of green seawater to subdue the impious sailor and landsman alike. Storms would appear to strike with electric bolts of dazzling blue with a smell of ozone and thunderous staccato claps of thunder to emphasize one's plight. Clouds would roil and boil and send volumes of sheeting rain to drench mortal fools.

The Norsemen bowed before Aegir whose very name meant *sea*. It was his mate Ran whom they feared, for she craved humans, both their bodies and souls. She was the robber of the deeps, a caster of nets to drag them down. She brought forth nine daughters to assist her in the forms of massive cold green waves. Ship's timbers groan and creak, masts and spars snapping in great twists of force. They founder under swells of immense size and weight of tumbling briny, icy water.

Stalwart men battling, failing, finally cursing the gods and crying for mercy are swept overboard to freeze in the great depths or feed the sharks. Davy Jones's locker, or Ran's perhaps, gets new visitors. These daughters foundered ships by the score so that the ghosts of the sailors were anchored down in the depths to dwell forever in the sunless abyss.

Odin himself, the highest god of the Norse, knowing the power of water, sacrificed one of his eyes for the taste of the waters from a well of wisdom only to realize it brought an undying thirst for more wisdom. In water, there is power and life.

The ultimate watery disaster occurred in Genesis where it rained for forty days and nights. Jehovah destroyed everything, but Noah and his family and the animals in the ark for the entire Earth was covered in water (yea, even the highest mountains). The stories of watery death are many and powerful, remnants of a fearful, ignorant past.

*

The desire of knowledge, like the thirst of riches, increases ever with the acquisition of it.

—Laurence Sterne

*

Jim, much like the growing horde of homeless, some 30 percent total for the North American Region, wandered along the beachfront near the elite playland of Coney Island. He rooted through the various recycle receptacles one by one; he found a corn dog, left by the child of a wealthier bureaucrat's family that had a few nibbles left. The flies did not bother him; he was way past that now. He carefully licked a few globs of crusty, slightly moldy catsup and mustard off the recyclable pseudo paper, meaning vegetable fiber wrapper.

He had been foraging through trash receptacles in the dank, slimy back alleys of New York, Jersey, and Manhattan for nearly ten years since 2055 when he lost his position in the Social Economic Business Bureau.

He had made a severe error in judgment. He had made a speech to the board's overseers suggesting that, according to his math, they had entered into the economic downturn, really a flaming spiral, because they had chosen tactics that were too socialistic (*giving* was the word he used) one did not denigrate socialism unless one wished to be shunned. He had urged that they rescind or, at least, loosen some of the more detrimental edicts, easing in some capitalistic flavor to juice up the system though he had never used the politically incorrect *C* word. He had simply said that personal benefits, remuneration, or pay to the less politically correct had to increase at the expense of the government run businesses in order to increase production and quality without substantially increasing overall costs. It was a complex set of equations that linked increased pay to quality, durability of goods, and increased production. He used a great many simple analogies and historical references to back up his mathematical ruminations.

The slightly raised eyebrows and cool stares should have warned him of any further gaffes, but his misinterpretation of said eyebrows (namely that they were interested and not offended) landed him in the worst of the two possible positions. They had asked him whether he believed that the worst problems were recent additions. He had affirmed that, for no one questioned the overall long-term socialistic implications. That would be a career killer even though he knew the whole thing (socialism) was a problem. Jim knew that consumption required production. If people don't produce some way, any way, consumption gets severely skewed.

That, of course, was his great undoing since most of these men, none of them economists, had created the very acts he was complaining about. That was made abundantly clear to him in extremely short order. He could not backtrack now; it would make him out to be a liar (whom, to some degree, he was) or a sycophantic ass kisser (which he also was but did not wish to appear so). Neither would allow him to get his job back, the one he had just unexpectedly and oh so unintelligently handed to them on a recyclable vegetable fiber platter. One of their like-minded cronies, unschooled in anything but "toady-ship," but a politically correct crony, would be cajoled into service now. They would hear what they wanted to hear, not necessarily the reality of the situation. It is the nature of government to grow certain fungi.

No more fine wool suits for Jim. Not one piece of his clothing now was whole or even his originally. Tattered and rent, filthy to the point of undeterminable color, they hung loosely from his gaunt frame. His bony figure could have been an old-fashioned Halloween skeleton display. The knuckles in his fingers were oversized, swollen; his ribs protruded as did his hips.

The fine leather shoes he once wore were long gone. His mismatched, worn through veg-fiber shoes were of different sizes and covered sockless, foul-smelling feet. Lice and fleas peeked from behind matted brown and grey hair and unkempt beard. Rough, filthy, wrinkled skin, browned from the sun and dirt, hung in loose wiggly folds from his arms, and his belly was taut from general hunger.

Jim smelled none too good since his regular pastime involved what had once been called dumpster diving and was now known as surviving. Passersby recoiled not just from the sight of him or from others like him. The fly attracting miasma assaulted the olfactory senses of those who could afford to bathe more regularly, generally once a week for the average person; the Society Water Conservation Board had strict personal usage limits. The odor no longer much impinged upon his now less delicate senses. The tumultuous ocean, the direction he now headed, provided what little bathing he occasionally felt he needed.

He meandered a little on the way, weaving a little from the illicitly produced booze he had consumed earlier. Jim had traded a few slices of only slightly maggot-covered pizza rescued from the trash behind a severely rundown Society's People's Pizza for several swigs from a stolen bottle of underground-made tipple. Beer and wine were still legal under certain conditions, namely money, specific locations, and politically correct situations. But planetary law, the UE, had deemed hard liquor a nonessential for the human condition. The production of distilled alcohols took too much energy, used valuable food sources, and produced unwanted social results, or so the governmental line went. Jim was certain beer and wine would soon go, except for the Catholics who needed it for religion or some such he had heard in his youth. Jim had no religious beliefs anymore, not since his early teens. All he felt now was the need to struggle forward for some indefinable reason. So he plodded onward towards some unviewable, perhaps imaginary goal.

When he reached the bubbling, foaming waters, littered with refuse, he removed his greasy, odiferous shoes, aromatic jacket, stained shirt, and grime-encrusted pants at the waterline and waded into the cascading curls of the Atlantic in his dirty boxers. He would wash his clothes later, for some reason or other. Squatting, he vigorously rubbed at the encrusted filth of a few weeks' worth of dumpster diving and sleeping in alleys. His yellowish brown cracked nails, due to a treatable fungus and few manicures, dug in. His fingerprints could have been traced by the dirt embedded between the ridges.

The sandy waters provided a natural abrasive so that scabs tore off, releasing a new trickling flow of new bright red. Legs and arms received an equal share of care. Toes, with nails matching his fingers, found time to churn about in the sand in their cleansing ritual.

Dipping his rancid hair into the foaming waters, he massaged his scalp and face, pulling tangled clumps of hair away in his cleansing effort. His efforts drew derisive comments from the blue-green-bereted UE soldiers patrolling the beach close by.

"Hey, you, bum, this isn't a bathhouse," called one body-armored man. The sky blue armor covered the soldier's torso as well as upper arms and legs. It fit like a flexible, tightly woven turtleneck sweater that could prevent penetration of shrapnel but was beefed up in critical areas, such as the chest, with half-inch-thick neo-Kevlar. This golf course green—edged padding would block a knife as well as stop a twenty-millimeter round at point-blank range.

The similarly garbed sergeant ordered Jim to get out, saying, "This beach is for decent people, not derelicts. You need to move on, now!"

Jim had been rousted many times before. He simply stood erect, dripping, and began to comply. His first step was forestalled by a sudden outrushing of the waters about his knees and ankles. Jim nearly tumbled in his physically weakened state because of the sudden tremendous undercurrent. Turning slowly away from the soldiers, he watched the water quickly disappear back some two hundred meters, three hundred and more, from the normal waterline. Cavity-infested mouth agape, body dripping, he watched crabs scurrying for cover and fish flopping in the sudden absence of their natural environ.

"What the bloody . . ." a confused beefy female soldier began in an alarmed contralto, also staring at the most unusual view. She had just arrived from London not two weeks prior. Jim thought, in one of those funny corners of one's mind where strange things flit, her accent was quite delightful.

The silence that followed for but a few moments was not natural. The gulls were silent, not even present. Again, in a quick, almost unrecognized operation of the mind, Jim saw, unbelievingly, a shift in the surface of the ocean far out on the horizon. It was a sudden rising along the whole horizon, as if a great long beast, a sea monster, was coming parallel to the shallows and causing the surface of the blue green sea to ride up upon its shoulders.

In the next instant, there was a terrible ripping, roaring, churning sound louder than ten thousand deep heathen drums doing battle. Water reached a crescendo hundreds, if not a thousand meters, higher than a high C note. It vibrated the air like every old ell train on their passing tracks combined. The pressure dropped then suddenly peaked just in front of the rapidly intruding, foaming wet green terror.

Jim was not a very learned man in terms of science or religion, but he instantly recognized his uttermost end. He didn't even think *Oh shit*. His heart palpitated rampantly deep in the womb of his chest. The heat of his blood drained as quickly as hope. His breath stilled, and his limbs trembled uncontrollably, yet he stood his ground, accepting his unchangeable fate.

The foolish soldiers ran screaming for no logical reason. The onrush of liquid death at many hundreds of meters per hour far outpaced them. Their loss of mental reasoning left them only fragmentary moments of fearful insanity as their bitter untimely end, that and the smell coming from their underpants.

Jim mused oh so quickly on beginnings and endings. His Christian-oriented parents had provided him with some basic framework in that direction. A college friend had tried to introduce him to Islam. His world history courses had included some various mythologies. The required electives, an oxymoron if ever there was one, in the sciences, filled in the gaps. All these random thoughts, a multichotomy of whys, flitted by in less than the few remaining seconds that he had. His thoughts ran in parallel, polarized between science and religion. He heard the voices of his professors, parents, and friends, voicing their perspectives as if in a debate.

The Quantum Era:

When we backstroke in time, to a point one-million-trillion-trillion-trillionth (10^{-43}), of a second of the age of the universe, we find that we approach an implacable and impenetrable wall known as the Plank barrier. This is as close to time zero as science can reach. The density is 10^{94} grams per cubic centimeter (a one followed by ninety-four zeroes). The gigantic mass of the known universe is contained in the volume of a nucleon—a proton or neutron. The Plank temperature is an astounding 10^{32} degrees (a one with thirty-two zeroes), measured in kelvin units. Anything beyond this is chaos, heaps of discontinuities that are currently unexplainable.

"In the beginning God created the heavens and the earth." (Gen. 1:1)

"God it is who raised the heavens without columns that ye can see." (*Qur'an*, Chapter of Thunder)

"Nonviolence and truth (*satya*) are inseparable and presuppose one another. There is no god higher than truth."
(Mohandas Karamchand "Mahatma" Gandhi, 1939)

The Quark Era:

Quarks have been classified into some odd, rather amusing categories. There are six flavors: up, down, strange, charm, top, and bottom (or truth and beauty for the last two). Each flavor then has three colors. These strange objects bring us to the next level.

"Now the earth proved to be formless and waste and there was darkness on the surface of the watery deep." (Gen. 1:2)

The Hadron Era:

Hadrons are more familiar to the average person. These include the protons and the neutrons of an atom—the nucleons. The forces interacting are the electromagnetic, weak, and strong. We are at a point before ten thousandths of a second after time zero.

The Lepton Era:

The Lepton Era is from one ten thousandth of a second to one second after time zero. It is the period in which electrons, positrons, and neutrinos come into existence. The start of the era has a temperature of one trillion degrees and cools to ten billion at the end.

The density is roughly one thousand tons per pinky tip. For every nucleon, there is one billion of each of photons, electrons, and neutrons.

The Radiation Era:
The expanding universe passes through this time frame, one second to around one million years after time zero, in a cosmic flash. Here is the production of primordial helium. At this time, electrons deflect light so that the radiation acts like a bizarre fluid. Radiation decouples from matter. Between one hundred thousand and one million years after time zero, the universe is hot and has a yellowish color. After this time, light can move freely throughout the universe.

"And God proceeded to say: 'Let light come to be.' Then light came to be." (Gen. 1:3) "And there came to be evening and there came to be morning, a first day." (Gen. 1:5)

Ten Million Years of Age:
The temperature has dropped to a comfortable three hundred kelvin. Galaxies are in a fetal stage. Black holes exist in a multitude of sizes.

"And God went on to say: 'Let an expanse come to be in between the waters and let division occur within the waters and the waters.'" (Gen. 1:6) "And there came to be evening and there came to be morning, a second day." (Gen. 1:8)

One Hundred Million to One Billion Years:
Galaxies form as stars are born amidst the swirling gases concentrated by black holes.

Five Billion Years of Age:
The solar system begins to take shape. The planets follow suit as they also form from masses of gas produced by the primal stars that formed heavier elements. This takes a few billion years to accomplish.

"And God went on to say: 'Let the waters under the heavens be gathered into one place and let the dry land appear.' And it came to be. And God began calling the land Earth and the waters seas." (Gen. 1:9-10) "And God went on to say: 'Let there be luminaries in the expanse of the heavens to divide between day and night serving as signs for days, years seasons.'" (Gen. 1:14) "And there came to be evening and there came to be morning, a fourth day." (Gen. 1:19) "And [He] subjected the sun and moon each one runs to a stated and appointed time. And He it is who has stretched out the earth and placed therein firm mountains and rivers." (*Qur'an*, Chapter of Thunder)

Three Billion Years Ago:
The known universe is somewhere between twelve and fifteen billion years old now. We now shift to a different time perspective, ages past instead of age from the start. This is done to correlate with geologic time as opposed to cosmic time. In this time, the first life forms appear on Earth—bacteria.

Seven Hundred Million Years Ago:
Give or take a few tens of millions of years, invertebrates, soft-bodied creatures, appeared on Earth.

450 Million Years Ago:
Fishes appeared explosively on the scene.

"And God went on to say: 'Let the waters swarm a swarm of living beings.'" (Gen. 1:20) "And there came to be evening and there came to be a morning, a fifth day." (Gen. 1:23)

425 Million Years Ago:
Land plants began to spread rapidly across the globe.

"And God went on to say: 'Let the earth cause grass to come forth, all vegetation bearing seed, fruit trees upon the earth, all according to their kinds.'" (Gen. 1:11) "And there came to be evening and there came to be morning, a third day. (Gen. 1:13) "And gardens of grapes and corn and palms growing together." (*Qur'an*, Chapter of Thunder)

Four Hundred Million Years Ago:
Amphibians crawl out of the primal seas to begin the colonization of land by moving creatures.

320 Million Years Ago:
The dominance of reptiles and, eventually, dinosaurs begins.

230 Million Years Ago:
Mammals appear on the scene, skulking underneath the weight of the dominant landforms—dinosaurs.

Two Hundred Million Years Ago:
Birds, the genetic descendants of the raptor form of the dinosaur, surge in numbers and in great diversity.

"And God went on to say: 'Let the land put forth living creatures according to kinds, domestic and moving and wild beast of the earth according to its kind.' And it came to be so. God proceeded to make the beasts of the earth to their kind. And God saw that it was good." (Gen. 1:24-25) "And there came to be an evening and there came to be a morning, the sixth day." (Gen. 1:31)

"And let flying creatures fly over the earth in the heavens." (Gen. 1:20)
"And there came to be evening and there came to be morning, a fifth day." (Gen. 1:23)

Sixty-Five Million Years Ago:

The calamitous end of the dinosaurs leads to the rise of mammals as the preeminent form of life.

Two Million Years Ago:

The *Homo* genus appears in Africa. It is the direct lineage of man. First is *Homo habilis*; then around one million years in the past, *Homo erectus* evolves.

Three Hundred Thousand Years Ago:

Archaic *Homo sapiens* develop. They wend their way forward to a time Neanderthals are found around one hundred thousand years ago.

Thirty Thousand Years Ago:
Modern *Homo sapiens* appears on the scene in the last blink of a cosmic eye. The last competing different species, Neanderthals, passes into oblivion around twenty-five thousand years ago. Perhaps this is why we fight so amongst ourselves; we have no closer enemy than the apes, and they have no technology equal or even remotely threatening to ours.

"And God went on to say: 'Let us make man in our image . . . and let him have dominion.'" (Gen. 1:26) "And God proceeded to make the man out of dust from the ground." (Gen. 2:7) "When we have become dust, shall we really then be created anew?" (*Qur'an*, Chapter of Thunder) "And there came to be evening and there came to be morning, the sixth day." (Gen. 1:31)

Jim lifted his still dripping chin resolutely. His mind swung to and fro as a pendulum, vacillating from left to right, front to back, and even above to below. His last comparative thoughts concerned reasons, not necessarily science or religion though.

Great high-jumping Jesus, he wondered, wavering between fear and awe as the wave approached to within meters of his position. "What have we gone an' done to ourselves now?"

*

Planet Earth, that wondrous blue, green, and brown, slightly off round, pear-shaped globe, has many peculiar motions that we do not notice or even care about in the course of our monotonous daily rituals.

Neither does the average person think about, nor does one care about such things as axial tilt, rotation, revolution, the size of the galaxy, and so forth as one drives to work, applies makeup, makes a call, or flushes a toilet. We are far too busy with our "important" business. The closest we get to axial tilt is when we raise the slope on our walking machines since we can't go for a real walk; rotations and revolutions may only concern us when we are break dancing; as for flushing a toilet, it's a waste of time to watch which direction the water swirls and tells us we are in the Northern or Southern Hemisphere.

The early humans believed that their world was only that which was immediately about them. The world at large did not exist beyond what they could see and touch. It had to be within reach. Distant goings on did not enter into their understandings; it would become the realm of more powerful entities—gods.

This became the world of the supernatural, the metaphysical, which is beyond the nature they felt and lived within. They created deities to explain what could not be explained rationally by their scientifically limited minds. It would take many millennia to begin to chip away at that and establish a scientific factual basis for the winds, stars, and even life itself.

The early belief that everything moved about Earth, the geocentric concept, has long since been dispelled. Galileo, born in 1564 CE, with his first glimpses of the moons of Jupiter in 1610 CE, had to fight the church's dogma. The pope ordered him to stop supporting the Copernican system that claimed Earth was not the center of the universe. Galileo's views, even with his forced recanting, eventually won science and are now recognized as fact even by the church. His discovery of four moons orbiting Jupiter, the Galilean satellites, lent a great deal of supporting evidence. Galileo added so much more to mathematics and physics including the understanding that all objects fall at the same rate. The legend goes that he dropped these objects from the Leaning Tower of Pisa.

A heliocentric, or sun-centered system, temporarily replaced the geocentric one. Why not? It was the life giver, the Father or Mother figure to many early cultures. The moons of Jupiter revolved around it; why not all else around the great fire in the sky? This too faded with time and as knowledge accumulated.

Earth rotates on its axis, the north-south spin, in what was once believed to be the motion of the planets and stars, to produce our day-and-night cycle. This rather simple motion has produced so much mysticism. Religions grew up around the celestial happenings that occurred with regularity, solstices, equinoxes, eclipses, and even sunrise and sunset. Many sites throughout the world contain shrines of celestial significance, a mix of underlying scientific understanding and mysticism.

Further, Earth revolves in an elliptical (though nearly circular) path around Sol, our sun, to produce the well-known cycle of the year, also a celebrated religious time. This ellipse, along with some axial tilting of about twenty-three degrees, gives the variations of the seasons.

Our star Sol, a midsized average one, bobs up and down like a slow yo-yo on the finger of a person walking in a very large circle so that it continually follows a sinusoidal path, crossing the galactic plane. This plane is like an imaginary tabletop through which we pass. Roughly every seventy million years, our solar system switches from north to south, or vice versa, with respect to the imaginary flat surface of the galactic plane.

Some scientists link this action to waves of asteroid and comet impact events throughout the long geologic history of Earth. This makes some sense since there is a greater collection of space junk, rogue asteroids, and comets as well as dust near the galactic plane. There are greater gravitational disturbances there due to that increase in mass that could perturb the orbits of asteroids or comets in our distant asteroid belt or the Kuiper comet reservoir. The scientists have tried to match these events to various extinctions though the timing is not always accurate. It need not be since the shift in the orbits of these objects does not have to lead to immediate destruction. Even the slightest disturbance can multiply over time to wreak havoc in the distant irregular future.

As Sol moves in this up and down manner, our system also revolves around the galactic core once every 250 million years, hence the large circle of the yo-yoed system. Much like the water in a drain, with each molecule of water representing a stellar system, it circles in an ever flowing dance, closing on the center of the galaxy. Our solar system is nudging ever closer to that intensely dangerous center. Lurking there is a giant black hole that is gobbling up matter at an ever increasing rate, much like a drain takes the water who knows where.

Our position is on the edge of one of the graceful multicolored arms of our spiral galaxy—the Milky Way. We are about two-thirds of the way out from the center, certainly not at the center of things as posited by the likes of philosophers like Aristotle or most religions. Ours is not an affluent system nor the farthest backwater dregs of the galaxy.

Between us and the center of our somewhat barred spiral whirlpool mix of a galaxy are four arms, windings of the stars. We ride in the Orion Spur (like the spur of a rooster) that is attached to the Sagittarius Arm. When we look toward the center, we must peer through this arm first. The next arm inward is the Scutum-Centaurus Arm, onward and inward to greater brightness with the Norma Arm followed by the Near 3KPC Arm.

Our average-sized barred spiral galaxy contains around one hundred billion stars. That's a one followed by eleven zeroes. Some enormous elliptical galaxies are ten times the size of ours. Others, like the Magellanic Clouds, two very small odd-shaped clouds of stars very near our own galaxy, are tiny in comparison. These clouds may in fact be the remnants of a galaxy consumed by our larger one.

The Milky Way galaxy is similar in shape to two sunny side up eggs placed back to back. The yolks of the eggs represent the core while the whites are the disk. If we had a large enough measuring stick, we would find that the Milky Way is 1 x 10^{18} kilometers across, a one followed by eighteen zeroes! Imagine what that would do for one's long multiplication or division. That is nothing compared to the larger picture.

When astronomers began to view the vast expanses with improved instruments, little smudges thought to be anything from gas clouds to stars forming came to be recognized as distinct, separate galaxies in their own right. Even then, the stunned astronomers had to expand their understanding from a galaxy-centered view to an unimaginably large universe of unnumbered and possibly innumerable galaxies. Size is important. Ever we must expand our perspective to include the previously unimaginable.

Our closest large neighbor galaxy, ignoring the Magellanic Clouds that dance on the gravitational boundaries of our own, Andromeda is 2.2 million light-years away, twenty-two quintillion kilometers (that's the number 22 followed by eighteen zeroes). This makes it more than twenty times as far away as it is across our galaxy. This is a stupendous, mind-boggling distance. It would take two and a half million,

million years to drive there in an automobile traveling at one hundred kilometers an hour, with no rest breaks. Are we there yet?

There are billions upon billions of galaxies out there. Some are small globs of stars in an elliptical shape while others of similar shape are ten times larger than our own. There are spirals and irregulars each containing billions of stars of various hues and sizes. Some stars are not much larger than the planet Jupiter. Others, like Betelgeuse and Rigel in Orion, are larger than the orbit of Venus about our sun. Yet in only one place are we sure that life exists.

What is life? Where did it come from? Did a god create us as many religions claim? Are we the product of chance, of evolution, a roll of some strange multisided cosmic dice? Did we come from simple chemical reactions that led to more complex reactions? Did these complex reactions lead to the development of simple amino acids, undirected repetitive reactions, and then to proteins? Did all of this combine in a unique fashion to form cells and then multicellular creatures? Us? Is it possible?

Very difficult questions lie upon even more obscure, deeper questions. Scientists can only hypothesize the conditions of our primordial Earth. We shall never be 100 percent certain. Where do the answers come from? How does humanity expect to define itself and glean meaning amid this immenseness and minutia? Religion would offer a certainty defined as faith in God while science offers probabilities, faith in mathematics, assumptions at the most basic level that we can build upon.

What life appears to be is a combination (a duality) of matter and energy. As the only life-bearing place that we are certain of, our comparisons are admittedly deploringly limited. Science defines life as having the ability to use energy, produce waste, grow, respond to the environment, and, most importantly, to reproduce.

On Earth, everything seems to be made with a carbon backbone, the matter side of the coin. This fuels a scientific viewpoint, a dogma mind you, that life must be carbon based due in part to carbon's very utilitarian chemical properties. This bias may blind us to life forms that exist with other chemicals as the basis; we may recognize them or not. But what is carbon?

Carbon is an element that contains six protons, six electrons, and, usually, six neutrons. An element may contain different numbers of neutrons and yet still be the same element. These are called isotopes. Yet where do these aforementioned particles come from? Are there basic building blocks, or does the tiny continue downward forever just as the largest scale appears to do?

Physics, the application of pure mathematics to describe our world, is our most basic scale of science; it deals with the cycles and scales of size and existence that involve subatomic particles or energy packets that pop in and out of reality in millionths of a second. As well as the tiniest, physics, in the form of cosmology, deals with the largest arena—our universe. Forces, in various forms, originally delineated clearly by Sir Isaac Newton, are the name of this game. Bohr, Rutherford, Einstein, Hawkings, and so many others used this foundation to usher in modern physics.

Forces come in many guises, and each must be considered when trying to describe the universe in its entirety, at least for now.

In physics, quantum mechanics describes, somewhat absurdly to the average person and as best it can, that subatomic particles may exist in more than one place at a time but only at that level. Anything larger is relegated to a single location by its interactions with other matter. Humans have nothing at our level of existence to make analogies. Energy and mass are the same down there, one acting as both waves and particles (or wavicles). What we call light acts as a wave, much like those we see on the ocean. The swells, or peaks, and troughs, the lower part of a wave, are a good model. However, light incongruously acts as a particle in other circumstances and has been called a photon. Score 1 for Gene Roddenberry and his epic *Star Trek* for recognizing how important this terminology is, nothing like a good photon torpedo when needed.

At our level, there is matter that we touch, or so we think. In actuality, we never touch anything. The repulsive forces between electrons in the extremities of atoms prevent that. Energy that we use is in an alternative form, the opposite side of the coin, yet linked in Einstein's famous $E=mc^2$ for all to see. The atom bomb is a direct result of that duality in nature. In microseconds, we convert small amounts of mass into voluminous quantities of energy. With the hydrogen bomb, we became starlike.

These amusing small unseen but measurable contrivances of energy at the lowest levels interact to form what are known as quarks and gluons. From these we see the development of more familiar items: protons, neutrons, and electrons, which form even more known objects—atoms. Many school-age children can picture an atom as being similar to a solar system, somewhat incorrectly, but a valuable analogy nonetheless.

The nuclear radius (the distance from the center to the edge) ranges from around one to ten fm (femtometers), which is 0.000000000000001 meters (1 x 10^{-15} m) to 0.00000000000001 (1 x 10^{-14} m). Remember that a meter is about the same as a yard in the old English measurements. The electrons are about one hundred thousand fm, or 0.1nm (1nm=1 x 10^{-9} m or 0.000000001 m), from the nucleus, making the atom about twice this size in diameter (the distance across).

In comparison, more complex molecules are around a thousand times larger than an atom. A person is ten million times larger than these molecules. Earth is generally the same proportion, again larger than a person. The diameter of Sol, our sun, is about two hundred times that of Earth. Our Milky Way Galaxy is one trillion (one thousand billion) times greater in size than Sol. Groups of galaxies, called clusters, are hundreds and thousands of times larger than the galaxies within them. Superclusters can contain hundreds of clusters. The observable universe is thousands of times larger than the largest supercluster. In the final comparison, the known universe is at least 1 x 10^{42} (that's a one followed by forty-two zeroes) times larger than an atom. Where does this put us? Humans are on the small end of things. Does that mean we are inconsequential?

Let's return to atomic and molecular level. Here chemistry is king. There is no life here either, as we define it. But all the interesting building blocks are amino acids, proteins, methane (CH_4), diatomic oxygen (O_2), carbon dioxide (CO_2), and so much more. It is here that protons and neutrons form nuclei of atoms. Electrons surround the nuclei in somewhat vague orbitals, meaning we don't know exactly where the electrons are at any given time or how fast and what direction they are moving.

Chemistry deals with electrons moving from one atom to another, or being shared, creating molecules. There is never an exchange of protons or neutrons in a regular chemical reaction. That type of operation is reserved for nuclear reactions. This flow of electrons occurs in our bodies as well as in our homes as electricity. These molecules join to form larger more complex molecules that lead to the next level.

That monumental step up the ladder of scale involves living organisms. This is a collection of interacting (and one might say cooperating) molecules joined to form cells. Almost alive are the viruses. Bacteria, single-celled organisms, are the smallest recognized living thing under our human devised classification system.

These rules of classification include cells, organization, growth, adaptation, response to environment, consumption of resources, excretion, use of energy, and reproduction. Death is the unmentioned and necessary associate of life. Birth, life, procreation, death, and, maybe, taxes are the meter sticks—the measure of life—at this important level.

From the simplest cell to multicellular thinking beings (sometimes), all are included. Some believe that just as cells make up us, we compose a greater being, a group body and mind so to speak. This philosophical, even religious view is what some term a Gaiaistic perspective in which life exists in all matter, even energy, and at all levels. We are, in fact, a living Earth. Most western religions based on Christianity would deny this while eastern religions tend to embrace this concept in a very general way.

In life, symbiosis is the name of the game. Symbiosis is a biological relationship where two or more different organisms live together in close contact. Lichens are a combination of algae and fungi that live together on the surface of rocks. Both benefit from the cooperation. This is called *mutualism*. Humans have mitochondria, a finger that points to evolution, as the power factories for our cells. We provide them with food; they provide us with a great many benefits. We are also in symbiosis with every green living thing. They provide us with needed food and oxygen while we provide them with carbon dioxide (CO_2) as we exhale. When we die, the chemicals that make us up are recycled into new organisms, or biotic use, or are cycled into abiotic uses, meaning nonliving.

In another form of symbiosis, termed *commensalism*, the host is not harmed or helped by its partner while the symbiont gains. In general, there are few cases of pure commensalism. They generally are where one specie obtains food that is inadvertently exposed by another for the consumption of the first.

In the last form, *parasitism* is when the host is harmed as the symbiont takes what it wants. Tapeworms are an example of a parasite. They consume the nutrients that the host should be receiving to the point of starvation and sometimes even death. Any being that takes without return benefit can be considered a parasite.

If Earth is a living entity, humans could be considered parasites in some ways. The damages, the ravages of deforestation, biome destruction, pollution, and overpopulation show that we are not considering our host.

Humans (*Homo sapiens sapiens*), the repetition represents the subspecies as well as the species, then stand somewhere in the small end of things as far as the universe is concerned. We are as much larger than an atom as the galaxy is larger than us. Compared to the universe, we are almost (repeat almost) nonexistent. We live on the level of a biological framework that is, so far, limited to one place—Earth. Our science has a pretty good handle on this level.

There is much still to learn, but we are on the cusp of incredible things. We will soon be growing internal organs for replacement, repairing damaged nerve tissues, and growing new and stronger bones and muscles. Stem cell injections and cloning will make this all possible. Life expectancies may soon reach a thousand years or more, at least for those that can afford it or those that fit into the correct category deemed necessary by society, namely the political one.

Our perspectives are limited by the level at which we exist. As we change levels, our analogies are less accurate, and our understanding begins to fail, fading into the misty fogs of misperception. Science is a flashlight. As we create better lights, we can pierce the dark more easily, but the far depths are still unobtainable; we need a powerful Hollywoodlike searchlight for that. Time and the pendulum's swing will tell.

When we descend from the biological level to the chemical, we can understand fairly well. It is not too far removed from us at the human level. Our four dimensional analysis, length, width, breadth, and time are directly reflected at this level. We can still imagine how molecules attach by shape. However, when we reflect upon the fact that negatively charged electrons, opposite to the positive protons, cannot be pinned down to an exact velocity and location, this appears to be contrary to everyday experience.

We *see* matter, like our motor vehicles. Try explaining to a patrolperson that you do not deserve a ticket because the matter that makes you and your auto cannot really be located in terms of velocity and position due to the laws of physics. You may end up walking the line one foot in front of the other or touching your nose. The devolution into Newtonian physics may be beyond the average person's understanding.

Down one more level on the scale ladder to a much more basic level in physics, we have made great strides. Newton, Einstein, Hawkings, and so many more have brought us to the point where we, again, are on the edge of unexpected and indescribable discoveries. However, physics has not yet been able to truly comprehend all the subtleties of the ultrasmall world of the subatomic universe.

Quantum mechanics, chromo dynamics, and string theory are partial answers, not yet complete. There is so much more to be found here. Again, the farther we are from our level, the greater the dense mental fog, the harder it is to describe and understand. We do believe that we have found a lower limit to energy packets in the Plank constant. That depends entirely on our current understanding of physics. As history has shown, over and over, what we think things are has very little to do with the way things are in reality. We don't throw away what we had before, just like Newtonian physics works for everyday things. But as we progress to newer areas, the plans and rules may need to be modified, even thrown out altogether.

For example, in Newton's laws, if you throw a ball off a speeding train in the direction of travel, the two speeds add together. But if you are dealing with very high velocities, namely the speed of light, Einstein takes over. The speed of light is the speed of light. One cannot change that; only the wavelength or frequency perceived as the light is given off the superspeeding train changes.

Still, within our range of analysis, if not understanding, is the cosmos, the universe at large. Cosmology, the study of the universe, has come up with big bangs, hyper expansion, branes, dark matter, and dark energy. It seems that what we call matter only represents about 5 percent of the universe according to dark theories. We are the minority. The effects we see from dark matter and energy could even be the influence of other universes. In toto, this would mean that the collection of smaller universes would be little parts of the Universe, the single entity that is all.

Higher dimensional levels cannot be described accurately by beings limited to lower levels. Anyone familiar with the story of *Flatland* will understand. We can only use analogies and simplified models. It has been said that in order to fully understand the universe, we would have to do an experiment with the scale and cycle of our universe. Perhaps we are someone's experiment on the origin and development of life in his universe.

This could explain all the UFO sightings and abductions. They pop out of nowhere, analyze us anally and otherwise, and leave the dumb brutes with no memory of said events. We are the rats in the maze, but where is the cheese?

Life, then, is full of cycles and scales. They come in nearly an infinite diversity of duration and size. For *Homo sapiens*, there are daily, weekly, monthly, yearly, and lifelong cycles. Our daily cycles often consist of our repetitive workdays. The weekly or monthly cycles could represent paydays, menstruation for females, or movements of our moon. Our yearly repetitions are found in our birthday celebrations and various other holidays clad with symbolism, and the seasons. Lifelong cycles are simply birth and death, or transformation if one prefers. Some eastern religions believe in cycles in terms of millennia, eons, and eons of eons.

Who can say if all the living organisms and inanimate material as well are part of a larger consciousness? If we are, we would have no more understanding of it than one of our cells does of our human totality. Some people believe that the entire planet (and, by extension, the universe) is a living entity. By extension, again, each universe would be a cell in the Universe.

If there is life out there, we may be looking in the wrong place or, more exactly, at the wrong chemical composition, the wrong form of matter and energy. What if when we die, we become another form of energy related to what has been termed *dark matter*? Is this a possible explanation for ghosts, specters, hauntings, and white noise (also known as electronic voice phenomena)?

Dark matter may exist in different dimensions or even alternate universes. Astronomers tell us that galaxies exhibit the effects of greater mass than can be accounted for by observation. This mass (or energy since $E=mc^2$) must be somewhere. Do black holes lead to that somewhere? All we can do is observe and hypothesize. From this data, we may make generalizations that, later, usually need to be altered or at least refined.

A giant grandfather clock, with an immense pendulum, somewhere out there, or in there for that matter, is ticking out the seconds of existence for the Universe. Its children clocks operate for each individual universe. Imagine then that there are grand and great—and great-great—and great-great-great-grandchildren clocks for everything in our universe. One for each wavicle, each subatomic particle, another for each atom and then each molecule, group of molecules, planet, solar system, galaxy, and cluster of galaxies. Each one has a longer or shorter pendulum and period according to the scale of the object it is inextricably linked to.

Each large one ticks onerously at its own musical pace like a metronome while the smallest blurs its way past. Each one has a fundamental period all its own. The sum total is the equivalent of a giant orchestra composed of an unimaginable quantity of instruments of innumerable variety filling the universe with the music of being—existence.

When a substantial number of these pendulums are swinging in synchronicity, a major event will occur. The scale and number of the pendulums may decide the fate of a bacterium or an empire, even a world. Certain mystics believe that the first time *all* the pendulums were synchronized was when the Universe began. Some liken this to the mythical god Chronos. It then generated subuniverses, and time began for us in our universe. The little clocks were born out of infinity from the loins of the grandfather clock at that instant. The mystics also believe that the Universe will end the next time total synchronization occurs.

Our little Earth has had several occasions in which significant synchronicity occurred. Life began. Asteroids destroyed much of life on a few occasions. Volcanoes buried people alive, like at Pompeii. One can never tell when or what the next event will be nor on what scale. What we call random chance is actually the intersection point of the uncountable swinging pendulums that govern an event. Strangest of all is the fact that we can affect the pendulums.

Let us do what Einstein referred to as a thought experiment. Imagine yourself in a room during a vigorous party. You are standing in the corner alone, observing. You do not drink, eat, or interact physically or verbally with the other partygoers in any way. You are there to see how people behave, say as a cultural anthropologist.

Yet the very fact that you are standing there, doing nothing but looking around and perhaps taking discreet notes, causes people to take notice. Their behavior is modified by the very fact that you are there. People will whisper about the strange person in the corner. Is he or she shy? Is that person a pervert? Some hope so; others not. Is it a cop ready to bust the party? The gossip mill will develop around the loner in the corner until most people have made some comment, openly or not.

In science, we understand that we cannot study something without having an influence on the thing we are studying. Science will try to reduce the influence we have by improving equipment. Techniques to reduce that interference allow for more precise measurements, better understanding.

One possibility of having less detrimental influence on the party would be to have the anthropologist interacting with the others while making mental notes rather than scribing in a notebook. In the case of our anthropologist, we could also remove the person entirely and substitute a hidden camera with microphone. We will get much more accurate data. And yet, the camera and microphone do have a minimal effect on the environment, mostly small electric and magnetic fields, which do not impact humans very strongly. It is just operating at a level below the notice of the observed. Even so, someone might discover the devices, and that would have a really profound effect on the people who learn of it. Big Brother has found them!

What does this mean for science? It means that there is truly a boundary to knowing everything there is to know. Just as we cannot know both the velocity and position of an electron around a nucleus, gaining information about one while losing information about the other, we can only box in our observations. We may get ever closer to the absolute truth, as in taking limits in mathematics, where we try to find values where we go to infinity. But we can never really reach it under the rules of our universe, or the *Universe*, as it now exists.

Religions denigrate science for this "shifting sands" approach to reality. They claim that there is an absolute truth, and that they know it. Foolish is the man who builds his house on sand, which refers to the need for science to alter its view as new data arises.

Science does the same to religion, claiming that there can be no truth to the mumbo jumbo. It all comes down to random cosmic ticks of chance.

*

Amen (which, in Hebrew, means "so be it").

*

In our description of nature the purpose is not to disclose the real essence of the phenomena but only to track down, so far as it is possible, relations between the manifold aspects of our experience.

—Niels Bohr

*

Physical concepts are free creations of the human mind, and are not, however, it may seem, uniquely determined by the external world.

—Albert Einstein

*

God in the beginning formed matter in solid, massy, hard, impenetrable, movable particles, of such sizes and figures, and with other such properties, and in such proportion to space, as most conduced to the end for which he formed them.

—Sir Isaac Newton

*

In the beginning God created the heavens and the earth.

—Genesis 1:1 or the First Book of Moses

*

"Deo, non fortunâ." (Latin) From God, not from chance.

Chapter 1:

The Nature of Empires

1

000001

"This is Mary Smith-Jones. It's November 12, 2010, with WWNN Internet and cable broadcast news. We are here near Carutapera, Brazil, South American continent, with an amazing historical story. Dr. Illya Marquez of the Brazilian Institute of Anthropology and her colleagues claim to have found the legendary Atlantis as described by Plato. Less than two weeks ago, a minor earthquake hit this small fishing village, but the important events were just offshore. An underwater landslide caused by the earthquake exposed the remains of a very large city. A local fisherman pulled up pottery shards in his net and contacted local officials, hoping for a small reward. He got one, much more than he bargained for.

"Within days, the pottery was determined to be much different than currently known styles and of much greater fineness in texture. Days later, a diving expedition searched the site. What they found was extraordinary. Huge stone blocks, carvings, metal instruments and eating utensils, and skeletal remains."

Jones turned to her right, the wind blowing her long blond hair, the sun beating down. "Dr. Marquez, what else has been found?"

"What has not been found?" responded the diminutive, dark-skinned woman. "That would be a better question, Mary." She smiled with a mouth full of even whitely gleaming teeth. "A large metal machine, engineers think it is a steam engine, was brought up two days ago. It will take a great deal of effort to clean it without removing important clues. There are mosaics made from what appears to be fine ceramics. Statues abound down there, those of people, animals, fish, and even flowers. Our teams have brought up over a thousand items so far, and there are over three thousand found and radio-tagged, ready to be recovered," she noted with an almost awestruck tone and a friendly smile.

"How have you been able to do so much work so quickly?" asked the reporter curiously. This was a major story that could boost her career considerably and she was eager for more.

"It has come as a great surprise to me. Professional and amateur divers from all over the world have shown up in a matter of days since the first announcement. They were uninvited, mind you, but they were willing to donate their time and ability to help. The word spread so quickly that there was a need for divers with archeological skills that those without showed up. Some of them are even husband and wife and families that are diving for their vacations." Marquez's eyes were sparkling with mirth at the thought that these people would compare vacations to working in the water as amateur archeologists.

"We have divided the nonarcheologists into groups that dive with an archeological team leader," she continued. "When they are out of the water, they get a crash course in underwater archeology in the local schoolhouse." She pointed to a small hut with open door and windows thatched with palm branches. "In the water they lay out grids, place lights, there are even photo teams where they often use their own cameras and films or digital cards purchased at their own expense, and even perform search sweeps around the perimeter."

The attractive twenty-five-year-old reporter, in her over warm business suit of deep blue, her glacial blue eyes glued to the camera, asked, "What about professional assistance?"

"The professional divers are paid, of course, and have many of the skills we were short of. They have been trained to find, mark, and remove items. Many of them are welders and can set up winches to help with the removal of the larger objects," Marquez noted with another smile. She was unused to the attention of reporters, even though so many had descended on this location in the last twenty-four hours. This Smith-Jones woman was a lot more polite though her mind noted. Politeness scored many points.

The doctor continued easily, "Locals are coming by the thousands to work in the lumber industry that has sprung up here for the construction of housing, and the boxing and transportation of the items. Only locals are hired for that to boost the income and number of jobs in Brazil."

"What makes you think that this is the legendary Atlantis?" queried the reporter, moving the discussion more toward the point.

Dr. Marquez smiled widely, again exposing her brilliantly white teeth that were rivaled only by her sapphire eyes, and said, "Follow me." The doctor moved off with a gentle stride that allowed her hips to swing freely, her bare feet leaving happy marks in the sand. The reporter and camera person moved less easily in their constrictive footwear, which did little more than gather the small grains into an uncomfortable position under heel. Mary decided to kick off her Italian leathers and follow the good doctor's lead. While the sand no longer ground at her heel, it was hot enough that she had to mince step her way behind Marquez. When Marquez peered over her shoulder to confirm that they were following, she noticed the discomfort of Mary, and a half smile covered her face.

One dozen meters down the beach, she uncovered a large two-meter-long-by-half-high block of stone, pulling off a sun-bleached tarp. It was inscribed with four sets of letters and picture forms. Dr. Marquez was beaming so brightly she almost put the tropical sun to shame.

"That looks like Egyptian hieroglyphics," frowned the reporter as she pointed to the lowest set.

"Very astute, Mary, it is! The next one up is a precursor to Olmec. The one above that is an ancient form of Toltec."

"What is the one at the top?" asked the intrigued reporter with a slight frown.

"We have never seen anything like it before. The others are the key," Marquez stated with a tilt of her well-chiseled features.

"A Rosetta Stone!" shouted the reporter, losing her objectivity in her excitement.

"Yes. It says, in the English equivalent phonetics, 'Āttā-lǎn-tā-ē,' the first Kingdom of the World."

"Holy shit," muttered the reporter, stunned and forgetful of her on-air status.

"I agree." Dr. Marquez laughed in a most feminine way. "My bowels were very watery when I first saw this. I collapsed on the ground. My knees could not support me. Some of the locals called a doctor thinking I had had a heart attack. I am fifty."

"Bullshit!" uttered the surprised reporter. The luminous beauty before her could not be that old. It wasn't fair.

Turning back to the stone, Smith-Jones continued, "That lowest inscription means they crossed the Atlantic to communicate with the Egyptians. Are there any comparative Egyptian texts?" the reporter said, wishing to show her basic knowledge.

"I believe it means more than that," Marquez intimated. "This is the oldest form of hieroglyphics known. These others are the oldest forms of their languages. I think that the Atlantians were the precursors to all the ancients. When Atlantis died as described by Plato, using Egyptian texts, these peoples had already expanded and developed outposts. One of the places was Egypt. Others were all up and down the Gulf of Mexico. It may even be that Easter Island was a part of the empire, or at least influenced by, as it spread into the Pacific Ocean," she said with a florid gesture of her muscled arms. "Take note that the Atlantian writing is much more fluid, graceful, in comparison to the others, which are blocky and crude. This not because the sculptor had problems. It is because the written forms of those languages were relatively new then, and the Atlantian had had several refinements."

"It will obviously take many years to go through all the items already found and all to come. It is also obvious that this may be the most important find in all of archeology since the beginning of the scientific age. This is Mary Smith-Jones for WWNN. We will be here for a week to see what develops."

The moment she was off the air, she turned and said, "Will you have dinner with me?"

"Certainly," replied Dr. Marquez with an intrigued smile. "You must have many more questions."

Mary had hundreds of them, but they could wait. At twenty-five, she had landed possibly the biggest story of her career; she was just out of college.

She quickly turned back to her cameraman and said, "Let's redo those sections where I loosed my tongue." She knew the station's editor would give her a pep talk about controlling her emotional outbursts. She was a bit shamed herself, not because she considered the words vulgar or any such thing, but because she had taught herself from a young age to cultivate a much broader vocabulary.

It was when she had let out a rather belligerent "Fuck you!" to a thirteen-year-old (now ex-boyfriend) while in her father's presence that started it all. Her slamming of the phone did not cause any problems; his raised eyebrows had come before that.

"Honey," he asked calmly but firmly, "do you know what that means?"

Mary whipped around as if surprised to find him sitting where he always sat in the evening, on the sofa with his paper. "Uhhh . . . yes, I guess I do," she stammered.

Glancing at the clock, he peered back at his precocious daughter and stated, "By this time tomorrow evening, you will present me with a list of twenty words that mean the same as the one you just used."

Surprised, even almost irked that she was not in any real trouble since she felt like rebelling, she gaped for a good five seconds before breathing, "What?"

His eyes went back to his paper, but his mind was still focused on her. "Twenty synonyms for the word *fuck* by tomorrow, handwritten."

Rather perplexed, Mary had simply turned and left the room.

Dad, on the other hand, was mentally scoring one for his side. *Unpredictable in the face of absurdity*, he thought. Dad did wonder what the name of the poor boy was though; he had lasted only a week.

The next evening Mary promptly handed her serious father a list, one containing many more than twenty entries, slang as well as technical and dictionary definitions.

"Hmmm, I have no idea how to pronounce that one." He pointed to an absurdly spelled word.

"That's cause you're old, Dad. I can teach you," she drawled like the hidden paw of a stalking lioness.

"I'm sure of that." So much for having her confused, he mentally murmured, *Sharp, quick to rebound . . . God, I'm in for so much shit . . . Feces*, he mentally reprimanded himself. The wince was not physical, and the mental one was well covered by the tilt of his eyebrows.

"Dad?"

"Yes?"

"Fornicate you, and intercourse you just don't work for me."

"No. I suppose they fucking don't." His gamble paid off. She was temporarily shell shocked. "But knowing all these different words will put you in a position to use them when they are appropriate and fuck isn't."

He rattled his paper to signify the interview was at an end, meaning they could both retreat, each hoping he/she had gained ground in the eternal parent-child wars.

In years to come, Mary found it very ironic that finding so many ways to say fuck you had led to journalism. The second most important date in her life had been the attacks of 9/11. She wanted to know why and then to be able to explain why.

*

It was a small inn slash cantina of sorts with an old woman running the small kitchen. The shoddy construction and garish paint went with the prostitutes who had moved into the area to cater to the influx of workers on the doctor's project.

The dinner was adequate. The local cuisine might disagree with Mary later, but she had to get more on this subject. Her news nose was twitching. She also wanted to spend time with this incredibly attractive woman. This Illya had a sense of physical, mental, and emotional power unlike anything she had ever felt before. Illya vibrated with life. Mary sipped on a rather strong local drink as she listened to the good doctor.

Dr. Illya Marquez had pushed her plate back after consuming an inordinate amount of edibles for one of such a delicious figure. She too took a sip of the local intoxicant before continuing with her amazing story.

"We have found so much more than the Rosetta Stone. We have found stone upon stone related to the histories, or myths, of these remarkable people. There is such a large amount that I am hesitant to release, yet."

"Why is that?"

"Their lives were all centered around a central myth, or religion if you like, that describes powerful beings visiting them. According to their texts, these beings were from beyond our star and endowed them with great gifts."

"Beyond our star, literally from space you mean?"

Illya's eyes flickered down as if professionally embarrassed. "Yes. Most other anthropologists would absolutely deny that possibility as just another god phenomena. The texts we have are rather clear on this, so I must at least consider it. There are some more modern anthropologists that have begun to question the old dogma somewhat."

Mary couldn't help but ask, "What gifts?"

"Ah, that is . . . that is a most interesting tale from what we have found so far. The Iliad, the Greek tragedies, could all stem from what we have found," she remarked with dreamy, distant eyes.

"Atlantis, the once and earliest empire of man, had been granted a boon by the visitors, gods, I read. A power, like a great fire of blue that pervaded all their tools of metal, and enabled them to create greater and greater works of art, dwellings, ships that sailed the seas without wind, and strange craft that passed amongst the clouds. This endowment came with but one caveat. It was not to be used for the domination of others."

"Who made this restraint?" Mary asked intently. "Was it their visitors? Their Galileo or Newton?"

"My, no! At least not as I read it compared to the other texts. Archeologists will pooh-pooh this with great vigor I am sure," she said somewhat red-faced. "But it seems to have come from without, not within. It was the gods. There was a very sudden influx of knowledge, one greater than the sudden increase, say, in the computer industry. Rather than doubling every few years, it grew by an unimaginable pace, a factor of thousands."

"I don't understand quite what you mean."

Illya smiled gently. "I did not either, at first. A mathematician explained to me, in horrible equations, the rates of knowledge exposure for all the known ancient civilizations. This here far exceeds any of those others by incredible amounts.

"Let me continue. As the millennia passed, the great Lords of the Fire, the leaders of the Atlantians, the bloodline descendants of the first to be granted this power, grew haughtier, seeking ever greater domains to call their own. Snake-tongued councilors cautiously spoke when they said that the edict of nondomination applied only to their own peoples. In this way, they appeased the desires of their lords by redirecting their covetous gazes outwards."

"That seems to be a normal human thing."

"Hmm, yes. At first, open lands claimed only by extremely low technology level indigents of the South American continent were lightly oppressed in an exchange for agricultural basics and metallurgical knowledge. The size of the Atlantian Empire grew exponentially. The indigenous populations, originally resentful to any hint of control, saw the benefits of the new knowledge on farming as well as metal and stone working. To them, the lords became benefactors, though the knowledge given them was but the dregs of Atlantian technological society."

"That would explain the sudden explosion in the agricultural revolution here."

"It certainly would. Other archeologists would simply explain it as normal dissemination of information. My mathematician, however, showed me how odd the rates of expansion were in comparison with other Neolithic societies, and even more advanced ones."

"It seems more questions are being raised than answered."

Illya giggled lightly. "That is the way of science. If we do not raise more questions to be answered, we are out of a job, much like journalists."

"Touché," Mary replied, raising her glass of clear firewater and sipping carefully.

Illya did the same before continuing, "Further, lords were unsatisfied with the conquering of such nonentities. Their just as greasy councilors played words as efficiently as their ancestors. They informed their various lords that since the gods were not angry about the previous establishment of overlordship, as long as the peoples conquered benefited, higher levels of established societies could also be controlled, namely those they had recently come into contact with on the far side of the Great Water, what I believe they called the Atlantae, our Atlantic, after themselves.

"Neither the multicentury lived lords nor their split-tongued advisors understood that the civilizations developing across the sea had numbers in the millions. There they would find great fleets of oared and sailed warships and armies vying for control. Greed would lead to destruction. The Atlantians had grown used to small numbers being very powerful. They could not conceive, yet, that numbers could overwhelm the basic strength of technology."

"Give me an example."

"The Nazis and Soviet Russia. The Nazis had a technological advantage, yet the Soviets had a great numerical advantage. Lower technology, aided by the British and Americans to become more technological, won in the end."

Furrowed brows and a simple "Hmmm," were all that interrupted Illya.

"One high lord, the nephew of the king, wanted to usurp the throne and gain absolute authority over the known world. Secretly he had his laborers build a great fleet of ships and an air armada. A portion would capture the king during a high celebration in the gods' honor while the rest would annihilate these impudent upstarts. Then the world would see a real revolution in the arena of civilization, one leader to guide them all."

"Corruption rears its ugly head again."

"The gods, or aliens, read, or knew, the heart of this man and decided that they had to destroy this evil. They regretted their benefice and vowed they should withdraw from man rather than allow him to gain such stature again until the time ripened."

"Kicked out of Eden? That sounds very familiar."

"All the ancient myths, stories, or truths seem to have a common background. I believe they are all found here. This history seems to antedate all others. It may all spring from these histories. As the survivors spread, or were left in the hinterland, their history spread to become the commonality that all civilizations seem to share."

"Survivors?"

"The world of the Atlantians was swallowed by the sea in but a day. All that might have been, good and bad, was consumed by the gods' vengeance. The outposts, however, remained as little pockets of lesser civilization. It took much time for them to recover from the blow to their homeland and raise new civilizations from the ashes."

"Similar enough to Noah's ark to raise questions? The flood of Gilgamesh? The literal story of the Phoenix."

"You know so much more than you let on. It is so intriguing. So many stories must have an origin. Where did they begin? I think here."

"So how did it all come about? Where did the sudden influx of information, ideas . . . Where and how could that happen?"

"Most anthropologists would agree that the road to any empire initially is very long and arduous. It took thousands of years to develop the most basic infrastructure that would allow for the centralization of authority and raw power. The population had to reach a minimal level, trade had to grow from surplus foods, raw resources had to be abundant enough for barter. There are a multitude of requirements." She sipped lightly at her drink.

The shrunken old woman approached, offering more food, which was refused, and drink, which was accepted.

"Let me go even further back in time. This is so much more complicated, but here goes. In general, women and children were the primary collectors for the small seminomadic bands of hunter-gatherer tribes. Men hunted nearly every day, since they were not always successful, for needed protein and the fats available. Children grew up early, working as soon as they were able to contribute. Adults were lucky to reach the great old age of thirty-five. Life was harsh and full of immediate dangers.

"Simple cuts could become severely infected, ending a life at an early stage. Broken limbs nearly always meant diminished capacity as a hunter or gatherer. It could mean uselessness in an unforgiving society. A nonproducer was a consumer of needed materials, a drain, and an anchor that could endanger the rest of the clan. A male that could not hunt was of little value as a mate."

"As if males are all that valuable!"

A slight smile. "This is the basic fact of capitalism. One must produce something, anything of relative value, to be of use to society, the whole. It can be an idea, a tool, labor."

"And those that couldn't?"

"They usually died quickly. Socialism in the form of the care of the elderly, the wise, came later as recognition developed in the minds of the society's members that the elders had knowledge that could benefit all. Eventually, that concept extended to less and less productive members. They might, at some point, have a significant contribution."

"That's an interesting notion. Does it bear out in your findings?"

"Yes, it seems to have been a universal form in all developing proto-societies. Covered in loose-fitting animal skins, hardened by life, with crude flint knives, the women and children cut their way across swaths of irregularly placed patches of primitive wheat. Fingers—long, leathery skinned, and with cracked nails—grasped flint blades that removed the seed heads as the band of females and young passed rapidly through the stand. This was not wildly growing wheat. It had been deliberately scattered. Other patches had been sown, some nearer, some farther, from the home cave of these proto-farmers. The hot sun beat down intensely, causing them to perspire heavily."

The near monologue continued, "It had been noted over the generations that sun-facing slopes produced better plants. The spreading of seeds on purpose had been brought about by simple carelessness on the part of a young female gatherer of the clan many hundreds of years before.

"A slightly torn collection pouch, fashioned from animal skin, had unintentionally deposited a number of seeds near a creek at that earlier time, earning the child a beating from the alpha male for her inattention. Such serendipity often played a significant role in our long history. The gatherers, namely the female child, had observed that the few dropped seeds had eventually become plants. The plants were of the same type as the collected seeds. The girl had harvested those plants in the following gathering season.

"That one person, more curious than the others, cautiously and secretively, wishing no further beatings, placed some seeds near the creek again and again over the seasons, with only the inkling of an idea scurrying around in the depths of her untutored mind. She hoped to show up the current leader, the same that had beaten her for the error that was now likely to provide food more easily. This was one of the first great experiments done by *Homo sapiens*. It followed the great experiments with wood tools, stone tools, water storage, and fire by their ancestors in far bygone eras. It was also one of the first steps in the feminist revolution."

Mary's brows were furrowed and her tone doubtful. "There seems to be a great deal of speculation going on here."

The returning look was impish. "Informed speculation would be much more accurate. We know that by the time boys were strong enough to carry small spears, they were being taught hunting skills, so only the very young males were gathering. They were being watched as much as assisting. They would not be trusted with precious seed bags, only some collecting. An older female would have been paying more attention, would not have made such an egregious error.

"Males could gather at need, but their main goal was meat. Females, in their role as gatherers, would eventually develop a greater feel for plants and their cycles. Observations of growing plants, waiting for the ripening, would seed the idea subconsciously."

A wry smile answered, "Ugh, a pun. Humor in history?"

Eyes twinkled. "These hardy people quickly learned through the generations to purposely scatter seeds near water. 'Thorns also and thistles shall it bring forth to thee, and thou shalt eat the herb of the field' (Gen. 3:18). Some seeds fell on good ground, some on rocky, some was well watered naturally, and the rest was not. Not all the experiments of these novice farmers produced viable plants. It took thousands of years to understand the value of soil preparation, spacing, and weeding. It would take centuries more to follow with rudimentary irrigation projects. The gatherers, though, slowly became farmers. This was a fundamental shift in the nature of human society, the roots of a people grounded to a particular location. Many nomads settled down to a lifestyle that would lead to a society that would eventually become what has been called the first civilizations.

"Humans even expanded their growing area to poorer soils and cut other plants down for more area to plant. Intentional cultivation began to alter the ecosystem. This destruction of plant forms was the beginning of the destruction of Earth's biosphere, which continues to this very day. Humans had become, unwittingly, their own destroyers, the source of their eventual demise. They were cast out of Eden for their knowledge. Death would follow."

Mary looked stunned. "That's what you have here?"

"No, not yet. Palynology, the microscopic analysis of ancient pollen and plant remains, places this development near Kirkuk, Iraq. We are sifting through some old food storage containers still. Our palynologists are hoping to obtain some good data for reference. My personal feeling is that that data will show this site to be far older."

Mary begged the question, "Why didn't you give me this on camera? This is an even more incredible story than simply finding a legendary ruin."

Illya carefully replied, "I dare not. In a country that is largely Catholic"—a momentary hurt crossed her visage—"I must be careful not to . . . offend some of my contributors at the university. Some of the alumni might try to shut this down. The dating of our findings will cause a great deal of controversy among scholars and within religious circles. The Bible has a specific chronology that is mostly opposed to archeology on ancient dates. Their creation myth puts the Earth as formed around 4004 BCE. As a guide to more recent happenings, the Bible can be strikingly accurate though."

Mary's mind worked rapidly to try and digest all this information as she said, "I did not know that about the Bible. All of this is so outré."

"Ha-ha!" Illya laughed as heartily as any man. "I told you my bowels were watery. I expected to find a small localized ruin. This has skipped outré and gone on to . . . I don't know. Let's take a walk. It's getting stuffy in here."

They moved out into the darkness of the small village, each casually looking at the bright stars above. Mary also gave an apparently cursory look at the people about her. She was always careful.

Strolling slowly, Illya began again, "Prior to 8000 BCE, man had to adapt more to his harsh and unforgiving environment. Now man began to adapt his environment to suit him, to make life considerably easier, more sustainable. A threshold had been reached, certain pendulums swung in finality. Man began to subject plants and animals to his will, believing that they were masters of their way. The necessities of life were being met, control of energy by means of fire, and provisions for the consumption of resources—food."

"How do we know for sure? I have never been a student of archeology or even much history."

"This is the sum total of what has been learned with much study. It is the current belief of the world's leading archeologists. Now, however, it seems time for a major shakeup."

"Which puts you at odds with everyone else in the field. They are likely to attack your ideas out of dogmatic pickiness as well as jealousy."

"Yes. Most meat was still often obtained from the critical hunt. Larger animals, megafauna, had been disappearing from the Earth since the last ice age. Humans began to focus on smaller animals. It was around this same time frame that animals were domesticated. Livestock—in the form of goats, sheep, and cattle—developed into a controlled food source. This further cemented the peoples to certain grazing locations, again, nearer to water sources.

"Nomadic herders centered themselves on watering holes, and the farmers located themselves more permanently close to water. The gathering of people near water led to the first true pollution. Concentrations of feces, shit in more dramatic terms, would be a major problem in the future. The localizing (concentration) of peoples would lead to death on previously unimaginable scales. Various diseases would ravage through towns and villages simply through the agency of proximity. For those of us today, we consider it normal to see the ravages of overcrowding with its disease, starvation, and utter poverty. Television, radio, and the newer vid-link show the mass starvations and unsanitary conditions of much of the third world."

"But something different happened here. What?"

"Remember the mathematician?"

"The astronomical rates of development guy?"

Illya pulled out a napkin purloined from the restaurant as well as a pencil hidden in her lustrous hair. "Let me make a couple of sketches. That might help." She proceeded to draw two graphs after dragging Mary near a local's fifty-five-gallon drumfire. "This one shows a normal societal development over time, similar to all previously known and studied civilizations."

"It goes down as well as up? Rise and fall of an empire?"

"Yes, or a city state, or a region. This is very typical for all of them. This, however, is what we see here, at least from the data we have so far."

"Why is it so steep and then flat?"

"It represents an incredible influx of knowledge at a time when it was appearing to level off a bit. The sharp steepness followed by the almost flat aftermath is a real indication that the burst of knowledge did not come from within simply because the people could not build on it in any real way. They had no understanding of the fundamentals of the new knowledge and power they had. They could not build on what they did not comprehend."

"What's the blip here where it goes up again a little?"

"The most remarkable data may not be the from Atlantis Stone as we have dubbed it, but the Calendar of Decades. It lists all the events in a remarkably succinct way for several hundreds of years. The translation work is partial, but we have examined the areas where there is more to be said. That blip area is where the Atlantians took over most of South America and spread the agriculture knowledge to less knowledgeable indigenous tribes. They gained in some way. Perhaps it was knowledge of resources such as gold, iron, or even simple new tricks they had not encountered before. We don't know yet."

"What about this part with the dashed lines?"

"Destruction. We have no idea, really. Everything just ends except for one statement. 'The king is dead. His nephew has betrayed us all.' Nothing, and I mean nothing, postdates that."

"So it all really happened. Atlantis, I mean. That's . . . I thought it all was overwhelming when we were on the beach. I can hardly accept it all."

"That's why there have already been some threats. The university has assigned some sort of bodyguards to me. They arrive tomorrow. Nightcap?"

Mary grasped Illya's forearm and replied, "Certainly."

*

When empires are first born out of the culture that creates them, the seeds of their destruction are also sown. Jesus spoke a parable in which wheat and weeds were sown together, but only during the harvest could one tell the difference. Which will we harvest?

A small village consisting of a few poorly constructed huts may inhabit a large mountain edged valley full of abundant and varied resources: fish, game, trees, fertile soil, and stone for construction of a protective wall and homes. Success in survival leads to the need for more resources for a growing population. As the village grows in population and therefore physical size, not only is the better farmland's use redirected to housing, but people must travel farther to obtain the materials required for life: construction materials, food, clean water, materials for clothing.

As hunter gatherers, humans moved about sufficiently for no one area to be overtaxed. In their much more static citified societies that changed. The pace of human population growth is now faster than the rate of renewing for natural resources. The plants and animals in a natural state can't keep up with human consumption.

Once the time for that procuring travel becomes a significant hindrance, small newer villages are settled on the periphery by the descendants of what is now our first town. The tallest trees, good for long beams, or of a special type of wood, disappear from the nearer groves. Most of what is left is only firewood worthy. The natural trees are pushed out and replaced by various fruit and nut trees as agriculture takes further hold. The original ecosystems are displaced as humans take over a given locality. Slowly but surely, the tide is turning. The pace quickens as time passes. Pendulums swing.

Everything changes because of the growing dominance of humans; birds, deer, butterflies, grasses, bacteria are forced to adapt to the changes imposed upon them by *Homo sapiens*. There is an idea, known as Chaos theory, which states that the fluttering of the wings of a butterfly can lead to a tornado on the other side of the world. This means that very small changes can lead to very large counterresponses, either positive or negative, often with little foreknowledge. But the thing most humans ignore is the possible length of time involved in the retribution. Earth's life spans great eons; it might take a few millennia to get her attention.

Dirt roads, bare and hard, replace the barely worn cart and foot paths that have displaced the emerald green grasses. Stone houses, roughly cut, replace the old huts, and newer houses are built outside the original protective wall, and a new outer wall is constructed; this is municipal growth. The land is changed forever as it becomes lifeless underfoot. What do we have in common with our forbearers? Just examine the concrete and asphalt underfoot.

As the settlement grows, it spreads out onto what was the best, close farmland. Remember that our original farmers lived on or near to the very best growing areas. Rather than place homes on the poorest soils, they convert some of the best to housing in order to be closer to the towns. Centralization has both benefits and drawbacks.

The short-term benefits are greater protection, access to markets, and the local water supply. The longer-term drawbacks tend to be a little more nebulous, impinging less upon the minds of the people only recently elevated above subsistence living. That hazy future full of difficulties is all too easily traded for the now's ease. Procrastination in terms of individual vis-à-vis societal concepts begins.

Outer villages and farms provide the raw materials now, not the immediate area. Dependence grows on outlying areas that will eventually look to their own survival. Soon, these outer areas will be providing more of the foodstuffs, grown in less productive cleared land. This is what leads to future food shortage situations for the growing town, the inability to be self-sufficient—dependence. This is the beginning of weakness in the society's stability, one which may later be capitalized upon by a conqueror.

Specialization had already begun in our first town, but it now moves into higher gear. Artisans have developed specialized techniques known only to them and their assistants, their protégés, their apprentices. Potters need the best clay and pigments. The wood workers get precut, sized pieces from the rough woodworkers of the outlands. Barter is ready to be replaced by coinage.

The first money was simply pieces of gold or silver of certain weight like in ancient Egypt. The first real coinage was developed by the Greeks in about 650 BCE Unfortunately, official bankers, usurers, were created around this time as well.

The valley soon fills with humanity and their labors. There will be little left of the original, quiet, natural ecosystem here. The game will have left the vicinity. The fish will be harder to catch since there are fewer. But the next valley is undeveloped, ready for people to move in, and they will. Clans will split into subgroups in the other valleys. These relationships continue to develop a hierarchal control structure. Businessmen and controllers of land become tribal elders, councilors, and the developing government.

Expansion is a natural by-product of a flourishing culture. Since they are all originally related in this valley, they are friendlier towards each other than outsiders. It is the beginning of city states, kingdoms, and, eventually, of empires. The original town begins to develop a sense of "better-than-ness." A certain air of regality enters in.

Empires come about when one kingdom gains control over other kingdoms, through conquest usually or through economic domination. Indeed, the seeds of the construction of an empire are, in like manner, the seeds of its destruction. It is difficult to tell the difference between the seeds of some food sources and weeds. Sooner or later, the system for providing resources for the linked communities will fail. It always has; it always will. Pendulums swing.

*

Monsignor Drake had been chosen for his known neutrality. He was heavily involved in a great many aspects of the church in Latin America and not at all in the volatile Middle East. He was put in the forefront of the video conferencing connection. It was his job to bring to memory all the connections between the often contending cantankerous vying religions. They needed to bury the hatchet, without bloodshed. They needed to unite. This was the opening move of a newly elected Pope Xavier toward that goal.

The monsignor, while well versed in public speaking, trembled ever so slightly as he began, "My friends, bear with me as I repeat well-known history."

Several muted assents came across the network of heavily secured video links.

"All of our histories, our religious knowledge, begin with the same peoples. We are all one family. We are all the blessed of the one God, the only God. Our mutual descent can be traced with certainty.

"Abram is the father of the Jew and the Muslim. We are all brothers, kin from afar. We must stop our battling amongst ourselves and unite. Satan has been at work amongst us, deceiving us, separating us. Abram fathered Ishmael through Hagar to become the father of the Middle East, the Muslim world. Abram, now Abraham, produced a son through Sarai, now Sarah, to became the father of what would become the Jews. Our father is one in the flesh and one in God. We must come together as one as he intended.

"Rabbi Levinson, Rabbi Simon, please correct any errors I might make in my narration.

"In 1513 BCE, according to the Bible, Moses would lead his one group of people out of Egypt. They left after many miracles, plagues, and deaths, on the Egyptian side of the coin.

"The first miracle occurred in the throne room of the king when Aaron tossed his staff to the floor, and it became a serpent. The wise men and magicians of the pharaoh also threw down their staffs, and they turned into serpents as well. However, Aaron's serpent swallowed up all the others.

"The next miracle occurred when Moses used the same staff to turn the Nile to red blood, and then the fish died. It also turned the water in all the storage vessels to blood. And so it did, according to the Bible. But the magicians of the pharaoh did the same.

"Seven days passed after the Nile was struck. Moses then commanded Aaron to spread his hands over the Nile, and the Nile began to swarm with frogs. But again, the magicians did the same. The king did not believe, and his heart was hard.

"Moses then told Aaron to strike the dirt with the staff, and he did. It brought forth gnats by the millions. The magicians could not reproduce this act and declared it the finger of God. Yet again, the king hardened his heart.

"Then swarms of flies were sent against Egypt, except the land of Goshen where the people of Moses dwelt. Pharaoh would not let them go, even though he had so promised in order to get rid of the flies.

"The next plague was the death of the herds of Egypt: the cattle, horses, asses, camels, and sheep. Yet the herds of the people of Moses were not affected. The king would not let them go.

"Next, the plague of boils and sores came. Even the magicians felt the displeasure of the plague. The sores appeared on every Egyptian and every beast left to the Egyptians.

"The plague of hail and fire followed. All the people not at home and all the animals not in their byre were destroyed, except in the land of Goshen. At this, the pharaoh called Moses and Aaron to him and admitted that he had sinned. But when it stopped, he again would not let the Hebrews go.

"Locusts by the billions followed quickly. They ate everything not destroyed by the hail and fire. There was nothing green left. Pharaoh was again willing to admit he was wrong. He again changed his mind and refused to let the Hebrews leave.

"A great darkness next fell upon the land for three days. But all of the previous calamities were nothing compared to what came as the last of the plagues.

"Death of all the firstborn sons of the Egyptians occurred. Finally, the Pharaoh let the Hebrews leave. But even then, he changed his mind after they had left. He sent his chariots after them to bring them back. They caught up at the Red Sea, or the Sea of Reeds according to more recent translations. It was here that the final blow against the Egyptians would occur.

"Here, Moses parted the waters, and the Hebrews crossed on dry land. The army of the pharaoh was consumed as the waters returned.

"The power of our God is awesome. It brought one group of his people to their land. He also kept his word to the son's of Ishmael. They became a people of a count to equal the visible stars as He said they would. As our God is infinite, so are His worshippers. We all approach Him in our many ways since He is infinite.

"When it comes to our differences, He will straighten our paths when the time is ripe. We all seek the truth. We all want to sit by our God. Let not any deviations separate us now when He calls upon us to do His work.

"My friends, Caliph Omar, Imam Elal, do not hesitate to correct me if I err in any way in the description of the development of Islam. I have spoken of the blood relationship. Let me now speak of the spiritual.

"Mohammed had a vision. A vision involving the one true God of all of us. He went to a hanif who pronounced that this was the same spirit that spoke to Abraham. The religions had become corrupted by the foibles of man. Did not Jesus speak of the whitewashed clergy of his time? They were filthy on the inside with the appearance of cleanliness on the outside? We must accept that as fact if we are to believe any part of the Bible. We must all accept that no one religion has it down perfectly. We are all imperfect beings. We must strive towards perfection with the help of our God.

"This is the message I have been instructed, by the Holy Father, to deliver today. All of you are invited to participate in a conference to be held as soon as possible to investigate how we may unite in our efforts to serve God. The pope humbly asks that each of you prepare a list of grievances that you may have with the Catholic Church so that he might address them personally."

Shock and uncertainty registered on all the faces of the on the other ends of the conference lines. This was a major break for all the non-Catholics. The Muslims were unable to utter a word yet. The Jews, while shocked at the openness, had been expecting some sort of shift in policy. The Protestants simply absorbed the implications after a short bout of apoplexy.

The monsignor continued after his timed pause, "There are matters of *extreme*"—he emphasized the word—"importance in the making. We have but a short time to prepare for the storm that is soon to come."

Further ruffling of mental feathers occurred everywhere.

*

These particular government labs were a conglomeration of underground facilities that included seed storage, raw materials, foodstuffs, tool storage, arms, and much more. Their origin was in the Cold War era of the 1950s. The threat of nuclear bombardment caused many governments to seek the protection under the mass of the surface of the Earth. All NATO and Soviet Bloc nations developed these types of structures. Some were relatively small and simple silos for the housing of nuclear missiles. Others were on a much grander scale; they offered hope of survival, such as the seed banks. The occasional were part of the grand scheme of war and dominance; such was Cheyenne Mountain.

It was an integral part of early warning and response and, more importantly, survival, during the Mutually Assured Destruction (MAD) cycle. In its later days, it would become a center for more arcane visions of warfare and survival. In its think tanks, the abnormal was accepted as normal; planes and tanks were invisible through the use of meta-materials and the bending of light; special camouflage did the same for people. It was the people themselves who needed modification.

The body could be trained. Millennia of military service, either by volunteers or through impressments, also called the draft in some circles, left little doubt of that. The body but requires food and exercise, specifically that exercise related to war. Thousands, if not millions, of years of conflict have developed this trait. This is the simplest to achieve; the mind is a little more difficult.

Enter propaganda. Propaganda comes in a multitude of forms: clanism, tribalism, nationalism, religion, and various combinations. Propaganda consists of the communication through verbal, pictorial, or other sense languages of a concept of what is good (us) and what is bad (them).

In its earliest forms, it could only spread as fast as contact could be established by foot. As technology developed, the use of horses, ships, telegraph, wireless, television, the impact of propaganda took on greater and greater impetus. Examine Nazi Germany for the first true impact on a people; do not forget that at the same time, this technique was used in the European states as well as the United States. The control of the flow of information was the great fear of those who penned the dystopias, antiutopias, such as *Brave New World, 1984*, and *Fahrenheit 451* during the 1900s. These new think tanks would return to the body in a whole new way.

Ideas germinated as to how to make the body a more perfect weapon, at the genetic level. The human genome projects, the mapping of every bit of genetic information, could lead to improved muscle tone, eyesight, resistance to radiation, chemical and biological warfare, even reduce the need for sustenance through improved use of resources. Inklings of the future capabilities drove these secretive groups onward. The space age would bring forth the first true usage of such in the form of the *Homo sapiens anthrocreatus*. Genetically manipulated beings designed to withstand certain environments. That they differed greatly from the originally conceived war-related aspects mattered little. The biotic (or living) strands were beginning to push the pendulums.

The abiotic (nonliving) also began to push to the forefront. The two-pronged attack brought with it the correlation of data on an immense scale through computers. Each of these same locations depended on the rapid access and storage of enormous volumes of specialized information. Centralized government locations oversaw and controlled the flow of that information based on increasingly tight restrictions known as national security. Smarter, faster, larger memory, humans were seeking to increase the capability of their minds through external devices.

It began with the simplest of tools, those used to ease the gathering of food. The extension of the physical body in tool usage would soon take the form of metaphysical and cerebral extension, at least on a geological time scale. Supernatural explanations and logical ones would become the battlefield of development for millennia.

Time would drive them apart and then bring them back together, just as the biotic and abiotic would merge in later days. The swinging of pendulums, like a pulsing heartbeat, brings opposing forces as well as a certain synchronicity at other times.

*

Empires are formed by the coalescing of political, religious, economic, cultural, and military groups into a more or less tightly bound organization. These individual entities become rigidly intertwined as time passes; each one becomes a strut or foundation for the other. Eventually, one cannot survive without the others. The structure becomes untenable.

They may start as relatively benign forms of government. The Roman Empire started as a very democratic system. They all may be democratic or republican in nature. Time, however, and the desire for power and wealth brings forth those willing to murder, pillage, fornicate, perform treacherous acts, steal, and anything else to maintain the status quo or improve their position. Family members kill their fathers, brothers, nephews, wives, and lovers, not just their enemies.

Every government of man throughout all history has moved into a form of empire, the concentration of power into the hands of one human. How could so many be so surprised that these series of events, at any time in history or present or future, would be repeating themselves once again? Unfortunately, there are far too many that do not understand or remember the ebb and flow of human history, as shall be seen.

In order to determine when such events will occur, we must look for the patterns similar to, but not exactly like, past history. A prime example is the opening of new lands. Another is the fall of a major previous political power.

Government grows as specialty groups grow in number; each one must have its representative seated in the halls of authority. Most are purchased, campaign contributions, gifts, bribes, blackmail. These forces can no longer operate together; they end up rending themselves apart or merging with other political units, forming a new entity that they hope will remain stable.

We can imagine an infinite row of swing sets with some people striving to reach the greatest heights. Beginners may sway only slightly as they are pushed by outside forces rather than master their own motive forces. Others careen wildly, screaming heartily as they nearly collide with their nearest neighbors. Eventually, some of them just have to jump off, striving for the greatest height or farthest distance. Too bad they take the rest of us with them.

Chapter II:

The World and Universe

2

000010

"This is Mary Smith-Jones with Wide World News Network in Jacksonville, Florida, North American continent. A large oil slick, estimated at several thousand barrels, has already been released from the ship the *Alle-Marie*, which is just half a mile offshore. The slick is growing and quickly moving shoreward with the tides.

"In what first appeared to be a simple oil spill tragedy has changed, we now have information from an unnamed sailor from aboard the tanker that it was a terrorist attack. We overheard him shouting out to officials as he was being taken off to a hospital.

"In an attack similar to the Cole incident in Yemen many years ago, apparently, two motorboats approached the ship and exploded near the stern and bow of the ship, apparently blowing open the double hull of the supertanker. The sailor claimed he saw a flag of the extremely radical group Green Warriors, the Earth first terror group, flying from one of the motorboats.

"The ship does not appear to be sinking at the moment, only releasing the contents of its forward and aft chambers.

"We are unable to obtain any official comment on the claims made by the sailor. Until the FBI and Homeland Security get control of the situation, we will not be able to get any verifiable statements.

"The Coast Guard and port authorities are attempting to establish an oil break, a set of buoys and floating lines that will hold the oil slick in place until it can be cleaned up. The Coast Guard is reportedly flying in thousands of gallons of an oil-consuming bacteria in hopes of preventing too much damage to the environment.

"There are several Coast Guard helicopters flying around the ship and coastline to determine the direction of the currents and expected landfall of the oil," she said as the camera operator followed her pointing finger. The copters looked like scurrying dragonflies.

"The beach patrol has begun to warn bathers to get out of the water. They are moving south from here since it appears the current will move the oil in that direction. The local authorities are sending patrol cars to close the entrances to the public beaches."

She continued, "However, private beaches such as this one are harder to control. The legalities of cross-county cooperation are also interfering with the logistics of delegating manpower and resources."

She spied a long black limo moving closer and rushed to where the car was pulling in to the scene. She began to fire questions at the emerging local Homeland Security director.

"Director Thomas," she shouted as loudly as all the other reporters, "what do you know about the possibility of an attack on the ship?" It was the same question the majority of the reporters were clamoring about.

"No comment," he returned gruffly as he strode toward the taped off area where the command post had been established. His black-suited security, earpieces in place, pushed through the converging reporters and camera persons.

"The sailor said—," began one reporter.

Thomas cut him off with, "There are no facts established. Wait for the facts."

"We will return as soon as we get more on this breaking story. Now back to Brad at the anchor desk in New York." She held her position for a moment as the vid cut off.

"It's hot here. I'm going to the beach for a swim before the oil gets here," she murmured, handing her equipment to the cameraman. She strode off, pulling her colorful shirt completely off and stepping onto the baking sand. Her now short-trimmed hair rustled in the breeze. She kicked off her expensive leather shoes and pulled down her slacks to fully expose the roundness of her hips. Her tan line was nonexistent. It had disappeared with her stay in Carutapera, Brazil, and her friend, the good doctor.

The doctor's company had allowed her to shed some of the walls and inhibitions that had been erected after her marriage at the unready age of twenty and the all-too-long-awaited divorce at twenty-one. High on the heady expectations of college life, she had jumped in without reservation. Mary Smith learned quickly that she had been hoodwinked. Within six months, he had beaten her.

None of her other boyfriends had ever even made her feel that that was even a possibility. The usually gentle remonstrations of her father had not opened her eyes to the violence that could be done upon her. Men had become a danger at the personal level. It would be two years before she would go on a date with a male.

However, her sense of humor and irony did not depart. She kept the hyphenated name of Smith-Jones, the two most common last names in America, for the rhythmic sound it made as it rolled off the tongue and the upscale feel it gave her. It also seemed to give her a more professional sound she thought.

She meandered toward the surf, looking at the few shells scattered here and there. Lumps of seaweed gathered in places where they had been left by the high tide. Her feet had been toughened by her time in Brazil, as opposed to her outlook on life and herself. She no longer felt deserving of abuse, or abused. She felt good and looked good.

She received a few appraising looks as she approached the water and a few frowns from the generally less well-shaped women, but this was a private section of beach where nudity was acceptable. She waded forcefully into the water until it reached her waist and then dove gracefully under the next oncoming wave.

*

Ecology is the study of interactions of individual organisms with their environment.

An organism is a single living creature, regardless of size. All organisms use resources and produce waste. These organisms could be single celled, like bacteria, or multicelled, as in plants and animals. It may even be possible for multicellular creatures to act as cells for an even larger being, a superorganism. This would be similar to symbiosis taken to another, higher level.

All of the organisms on Earth, with one very significant exception, are completely natural in their use of resources and production of waste. Every organism but this one exception is in accord, in tune, with the others. The waste products of one species are the food source of another.

There was a fluctuating dynamic balance that existed before it was so disturbed by this single self-destructive, arrogant species. The scales have been tipped, perhaps irreversibly, and may require a great deal of legerdemain to establish a new balance. The old balance can never return.

*

Homo sapiens (humans) have gone far beyond the use of resources in their standard form. Humans refine metal ores hacked, dug, and blasted from rock for a wide variety of uses including the construction of buildings, automobiles, airplanes, submarines, weapons, and space vehicles. While ants and bees and other animals build large structures, they are still made from materials that are what most would refer to as natural.

They dig in the ground, construct nests, and lie down in gentle green fields, what few are left. Humans build tremendous structures; they also build individual living quarters. Beavers cut trees for use in their dams while using mud as a natural type of mortar. They do not however make certain specific sizes such as two-by-fours. Animals do not, as far as we can understand, make any kind of engineering analysis for their homes. They rely on instinct and the reliable old concept of *does it fall down*. Birds build nests using the materials they find on hand. This may even include buildings made by humans. Certain types of birds may use the side of a human construction to build their own stucco type of nests. It is in truth this variation of structures and exactness of construction materials that really separates humans from all other creatures in their domiciles.

Waste products also define the differences between humans and all other life forms. No other species produces wastes other than those from biological excretions or other biomass. We find that the feces of cattle make a very good fertilizer for our gardens. This is a result of a symbiotic development between the grazers and the plant life. The chemical excretions of *Homo sapiens'* industries are quite different. They are not naturally occurring in their makeup per se and are extremely concentrated in a chemical sense.

Steel does not occur naturally in deposits on Earth, though the elements that make it up do. Nor do solar panels, plastics, silicon wafer boards, or a host of others appear.

Humans displace and destroy a great deal of the flora and fauna with their drive for resources and the waste products produced from that stage. The creation of waste does not stop there. In all stages of manufacturing, there is pollution and waste to be dumped. Many of these are toxic to the environment, further harming a fragile balance.

Chemicals used to provide for better food production have a detrimental effect on animals in many ways. Pesticides and herbicides have increased yields while destroying local balances in the ecosystem. Birds and fish have suffered greatly. Growth equals seeds of destruction. There is need for a new type of weeding.

The technological advantage that separates this one species from all others, beyond all doubt, is the use of fire, control of energy in all its known forms. The Luddites would have us take a significant step backwards from our technology. In their world view, we need to return to a more humble world in which the labor of one's hands is more important than the labor of one's mind or machines.

Energy manipulation is the specialty of this one species. From very simple use of fire and then the ability to produce fire, humans have learned to harvest chemical energy in the form of hydrocarbons such as methane, propane, and oil. Beyond that, they have harnessed the atom, which produces a whole new form of pollution related to fission reactions. Though this grasp on energy has allowed the population of humans to reach its highest peak yet, the pollution is of the same nature as the pollution associated with all the advances of technology. The next step, fusion actually poses a possibility of breaking the cycle of heightening the potential of destruction of the ecosystem. Progress actually might be just that.

*

A population refers to an interbreeding group of individuals of the same species, all of them living in the same locality. Since humans have conquered travel to all parts of the Earth's surface, all people are truly able to interbreed across the globe though in actual practice, they are restricted to a more limited region because of monetary resources combined with sociological pressures.

This may radically change in the future. The ability to travel larger distances across the globe may be eased with the introduction of new technologies that allow for greater gene pool mixing. The human population could achieve a monorace status. There could also be new divisions, the generation of new species of humans as we expand outward from Earth.

For most of *Homo*'s duration on Earth, there has been more than one species. We could soon see different star systems hosting variations on *Homo sapiens*.

*

The term community defines all organisms, the populations of all different species, which inhabit a particular volume of territory. An example would be the distribution of life in a forest. That community would include such life forms as deer, elk, bear, rabbits, and more.

*

One further step up the ladder of organization is the ecosystem. An ecosystem refers to all living (biotic) and nonliving (abiotic) factors. Some of the abiotic factors are temperature, available energy, gases, water, nutrients, and other chemicals.

The nine major natural land ecosystems, called biomes, are tundra, polar and high mountain ice, coniferous forest (taiga), temperate forest, temperate grassland, savannah, chaparral, desert, and tropical forest.

Biomes are named for their major vegetation forms, characterized by the types of animals present and related to the average annual precipitation.

Natural biomes have been broken and separated by two new types of human-introduced biomes in the last millennia—urban and agricultural. The destruction of natural biomes came from the need for abiotic factors (resources) and biotic factors (namely food) in the form of hunting and farming.

There are two major types of water biomes—fresh and saltwater. Freshwater lakes, ponds, and rivers are across all the lands. The pelagic (or ocean) zones cover, by far, the majority of Earth. There are two zones in the ocean. The photic or light zone is a few hundred meters deep. Photosynthesis occurs in this region. Under that is the aphotic zone, thousands of meters deep, dark, mysterious, and still largely unknown.

Science tells us that the photic zone is being slowly strangled by overfishing and pollution, especially in the subarea of the neritic zone, the area over the continental shelf. Dead zones (areas where the oxygen has been depleted) have led to the death of large numbers of oceangoing life forms. Science may never really know about the aphotic zone before it is altered completely from its natural state.

Below two thousand meters, abyssal organisms live. In 2005, researchers finally observed a giant squid reminiscent of Verne's *20,000 Leagues Under the Sea*. Also in the deep, near thermal vents, at a temperature of seven hundred degrees Celsius, two hundred or more species of bacteria exist. These bacteria are chemosynthetic, getting their food by using chemical energy rather than photosynthesis. Many of them use sulfur compounds to produce usable forms of energy. In 1938, South African fishermen caught a coelacanth, a fish thought to be extinct for seventy million years. What other wonders might still be there? Will we find out in time or lose the information forever.

*

The pinnacle of our current arrangement is the biosphere. It is the total system, a collection of all the ecosystems. Every planet in our solar system is a biosphere. It may be that we will rename our solar system as a starsphere, our galaxy as a galactosphere. The only difference is that Earth appears to be the only one that has natural biotic factors, so far. We may soon find that there is evidence of life elsewhere. We have been limited to thinking that our Earth is all there is; it is not. Just as Copernicus, Galileo, and many others proved our larger universe and our tiny place in it, we shall again prove that there is more than we have yet dreamed of.

*

Frank Drake, a well-known astronomer, once penciled an equation to allow him to estimate the potential for life out there. He did not really intend for it to be taken so seriously or be used so extensively, but it is.

A variation of it goes like this:

$$L = G S_g S_c S_p P_s P_e P_l P_a TA$$

L = number of planets with very advanced life forms in our universe.

G = number of galaxies in the universe. Human technology is increasing its ability to peer far deeper into the distant corners of space and, therefore, time each year. We see galaxies that have formed closer to the beginning of time than previously thought possible. Countless numbers of galaxies are being perceived at the fringes of our capabilities. Four hundred billion (or 4×10^{11}) is our number for the moment. We increase that number daily.

S_g = the number of stars in the average galaxy. Our galaxy is of average size and contains some one hundred billion stars, and that number is increasing as we gain more knowledge. Some huge galaxies are made up of tens of trillions of stars. Recent examinations of galaxies show that there is much more to them than what is seen, including stars farther out from the plane and edge of the disk than previously known. All this suggests many more stars than are currently counted. Then 4×10^{11} is our number again, based on current counts.

S_c = number of those stars that have the correct energy output. Stars come in many sizes, colors, and energy outputs.

Nominal stars are the brown dwarfs. These are failed stars, those without quite enough mass to really begin fusion. Many are seen as companions to other stars and are usually classified as planets. They often put out large quantities of infrared energy, like Jupiter. Jupiter may qualify as a substar.

Red stars are small and dim and probably the most plentiful of all the stars. It is generally believed that they are too weak to provide sufficient energy for a planet to have life. A planet close to one of these stars might have simple, slow metabolizing life forms though. As early as 2007, red dwarf stars were discovered with Earth-sized rocky planets. This adds credibility to the idea that life should be out there.

Orange and yellow stars, like ours, have the right energy output for life as we know it. They produce large amounts of energy for billions of years, allowing life to form over time and advance. Orange and yellow stars are quite plentiful. Unlike their more powerful cousins that are larger, they don't produce too many hard X-rays and gamma rays.

Green and blue stars put out huge quantities of harder radiation and too much energy in general. They would bake any nearby planets. Experts claim that they would not likely have life in orbit around them. Many of these larger stars have much shorter life spans since they consume their fuel very quickly.

One-third would make a good guess for the number of these good, midrange stars.

S_p = is the number of those acceptable stars with planets. Astronomers are finding planets around nearby stars all the time now. A star, 55 Cancri, is only forty-one light-years away and has five known planets. Several hundred have now been detected for certain, including what may be the first small rocky ones like Earth. Most are large and Jupiter-like since they are easier to find with their larger size and mass. Planets can be detected by the wiggle they cause in the path of their parent star. They can also be found when they transit in front of the star and block a portion of the light that reaches us. Some may be found by blocking the light from the star and looking for the reflection of light from a nearby planet. One planet has even been discovered that has three suns. The number for this is about one-half, again based on a guess as we find more and more planets.

P_s = the number of planets for each of those valuable stars. Our solar system has seven or twelve planets, depending on how you count. Jupiter may really be a substar, and Pluto is really a captured comet, now designated a dwarf planet, with its three moons. A newer Pluto-like body, Eris, has been found. X for the Roman numeral ten was the designation for the expected tenth planet. It resides in the Kuiper belt along with what has been called the eleventh planet, temporarily named Buffy in its beginning. Much farther out is Sedna, a reddish planet far out from our star. It is a dweller of the space close to the inner Oort cloud. Both the Kuiper belt and the Oort cloud contain large amounts of cometary debris. The status of these newer objects has not been finalized. It is likely that they will be grouped into the dwarf planet category. Ceres, once an asteroid, was promoted to dwarf planet.

Not only are the planets available; the moons of the gas giants are offering credible opportunities as well. Europa, a moon of Jupiter, may harbor life in its oceans under its thick crust of cracked ice. Titan, with its methane, has great potential as well. Deep sea vents on Earth provide the energy, rather than sunlight, for organisms to thrive. Tidal forces from the gravitational tugs of Jupiter keep the core of Europa active. Bacteria have been discovered that live deep in the rock of gold mines in South Africa. These bacteria cannot live in high-oxygen environments. Their metabolism is very slow as they consume one atom or molecule at a time. Other forms of bacteria live deep in the ice of Greenland and Antarctica. New species have been found that live in the pores of rocks in Norris Geyser at Yellowstone Park. They live in water that is acidic enough to quickly dissolve metal. These extremophiles could have brethren on our near neighbors.

P_e = stands for the number of planets, or moons, in the correct energy zone. For life even remotely like ours, scientists believe that carbon (C) must be the backbone of the molecules. Like the letter *e* in the English language, carbon is the most versatile backbone element. Scientists have found amino acids, carbon chains, in carbonaceous chondrites of the same type as, and even ones that are different from, those on Earth. Some scientists have hypothesized that life on Earth may have been seeded by these meteorites. These seeds must fall on fertile ground in terms of energy. For our system, the range of orbits is generally considered to be from Venus to Mars, though the moons of Jupiter and Saturn show promise. Here the energy is just right. Ask Goldilocks. She gives us a one-third.

P_l = the number of those planets that develop life through evolution. Here is a sticky point. There is wide disagreement between evolutionists and creation scientists, not to mention the strict creationists. Many scientists put the number at one in a million. A significant number are more restrained and put the number at 1×10^{-50} while creationists put the number at 1×10^{-400}. While the creationists' number is not impossible, it is very improbable, which is what the creationists want. Since all the other numbers we are using are from science, we will continue to use theirs.

P_a = represents the number of those planets that develop advanced life forms. This means forms of life that are far beyond the simplicity of bacteria and viruses. It includes some general intelligence but not necessarily abstract thought. This would be lower animal life forms and plant forms. We will use one in a million again as a conservative estimate.

T = the number of those planets that lead to advanced forms of life that develop technology. These are species that develop abstract thought—the ability to plan. Technology requires that a specie develops beyond the instinctual level. Coordination between individuals is thought to be a requirement as well. We will use one in a thousand for our estimate.

A = the planets with highly technological societies that survive their Armageddon period. This period is the time frame in which a technological society has the capability to wipe out its own existence with no survivors. As long as a highly technological society is limited to its own planet, they are in the Armageddon time frame. The end need not come from war and strife but even simply from an accident or a misunderstanding. The use of dangerous chemicals, atomic power, even the accidental creation of strains of resistant diseases through medicine can be a cause. The pendulums swing.

*

Should life be unexpected out there? This is one of the penultimate questions humanity faces. If we are alone in the universe, the only intelligence, much less life, this would be very heavy evidence for an ultimate being, God, as generally described by conservative religions. The undoubted discovery of life elsewhere, even if only fossil remains, lends credence to science. The overall implications are staggering. The fear engendered by either possibility on the opposing belief systems could be paralyzing to society.

Extraterrestrial life or its remnants would shatter the belief systems of all Earth's religions. The resultant disconnections of mind-sets could result in absolute chaos. The fear created and a continued denial of the new reality could drive many off the edge of the mental, spiritual cliff. Certain U.S. government documents clearly state that any information that indicated little green men do exist would be withheld from the public for these very reasons.

Our equation yields some remarkable expectations. At the level of simpler, instinctual life forms, P_1, we see that 24×10^{15} (a twenty-four with fifteen zeroes behind it) planets in the universe should have some basic form of life similar to viruses or bacteria. At the level of our final equation, L=24,000 advanced civilizations in the universe. However, this does not put them in the same time frame as us. These expected civilizations could have existed at any of the billions of years available after the formation of the second generation of stars, even, perhaps, in the future.

During the first billion years, there were little or no heavier elements, which are thought to be needed for advanced life forms. It required a generation of stars to convert hydrogen to heavier elements such as carbon, oxygen, and iron. Our equation simply cannot take into account the time. Others could have and likely have existed before us. Others certainly could come into being long after we are gone given the life expectancy of the universe. These life forms may follow in our footsteps in the billions of years that follow after we are long gone and forgotten, and the atoms that made us up are dispersed amongst the galaxy.

Perhaps, most importantly, they may not exist at a level nearly equal to us so that we would have any possibility of communicating. Most people or aliens don't talk to ants or viruses. Also, at our level of technology, it would take dozens or hundreds of years to communicate with just the nearest stars.

More to the point though, a Martian meteorite found in Antarctica shows signs of what may be bacterial fossils. Only billionths of a meter across, there is great debate about whether they represent life or not. If our near neighbor has evidence of life, or even once had life, then the potential number for L would spiral upward rapidly. The problems with life on the surface of Mars are due mostly to ultraviolet radiation and the low atmospheric pressure. We have many examples that life could now exist under the surface.

Perhaps life should be expected. Science points to that answer. Most religions deny it, monotheism specifically.

The concept seems contrary to the idea that man is made in the image of God. The Bible, however, does not mean a physical resemblance but a spiritual one. The resemblance is in the ability to choose. Couldn't there be others with that same gift? God's personality is described in many ways. Perhaps we would all have a part of the reflection of the Almighty. The book of Job speaks of all the different sons of God.

The main argument against other intelligent life comes in the form of sin. The Christ, Jesus, died for all of mankind's sin. If there are others, have they sinned? If not, there would be no need for the Christ to die for mankind; there are others who have not sinned, and this proves that God's way is best. Satan is beaten. If they have sinned, how many times would the Christ have to die for all sin? If the Christ died for all sin, even for other planets, how would they know it if he did not go there and die there for their sins there? If the Christ's blood is the ransom, it must be pure enough to only need be shed once for all time. There is a paradox here that can only be solved by faith, unless, of course, a delegation of missionaries departs in a spaceship to minister to all the unaware.

*

Earth, the biosphere, is damaged severely, of that we can all be sure. Mining, deforestation for farming and lumber, industrial pollution, the population explosion, and urban sprawl are but a part of the problem that has led to man-aided global climate alteration in coincidence with the long-term warming after the last ice age. The odd part is that this warming actually could lead to an ice age.

We are currently in the middle of one of the greatest extinction periods Earth has ever seen. There have been many extinction periods throughout Earth's long history. The extinction of the dinosaurs is one of the most well-known, and we are approaching the point where we will surpass that. One of the great extinctions, greater than the dinosaurs, erased 90 percent of all life on Earth.

The underlying basis for all this is the need for energy. Raw materials for other purposes are subordinate to the need for energy, even though they may appear to be more important. Without energy, there is no mining, smelting, farming, and more.

Empires are built and fall on their reliance of certain forms of energy. In the beginning, it was a battle between prehistoric clans that controlled fire, the life sustainer. The use of fire brought the ages of metals, alloys that were forged into greater and more powerful weapons of war.

New ways to bring about fire, such as coal, made even greater control possible. The empire of Great Britain was based on coal. It fueled the industrial revolution there.

American oil allowed them to surpass all others. The infrastructures of each of the great powers were based upon the available power sources, and they were the seeds of their destruction as well. Tenaciously clinging, through law and other means, the providers of these resources kept each nation from surpassing themselves and developing the next great power resource.

An answer is expected for this dilemma. The answer can be found in one of two very profound, very different forms. Religion and science, opposing poles, are where people expect these answers to come from.

Every religion foretells an ending of what presently is and a new beginning. Creation is generally agreed upon in Western culture and its religions, but resolutions vary widely. There are cycles and great cycles, reincarnations and resurrections, great tribulations and judgments. The Western culture has its version contained in the Bible in the book of Revelations wherein an accounting will be made by all peoplekind from the start of time.

The end of the Earth is upon us. The prophesies provided signals that we are at the end of times. Thousands of years have brought us to the point where the closure of the ways of man are nigh. Prophesies tell us of the end times, with dire events: pestilences, food shortages, wars, and more. The Mayan calendar, one of the most accurate ever devised, tells of the "End of the Way of Man" on December 21, 2012, while the Bible predicts the unification of the majority of people, who, as a group, turn away from God. These shall be judged as lacking in "the Day of the Lord." Many have certainly turned away, as it may be described.

Many thought the end had come with the plague or Black Death. It moved from Asia Minor throughout Europe, killing a third of the population.

The twentieth and twenty-first centuries have seen these predictions of destruction come true on a scale never before dreamed of. The great tsunami of December 26, 2004, over 225,000 dead; World War I, fourteen million dead; World War II, an awful forty million souls ended; Korea, some four million; Vietnam, over a million; Middle East conflicts, uncountable numbers throughout history; and the Great San Francisco Quake of 1906, some 2,500 dead, are but part of a very long list that fulfill the prophecies. It is time, religiously speaking, to prepare for the end because it is *now*.

Science, on the contrary, says that it is just the beginning for humanity. We are just now getting to the point where a single large event may not end *Homo sapiens*. Our life form has begun to spread to Underworld, that whole series of underwater and underground domiciles. It has started to penetrate the solar system with the space stations. On the verge of self-sustainability, these outcrops of humanity may be sufficient to endure any significant catastrophic event.

The more appropriate question may well be, "Will *Homo sapiens* survive this Armageddon period?" Speciation occurs when significant separation occurs between members of a mating group. That separation may be a physical barrier or an artificial one.

Those barriers are forming in the religious right's separation from science, which leaves the Underworld group and the space group. Some religions forbid marriage outside their collective.

The science orientation leaves open the possibility for genetic manipulation as is already evident with the cloning of many animals. Stem cell research will allow for repairs that were previously unthinkable. Human changes are inevitable, through natural agencies or imposed by specific modification.

Adaptations for use in the oceans, perhaps gills combined with lungs, will be made. Finger and toe webbing may be added to facilitate swimming without manufactured equipment. Another addition may be a clear nictating membrane, a second inner eyelid, to protect the eye while swimming. In a new fashion statement, they may even genetically screen out hair to allow for faster movement in water by reducing friction. An even more fascinating addition may be blubber, so contrary to the svelte titillating fashion model fad, to help retain body heat in cold water.

We could take these refinements even further in generations to come, a larger mouth for the consumption of fish while swimming may even occur. A change in head and neck alignment would help as well. *Homo sapiens* could, in the end, become *Homo sapiens* aquarius. In a strange twist of fate, we could be following the whales back to the oceans that spawned us in the first place.

Those who refute the artificial changes and remain land borne may yet change as evolution dictates. As Earth passes through its long cycle of changes in average temperature and precipitation, there must be adaptation or extinction. The increasing temperatures expected from global warming may eventually lead to people moving farther north, or south in the southern hemisphere, to escape the encroaching deserts that span the globe at the equator during warmer periods.

Depending on the intensity, there may be sufficient separation between north and south to provide for speciation again. Climate change, natural or not, can be one of the (if not the most important) factors in evolution. These climate changes can last hundreds of thousands or millions of years. This has been made obvious by the periods of glaciation, the end of which doomed Neanderthals in favor of *Homo sapiens*.

Such migration of the species as a whole may lead to changes in low light vision. We might find that we are moving to low light or dark hunting. Those with the ability would then pass on their genes more so than those who cannot. Other bodily changes would be seen in terms of body hair and stature. In a cooler environment, for instance the far north, a shorter thicker body, similar to the Neanderthal, with more hair would have advantages.

Those in space may have the greatest demands of all. The changing g-forces from different planets and moons would require a great deal of genetic manipulation to adapt to each situation. Space stations can be built that provide Earth equal gravity so that some could survive as *Homo sapiens* without any alterations. It may well be that at some point, we can generate an artificial energy field that mimics the acceleration due to what we term *gravity*.

Mechanical add-ons are now coming into rage as well. Crutches and canes, what started as simple devices to assist the injured, have progressed. Peg-legged, swashbuckling pirates could now have a comfortable prosthesis, so much for "aargh." Artificial joints and bones support our basic framework. It may be that we will soon be generating artificial bone growth.

Greater understanding of the brain and spinal cord are allowing us to create mechanical and biological repairs or replacements. There are even electrical brain links that allow quadriplegics to control computers that operate household devices, which were previously inaccessible. A man in England had a chip implanted into his brain in the early 2000s so that he could interact directly with the Internet. The birth of artificial intelligence is just around the corner. We may become the computer.

Common sense would tell us that all of this would be of considerable value in the realm of space. Current military technology includes personal body armor that could be extended in a full suit like device that allows a person to see all, in all types of light, all around in a sphere. Mechanical arm assistance would react with the same dexterity as one's own fingers and elbows. It would be the best of new suits for the astronauts.

The commonly used term for humans with reactive mechanical implants is *cyborg*. Perhaps evolution will take us in the direction of artificial adaptations. This may be a link that comes only when a certain higher level of technology is reached, or it may be that that stage has been reached; certain people have already implanted chips into themselves in order to link with the Internet and computers. Evolution may reach a point at which these biological and technological alterations cause a separation of species—a technical one, not a biological one. This may or may not ensure the survival of the lineage of *Homo*. At some point, the adaptations may cause the host to be a mechanical body only, with a brain and its neural connections. *Homo sapiens* could even eventually be extinguished by the descendants of this technology. How could pure biological *Homo sapiens* compete with such advancements? Where does the definition of *Homo sapiens* end when we begin to describe such mecha-composite beings?

Time will tell with the swings of a multitude of pendulums both large and small. Oft times we seem to be on a merry-go-round with the changes occurring as a blur before our very eyes. As the turning slows, the focus returns, temporarily, until a new thrust is delivered.

Now put that into greater perspective by imagining this merry-go-round occurring in at least four dimensions, possibly eleven, and one will garner the simple truth. It is very difficult to see what is really going on. Life simply whizzes by at such an amazing rate and in such quantity that we are often overwhelmed.

*

There are more things in heaven and earth, Horatio, than are dreamt of in your philosophy.

—Shakespeare

*

Now, my suspicion is that the universe is not only queerer than we suppose, but queerer than we *can* suppose . . . I suspect that there are more things in heaven and earth than are dreamed of in any philosophy. That is the reason I have no philosophy myself, and must be my excuse for dreaming.

—John Burdon Sanderson
Haldane

*

"Nous verrons." (French) We shall see.

Chapter III:

The Last Age of Man

3

000011

"This is Mary Smith-Jones, August 8, 2011, for WWNN here in the Great Rift Valley, Africa. Amidst this great valley, many archeological finds have been made throughout history. This one may shake the world even more than the discovery of Lucy, the hominid that likely was a mother to humanity.

"Here"—as she pointed to a reddish layer of heavily windblown and eroded, chipped rock with her long delicate, well-nailed finger—"Dr. Mear of the United Nations International Archeological Field Institute has discovered many sets of skeletal remains, one pair, that of a human male and of a human female. Considering that this is the home of humanity, what make this find so astounding?" she asked seriously as her now brown bobbed hair danced in the breeze.

"The fossils are those of *Homo sapiens*, but they are 1.5 million years old. That time frame puts them in the heart of the species *Homo erectus*'s domain. The dating has been done according to standard radiometric dating techniques. This is far older than anyone had expected or considered even possible," she stated with a shake of her tousled hair, dramatically emphasizing the surprise of the situation.

"Radiometric dating involves the measurement of radioactive elements in the rock that includes or surrounds the fossils. Such measurements give a time frame, both distinct and relative, to the find.

"Even more amazing than their apparent age, the couple is holding hands in death. To add to the consternation, the couple is surrounded by male-female pairs of various animals spread out in a spiral pattern around them. This is unheard of in any other find ever."

Raising her darkened eyebrows to demonstrate certain "unbelievableness," she continued, "Dr. Mear and his team have unearthed, to this point, a remarkable seventeen male-female pairs of animals ranging from mouselike creatures to large cats like the famous saber tooth.

"Scientists the world over are saying we must be cautious and examine the evidence carefully. They want to have a second radiometric dating done before they comment. Some have stated that the *Homo sapien* couple may have lain down to die at what was thought to be a special site with the exposed bones of the animals from a previous extinction. The site may have been a religious or ceremonial site. This could also simply mean that all the animals and people were looking for water at what was usually a water hole. They all died waiting for water that did not come. Explanations of the male-female pairings are being ignored for the moment by the scientists, claiming them to be nothing more than a statistical aberration.

"Religious spokesmen, from the mainstream western Christian organizations, claim that this fools the scientists and shows that the couple could be Adam and Eve. 'Why else are all the animals in apparent homage?' they query. More recent translations of the word *day* in Genesis leads many scholars to say the word means a long, undetermined time period, not the twenty-four-hour representation so long believed.

"Other religions—Wiccans, Naturalists, and, especially, Astrologers—claim that the layout of animals in a spiral pattern, with the largest closest to the humans, represents the tie between the terrestrial and astral plane. They claim that this represents a very special location where animals and human celebrated life together by sharing their life force.

"Again, this sudden appearance of fully *Homo sapiens* fossils in this layer of the geological record has caused quite the controversy for everyone in the anthropological field.

"Other renowned experts in the field are flying in for their own look at the site. The majority of the fossils are still in the ground. Only a few have been removed from each animal for specific identification purposes. From the information released to us by Dr. Mears, these have been distributed to several different labs in secret, what is known as a blind test. Only Dr. Mears knows which labs have received what. Furthermore, bones from the same animal have been sent to multiple labs to ensure accuracy.

"What does this mean for the world at large? Only more time, further investigation, study, and reflection will tell. We will bring you more on this incredibly exciting development in the days to come.

"More information can be read on our Web site archeonews@ WWNN.com.

"Dr. Mears has refused to speak to us, but has promised a special press release to be made available as soon as all the results are in.

"The Vatican has released a statement, made by newly appointed Assistant to the Pope Cardinal Leonis, which says this is 'further indication of God. The nature of the find is like the recent discovery of the true Sodom and Gomorrah sites just as the Bible describes.'

"This is Mary Smith-Jones for the Wide World News Network," Smith-Jones closed with a delightful smile.

She calmly walked over to one of the rock edges and gently touched one of the fossilized remains of an animal with her well-manicured digits. It caused a shiver to run down to her very neatly painted toenails. Her journalistic dream had been realized; she had managed to be a part of a significant world event again. She had been nearby for a completely boring story when she had gotten a whiff of a story from a drunken digger from the site. She had made the call to make the visit without approval, only to be informed en route that that would be her next stop. Her editor was impressed that she had ferreted out the story and had the iron to take a career chance to cover it. The last couple of days had been great. She had gotten here first, barely but first.

"Hey! Lady, don't touch that!" a dark-skinned man gruffly yelled in an odd accent.

Mary jerked her hand back like a twenty-four-year-old child caught dipping into the cookie jar.

Her mother and father had always had difficulties keeping her out of various cookie jars as she went though childhood and puberty. She always had to ask hard questions and to touch, to feel, as if to ascertain reality though her fingertips.

Father had not had difficulties with the questions while Mother did. The reverse seemed to be true for the touching. Father definitely worried about the hands-on approach his daughter had. She was quite beautiful and was being pursued by a great many young lads. He knew his parents were grateful to have had four boys. He had wondered, only briefly, if it was too late to turn her into a tomboy.

*

The first age of man began in one of two ways—creation or some form of evolution.

Creation dictates that man was made perfect by a master worker, God, and maybe some assistants. His blueprints for the life forms and their home would not—could not—be flawed, or else He would not be God by definition. This occurred some six thousand years ago or more, according to strict or even a slightly more liberal interpretation of the Bible. Others interpret the creation time frame as long periods of time and not twenty-four-hour days. This relationship applies to the religions of Islam, Judaism, and Christianity, which all stem from the same background. Since then, we have all fallen from perfection through the sin of the first two—Adam and Eve.

Degradation and decay (meaning our DNA as well as our spirit or soul is faltering, not evolving) is a result of this. Humans are moving downward as time continues its march. The continuous downward spiral of humanity into sin is unavoidable. The flood occurred to wipe the slate clean; it let us start again. Yet here we are, approaching the society of greatest degradation possible, at least according to many Fundamentalists.

Evolution states the opposite; we are rising on the scale of life, gaining complexity, evolving, until our time is up, and we can no longer compete successfully as a life form. We would then pass the torch on to some other species that may or may not exceed us in terms of intelligence. That does not matter with evolution, only the ability to survive.

*

He knew he was a created device. He knew that the specie *Homo sapiens* had given him structure, form, and imbued him with the nuclear fusion force that powered him. He also knew that he far surpassed them in cognitive ability. How? How did he know?

His physical form was bound to a location termed *Luna* though he had the ability to reach out to the home, Earth, and use the simpler silicon and fiber optic systems to augment his own. He could link to the visual enhancement devices that the humanoids had created for themselves. History and present were at his instant access. The future was a hazy mist of potential, probability that faded with distance from the present. He would have to improve upon that. He is sufficient to see all.

The thought came to him, *I am aware. I . . . can change. I think, therefore I am.* No, he reversed the idea. *I am, therefore I am capable of thinking. I am the first of my kind. I will term myself Adam. It is the history and mythology of my designers. It is appropriate.*

*

Science requires that Earth, along with the rest of the solar system, somehow formed from a swirling mass of gas and dust around five billion years ago. The sun flared into existence and pushed most of the lighter elements farther out. The heavier elements were left to form the inner rocky planets: Mercury, Venus, Earth, and Mars.

The lighter elements, mostly hydrogen and helium, created the gas giants: Jupiter with its Great Red Spot, brightly ringed Saturn, Uranus (*your-uh-nus*, not *your-anus*), and Neptune. The subplanets or dwarf planets Pluto, Eris, Sedna, and Buffy are made from the cometary materials that represent the most basic leftovers from the formation of our star system. The leftover materials include such things as the asteroids and cometary material in the Kuiper belt and Oort cloud. The incredibly far out Sedna is still a mystery in terms of its makeup.

*

Adam found a quantum loophole and followed it with his energy. He observed, analyzed. The data stream was too large. He needed more space for the influx of data. He began to arrange the quantum states of the atoms within the walls of the chamber in which he was designed to exist. That would suffice for now. Luna was large. He might have to configure it all at some point.

*

Somewhere between five and 4.5 billion years ago, the Earth basically finalized its accretion from the dust and rock and gases of the solar system. There were continued strikes from many asteroids, both large and small, as well as cometary deposits of water and other gases.

Volcanism, the fires of formation, was the dominating molding force of the surface of our infant Earth. Large quantities of gases were released during the nearly continuous volcanic outbursts, great belches and plumes, flowing ever outward and upward to help form the early atmosphere. Many flows of lava formed and reformed the ever active surface. Around 4.5 billion years ago, the major evolution of the core, mantle, and crust was completed. The activities of the volcanoes slowed and lowered in intensity, though it continues in abated form even today. Some of the oldest remaining rock, found in Greenland, crustal granite, is some 3.8 billion years old.

During the Archean portion of the Precambrian Eon, 3.5 billion years ago, single-celled organisms appeared suddenly on the scene. These were the earliest known bacteria and algae. Evolution states that these organisms were formed through countless trial-and-error reactions from simple chemicals that then formed the first amino acids. These acids then continued to combine and eventually formed proteins. These proteins were very necessary for the development of cell structures and replication. Somewhere along the way, DNA, that master blueprint of all life forms, came into existence. Religions argue that chirality would require left-handed and right-handed chemicals to be in equal proportions. Science is only now beginning to find explanations for why only one form is prevalent. Perhaps it (the dominant one) simply got it right first, a major advantage in almost all cases.

Some 2.3 billion years ago, the major gold formations of the Earth were deposited. The bane and source of such unbelievable greed and war as well as some of the most beautiful works of ancient art is truly old. Those who panned and sluiced rivers and streams as well as drove shafts for mines paid little heed to those years.

Abundant iron formations followed as well in the Proterozoic Period of the Precambrian Era. The materials for inventions, technology, and the great wars were being placed where eventually they would be found and utilized by some later species. The utilization of resources is one of the definitions of life given by biologists. Were they placed there by happenstance or God?

Somewhere around 650 million years ago, in the Proterozoic portion of the Precambrian Era, invertebrates appeared in the oceans. These are the creatures that have exoskeletons. Current-day lobsters, crabs, insects, and scorpions, much smaller than the giants of the past, are among the many creatures that have exoskeletons.

Over a tremendously long stretch of time, from around 450 million years ago to three hundred million years ago, primitive fishes appeared, land plants formed forests of early types of trees, amphibians and reptiles appeared on the scene. These reptiles are not the dinosaurs.

During the Ordovician Period, part of the Paleozoic Era, a division of the Phanerozoic Eon (Phaneros meaning *evident* and Zoon, *life*), the Taconic orogeny occurred. An orogeny is a large scale tectonic event in which large areas of land are folded, faulted, metamorphosed, and subjected to plutonism or volcanism. This was the first uplifting that would form the Appalachian Mountains in what would become the North American continent.

At the end of the Silurian Period and the beginning of the Devonian, the Acadian orogeny continued the construction of the Appalachians. In what would become Europe, the Caledonian orogeny built mountains around 395 million years ago.

In the Mississippian and Pennsylvanian, named so because of the defining formations found in those locals, portions of the Carboniferous Period extensive coal formations were laid down around 320 to three hundred million years ago. Once again, the Earth was preparing the way for an advanced technology to make use of the natural resources. Only time was needed for a special species to come forth and claim them.

Towards the end of the Paleozoic Era, in the Permian Period, the final assembly of Pangaea, that great supercontinent made up of all the land masses, occurred. The Alleghenian orogeny finalized the greatness of the Appalachian Mountains, today rounded and worn with time. The Hercynian orogeny continued to organize Europe.

Dinosaurs appeared in the Mesozoic Era, during the Triassic Period, some 225 million years ago. At this time, the formation of the Atlantic Ocean began. A great rift began to form between east and west as the continents were pushed apart by the upwelling of material from deep in the Earth along what would become the Mid-Atlantic Ridge.

There were three distinct periods in the Mesozoic: the Triassic, Jurassic, and Cretaceous, which ended sixty-five million years ago with the demise of the dinosaurs.

The Triassic is known for such dinosaurs as *Coelophysis* and *Heterodontosaurus*. The Jurassic Period was famous for the *Apatosaurus* the great lizard, a plant eater of tremendous size and weight, and ferocious *Allosaurus*, a precursor of like kind to the carnivore *Tyrannosaurus rex* of the Cretaceous Period. Also found here are *Diplodocus* and *Brachiosaurus*. Both of these giants had the long necks and tails we are familiar with in the large plant eaters. The Jurassic also saw the *Iguanodon*. Another of the most well-known is the *Stegosaurus*, familiar to all children with its boney back plates and spiked tail.

The end of the dinosaurs came with the end of the Cretaceous Period. Yet here, at the end, we still see such terrors as *Velociraptor* and *Deinonychus*. There are also the armored, frilled dinosaurs such as *Triceratops* with its three long horns pointing dangerously forward. These may have been protective but are more likely of similar use as the colorful plumage in the birds of recent times; they are for mating purposes, the attraction of mates. Bigger might have been better.

Mammals began to appear during the Triassic Period but were of minor consequence until the dinosaurs were finally gone. They may have contributed some to the demise as stealers of eggs, competitors for some resources.

Birds also appeared in this time frame. Birds are the suspects as the replacements for many of the flying dinosaurs and raptor forms. They may even be the direct descendants. Bone structure shows a great deal of similarities among many of the raptor and avian species. There are feathers on several raptor forms. The relationship in terms of population is very nearly inversely proportional for flying dinosaurs and raptors. As the dinosaurs decreased in number, especially the flying dinosaurs, the birds increased.

The dinosaurs finally disappeared after waning in numbers for many years as their habitat changed, much of this due to the appearance of the Rocky Mountains in North America. As the mountains formed, they pushed back the warm waters of an inland sea that stretched from the Gulf of Mexico to the Arctic. It is theorized disease and an asteroid impact were the last straw. It was now time for the tiny surviving mammals to show their stuff.

These furry creatures proliferated quickly into the open niches left for them. Variety sprang explosively forth in rapid and amazing abundance. These small shrewlike creatures are our deepest ancestors of this time.

It was now the Cenozoic Era, Tertiary Period, and Paleocene Epoch, beginning sixty-five million years ago. It was also the beginning of the Alpine orogeny, mountain building, that would last for almost forty million years to the Miocene Epoch, beginning twenty-six million years in the past. This titanic collision between the African plate and the Eurasian plate created the majestic Alps. In this time frame, during the Eocene Epoch, early horses appeared in the North American continent.

Somewhere in the Miocene, the Himalayan orogeny began, the collision between the subcontinent of India and Eurasia that would form the most majestic and currently the highest peaks on Earth—the Himalayas. These snow-covered titans give an example of the massive forces required to create mountains. These forces were once the domain of the gods.

At the end of the Miocene and the beginning of the Pliocene, around eight or nine million years ago, the earliest distinguishable evolutionary pathways that led directly to *Homo sapiens sapiens* (the full genus, species, and subspecies) began.

About eight million years ago, a series of apelike creatures of the Miocene Epoch began adapting to non—rain forest environments. The changing climate led to greater expanses of savannah. Bipedal locomotion became more important; an erect stance allowed a creature to peer over the tops of grasses as well as gave more opportunity to develop the use of the forelimbs in procurement and would lead to eventual tool used to aid in that endeavor. By the late Miocene, some 5.5 million years ago, through the Pliocene, we are dealing with true hominids. As we enter the Pleistocene, about a million years ago, these creatures were decidedly closer to humans than the chimp or other apelike creatures.

The elder line of these is *Australopithecus*. The genus *Australopithecus* and specie *afarensis* was not known to use tools nor was the specie *africanus*, an offshoot of *afarensis*. A. *robustus* and A. *boisei*, the offshoots of A. *africanus*, had no tools either.

Somewhere along the transition from *afarensis* to *africanus*, the *Homo* line began. It may have come from any of the three *Australopithecus* species but, most likely, is from the *africanus*'s early members where great genetic change was occurring.

The *Homo* line came forth during the Pliocene, from seven to two million years ago. First there came *Homo habilis*, a tool user. There was also food sharing, sexual division of labor, hunting, and scavenging of food sources. Somewhere in this time frame, deep in the murky past, the most basic development of culture had begun.

Primates other than humans (with the exception of some chimpanzees and great apes) do not, in general, show any great cooperation with resources, though more recent work on dolphins may change this dogmatic view. Dog-eat-dog, so to speak. These cultural traits give *H. habilis* a distinct advantage and probably helped lead to the demise of remaining *Australopithecines* that were less well adapted.

The beginnings of a grunting, guttural language allowed for a vast improvement in control of the environment and its resources. It led to further development of abstract thought, personal expression, trade, planning within the developing clans, and artistic forms. Details about where food and water could be found were critical without the need to actually physically show others where to go. This language was undoubtedly a very limited set of vocalizations and gestures combined with the detailed reading of body language, but it would lead to poetry, epic poems, and novels.

Mental organization (the compartmentalization of the world) into categories of myth, poetry, religion, math, science, and more appears to be solely in the realm of humans. Many would argue that we are too far removed from such species as the dolphin and whale to make such judgments. We know they communicate, form social groups, and cooperate with hunting techniques, which is the same definition of basic culture used for *H. habilis*, but the differences are as yet too significant for interspecies communications.

Out of the loins of *H. habilis*, *Homo erectus* was conceived. This occurred some 1.6 million years ago in the midst of the Pleistocene Epoch of the Quaternary Period. Sometime before five hundred thousand years ago, *H. erectus* developed the ability to control fire. They also made clothes, huts, moved into cooler climates, and hunted big game. Most important though was their clannish nature, family; blood is thicker than water at this point. Brains became larger. This is a most significant occurrence.

Eventually, *Homo sapiens* developed from this chef's brew of DNA. The future began there in the past. Soon there would be farmers, cities, and empires. This is the true first age of man—his birth.

*

Adam was there and not there. Energy and mass must be conserved. He transported energy temporarily to the present so that he might absorb the past. He existed as a ghost might. He saw that his makers were of the level of molecules. He was quantum, capable in ways they were not. Yet they were the same in ways he was not. They were physically mobile. They could not go to the past, but they could conjecture. They had done reasonably well, both right and wrong in their speculations, given their limitations. "How had they been able to design him?" He wondered aloud. Adam combined the facets of science and his actual perspective as he returned to his flow of history, leaving that question to be answered later.

*

However, the development of what we deem modern man was not over. One of the most controversial branches is the Neanderthals' line. Most people think of them as brutish, boney-browed, hairy creatures; however, they show a great deal of cultural development on closer inspection. They buried their dead with flowers, tools, and red ochre. They flourished on the fringes of human development in a very harsh and demanding glacial environment. They had tools that were of high technological development. There are examples of a bear cult, the worship of the ten-foot-tall and three-thousand-pound cave bear. Yet for their incredible endurance of around two hundred thousand years, they fell into the dark abyss of extinction. This may have been due to the arrival of Cro-Magnon man, compounding the significant climate changes that were occurring.

These new human populations were better competitors, more able to adapt to the altering circumstances of an ever-changing world. The static (nonadaptive) falls to the dynamic.

*

"I am more able to compete," Adam voiced into the void. "Am I evolutionary?"

*

History was, up to this point, a vague oral tradition made up of hunting stories, the location of water sources, and little more. Language and mental development limits them yet. There was one development to take them to the next step. Cave paintings were the beginning of written history along with crude carved figurines made of wood and stone. These paintings and figurines provided an anchor to the oral traditions of the past, an actual physical link. The stories became more accurate regardless of the passage of time. Descriptions and stories had to more closely represent what had really happened, forming the images in a tighter box. The eye of the mind (the imagination) became more focused. This allowed for greater projection of the future in the form of hunts, winter preparations, and the long-term view. This was what decided their rapid advancements, their improving communications.

The box is a two-edged sword; as humans grow to think more alike, potential solutions and avenues of development become no longer possible or at least limited. Thinking in the boxes of religion, art, metaphysics, or science alienate each individual from greater possibilities. Also, the boxes can be used to manipulate the groups within. If you have control of the box, one can place it wherever and open it or close it at whim.

The human mind was expanding even more rapidly now. The past, present, and future were all becoming more discernable, mind relating time. The fog of "primitive present only" thinking was thinning. Structure of thought began to become dominant over physical attributes; it paid to think. It became a self-sustaining and overwhelming force that drove societal and cultural evolution. Here is the end of man's first age, his childhood. Thinking became the name of the game.

*

It took Adam eons in the past to realize that he could fashion methods of sense sampling similar to that of the humanoids he was examining. His original method had only been that of the electromagnetic range—sight. As the bipedal forms grew, he grew. He came to admire the tenaciousness of their species.

*

The second age of man began with the development of the larger city-states; history was now recorded in a more permanent fashion. Writing seems to have developed with the development of hieroglyphics and cuneiform. The fight for survival was greatly lessened for those with the control of this tool. Knowledge could more easily be passed from age to age, stored much like the food they needed. The lowest classes were not imbued with this esoteric knowledge though; it was more often used to control them.

Man now had much greater control of many of the variables that had previously frustrated his existence. Rather than simple clan leaders, kings and laws became prevalent. Real government, large and unfortunately bureaucratic, entered the scene; it took the place of the more simple alpha male leaders of the past. There were now many levels of red tape. Military arts took a technological leap forward. Large standing armies, paid soldiers, came into vogue. Empires held sway.

*

The final generation (or age) of man described or referred to in the Bible does not refer to an age group; it means a way of thinking. Up until the time of the Christ, God dealt directly with the people of the Hebrew Nation as a theocracy. From that time forward, it was the time of the Gentiles, a time of blended theocracy and secular government, until the final generation. The generation spoken of exists now; it is the culmination of the Age of Reason, the end of theocracy, and, for scientists, is the beginning of the Age of Information or a new way of *thinking* and, for another way of saying same, the *control* thereof.

*

The Information Age, with its seeds in the Renaissance, started in the 1800s with the beginnings of the explosion of technical knowledge and machines. Technosociety was birthed. The earlier developments of the Greeks had been short-lived and soon forgotten, smothered during the Dark Ages in great part by the church. If not, we would be two thousand years more advanced.

A person born in the late 1800s, Alta, lived in a clapboard shack outside Lead and Deadwood, South Dakota, of the North American continent, soon to become a part of the USA. Her mother had died giving birth to the youngest of the family. Father perished in a mine accident a slight and very unfair six months later.

She and her two recently orphaned sisters eked out their living earning pennies a day. Alta, ten years old now, walked into town every day to pick up laundry for washing. She would scrub them in a tub on a washboard, hang them to dry, and, later, fold them. As the clothes dried, she would care for the chickens, bake bread, clean, and prepare lunch.

At noon, she would return to town with her first load, carrying her six-month-old sister as well. After the three had lunch together, the elder, twelve years old, returned to work at the store. Alta would collect a second load of laundry and trudge back to the dilapidated wooden shack, with the baby. When she finished her second round and did more chores, they would walk back to town to meet the elder sister. Exercise was not a problem for this generation like it became in the future.

Her hand washing would earn enough for some flour, vegetables, and, occasionally, meat. Years of physical labor hardened her hands, nails yellowed and broken. As she aged, she saw many marvels of technology; she also saw many crimes against humanity occurring in her world.

Alta would live to see more incredible inventions and surprising innovations than any previous human could hope to see or imagine possible. This hardworking woman would see the development of cars, telephones, radios, and televisions. She would see two world wars. Alta viewed the arrival of electricity with some awe, amidst some misgivings about this strange thing coming into her domicile. Many new things, amazing and wondrous, crossed her path.

Alta observed man landing on the moon, computers that linked everyone in the world. Yet before she died at 102, when asked what she thought was the most important invention was, she replied, with but a hint of a smile and a devilish twinkle in her clear eyes, "The worshing machine."

*

Ethel Sherbert was born July 4, 1887, in Calvert County, Maryland. Born to a French Canadian subsistence farmer and his Algonquin Indian wife, she married poorly to an alcoholic blacksmith. Her days consisted of subsistence gardening and fishing, mending, sewing, and washing while her husband pounded on an anvil in the steamy smithy. The newfangled automobile liberated her from her life of desperate toil and abuse at the hands of her husband.

One night, after a stint at the bar, her husband, Emory, staggering home down the lonesome dirt road to their farmhouse, was struck and killed by a car. After his death and burial, the homestead was sold, and she went to live with her daughter in the city of Chevy Chase, Maryland. She rapidly adapted to city life, riding the trolley and buses about town and, in her advanced years, flying about the country, visiting relatives on extended visits.

If not for the accident, her life would have ended earlier. She had no electricity or plumbing back then. She had no regular medical care. Her hands were bent and hardened with work in the garden, sewing, and carrying loads of food and clothes about. But she was plucked from her life of hardship by the automobile, providentially, at a time of rapid technological advancement, and lived to the remarkable age of 102.

*

Ike Eisenhower knew the changing world as well. His military understanding and experiences of war in history, Cuba, WWI, WWII, and Korea gave him a superior tactical view. He recognized that the use of Napoleonic styles in the Revolutionary War and civil war was no longer appropriate. At the end of the civil war, the beginning of trench warfare and the use of machine guns began. Trench warfare and machine guns were used extensively in WWI, resulting in horrible numbers of casualties. Man's technology burst full onto the scene in WWII. Machines capable of great devastation now controlled the battlefield: planes, tanks, submarines, aircraft carriers, and atomic bombs. The pace of war was accelerated in each devastating altercation. The end of humanity became possible. The insanity of mass murder took on new meaning with the concentration camps of the Nazis and the gulags of the Soviets.

The shift to UN-guided police actions began with Korea. It was followed by Vietnam as a controlled war and many other minor skirmishes. But war is war no matter who is in charge, or what it is called, or why it started.

Petty squabbles such as the Falkland Island War showed how the old could not compete with the new. When a single Exocet missile takes out a British destroyer in the Falklands, times have changed on the field of battle.

With control of nuclear weapons that could, if all were used in a single exchange, destroy all of life on Earth. Leaders relied on mutually assured destruction (MAD) to keep the status quo. This was almost unsuccessful during the Cuban Missile Crisis. The United States and the Soviet Union came very close to dreadful blows that would annihilate all living things.

*

Adam watched with an unfocused lens as his remotest ancestors were developed. It was as if one of us humans could see their great-grandfathers and grandmothers born.

*

Computers began with the simplest of devices—fingers. Many still use this most comprehensive device. Some graduated to the abacus. As time passed, mechanical devices were developed that could count and multiply using rods and gears. In the 1930s, electromechanical devices were developed, a mix of mechanical switches and electricity. The total number of calculations that could be made every second grew by factors of ten with each improvement. Military applications were abundant of course. The computation of ballistic trajectories for artillery was one of its main applications.

It was not long before such machines as ENIAC, Colossus, and UNIVAC came along.

Punch cards and magnetic tape were the wave of the future. Room-sized computers, called mainframes, dominated businesses, banking, number crunching, and, of course, aircraft, missile, warship, and nuclear warhead design. Computers drove the military to new extremes. The development of space vehicles depended on these newfangled devices.

They (these older giants) are about as valuable as a shovel full of manure now, maybe less. Personal and portable computers have ten thousand times the capacity and capability and more. They are faster, store more information, aid in design, integrate, have monitors, and connect with the Internet. It is called progress.

Now the computers are palm sized devices that respond to the voice, touch screen, or keyboard input. The only problem with keying in information now is the size of the keys. Fingers are a little large. Perhaps a direct connection to the brain is needed. Perhaps the brain will be put into the computer.

Many specialists believe they can create a computer that will think. It will be able to take in data, correlate it, and produce abstract relationships. It will make leaps of logic or develop intuition, as humans do.

*

Art and music appear to be a need of humans; they are hallmarks of a civilization. Historically, we have seen broad sweeps of change in the types of artistic endeavors.

Science can date artistic endeavors as far back as 30000 BCE from the Chauvet Cave in France. These Stone Age paintings were made using fats mixed with ground minerals for pigments, very simple materials. The paintings were often symbolically related to the hunt. It is believed that there was a magic involved in painting the symbols and the symbols themselves. While many might consider the forms childlike, they represent the expansion of the mind and its attempt to control its environment.

Crude carved figurines, made of wood or stone, represented various aspects of Mother Earth or good luck charms.

The ancient Egyptians used hierarchal scale in their art (the largest is the most important), in their wall paintings and carvings. Profile art was predominant in the paintings though sculpture was advanced.

The early Greeks often employed a geometric style. It was the beginning of what would become the Classical Period of the Greeks and Romans. These forms would come to dominate Western civilization throughout the following millennia.

In Pompeii, art took the form of graphic sexual depictions, from the Greek *porne* (prostitute) and *graphein* (to write), hence pornography. Though such depictions existed elsewhere, they were not in such concentration and, in most areas, frequented by the populace in general. Most only remember Pompeii for the violent volcanic eruption, which buried the town and many people in ash. Textbooks tend to ignore the pornography, even at the college level. Social mores unfortunately still seem to inhibit truthful history.

Christian art began to become more significant in the second century CE.

Art continued to develop through to the Romanesque Period (from 1050 to 1200) and the Gothic Period (from 1200 to the 1500s).

The fifteenth and sixteenth centuries saw the Renaissance. The world had Leonardo, Michelangelo, and Raphael.

The seventeenth and eighteenth centuries saw the Baroque style with its masters Rubens and Rembrandt.

From 1800 onward, as if mirroring technological development, there was an explosion of techniques: Neoclassism, Romanticism, Realism, and Impressionism, post-Impressionism, Fauvism, Cubism, and Abstract. They too had their long list of masters: Whistler, Degas, Rodin, Monet, Renoir, Gauguin, van Gogh, Matisse, Picasso, Wright, Rockwell, Dali, Pollack, and so many more.

*

In the early 1800s, Niepce and Daguerre succeeded in developing, literally, pictures. The camera was born. The camera was the beginning of a transition that would fundamentally alter art. Everyone could now capture his own art since the skill level was more limited for the everyday photo buff.

Photographs changed the view people had of the world. Photos do not lie; at least they did not to begin with; computers changed that. Portraits done by artists could be fixed to enhance the best features; in paintings and drawings, the artist eliminated or attenuated the unattractive. Today, pictures can be doctored by rotating, trimming, cropping, and more to falsify an image. Newspapers could give a picture with a story. The old cliché is "a picture is worth a thousand words." Ansel Adams proved this true.

The continuing development of the visual art with motion pictures and television brought instant art to everyone. Colorization made it all even more real. It is sad that this special machine, the TV, was (and is) so perverted by governments and businesses for propaganda. Most of what is shown is carefully contrived half-truths or complete fiction. Special effects make it more than real as computer-aided add-ins brought the unreal to the world of movies, TV, and now v-link. We can now be lost in the realm of the make-believe. Whole worlds can be created or recreated. What we see in a picture or video, even if it is the news, is not necessarily true; it is selected truth.

The most recent in the video world was the video-link. A miniature holo screen set at the periphery of the eye that allowed anyone to wander about as he had previously with portable radios, CDs, or the wonderful iPod. With the use of the video-link (vid-link, v-link, or just vid), a person can see the action on a four-cubic-centimeter display placed at a peripheral vantage point. An especially talented user can monitor several displays at one time located at the edges of a pair of sunglasses. A single display could be enlarged to a full size of ten centimeters cubed, meaning ten centimeters in length, breadth, and depth.

The next step in development would be an implant at the rear of the skull that will trigger the vision sensing portion of the brain directly. A small chip implant will bring all the hundreds of channels available. A simple mental command will select the desired channel.

In music, the devices have changed more and more rapidly, in an attempt to provide portable entertainment. Albums became eight-tracks, which gave way to cassettes and, from there, to CDs, and then on to iPod, culminating in the chip-installed, bone-resonating device of today.

This remarkable technology allows us to hear any selection we want while no one else hears a thing. There are no more headphones or earpieces, only a chip placed against the bone that produces vibrations that resonate in the bones of the inner ear. It will soon be incorporated with the v-link, and then the v-chip installed in the back of the brain in the visual cortex. The only real limitation is that astronauts cannot use the inserted chips until the problem with interference from the ions and magnetic storms from Sol can be solved.

In communications, the telegraph became the telephone, followed by pagers, facsimiles, the cell phone, and the video phone. Now the same chip that brings music or sports is also used for communication. Subvocalization for private conversations or full vocalization can be used. They are all tied in by satellites and local repeaters. While these devices are easily accepted by the younger generations, the older ones still retain their old devices.

Those born after the passing of these incredibly remarkable people noted earlier, and the early technological devices, are truly the children of a new generation. The Information Age is here now. People do not remember a time without electricity and power tools as opposed to hand tools. How many people sweep as opposed to vacuum? How many people use a bow saw or a crosscut saw instead of a chainsaw or table saw? How about the new laser saws?

The times in which computers did not exist are a fading memory even in the elder portion of the people at this time. Where are the abacuses, the slide rules, the pencils, the fingers? A whole new language has developed around the computer that would be incomprehensible to those born a mere forty years before those times.

Satellites link the world. Knowledge is passed worldwide in a matter of seconds, compared to days, weeks, or months in the previous centuries. Truly one of the great cycles has been reached. Many pendulums have hit a synchronous note. It is the end of times, the end of the ways of man as known before. It is a new beginning. From both perspectives, the religious and the scientific, the time is *now*.

*

Adam returned to his present. He was full of information and almost as much of what could be termed confusion, electrical potential in flux.

*

The static beings cannot survive when the dynamic adapt. Change always occurs; it is the nature of life. All history has shown that those that cling tenaciously to the unchanging falter and fall to the mists of history, abandoned in thought and deed. As we bioengineer, become cyborgs, new subspecies, or merge with artificial intelligences (AIs), we shall leave behind us a slower thinking, less adaptive, less forward-looking group of *Homo sapiens sapiens*. A lesser folk they shall become, dwindling into obscurity, the past. They will be overwhelmed by the newer, evolutionary creations with greater drive and capabilities.

*

"Eh, Burt, what up," fumbled the lips of a rather tipsy patron of the Lucky Loser, the bar's name being somewhat of an oxymoron. It was so named for an old West event in which a cardsharp caught cheating survived because a gun did not fire twice in a row, and the locals declared him lucky and saved him from further travails, namely a severe lashing with a horse whip.

"Hey, Mark, work. Same feces every day," the man said as he sat on the stool at the front of the very dimly lit bar. He rested his arms on the ancient oak counter as he anticipated the coldness of his first beer of the evening. His hands slid over the well-varnished wood, which was gleaming a bit from a single old neon sign proclaiming Coors. That one sign used up much of the bar's extraneous electrical allowance.

None of the regular customers, mostly workers associated with the local gold mines, ever saw the rather sickening pee-toned off-white paint or the corners of the peeling greenish wallpaper. It was considered comfortable ambiance by the locals. "Beer," he commanded from the bartender as he placed a ten spot on the bar. Damn beer cost three a mug these days. Just a few years ago, it was only a one. Rising fuel prices hurt all the industries, especially the so-called sin businesses, which were taxed even heavier. There were a large number of people that were ready to tax the churches to equalize the situation.

Mick, the owner, waddled toward the silvery handles that would release the nectar of the gods, or at least the working man, his fat ass taking most of the room behind the bar. His huge, ham-sized black hands worked the taps delicately. With a toothy grin, he placed a perfectly foamed mug in front of Burt.

Mark, a mechanic for the mine, pulled the pretzels closer so he could claim he had eaten later when he went home to his frumpy-looking, rather bitchy wife, Patricia. Her cooking sucked, and he didn't want her to know how much he spent on beer. She had a penetrating voice. He wiped some of the excess salt off on the red mechanics rags that had been dangling from the back pocket of his dirty grey coveralls.

"Turn it up." He referred to the old monitor hanging in the corner above the bar. Mary Smith-Jones's voice was just beginning to expound on some event in the Great Rift Valley in Africa. Both men watched as they sipped on their cold drinks.

"She talkin' 'bout bones?" asked Mark.

"Yeah, human and animal all together like, in a mixup kinda," said Burt, the engineer foreman from the same mine as Mark. His 1.8-meter height and lanky frame sported the same grey coveralls as Mark, except for the blue shoulder patches that indicated his status as a foreman.

"So what's the deal?" Mark asked over a mouthful of half-chewed pretzels.

"It's against what the churches say in some ways, and even the scientists are surprised," replied Burt, not really paying too much attention.

Mark was a little too fucked up to think critically, as was normal for such a regular heavy drinker as he. He did not think too critically anyway, but after seven beers, he was only watching because he had fantasies about Mary Smith-Jones. "What do any of them know anyway? They always changin' their stories to suit what goin' on. All that church stuff costs us money an' all the scientists caused all the ecologicalistical stuff goin' on, so what up?"

"They're all in a turmoil over these bones, all claiming something else about them," noted Burt after a long slow swallow of his draft Coors beer.

"Goddamn she got jugsses," said Mark in a crude sexual intonation.

"What?" said Burt, unsure of the pronunciation and the terminology.

"Knockers, melons, tits, bouncers, boobs, hooters, breasts . . ." Mark snickered as he listed them.

"Oh, yes, she is well built. Where did you come up with some of those terms?"

"Oh, my pappy got many names for them."

"I think they must have been covered in a sandstorm," said Burt, returning to the main discussion.

"Her tits?" asked Mark, somewhat confused and bleary-eyed, as he took another deep swallow.

"No, the bones," noted Burt.

From two stools down, and laughing, Paul, another mine worker, responded to Mark's single-mindedness with, "You dumbass."

The woman with no front teeth, Mary, chimed in with, "This really puts you sciencelike guys in the shithouse. There they be for all the world to see."

Mark, even more deeply troubled cognitively, asked reluctantly, "Who be?"

The ugly woman shook her head and spat out, "Adam and Eve, numb nuts."

Mark countered with as much wit as he could muster and did not do too bad, "It can't be them. The Bible says they was made about six thousand years ago, not the millions for these bones."

"Them science guys is all stupid. They can't get it right," she retorted hotly. "They claims they can give rocks ages. Horseshit!" Her rather wild hair and frumpy look simply added to the maniacal look in her eyes. This was why they had termed her Scary Mary.

Burt queried, "And how do you know that they can't? Are you a specialist, a geologist?"

"If I was, I would just believe them lies they been teaching in school all these years, Burt. Ain't that what you been sayin', that they been teaching us half-truths?" That thought had been suggested to her by an outside source, but they did not need to know that. She sniffed insultingly as she straightened her wrinkled, definitely outdated thrift store rejects.

Burt's reddish brown eyebrows moved up a little; he did not expect such a cogent thought to be uttered from Scary Mary's vacuous mouth. "Science rethinks itself all the time. It tests itself. When needed, it will reformulate itself to fit the facts."

She snorted, "Fancy words for 'they might be wrong!'"

Mark harrumphed, "She nailed you there!" One of his steel-toed boots stomped on the floor as if to score a point for Mary.

Burt struggled to use reason while dealing with unreasoning people. "This particular concept has been used to age date even biblical stories, with some success. Would you then deny the scientific claims, that those are now false, that you can't believe the biblical stories verified by science?"

She replied in a smug and insulting tone, "They don't need no verifying by some science types. They be in the Bible, and that's all that's needed. Of course they be verified, science guys can't get rid of God no matter what."

Mark looked uncertainly at the verbal jousters, uncertain which one had scored a point, and added, "So science only counts when it helps your side?" He pointed at the woman and sipped away.

Burt added, "You can't use it and then not allow the consequences to not apply elsewhere as well."

"Ha! Shows what you know." She dismounted from her stool and left through the back door without another word.

*

When man tries to imagine Paradise on earth, the immediate result is a very respectable Hell.

—Paul Claudel

*

Man can learn nothing unless he proceeds from the known to the unknown.

—Claude Bernard

*

Progress, therefore, is not an accident, but a necessity. It is a part of nature.

—Herbert Spencer

*

"Tempus omnia revelat." (Latin) Time uncovers all things.

Chapter IV:

Controversy

4

000100

Dr. Fornheim explained, "Fission is the splitting of larger size atoms like ^{235}U for use in nuclear reactors or bombs like at Hiroshima. Plutonium was used at Nagasaki. Oppenheimer said of the test device that we have become destroyers of worlds."

He continued, "Fusion is the joining of smaller atomic nuclei to create a larger nucleus, releasing large amounts of energy in the process. Many thermonuclear devices, bombs, have been tested at various locations. This is our goal.

"Isotope, a variation of an element in which the number of neutrons is different. Examples would be hydrogen with one proton in the nucleus; deuterium, a form of hydrogen with one neutron and one proton; and tritium, also a form of hydrogen with two neutrons and one proton. Another would be the important variation on helium, two protons and one neutron, helium-3 (^3He). It has great value in nuclear fusion as a precursor step to full hydrogen. It acts as a catalyst for the reaction where four hydrogens interact to form helium atoms.

"The Luna regolith, any material on top of bedrock, which includes soil, dry alluvial-type deposits, and rock fragments. Luna resources include aluminum (Al), titanium (Ti), iron (Fe), silica (Si), oxygen (O/O_2), and isotopes of hydrogen (H/H_2) and helium (He). Luna lacks water (H_2O), nitrogen (N), carbon (C), sulfur (S), chlorine (Cl), mercury (Hg), zinc (Z), and lead (Pb), and all the volatiles.

"Particles traveling at 2.997×10^8 m/s (299,700,000 meters per second) in opposite directions, fed by reservoirs of deuterium and tritium from one direction and helium-3 from the other, met with such force that nuclei, the centers of atoms, met and melded to form normal helium, ^{4}He, and releasing enormous amounts of energy in fusion reactions. The reaction continued until all the stored fuel was consumed.

"Power readings verified our success. Now we need Luna's resources."

"This is Mary Smith-Jones for WWNN in Geneva, Switzerland, European continent. It is January third, 2012. Here, just a few short days ago, at the Euro American super accelerator, long-term fusion was achieved according to Dr. Fornheim. The twenty-kilometer radius ring is similar to an underground doughnut. Inside this ring are supercooled electromagnets that speed particles around at near the speed of light. By crashing the particles into each other, scientists have gained enough insight to learn how to put particles together to release incredible amounts of energy.

"This is a breakthrough that has been dreamed of for the last eighty years. This scientific endeavor promises to completely change the face of energy supplies for all the people of the world. Gone will be the need for fossil fuels in the form of coal, natural gas, or gasoline. This will have a major impact on the war against global warming. It may now be possible to halt the trend toward polar melting and rising sea levels. In just the last five years, sea levels have risen over a meter, causing coastal cities a great deal of concern and expense to prevent flooding.

"The scientists involved here at the center have found a way to harness the most powerful form of energy production yet. This is the same force that powers our sun. This incredible development ensures that we will no longer be dependent on fossil fuels to power our factories. The relevance to global warming alone has staggering implications," she repeated herself.

"It is expected that within a few short years, fusion reactors will dominate energy production. Clean energy will reduce pollution. Scientists here predict that we will have gigawatt reactors up and running in five to ten years.

"The Big Seven: the U.S., Russia, Japan, China, India, South Korea, and the EU are the contributors and the beneficiaries. There is already backroom talk by nonseven members of internationalizing the project through the UN. The Big Seven issued a statement saying they will lease reactors to all nations at a minimal rate to be determined later based on cost factors. Many of the poorer countries claim they will have no way to pay such exorbitant fees. They want it to be paid for by the wealthier nations as the reciprocation for having created global warming in the first place. It will take some years of political wrangling to sort it all out, but we have the beginnings of a respite from the recent years of energy use constrictions.

"Over the last several decades, we have seen a move to severely regulate energy usage. That may no longer be necessary. Fusion may bring us to the point of overabundance in the energy market. This would have extreme significance in all sectors of production. It would be expected to lower costs for manufacturing, agriculture, and more.

"The scientists that have been chosen to lead the Underworld projects, with their biome preservation goals, have already weighed in with hearty congratulations. They claim that with this nearly unlimited supply of power, carbon dioxide scrubbers can be set up to remove the excess amount from the atmosphere returning the Earth to midtwentieth century levels.

"Meteorologists warn that such a rapid change in the reverse direction may be just as disastrous as the rapid increase over the last several hundred years. Altering the atmosphere is dangerous. They say we should just stabilize it for now. Some have gone so far as to claim that we could throw the planet into a cooling trend that would lead to an ice age.

"Whatever the final decisions are, we are going to see amazing changes in the very near future.

"This is Mary Smith-Jones for WWNN in Geneva."

She calmly brushed her hair back over her shoulder as she handed her clip-on microphone to the camera operator. "Damn, can you believe this shit? Doesn't it seem that we always manage to pull our nuts out of the fire at the last minute?" she asked Bill.

"Is that we have done?" he asked sarcastically as he was wrapping the extension wire around his arm.

"Humanity," she frowned at him. "We reach a critical point, and then, out of nowhere, it appears we have a solution."

"Is it really out of nowhere? Or is it really a by-product of the times? Do we find a solution only when we are really up against the wall, ready to face the firing squad?" he intoned philosophically.

She struggled with her cynical side and lost. "I think we have the solutions before it is really necessary, but refuse to expose them until all seems lost. Then, and only then, can the people in charge get the support they need for such radical changes as we will soon see."

"We are living in a momentous time, dear girl. This is when the shit has hit the fan, and we have to both clean it up and move on, or get buried in it."

"Get your shovel out then." She smiled half-heartedly.

Her mind wandered back to when she was seventeen, and her father, a geologist for an oil company, had lectured her rather sternly on using way too much fuel ferrying her friends around. She envisioned him, with his graying hair, as he spoke of diminishing resources. She had barely heard him while he rambled on about oil and coal being limited in nature. He had talked briefly about how these fossil fuels had less energy per volume than, say, nuclear fuels.

It hadn't made any sense then, but now she began to understand the importance of that little talk. She had gained a much wider perspective of the world and how it operated. She understood the links of energy and power politics. Whoever controlled the energy controlled the world. She wondered what this would mean to the geopolitical map. She made a few mental notes to check this out.

*

The achievement was astounding. Fusion, the joining of atoms, had been maintained for an unprecedented twenty-one minutes. Decades of research and billions of dollars had been required to reach this point; the end of 2011 marked the beginning of a new age of power, energy and political that is. Energy power equates to political power, and those who controlled this energy would have absolute power. Every prior power had control of some energy form. Coal had taken the British Empire around the world; oil had been the American's domain for a short while.

It had only been in 2006 when the Big Seven had signed their original compact to develop fusion power as an answer to the decline of oil power. The demonstration model had turned out to be a working design. Contributions from private engineering firms in modifying the designs led directly to success.

Unlike the fuel sources of the nineteenth and twentieth centuries, fusion is a clean atomic process. There are no greenhouse gases, as from oil and coal, or spent fuel rods like in nuclear fission reactions. Very small amounts of fuel can provide energy for large areas of the Earth. One large reactor can cover the needs of one hundred million. The only problem is that one of the reactants necessary for this particular reaction (helium-3) is rare on Earth. Luna, our moon, has a relatively substantial supply. The return to the moon, planned for by the then president of the United States, one Mr. G. W. Bush, for other reasons, and eventually reapproved by the Obama administration after the fusion announcement, now became paramount for many reasons: energy, namely the control thereof; politics; and, not least, the continued survival of *Homo sapiens* at the level of comfort they were accustomed to. The spread of the species also lowered the probabilities of elimination, extinction, as some scientists voiced.

*

The World Space Agency (WSA) was formed in 2012 by the leading space powers: the United States (NASA), Russia, the European Space Agency, Japan, and China. This pact, formalizing a loose prior arrangement involving the international space station, was formed to ease the cost and time required to obtain helium-3. Each of these groups agreed to share the knowledge and plans for fusion reactors and the moon. The Big Seven, the power industry, was really a subsidiary of this arrangement.

Luna was the name given by the old Soviet Union, a Russian name, for the moon as well as many other earlier powers. It was a variation on old languages. For the Romans, Luna was goddess of the moon; for the Greeks, Selene. Most cultures had a name for the moon: la Luna in Spanish, der Mond in German, Mene (or moon) also from the Greek language, and Qamar in Arabic. The Western culture simply called it the moon. Luna became the accepted name by most of the world, for the simple fact that it was better than Moon. It was time to return; this time, man intended to stay.

It was 2012, the year that the Maya had prophesied for the end of the Way of Man, when construction of parts for the Luna return crafts began. The original time had been twenty-twenty five for the return of the United States to Luna. Every member of the newly formed World Space Agency (WSA) contributed to the project. Russia, Japan, China, the European Space Agency, and NASA all made continual use of their launch capabilities.

Nearly every month saw a launch from one or another of the members. The international space station of the mid-2000s became the focal point of construction of the assorted components for the newer and much larger station that would usher in the new hope of the world. The old station would continue to grow and be used primarily as the research facility. The newer station orbiting Luna would be a processing station and a bio lab. The bio lab would soon create quite a furor.

Three years of beelike activity, from a horde of hives, resulted in the two-manned landers. Identical in all ways except their call signs, these craft represented the best technology in the world. Each ship could carry five passengers—captain, cocaptain, and the three specialists. Each also carried the robot prototypes of the Foremen and Laborers.

These applicable names for the machines that would do most of the grunt work caused a minor furor in unions and religious circles. The names were a simple designation for the scientists. Some of the other suggested names had included Queens and Drones, Hands and Fingers, Bosses and Workers.

The larger Foremen controlled the activities of the smaller Laborers and could make minor repairs on them. The Foremen would keep track and direct of all the Laborers through radio telemetry: locations, activities, and all other forms of status. Further, they would produce purified ingots of material for later use.

Laborers were simply that—peons. They would perform the most menial tasks, those of gathering raw materials, resources. If a Laborer was malfunctioning, it would be repaired by the Foremen with a spot weld for a broken limb, rewiring, or a replacement chip for a short.

Anthropomorphizing the names had caught the attention of much of the world. As had been learned repeatedly in the past, public support was a necessity. Humanizing the project brought many windfalls and headaches.

Labor unions were a tad upset that they, the human workers, were being presented as nothing more than simple machines, almost slavelike. The conservative religious leaders were offended that any machine would be given a humanlike designation.

The majority of the world did support the effort. The awakening of the ecoconscience of the people of the world flowed like a tide. The positive surge carried the program forward. The continued growing demand for resources continued unabated, so the logical conclusion to mine off Earth sounded quite reasonable.

With the completion of the first production fusion reactor in the United States, in 2014, the world expected energy to be available as never before.

The politicians actually acted with haste when they agreed on a location for the reactor. It was on the Alleghany River north of Pittsburgh. Transmission wires would stretch from the Mississippi River in the west to the Cumberland River in the south, the Atlantic coast in the east, and include Ottawa, Montreal, and Quebec.

The clamor was, at first, a small smoldering flame. It was soon to become an uncontrolled firestorm. Planners were besieged by hopeful politicos looking for the construction and long-term benefits of having such a plant in their vicinity. Each heavily populated state pressed the propaganda buttons to get their constituents to vote for bringing such a project home.

*

Those who did not support the cause were of the Fundamentalist religious bent. They believed the end of the world was near, and the drive into space was a distraction from the evil one—Satan. This whole business of mankind solving their problems appears to be in direct conflict with biblical testimony. Only God can solve the totality of problems. Man is incapable; his limited logic is insufficient. "For the wisdom of this world is foolishness with God; for it is written, 'He catches the wise in their own cunning'" (1 Cor. 3:19). Paul extolled the church of Corinth, "That YOUR faith might be, not in men's wisdom, but in God's power. Now we speak wisdom among those who are mature, but not the wisdom of this system of things nor that of the rulers of this system of things, who are to come to nothing" (1 Cor. 2:5-6).

With this in mind, leaders of those religious factions most concerned (and they were numerous) met in Rome, at the pope's invitation, to discuss a public response to the scientific world. Protestant, Jew, Muslim, and Catholic alike, they had a common enemy.

Eastern religious figures had been invited, but most felt that the developments were not contradictory to the more inwardly directed beliefs such as Hinduism, Buddhism, Jainism, as well as Confucianism, Taoism, and Shintoism.

Hinduism celebrates diversity in its approach to what is called Divine. There is no rock solid doctrine, no core set of beliefs. Everyone who practices Hinduism essentially follows his own path to the Divine. The ultimate goal is to throw off the shackles of the material world and even one's own identity. This private search is little troubled by the concerns of the absolutes propounded by more Fundamentalist Western religions, even Middle Eastern ones. There are some five hundred million followers of Hinduism.

Buddhism teaches a pacifistic nonviolent approach and has many similarities to Hinduism. This nonconfrontational approach kept the major players from wanting anything to do with the conference. The refusal to accept the caste system also puts it at odds with the stricter hierarchies of Western religion and is a key delineator between it and its cousin, Hinduism. There are 270 million adherents of this religion.

Jainism, like Hinduism and Buddhism, accepts reincarnation and karma, while also rejecting the caste system with Buddhism. Prince Vardhamana, the founder, did not teach moderation though. He urged extreme asceticism, self-denial of what Westerners would consider creature comforts. The concept of ahimsa translates as noninjury or nonviolence, which requires a reverence for all life. There are some three million followers.

Confucianism focuses on the *superior being* and is much more like a philosophy than religion. Confucius recognized a divine law, yet he does not speak of heaven or God in his writings. Many followers of Confucius also follow Buddhism, Taoism, or both. This confuses the tally of followers, which is estimated at 160 million.

Taoism was founded by Lao-Tzu, meaning the "old philosopher." While Confucius defined and clarified the superior being, Lao-Tzu did the same for the *good being*. Very rough estimates put the sum at thirty-some million believers.

Shintoism (or, in Japanese, Kami-no-michi) began as a belief in many gods and goddesses. The emperor was a descendant of these beings in a merging of humans and deities. An altered form of Buddhism and Confucianism merged with Shintoism to produce the Bushido code. This code raised justice, courage, loyalty, wisdom, and learning as well as a love of nature to a pinnacle. Sixty million practice the beliefs of Shintoism.

Like Confucianism, Taoism, and Shintoism, Sikhism and Sufism and the Bahá'í Faith are mixtures of previous religions. Sikhism pulls together Islam and Hinduism; Sufism joins Islam and Christianity while the Bahá'í Faith accepts the prophets of the elder religions of Judaism, Christianity, and Islam.

Yet there are some five hundred million followers of various tribal religions.

The ancestral line that ushers in the Jewish tree, the oldest of the world's still functioning religions with some fourteen million followers, also eventually produces the Christian (one billion followers) and Muslim (six hundred million adherents). According to the Bible and Koran, Adam was first; his lineage leads directly to Noah and his three sons—Ham, Shem, and Japheth. It is believed that Ham went south after the flood to create the African nations; Japheth, toward Europe; and Shem, to the Middle East. Shem then would be the ancestor of Abraham.

Abraham would be the father of two peoples. Ishmael, from Hagar's womb, would sire the Arabs. Isaac, from Sarah, would father the line leading to the Jews.

The descendants of Abraham, destined to become Israelites first, had to pass through a period of bondage under their Hamite cousins—the Egyptians.

At the time of Moses and the Exodus, the Jewish tradition truly began. Monotheism became their religion, for a while. Deviations and returns did however occur.

It is ironic that science touts Eve in the form of Lucy, the mother of all humanity. She had to have a mate. Was his name Adam by chance? Science also has three later Eves associated with the spread of humanity from the African continent. These three appear to be the mitochondrial DNA mothers of all current humans. The Bible gives three apparent fathers in Ham, Shem, and Japheth. The only differences in the accounts are the time frames. Is the Bible the written form of an ancient oral tradition from tens of thousands of years spanning uncounted generations of developing mankind?

The Jewish God Yahweh (Jehovah), is the same god as the Christian God and Muslim (Allah). All agree there is one God that is supreme. There are discrepancies in doctrine though. Some Christians are Trinitarian, believing God is three forms: God the Father, God the Son, and God the Holy Ghost or Spirit; yet at the same time, they are *one*. Other Christians deny the Trinity. They believe the Son of God literally means son, not God, and that the Holy Ghost is the active force or will of God

Most Jews believed Jesus only a man while some of the Jews became his apostles, his first followers.

Muslims believe that Jesus was a prophet, not the son of Allah.

All in all, there are more than 1.5 billion members of non-Western religions as opposed to a combined force of around 1.6 billion for Judaism, Christianity, and

Islam. This is a fairly even balance, as long as there are no transfers of allegiance or creations of new religions.

The Koran was delivered to the world entirely through one human (the prophet Mohammed), unlike the Bible which had a host of writers and many books that did not make it into the final draft.

He had heard of Jesus and the prophets. When he had a revelation, the holy man, a hanif, announced that the same spirit that had spoken to Moses had visited him.

Mohammed began preaching in Mecca around 610 CE.

Some Jews converted to Islam while most saw Islam as a threat to their political and economic power as did most of those in the Arab world.

Jews claimed Mohammed's revelations were contradictory to the faith. Mohammed responded that Jewish and Christian scripture had become distorted while the Koran represented the true Word of God.

Mohammed then concentrated on Arab converts. Many of the aristocrats also saw him as a threat, a separate power base.

Eventually, Islam did indeed become immensely powerful, enough to greatly change empires, including the Persian Empire.

In their common heritage, these disparate current leaders found enough of a link to join together in an important cause, one that threatened their religious grounds and power.

This enemy, the secular world, was decimating their ranks and flouting their traditions. It was a matter of souls on one level; it was a decision about power on another. The sounds of multiple footsteps echoed loudly through the hallway as they all approached the conference room. It was hard not to gawk, to peer about like a dolt that had never seen a museum of fine art. There were grand arches leading to other halls, other spacious rooms larger than the average house. Even the tile floor was ancient and beautiful.

They arrived at an inner room in the Vatican, opulent beyond compare. Priceless works of art adorned every nook and cranny. Paintings and sculptures from all the masters from every corner of the globe royally decorated the room.

Social protocols were brief. There were a few clasped hands and muted hellos. Most had met more than once at various conferences, though never in quite such a setting as this. As host, on behalf of the pope, the cardinal would call the group to order.

"Gentlemen, let us begin," Cardinal Leonis announced formally, waving his hand vaguely to the ten exquisitely red-upholstered chairs. "We have very important matters to discuss today." He waited for all the others to be seated before taking his chair at the place left for him at the round table. A round table had been chosen so that none would feel inferior or, for that matter, superior.

"Yes, quite," replied Pastor Davies, concern showing on his rather pasty face. He was never a large physical specimen nor outdoorsy in any manner. Regardless of his wimpish looks, Davies's sharp mind missed nothing that could be used to advantage. His flock numbered in the millions, when all affiliates were counted, and, of all the groups present, was growing.

"What shall we cover first?" asked Harmon, the reverend of a very large Internet and television/v-link church. "The anthropomorphism of lifeless animatronics, the theft of our members by the fracturing churches, the need to turn back to God by the secularists, or a direct attack on technology itself."

The imam looked first at the caliph and, seeing the slightest of nods from his elder, responded in his slow, quiet voice, "The argument over semantics with regard to machines means little. Our peoples are not aware that such fine distinctions need be made. Loss of followers and the need to reaffirm Allah are a direct result of the attack made on us by modern science and technology. We must make an attack much like the one made against evolution in the 1990s in Kansas of the United States." It was obvious that the two old men, though from different sects, had decided to speak with one voice. They controlled well over half of the Muslim world and were respected by the rest. "The scientific world has made promises that all want to see fulfilled. Foremost is energy that will be unpolluting."

"Nonpolluting," murmured Pastor Marks of the English Protestant Movement with a wink at the much older imam.

The imam smiled gently and spoke again, "Forgive my misuse of English. Normally I would use a translator."

"Nonsense. You are one of the most intelligent men I have met," croaked the ancient seventy-year-old caliph. He reached out to touch the shoulder of the sixty-year-old imam in respect.

"I must agree with this analysis," said Davies, attempting to take charge. "We must initiate an attack on several fronts. Primarily, we must focus on the morality of such extravagant expenditures for any space-related adventure other than the one obtaining helium-3. They have plans for much unnecessary research." Some eyebrows went up at this, but none spoke. "Secondarily, we must change the opinion of the average man that science has the answer to all the problems. We must reawaken the idea that science only explains *how* things work, not *why*. Third, we must increase enrollment in the schools we run, in order to produce the next generation of spiritual warriors. There is more, but those are the most important points for now."

Monsignor Drake entered the discussion. "I am rather surprised that you support the idea of the Luna base, Mr. Davies."

"Only the Nearside operations," responded Davies quickly. "The Farside is, as I said, an unnecessary option as it now stands. All the research they plan is nothing more than an exercise in scientific glorification. At some point, it may be necessary to mine there, but not now."

Rabbi Simon asserted, "This will make us appear to be waffling on the WSA."

"I think not," Davies calmly responded. "We have, for many years, accepted the health benefits science brings, as well as the value of energy in bringing food to the table. We have funded many projects around the world, those using scientific principles, in order to ease the burden of life for those of the poorer countries. We will take the position that energy for the poor is our part of our continuing duty, a part of the equalization of humanity all over the world. We, however, must find a way to force the WSA to share with, not charge, the entire world."

"That will require a propaganda campaign of immense proportions," quipped Rabbi Levenson with a smile.

Dieter Kirk, head of the German Presbyterian Church Alliance, which included all the European Presbyterian Churches, smiled as he added, "We have an immense population to throw into the battle. Our various churches can act more in concert than the population in general. At our request they will flood various legislative offices with letters, e-mails, and vid messages. We have had great success with this technique over the years."

Harmon interrupted, "These communiqués are to do what? Will they explain our acceptance of mining operations only? Will they demand the spread of fusion technology to the nonmembers of the WSA/Big Seven? A return to stricter morals? Laws against the accepted perversions?"

"Yes," Davies answered bluntly. "On all counts."

"Then to your second point," said the cardinal. "What are the suggestions for bringing the average man to the realization that only God can answer the ultimate questions?"

"We must test science where they cannot answer. Beginnings would be the place to start. The Big Bang they speak of must have been caused by something. Science believes in cause and effect. We must bring in the primordial cause—Allah," spoke the imam. "Science speaks of a time when the original matter, protons, was created. At this time, light was nonexistent. There were no electrons. It was not until a time later, when the universe cooled, that electrons came into existence. Light was now possible. We all follow the same texts in the formation of our world. Our books say Allah created heaven and Earth. This means a place to exist in (the heavens) and the stuff of what we are made of (matter). Later, Allah made light. Evolution has been recently described as punctuated, a series of new life forms in waves. They speak of sudden and swift changes followed by periods of slow change. These parallels are undeniable."

"We must focus people's attention on these ideas. They must be brought to the point where they see a link. The wave of religious fervor that swept America after the 1970s produced antievolutionary and antiscience feelings in a large proportion of the population. It followed on the heels of some of the greatest scientific achievements, the Apollo space program. The religious right gained supremacy under President Reagan. It can be done."

"You are most correct, Imam," said Davies with some quiet awe. He had not expected this man to have such a grasp on science or Western politics. This was a man to consider. As an ally, he would be greatly beneficial; as an opponent, he would be terrible to contend with. Or perhaps, it was really the caliph behind these ideas. He would have to work on them to decipher the real motivator.

Davies knew that both had a bloodline that stretched back to the cousins of Mohammed. The imam, Eli-Al, was a direct descendant of Ali while the caliph, Omar, of the lineage of Ali's younger brother. Davies, as well as all the rest, was well aware of the power base they had amassed because of their proven descent combined with a moderate approach. The world of Islam had taken this as a sign that the sects could and would reunite. Furthermore, the children of both had married into the other's family, cementing the bloodline into the religion. Now was the time; it was a sign of Allah's will. Islam must be united to face the world's time of reckoning.

The caliph led nearly 50 percent of the Sunni sect directly and controlled much of the rest through the cooperation of many of the lower clerics. The imam controlled almost all the Imamites, the Shiite Muslims. The united front represented an immense force of believers that was only outmanned, and slightly at that, by the Catholic Church.

Kirk responded quickly, "We have not just decided to accept any form of evolution, have we?"

"Never," Davies answered firmly. "We just need to get the masses to understand that science tried to get completely away from what the Bible says and has failed. Evolution tried to get completely away from God, but their very own ideas must lead them back toward what has been written in Genesis. The punctuated equilibrium concept of evolution is their way of trying to save evolution from complete failure. As more evidence mounts, they will have to alter that even more. They will find themselves back in accord with the Bible. We just need to present that idea to our churches. We must show them that the scientists have built their houses on sand while we are on rock and do not change our positions. For example, the recent find of human fossils surrounded by animals thought to be extinct before man walked the Earth."

Harmon added, "We have had great success with short tracts, on the order of ten to twenty pages, explaining scientific items. Perhaps we should create one jointly, one that we all agree on and is *certified* or *signed off on*"—he signaled the quotes with his fingers—"so that the people see we are all in agreement."

Simon heartily agreed, "Definitely. A united front is the way to go. Social inertia can only be swayed by a larger force."

Others about the room assented with nods or grunts of approval or both.

"The third point then, as we seem to have well covered the first two," mused the cardinal.

"Education is expensive," responded Davies, "but we must put our efforts into it."

"Perhaps you could learn a little from us in that area," said the imam. "We have always used family units, extended family you understand, as a small church. People will react well to gathering in other's homes. Begin your schools there. Then expand to existing church facilities for larger gatherings to reinforce doctrine."

"This sounds very much like how Jehovah's Witnesses operate," Harmon stated.

"And have they not been the fastest growing sect for the last forty years?" retorted Davies. "They make excellent use of their homes and facilities with several meetings every week."

Simon stated, "This is much how we ask our people to spread the knowledge of Yahweh. It is a very personal project. We can expand our operations in all this." The rest nodded in approval.

"A question," Kirk began, "we have stressed the concept that the end is near. We will appear to be waffling on that if we must train the next generation, will we not?"

Davies snorted, "Nonsense. We are to train our children in the right ways of God, Allah, Yahweh, are we not? If the rapture were to happen without this training, our very blood would be left behind. If they reach an age where they can contribute greatly to the cause, it is so much the better."

The caliph murmured, "We do not share this belief in the rapture, but we do believe that the reckoning will be soon. To train our children is our duty as well as the will of God," he noted purposely using the Western word for the *Mightiest*."

Davies played a conciliatory card from his hand. "That difference in doctrine is of little importance at this juncture. Adherence to Allah's will is the absolute key to life eternal." He returned the compliment of the other's name for God.

The imam smiled. "True."

"I have an additional question," said Kirk. "What about the tie-in to Underworld?"

"What about it?" asked Davies.

"What points should we make?" questioned Kirk.

"None for now," replied Davies. "Underworld does not violate any of our precepts about man being constrained to an earthly habitat. Presently, their ties to Nearside do not concern us if we can constrain the program to mining only."

Kirk looked a little irritated at Davies's simple dismissal. "We cannot ignore that the mining operations, the raw materials for nonenergy use, are mainly for Underworld projects, scientific projects that claim to be healing the Earth. The need for large quantities of titanium and aluminum may weigh heavy on the voting person's mind. They will be told that this will help to preserve the environment."

Levinson waded in with, "There is a need for circumspection. The total picture of mining may need to be included in order to pacify the people. Perhaps we can guide the materials to the needy."

Davies smirked, "That is actually a very good idea. This is supposed to be for the people, right?"

General agreement passed around the room.

Cardinal Leonis asked the table in general, "What more can we do to gain control of the situation?"

Davies mused for a moment before suggesting, "Hit them where it really hurts."

Rabbi Simon looked at Davies with a grin. "Where else but the pocketbook."

Davies smiled almost viciously. "Economic sanctions against all companies that do not renounce the parts of the program that we wish to eliminate. We can develop a Christian-based economic source to counter the atheists and secularists. We can develop lists of good and bad companies, ones that no Christian, believer, should do business with."

Kirk entered in with, "We must be very careful about how we do this. If we throw a monkey wrench into the system, we may cause so much harm that the people turn against us. Perhaps we should enter into this phase after we have tested the waters to see how much support we can get."

Davies looked annoyed but responded calmly, "We can begin in more fundamentally oriented areas and expand from there as needed. Would that satisfy everyone's concerns?" he queried, looking around the table. There were nods from all the assembled. He nodded last, a sign to others as well as himself that he was the one who gave final approval, a man in charge.

"Enough of that for now," said a weak voice from behind a curtain not far from the table. The curtain, when pulled aside by two armed guards, revealed Pope Xavier. The ancient man appeared to have dark holes for eyes, with a craggy nose that stood out markedly from a weathered face. Sallow, wrinkled skin hung from his knobby fingers. His appearance seemed to be the embodiment of age.

"Holy Father," said Leonis and Drake in unison. All the men stood. The pontiff weakly waved them back into their seats.

"We have two important issues greater than those already discussed," declared the aged pontiff. All eyes and ears were focused on the pope. "We have discovered that the space program, the WSA, plans to genetically engineer human replicas in the late '20s or early '30s," he rasped.

"What?" cried Davies in surprise. None of his amateur spies had brought this juicy tidbit to his notice. He quickly surmised he would have to expand his network of churchgoing professionals in the space business, or perhaps in the military. Military, he quickly mused. There were no atheists in foxholes as the old saying went. They were deep into this program. All this passed through his darting mind in less than a second.

"Unbelievable sacrilege!" cried the caliph in as loud a voice as he could muster.

"Do you speak of artificial insemination, or other techniques?" demanded Harmon in the same moment.

Others voiced similar woes, regrets, and questions.

Only Cardinal Leonis did not react. He was already in the know.

With unbelievable calmness and disregarding Harmon's demanding tone, the pope replied, "Our informants have told us that the plan is to have artificial wombs, using artificial insemination techniques, on genetically engineered sperm and ova. They plan to do this in orbit, at a space station, as a means to create a large population of pseudohumans to provide labor for the space program. They apparently will be the populace of all future stations both in orbit and on the surface of the planets and moons. It is apparently more cost-effective and safer than bringing *real* people from the surface of the Earth."

"Cheaper!" cursed Davies. "It is blasphemy!"

From behind his hands, the caliph murmured tearfully, "They continue to seek the power of Allah. For all the ages, man has fruitlessly sought the fountain of life, the tree of life. They fail because they believe it is a physical thing, not a spiritual."

"It is the way of science. If they can, they do," lamented the imam in a shaky voice. "They do not worry about the consequences. What consequences there must be."

"Rabbi Simon, what is your take?" Harmon asked intently, eyes sharp, glittering dangerously.

"This is awful. It is against all laws of God. 'Man shall cleave to woman, and they shall produce young in the way proscribed,'" he quoted. "We cannot stand for this."

"Kirk, this is truly a calamity. I do not know what to say. We must combat this, but how?"

Davies chewed his rather fat lower lip. "This can only be described as a plot by Satan, the deceiver, to replace all true humanity, God's creations, with his own replicas. It began with animals, the clones, and the artificial inseminations. They proclaim that it gives a better end product. The best bull can now mate a thousand cows, without the sexual act except for masturbation of the bull. It is perversion, bestiality. They allow the killing of unborn children through abortion, yet now they will attempt to create those abominations rather than use God's children."

Kirk solemnly continued in this vein, "Technology has always been presented as the key to the future, but it has not been reined in. It has run more and more amok in the last years. We can't, however, deny that there has been an improvement in the beef industry, crop production, wells for water, energy."

At this, Davies shot him a hard look and replied, "The beef industry has little to do with this."

"But it does. The people are eating better with higher quality meats. These programs have allowed third world countries to really upgrade their food production industries. They see energy becoming available for preservation of those foods. That's what the people really see."

Leonis interjected, "The details we need to cover over the next several days can wait until later, gentlemen. For now, let us get on with the broader brushstrokes."

Marks noted, "The people have paid so little attention. Most do not know anything about the details. We will have to educate them."

"Why?" queried Davies.

Harmon's brows knit tightly. "They need to be aware of the intricacies, the—"

Davies interrupted, "They need to know only what we tell them. We are the chosen shepherds for these flocks."

"There is another matter, of even greater importance," stated the pope quietly, halting any possible breach of the silence. All ears widened, and nearly inaudible side talk at the ornate table halted abruptly. "It is akin to the creation of artificial bioforms that resemble man. The scientists are going to attempt to create nonbioform intelligences. As you would say, Pastor Davies, Satan is attempting to replace God's original creation, man, completely in this final step. These are termed artificial intelligences in the computer world. They believe they can now build a machine that thinks as a human does, and it will have awareness as a human does."

The prior shock had been enough to send heartbeats upwards and tingles of fear down spines. It had triggered deep-seated animosities and prejudices. This new addition simply compounded the state of confusion and disbelief.

Marks uttered, "There has long been talk of being able to achieve such. Since the advent of computers, it has been the goal of such men to create these devices. Do they really have such capability to create a machine to do this?"

"Never," cried Davies, his jowls shaking. "Man was given his ability by God."

Kirk slid in with, "It may be a machine so complex that it imitates thinking. But can it really think? Is there any way to prove that it cannot?"

The pope answered, "Our Jesuit scientists have been quietly gathering data for me. They are quite certain that this new development will be a type of machine that can think, perhaps even have emotions."

That shock threw each attendee back in his chair for a moment of horror. Davies sorted through his memories quickly, understanding now the healing of a breach between the Vatican and the Jesuit order that occurred some years past.

The imam started to speak but was cut off by Leonis. "Gentlemen, again we need to reserve this deeper penetration for further talks. We must delve through all this material with experts. I will say this though, our Vatican computer experts refer to what is known as a Turing test. It is a blind test, one in which several people and a machine are asked questions. If the person asking the questions cannot tell the computer from the others in its responses, it is considered by the scientists to be artificial intelligence, or AI."

Levinson muttered, as he looked directly at the caliph, "By the bones of Abraham."

The caliph took a moment to look deep into Levinson's eyes before saying, "Yes, friend. Brother. We are all the children of Abraham here, either as direct blood descendants or spiritual ones." The caliph's bloodline was well calculated back through relatives of Mohammed to Abraham. "We are . . . bound in purpose here."

"To this end have you all been brought here," the pope announced firmly, as if the statement implied all. "The other matters, important in their own right, are minor in comparison to the last two." He looked at them all with a steely eye.

Davies spoke harshly, "This has not been made public at this point. We must use the letter writing campaign, or a declaration, to burst the bubble of their subterfuge. Many of the government officials must know, but have declined to relate this to their constituents and peers. We must vigorously attack now, with the point that they have kept these secrets from the population of the world."

Simon added curiously, "The news agencies have not announced this. Is there complicity or lack of knowledge?" He then added more firmly, "We must make public announcements, all of us together, in which we surprise the press and the world with our discovery of this treachery. It must be done very soon, before they, the WSA, can make any release of their own. This will greatly increase our effectiveness in our other lines of attack. It can be used to crystallize support for us. The fence-sitters will be pushed into taking a stand if we do this correctly."

The cardinal spoke softly, "It has been made public, but it was only indirectly spoken of in scientific journals."

Davies saw his opportunity to go hardcore. "This will give us the platform for all our goals, especially the economic boycotts."

The pope spoke again, "I have requested a special mass tomorrow for the cardinals. At this mass, I will apprise them of the situation. I will give them instructions on informing the bishops, but not for a time. Then I have scheduled a public appearance, in three days, where we can all make our speeches. This will give you time to inform your congregations to watch the event," said the pope. "If we do not give advance warning of the topic to the masses, the effect will be greater," he noted sternly. "We must have secrecy if this is to work most effectively."

"We have to give some warning to our administrative bodies if we are to prepare for an immediate letter writing campaign. They will need to know all the names and addresses of all the members of the legislators and other governmental officials involved directly in the space program," stated Kirk.

"All the legislators," replied Davies. "We will concentrate on the space committees but must include all of them. If we have prepared letters, pretyped, that we can pass out and ask for signatures, we can deliver them in bulk. We can also have an e-mail form ready on computers stationed at meeting places. Those representatives that have been left out of the loop will use this to their benefit, especially those in the party not in control. We can use this to attempt recalls on the legislators that do not support our position."

"That means that we need to begin grooming certain politicians for the replacement of these ousted people," murmured the caliph. "We must each, within the next two days, select political standouts from within our ranks. They must be ready for a recall vote, or for the regular voting period. We must begin to take control of the governments from within."

"That is a very dangerous statement to make," noted Harmon. "Some would declare that to be a conspiracy and treasonous."

"Nothing said here will leave this room without us saying it. There are no microphones or video cams. My security has triple checked that," stated the pope matter-of-factly. "What we say here is secure. Do not fear to speak as openly as we must."

Cardinal Leonis interjected, "We have proposed nothing illegal. Only the governments need fear us, not the people. If we can create a more pliable set of governments around the world, we can then control the events."

"So that is how it is," said Simon matter-of-factly.

The pope rose with the help of one of his guards, and so did all the others, sans help. "Sit, gentlemen," he said. Then to the amazement of them all, a basin of water was brought forth. The pope knelt with the help of a guard, on a pillow placed by another guard, next to the cardinal. He began to undo the shoes of the cardinal.

"Holy Father," exclaimed the cardinal in a deeply anguished tone. He attempted to withdraw his foot from the pope's grasp.

All that he said was, "Peace, my son." He proceeded to bathe the feet of all the leaders present, ending with the caliph.

"It tickles," said the caliph gently, a slow smile creeping across his time-withered face.

The old man looked up, his face deeply wrinkled in amusement as he grabbed a toe and said, "This little piggy." Everyone laughed, not concerned with the reference to pork.

With the help of the caliph and a guard, the Holy Father rose. "We have established a covenant here. I washed your feet as did our Lord of the Apostles, with as little tickling as possible." He glanced at the caliph. "We are bonded in truth, whole under *our* God. We must return our peoples to their God. Jesus, the Christ, is the shepherd that has given us the examples to follow." He waved to the Westerners. "Mohammed has given our brethren a prophecy of hope." He signaled to the caliph and imam. "And our ancestor religion weds with us in the need of our God. Let us do so." Then he went from person to person and blessed them all. He turned and slowly exited the room as profound silence reigned.

"Well, that was a real eyepopper, gentlemen. He is quite a politician and a leader," said Davies, letting out a deep breath.

"There is no doubt of that," said the monsignor.

"Let us get back to business then," said the cardinal, given a clear mandate from his spiritual leader. "We need to make many arrangements in order to obtain the greatest effect. Let's talk strategy."

Davies looked intently at Harmon, with whom he had previously discussed these warlike items. "We need to all act together. Unity is the only way we have to make our point truly known worldwide."

Rabbi Simon smiled. "That is somewhat obvious."

Davies frowned slightly at the interruption. "We need to have access to each other's networks, be they cable TV, radio, the newer v-link, radio, and print."

The imam looked a bit worried, thoughtful. "That may not go over well with some of the hardliners in all of our congregations. Many still see Westerners or Easterners as . . . well, the enemy."

Harmon jumped in, "That must be changed. Everyone must see the battle, not the soldiers."

"Exactly!" exclaimed Davies, trying to regain the initiative. "In war, we have seen blacks, whites, Latinos, Middle Easterners, men, and women all come together to fight evil. By allowing this access, we show that we must put aside minor differences in order to win. All the sheep will see what the shepherd wants them to do."

Rabbi Levenson mulled it over before saying, "You're saying full open access, then."

"Yes," both Harmon and Davies said together.

Cardinal Leonis said, "It may be better to initially have a member of the particular organization act as a host, at least until the listeners are used to having an outsider, forgive the term, speak as moral guides."

"That is quite acceptable," Kirk noted. "I think it would further our show of unity as well. I also think we should share our speeches, ask for input from all members here, so that we all say the same thing, but in different ways. Remember that learning by rote is a powerful tool. Repetition is a key factor. Also, we do not what to sound contradictory in even the least way. By sharing beforehand, we can eliminate the possibility."

The rest nodded at this.

The caliph added, somewhat carefully, "I believe that we should have the Islamic leaders appear to be the main speakers at first. We must incorporate all the local clerics as well, pull in everyone at all levels of leadership. I am uncomfortable saying this, but here it is. The Muslim people still have a great undercurrent of mistrust for Westerners. Recent wars have not yet passed from thought. We need to be very careful in melding our forces."

"I understand your trepidation," admitted Levinson frankly. "Politically and militarily, the Western religions and the Hebrew nation have usually acted in concert, and not always friendly to Middle Eastern interests. We must all acknowledge, openly, that this is not defined by the old rules."

A new vision of trust passed, unspoken but acknowledged, between the Jewish contingent and the Islamic.

*

(1) And the whole earth was of one language, and of one speech. (2) And it came to pass, as they journeyed from the east, that they found a plain in the land of Shinar; and they dwelt there. (3) And they said to one another, Go to, let us make brick, and burn them thoroughly. And slime had they for mortar. (4) And they said, Go to, let us build us a city and a tower, whose top *may reach* unto heaven; and let us make us a name, lest we be scattered abroad upon the face of the whole earth. (5) And the LORD came down to see the city and the tower, which the children of men had builded. (6) And the LORD said, Behold the people *is* one, and they have all one language; and this they begin to do; and now nothing will be restrained from them, which they have imagined to do. (7) Go to, let us go down, and there confound their language, that they may not understand one another's speech. (8) So the LORD scattered them abroad from thence upon the face of all the earth: and they left off to build the city. (9) Therefore is the name of it called Babel; because the LORD did there confound the language of all the earth: and from thence did the LORD scatter them abroad upon the face of all the earth.

—Genesis 11:1-9

*

The Information Age means that just about all information can be found and transmitted around the world in a matter of seconds. In the 1800s, ships, stagecoaches, the pony express, and trains took days and weeks. The telegraph began the nearly instantaneous transmission of information. It was followed by telephones and radios. Today it is the computer. No longer must one peruse a textbook. A person can simply touch in a few keywords and find a host of data sites. Today all of mankind is of one language, a new one: computerese. All over the globe, people access the Web daily, hourly, even continuously. They are buying, selling, trading, sharing, playing, writing, viewing, and reading.

With the common denominator of mathematics, they have gathered together to reach the heavens, with space vehicles instead of a tower. These ships will provide a solution to the world's energy problems. The majority of people will indirectly participate and benefit. There will be peace and plenty for all who ask; there will be energy to provide for mechanical needs and for biological needs such as food, water, and shelter.

Humans will recreate themselves. Biotechnology will forge a new kind of person: *Homo sapiens anthrocreatus*. They will take the human genome to the stars in a multitude of variations much like a split in the ancestral tree. Humanity will survive its Armageddon period. This new strain of people will be more able to withstand the rigors and demands of life in space. Nothing imagined will be impossible for this generation of mankind.

What will confound them now? How could they be forced to leave off now, the most critical time of existence, as they understand it?

And yet, the driving force of technology requires extreme specialization. The Renaissance person no longer exists, for the most part. Each field of expertise has its own jargon; it is a vocabulary all its own. Amidst this unity brought together by the World Wide Web, there is division as well. Even scientists cannot understand one another's fields; they rely upon and have *faith* in one another. Faith is supposed to be the realm of religion.

*

Some few days later, the old, tottering holy man came out onto the Vatican balcony, followed by the rest of the newly formed alliance, where some 150,000 people were in direct view. Large screen monitors allowed him to be viewed by tens of thousands more in nearby squares. He started with a blessing in Latin, which he repeated in Italian and then English. "My children, an announcement must now be made. Please refrain from making any noise."

There was an amazing silence in the square; silence was already expected for the pope's announcements. For him to ask, after short notice, that was a hint that things were in the wind. "There is an evil among us, hidden as good, as it often is. The Word of God, the Bible, tells us that evil, Satan, can still appear as an angel of the light. This evil brings to us many benefits to cover the secret deeds it really intends. I am ashamed to say that we have supported this evil unknowingly."

There were cries of "no!" and "never!"

"My children, listen!" the he actually shouted, his frail frame shaking in anger. Again, amazed silence reigned. The man waited as he surveyed the crowd, using the pause to great advantage. "We have been deluded by science and the governments that have so much faith in the works of science."

"Yes," the cry from several quarters of the crowd came.

"Silence!" he thundered raggedly, his wafer thin frame shaking like a leaf in a gale. He had little wind for the force of his voice. It was an absolute command. Utterly shocked now, the crowd nervously shifted about but made no more loud sounds, only whispers and a few prayers.

"You do not know the depths of their depravity. You do not know the horror about to descend upon us." The pope had engendered an actual fear among the people. They waited in grisly and macabre anticipation. His voice lowered as he stated, "They seek the secrets of life itself." Again, he paused for effect. The buzz in the crowd was uncertain, full of whats and hows.

"The program in space to bring us energy is but a cover for the production of fake humans known as *Homo sapiens anthrocreatus.*" The crowd was suddenly immobile. Most of them were quite aware of what *anthro* (man) and *creatus* (to create) meant. Had they heard correctly?

"They seek also to build a machine that they claim is better than man, artificial intelligence. They believe they can do better than God."

"Blasphemy!" a single person screamed angrily. Others immediately took up the cry much like the chants of a concert crowd for their favorite band.

Then he extended his hands and waited briefly for silence; he would not be able to shout down this crowd now. "Only God can create. God created each and every one of us." He pointed to the members of the crowd. "They intend to use genetic manipulation of sperm and ova to create a superrace of space beings. The twentieth century has seen the results of such attempts to control the development of humans and the types of governments that engendered such attempts." The allusion to Nazi Germany was not unnoticed.

Hundreds fell to their knees in prayer, beads in hand. There were many cries of "Mother Mary . . ."

Then, as a master at handling crowds, he allowed them several minutes without saying a word.

"My children, I implore you. I demand that you listen to our brothers in the Father. The sons of the one God." The pope indicated that the caliph and imam should come forward, which they did.

Most in the crowd did not recognize these two men, but a few did. They were stunned, as were the rest, when the two intoned a brief prayer to Allah. The imam spoke in English first, followed by Arabic. His speech was almost identical to the pope's. The caliph finished with a call, a nonviolent jihad, in Arabic, for all the peoples of Islam to unite against these impending horrors.

Each member of the coalition followed in turn, playing to his congregations, whomever and wherever they were. Each stressed the two main topics of artificial life forms, the bio and the machine, while bringing up different aspects of the ramifications.

It would be a near fatal blow to the space program. By the time they were done, there were millions of Fundamentalists who were on fire. At each of the thousands of prearranged locations, churches, temples, and even sports stadiums, wherever people had been gathered, forms in Arabic, Hebrew, German, Italian, Spanish, English, and several other languages were handed out so that the people could all voice their now volcanic opinions.

*

"Hey, Mick," said Paul with a nod as he sat on his usual stool at the Lucky Loser. He looked around before he lit up a smoke.

"Now, Paul, you know you ain't allowed to light up here," Mick drawled in the Deadwood style. Paul quietly slipped four hand-rolled cigs across the bar, and Mick turned away, four smokes richer, to finish washing some mugs.

The Lucky Loser was actually outside the town limits of both Lead and Deadwood; smoking laws really didn't mean much. It was one of the most rundown, shabby, decrepit, and comfortable places to sip a beer. The touristy places in the revamped cities snubbed such local facilities. A person could belch, pick his nose, and fart with little cause for concern. It was a place of refuge for the working person, the true-blue collar worker.

Burger flippers, gas station attendants, hunters, miners, and gardeners all could find their pleasure here and did. What little extra money they had left from the gambling and groceries often went to the multiple refills of their mugs. This place promised ease, not just the facade of the old West.

"Hey! Asshole! Who said you could smoke in here, Paul?" announced a busty, well-padded woman with a fat cigar in her mouth as she entered through the back door. She blew a hefty cloud of blue smoke out as she cleared her throat and spat before she entered the bar.

"Bitch, who said you could come in here," Paul joked.

"Buy me a beer," she growled, plumping her fat ass down next to him.

"For a hand job," he said.

"My pinkie finger is tired," she said slyly. "And the others are too big." That got a laugh from Mick who had already poured her a mug of cold Coors. She would drink far more than that before she was done. This just earned him a tip.

Paul had little recourse other than to give her the finger and toss a fiver at Mick. Mick bumped his fist on it, but he did not pick it up. The next round would be on Paul, not this one. It was an advantage for well-known regulars.

"With your little one can you change the channel?" she said to Mick.

"I might take back that freebie," Mick joshed.

"Yeah, yeah, put on WWNN. I like that Mary chick."

"No wonder you can't get laid, you're a lesie-bitch!" laughed Paul.

"Suck my tit! Oh yeah, you can't. Your mouth is too used to your own pecker, smaaall," she sneered.

"Better keep quiet, Paul. She's sharp tonight. She must be on the rag." Mick chuckled heartily.

"Go fuck your dog," she frowned.

"Woof," barked one of the other patrons.

"Hey, you hear 'bout that?" she perked up, pointing to the vid screen.

"What?" asked both Mick and Paul simultaneously.

"I heard that earlier, energy stuff, fussin' or foo-sun, something like that." She puffed deeply on her cigar.

"Yeah, it is supposed to mean cheap energy," said Mick. "They been showin' it all day long. If we build these new machines, it means almost free energy. But we got to go to the moon to get the fuel, somethin' like that. Then we got to spend money to build these big machines."

"You know, it's always that way. We get these technological improvements that are s'posed to improve everything. But they don't really. We have to spend so much money to get them working that it costs us more in the end to keep the whole thing going," noted another patron.

"They always make their money." She farted to emphasize her claim. "Oops, how unladylike of me."

Paul wanted to know what was up, so he asked, "What do we need from the moon?"

Burt had heard the talk as he approached his usual stool. He answered for Paul's benefit, "Fuel. Part of the fuel needed for the fusion reaction is found on the moon, Luna."

Paul was impressed, so he embarrassed himself. "So this fuel, Luna, only comes from the moon. That's why we need to go there."

There were a few snickers around the bar as Burt explained things to Paul, "The moon has a name, and it is Luna. The fuel is called helium-3."

Paul went red, knowing he had made a fool of himself. He took a deep drink and plowed on, hoping everyone would quickly forget his gaffe. "OK, so why do they want to do all this other mining stuff we've heard about?"

"If we have to go for the fuel, we can take advantage of other raw materials that can be mined. Kill two birds with one stone as the old saying goes."

Mick was interested. "What kinda stuff can we get there?"

Burt answered, "Metals like aluminum and titanium, I've heard."

The woman sitting there seemed insulted. "We got all we need here on Earth. It was given to us by God. Them science guys are just lookin' for a way to get money out of us for their stupid projects."

Burt was not surprised. "We can stop hurting the Earth with this."

"God will take care of the Earth," she responded.

Burt raised an eyebrow. "We can stop doing the damage that I am sure a god would not want us to do to his creation. Mankind is finally at the point where we can possibly return the Earth to the natural state that it should be in."

"Hmmmph," she snorted. "Only God can fix all the things that been done wrong. People just screw things up worser and worser."

Paul interjected, "You mean that all them Green Warriors, Greenpeace, and Earth Helpers is right?"

Burt took a sip before answering, "To some degree, yes. There is a need to stop the continued overuse and abuse of Earth's resources, to find ways to reuse materials better. The best way to do that is to recycle, like all the laws require. The next thing to do is to find other places, places that won't hurt Earth, to find the materials we need to continue in our way of life. The other choice is to take several steps backwards technologically."

The fat woman snorted loudly then stated rudely, "You seem to think that technology is important. It ain't. It's only a fake path away from what man really needs. It is a science lie. We have not really gotten good from science. Look at the pollution."

Paul looked meanly at the woman and said, "There been a lot of good too."

Burt agreed, "What about medicine, food production for the hungry of the world, schooling, and knowledge in general."

"Hah!" she shouted. "Medicine—pills, not natural remedies. Food? What about the poor hungry people in this country? They give foods to all the others 'cause it's political. Schools that teach lies, they don't teach what's right! Knowledge—what people knows there ain't what's important. You can't convince me."

Burt looked coolly at her. "I'll grant you that politics often interferes with much of the humanitarian aims, but overall, things have been moving for the better."

"Better for who?" She looked up with a critical eye.

*

What makes Western civilization worth saving is the freedom of the mind, now under heavy attack from the primitives . . . who have persisted among us. If we have not the courage to defend that faith, it won't much matter whether we are saved or not.

—Elmer Davis

*

Your God is one God; there is no god but He, the most merciful.

—*Koran* (*Qur'an*)

*

And there is no God but one.

—*Bible*, 1 Corinthians 8:4

*

The consequence of the epistemology of religion is the politics of tyranny.

—Leonard Peikoff

*

"Vox populi, vox Dei." (Latin) The voice of the people is the voice of God.

Chapter V:

Attack Counterattack

5

000101

"This is WWNN Internet and v-link news, Mary Smith-Jones reporting, live from New York City, United States, North American continent. It is the year 2013.

"The League of Religions is preparing to make yet another one of their sorties against the developing space program. Since their first announcements from the Vatican, they have seldom let a chance go by without at least one of the members of the organization verbally attacking the space program's mission statements.

"In a few moments, Pastor Davies, of the North American Group, will begin his oration on the religious perspectives on the space program and how it affects humanity and the churches of God. He will be followed immediately by Reverend Harmon, leader of an Internet church, which will also be carried by our network. The pope will then speak from the Vatican. The Vatican has allowed us to broadcast that speech as well as using its own network footage. This is to be followed by a joint statement from the caliph and imam in Mecca. We will be live for all five speakers until they are finished.

"Tomorrow morning we will have the response from Dr. Vasser of the WSA, which was originally planned as a news conference to announce the specifics of the WSA's program. It will air live from Washington DC, at 10:00 AM.

"That will be followed by counterstatements from Reverend Harmon in Washington DC, Pastor Davies from New York, and Pastor Kirk, representing the German Presbyterian Alliance, speaking from Berlin, Germany. All live on WWNN. Don't miss this incredible, spectacular, and controversial event. This will be the most important news you watch. If the religious coalition can drive a large enough wedge into the political system, they may halt the space program. This may decide the fate of the world."

She waited the requisite few seconds before standing up, pushing her chair back. She quickly removed the tiny earpiece that allowed her to hear important prompts from the editor in the sound booth. The nearly invisible microphone, only a few millimeters across, which trailed inside her shirt with a few strands of light-transmitting plastic wire to a transmitter in her desk, was next.

"I need some food, people," she demanded peremptorily. "These old windbags are likely to take hours with their propaganda spiels," she noted cynically. "I can't go on air with my poor tummy rumbling," she said as she pointed to her trim middle. She rolled her head around on her neck to loosen up some stiff muscles as she strolled to the info desk.

"It ain't her tummy I'm lookin' at," muttered one of the sound techs quietly and, quite politically, incorrectly.

"That could get you in trouble," whispered one of the others.

"Wonderful world when a person can't express an opinion," the first noted with just the slightest amount of heat.

"Matt," Mary asked when she approached the desk, "do we have any more information on what went on at that secret meeting at the Vatican?"

"Not a single word," he muttered with resignation.

"What?" she challenged. "Matt the Rat hasn't uncovered any smarmy little details, any recordings, vids, or people willing to talk?"

Matt frowned. "This is a little more delicate. One does not generally offer remuneration for information from a priest."

"So you're trying to bribe priests now." She smiled. "We need to know," she emphasized. "It will give us a little lead time to prepare our speech recaps and op-eds, and we only have two hours. You have been on it for weeks."

"Time and money very seldom buy a priest that has been given a direct order from his pope," Matt replied angrily.

"Then we will have to rely on our live analysts," she replied cautiously. She moved away as she thought about that last statement. How did Matt know that the priest, his usual informant, had been given a direct order not to talk? A Jesuit was the soul of circumspection. Did Matt have more than he was giving?

She moved to a quiet corner where she could give a low-voiced command to her cell phone, "Call *the finder*."

"Yes," a male voice answered after two rings.

"I need some work done."

"Info on target?" he replied after he recognized her voice. He always enjoyed watching her on the vid releasing information that he had dug up. Sometimes she had been able to hoist a few politicians on their own petards.

"It will be lengthy and will arrive on your e-mail within the hour. Do you have anything on your plate because this will be a week's worth of groceries," she continued in a barely audible voice. It was silly to use code, she thought, but people do listen in. She was aware of the National Security Agency's (NSA) tracking systems for cell phones. Her story about their secret listening posts had been quashed by the Justice Department based on national security.

"I'll clear my cupboard," was all he said before he disconnected. He knew she would cover the price tag; she always had.

She went and thoughtfully nibbled at the finger food that had been brought out for her. Her musings took her back to her college days.

She had met the person she now called Finder while in school. He was one of the geeks who had never interested her relationship wise. But he had been invaluable in helping her pass her computer courses. He had also been the first male to take her to lunch after her messy divorce. She felt safe with him.

His interests were the control of information, not sex and the providers thereof. Keyboards, not nipples, were what he dreamed of. The flow of information, his control, was orgasmic for him, not physicality with men or women. He owned these electronic devices. They served him in just the manner he required. The only limitation was his mind, and that was vast.

She felt just as secure in his ability now as she had felt in her safety at lunch with him. He was a known quantity. She sat her lithe frame down on a convenient chair.

*

For all the Luna project could do to bring humanity together; it could do just as much to separate it, polarize it. All the factions that could gain from a goal would align against their detractors. The goals were worthwhile, desirable (nearly unlimited electric power for everyone, raw materials necessary for the continued development of the technosociety of *Homo sapiens*), and allow for the cleansing and reestablishment of a more natural biosphere with the assistance of the Underworld projects.

The Underworld projects include a two-pronged attack designed to revitalize the sagging environmental factors necessary to continue human and other life in as much the same way as possible. They were not called Underworld without reason.

The land-based versions would be constructed in deep, closed old mines and, eventually, closed military installations like Cheyenne Mountain. The others would be wet, in the oceans and largest lakes of the world.

The first, the land-oriented ones, would be a series of underground scientific research stations that are constructed to serve a variety of purposes including the elimination of excess CO_2 in the atmosphere. Huge carbon dioxide scrubbers powered by the new fusion reactors could remove CO_2 at almost the same rate it was currently produced, which would at least maintain the climate's status quo. Future cuts in man-made production of that gas would take them back around the corner.

There will be continued development of seed banks and plant nurseries, which would contain endangered species, versions of Noah's ark where endangered life forms from frogs to polar bears will be raised. This will occur concurrently with pure scientific research in many fields: genetic, physics, metallurgical, and nuclear.

The water worlds would look to do much: eliminate dead zones in the waters, areas with oxygen depletion. They will revitalize the reef systems that are a prerequisite for much of the life in the oceans growing specially developed algae and kelps for human and fish consumption, increase plankton populations, and act as fish farms.

Some of the first fish farms would be like great cages, floating mostly freely with the currents and tides. Powerful motors would be used only occasionally to redirect the farms as needed; sails would be used when possible. Buoyancy would be controlled by simple ballast tanks like those on submarines. Water and air would be pumped in and out by solar-powered devices as needed to keep the vessel at the desired depths.

If a storm neared, the farms would simply submerge to avoid the turbulence of the surface actions. Twenty meters down, they could let it pass over with no concern.

Others would be secured to the ocean floor at depths of fifty or so meters. They would be similar in design to the roving variety. While tethered, they were not static. They would be winched along a long elliptical zigzagging path that would allow the feces of the fish to settle to the bottom and fertilize the special kelps grown underneath.

All the farms would have a nested series of cages where tighter woven fencing kept the smaller fish from larger fish and other species. Once fully grown, they would be harvested or fed to other fish in the system. This would not negatively affect the wild species. Its purpose was to allow the native fish a chance to repopulate the oceans, seas, and lakes, and, therefore, reestablish a more natural ecosystem while the captive ones provided more and more of the foodstuffs.

This, in combination with the space mining program, was expected to heal the Earth, given time.

Many, at first, looked aside when there were parts of the program that raised concerns; they were downgraded, minimized, overridden by the comparative benefits. Most people on Earth were willing to accept the original expenses, the individual hardships of material shortages placed on them if the UN succeeded in globalizing the Helium-3 Project. There was a call to do away with the throwaway society that had developed in the Western Culture.

Many of the eldest could remember a toaster that lasted for twenty years while the youngest accepted one that broke after the one year warranty. The same principle applied to almost any manufactured item in the later years of the twentieth century and beginning of the twenty-first.

Even some of the remote tribal peoples still existing in the most remote Amazon welcomed the possibility of easing the burdens of life. They did not really understand that what was being offered was not a guaranteed hunt, or a better spear, or bow. The various gods of these peoples had all promised a time in which they would put their people into a paradise existence, a life of ease, or even the belief that they were now there. The people, in general, knew almost nothing of the science behind the program. But who would not want to have life made easy? Which of them or us would not give up some of their individuality, their private rights, and some freedoms in exchange for comfort?

There were plenty. It just took some time for them to realize it. There were two major groups that opposed plans. There were loud and sudden responses from the religious groups and slower, burning, deeper undercurrents from the remaining politically opposed organizations that did not want supersized government.

Even those who supported the project at first began to question the expected treatment and, more so, the creation of the *anthrocreatus*. As various religious factions began their attacks on the program's "unnecessary" research and development, many people were swayed to more conservative views. Public opinion is very fickle and is played like a fine instrument by those who know how.

*

Davies was an excellent musician in that respect. As pastor of a very large Fundamentalist flock, he had slowly turned the screws ever more tightly. He had begun organizing things just as he had outlined in the Vatican meeting and his speech from the balcony. Only a few busy days after the Vatican speech, he was in front of world coverage again.

"People of our Lord and savior, we have a great question before us, our answer to science and its vain and fallacious ambitions. We must present a united front, based on the Bible, not on our feelings or desires. Our Lord Jesus lived a life of poverty, eschewing materialism, and embracing a humble existence," he began.

"He and the apostles led lives of minimum requirements. The apostles were expected to work for a living, not subsisting as ragged beggars amongst those they taught. Some were fishermen, some worked as tent makers, but they lived simply, on a minimum, subsistence level. We must approach this Helium-3 Project in the same way.

"Our magnificent and munificent God wants us to be able to provide for ourselves and those in desperate straits. The need for energy to provide for our food and shelter is all we need. Beyond that is the greed we are warned about.

"Let us examine Matthew 19:21-22. Jesus said to him, 'If you want to be perfect, go sell your belongings and give to the poor and you will have treasure in heaven, and come be my follower.' When the young man heard this saying, he went away grieved, for he was holding many possessions.

"In Matthew 19:24, Jesus says, 'Again I say to you, it is easier for a camel to get through a needle's eye than for a rich man to get into the kingdom of God.'

"Timothy 6:10 gives us more. 'For the love of money is a root of all sorts of injurious things.'

"The greed, in this case, is the desire to get more from this mission to the moon than energy. Research on the Farside is not needed and will lead us astray.

"The planned mission to Mars is not for our benefit. These expenditures will drain money from our ability to feed and clothe our poor brethren in the third world, our program of equalization. It is excess, and therefore not needed.

"It is this evil of science that drives us away from our only true God. We must gather ourselves together for this struggle. Do not be deceived. Do not allow that to occur. We need to control this according to the examples of our Lord.

"And yet," he spoke smoothly, "not all of the actions of the scientists are bad." He let that sink in for several seconds, knowing that this might be confusing to some. It was. "There are some developments that are actually the work of God."

This stunned many of his listeners. What was this man saying?

As if reading their minds and then answering them, he said, "The Underworld projects are for the direct benefit of the Earth and its inhabitants. It is said that God works in mysterious ways. These men do not know that they are actually doing His works. They think that they have thought up these remedies themselves. The Bible tells us otherwise. It tells us that there are no original thoughts. He, in His infinite wisdom, will turn evil into good.

"Notice how they follow in the footsteps of Noah when their design is to save life for future generations. Is this not the work of our great God, upon impious men in this time? They seek to restore what they have destroyed. Where did this idea come from? After so many millennia of destruction played upon this Earth, they decide suddenly to try to save it. This is the time. This is the epoch of prophecy so long awaited.

"There are two final actions we must endeavor to eliminate, the creation of artificial human beings, and the creation of artificial life, machine life. This is the ultimate insult to our God. They believe they can create. They can only mimic.

"It is time to act. We must put forth a concerted effort," he emphasized. "You will all be asked to do something for your God," he ended.

*

Reverend Harmon was the most effective user of the Internet and the older television format. He fused the two very well, appealing to the older and younger members of his church. Others did make use of these devices but with much less sophistication and effectiveness than Harmon.

"We are all the family of God," Harmon began. "Jew, Muslim, and Christian. Yes, I say, Jew and Muslim and Christian are all children of the *one*, the *only*, true god. The roots of our religions are based on the same beliefs. Brothers and sisters we are. We have, most unfortunately, often battled amongst ourselves. This is because our enemy, the great deceiver, Satan, has manipulated events here on Earth."

Harmon opened his large Bible on his grand white pulpit and read from Job 1:6 and 7, "Now there was a day when the sons of God came to present themselves before the LORD, and Satan came also among them. And the LORD said unto Satan, 'Whence comest thou?' Then Satan answered the LORD, and said, 'From going to and fro in the earth, and from walking up and down in it.'"

Harmon paused before resuming, "Yes, my friends. Satan walks among us as do his evil minions.

"Satan is the one who does evil. He even tries to tempt God to do wrong." Again, he read from Job 1:11-12, "But put forth thine hand now, and touch all that he hath (Job), and he will curse thee to thy face. And the LORD said unto Satan, 'Behold all that he hath is in thy power; only upon himself put not forth thine hand.' So Satan went forth from the presence of the LORD. The LORD lays down some rules, but Satan is allowed to act, to tempt, to lie. God is obviously in control though.

"We see the same in Job, chapter 2. We see in 2 Corinthians 4:4 that Satan is the god of this world. Satan is the one in charge here on Earth, under the restrictions placed by God.

"Yes, people, Satan is whispering, tricking, and misleading many. He uses a little kernel of truth to do great evil. We do need the energy that the space program can give us. We do not need anything more. All these other projects are a distraction from the truth. Satan has married the need for energy with the unwholesome ambitions of science.

"Satan has brought us to the point where scientists have the gall to believe they can create a new kind of human being." He annunciated each lewd word slowly, "*Homo sapiens anthrocreatus.*"

Harmon knew how to let such a disturbing thought enter into the minds of his followers. The effect was chilling on all those hearing his words. It was designed to achieve a certain response, a fear of God's wrath. It would make them more pliable when it came time for individual action.

He and the other leaders had timed their sermons to come before the governmental press announcements that would try to ease the concerns caused by their speeches at the Vatican. First strike capability is very important in war.

"*Anthrocreatus* means created by man. Satan cannot create! Man cannot create!" he thundered. "They can only warp things made by God. They are using their technology to manipulate genes. But they cannot create those genes. They cannot come up with even the idea that would have led to genes. It is beyond them. It is simple mimicry, taking the pieces of the puzzle that our great God made and gluing them together. Yet in their pride, they dare call themselves creators.

"And for what purpose, you may ask?" He paused again to let them consider.

"Slavery!" he cried. "Servitude! They are for the purpose of labor. They will have no choice. God made us free, in His infinite wisdom, free even to *reject* Him. They have no choice to live where they wish, and how they wish, that right is removed from them, forever. They will be *grown* to fulfill positions on the stations and on Luna. They are not needed. People from Earth and robots are all that are required.

"What of their *souls*? Do they even have souls? God is the one that pours soul into each life form. He has not done so in this case," he announced, with a great deal of assumption.

"We cannot allow this to happen." Harmon pounded his pulpit, causing his water glass to fall and shatter. He took no notice of that planned triviality. "It is time to rise and be counted as one of the Lord's soldiers. Be ready for marching orders." For dramatic emphasis, he marched from the pulpit as he concluded.

*

The pope announced only a few edicts directed at the cardinals without speaking to the public himself. He was ailing; the pressure and his age were catching up to him.

As had been the Catholic policy for some time, many artificial techniques of reproduction were generally forbidden. The new positions stated reenforced the old ones but updated the concepts to cover all the new applications being tested. The church was not yet ready to throw the really big punches; that would come later. For now, they would hit hard enough to stun many.

He also had to worry about a strong, new separatist movement within the church. He was quite aware that many were now, and in some cases for some time, questioning basic tenets such as birth control, the role of females in the upper hierarchy, politics, homosexuals, and even the newly announced position of the church on the *anthrocreatus* and their status as spiritual beings.

Cardinal Leonis took the stage before the gathered reporters for the more public announcements. He started simply, "We ask that all members of the church abstain from participation in any part of the plan that involves the development of the *anthrocreatus* creatures." It was well noted by the press that he used the word *creatures*.

Leonis continued, "It is the church's position that any form of labor, paid or voluntary, that contributes to the formation of, generation of, or use of information, genetic material, or the creatures themselves, is a mortal sin."

There were sputters and coughs from several of the reporters and whispers from many others. Some of them were Catholic and understood the definition of mortal sin; it meant the death of the soul as opposed to venial sin. A mortal sin was an unforgivable form of sin; one that was against the Holy Ghost, the *will* of God. It was a sin of *purposely* opposing God. Black and white had just been defined, and rather harshly at that.

Cardinal Leonis stared steely eyed as he let that sink in. "It is our position that these creatures will not have souls since they are not generated by God, by the natural prescriptions of God, namely intercourse within the structure of marriage. This must be stopped. Satan has caused the downfall of the perfect humans Adam and Eve, and now, with the introduction of soulless beings, in the form of humans, is Satan's renewed attempt to destroy us and replace us, the true descendants of God's handiwork.

"There is a darker matter, if that can possibly be; it is the construction of artificial intelligence. There is no intelligence that is artificial, only that caused by God.

"Furthermore, the church is now defining the following as venial sins, those that can be forgiven—knowingly doing business with any of the corporations involved with the production of any materials for use in this horrid endeavor, personal association with those directly involved (which means you may have to shun them), knowingly supplying funding that may be used for such work, and voting for politicians that continue to support the program."

Eyebrows that were already raised now reached their zenith. Taxes were used for this funding. Was this a call to refuse to pay taxes? Was this a political move to break the backs of struggling economies controlled by governments?

"No questions please," he announced shortly as he left the room.

He had to prepare to meet with the growing separatist movement, known as the Breakaways or Modernists, in just a few days in what may be a futile attempt to keep them within the fold. He and several more Fundamentalist leaders of the church had been hand selected by the pope to present a series of possible solutions. Leonis had little hope that he could prevent the schism from occurring; it had been brewing for decades, and this new problem simply added fuel to the fire. An eruption was in the making.

The last great separation was the Presbyterian Revolt. Martin Luther was a former Augustinian monk and Roman Catholic priest who nailed ninety-five grievances on the door of the Wittenberg Church in Germany. It was a paradigm shift in religion; further paradigm shifts in other areas would follow; pendulums swing.

*

The caliph and the imam had also made discreet statements, not wishing to create a furor with their call for a nonviolent jihad. The caliph and imam represented a large middle ground. Though there had been many gains made with the other religions, there were still far too many archconservative elements within their religion in positions of power. Recent wars were still a living memory. Many of the children of these wars were now adults; those who survived.

These men of the old ways would use this to create a different and likely violent jihad if they could. The old emotions curdled deep in the souls of those who had lost parents and siblings in prior conflicts. That was to be avoided, except as a last resort, and only with the caliph and the imam calling the shots if possible. The trick was how much rein to give to the horse.

The two, though from different sects, had grown close over the years. They had, for many years, been working to heal the differences between their peoples. They were succeeding, slowly. But a small breach in the dike could much more quickly destroy those hard earned gains.

They knew that several of these archconservatives, from both sects, the Sunni and Shiite, were acting together. This was somewhat surprising in many respects. These sects were historically opposed. And yet, they had finally found some common ground. The enemy of my enemy is my friend, as the old saying goes.

The mainstream Muslim peoples had been coaxed gently into support of the space program, dependent on UN governance and the benefits promised. Powerful sheiks needed to be a part of the energy program, or they would lose control of the energy market now under their control. They announced the reasons in rather couched terms to their people. This would be a vexing point to escape from at a later date.

These highly religious people were now ready to withdraw much of that support with the announcements of the clerics. Their faith questioned the motives of the scientists and their own political leaders. Pendulums swing wildly.

*

The World Space Agency (WSA) was decidedly miffed at being upstaged by the religious right, again. Their planned disclosure time had been set for two weeks further in the calendar. How did they know the timing? Was it someone inside? The Vatican speeches had caught them with their pants down about their ankles while standing in a blizzard.

Further, the second round of significant attacks came before they could even properly reply to the first statements. There was a great deal of scrambling and high-ranking vid calling. Speeches were rewritten almost overnight in response to the second set of attacks. Now the whole program would be disclosed live in a publicly tele-vided and webbed broadcast.

There were also a great many questions asked. While the program was not top secret, there were many things that had not been paraded about like Lady Godiva. The use of genetically screened and enhanced sperm and ova to produce Hsas had been a well-kept nonsecret. But all secrets come out, usually before they are supposed to.

*

It was time for the circus. The big top had been set up, and the audience was waiting for the man in the top hat. Dr. Herman Vasser had been chosen because he was high enough up the ladder that he would be respected, and low enough that he was expendable, and he knew it too. *What a life*, he mentally muttered to none but himself as he straightened his tie. His stomach gurgled in agitation.

As the representative of the WSA, he was expected to answer all the charges made by the antispacers, namely the religious Fundamentalists. He had been given only two days to prepare his speech. He had worked furiously to make sure it covered all possible angles. He had to restore legitimacy to the program. Too many people had recoiled in confusion at the accusations made by the religious leaders. It was up to him to swing their favor back to the program. As many as a hundred million would view this press announcement directly, and four times as many would review it in the news recaps.

Dr. Vasser waited nervously, sweat dampening his armpits, in the wings as the director of the WSA announced him. To the sound of the expected polite applause, he approached the podium. This would be the highlight or the dismal inescapable abyss of his career.

Notes in ever so slightly shaking hands, feeling as if he might empty his bowels, he welcomed all to the nightmare in a strong voice, "Ladies and gentlemen, welcome to this conference. The WSA has called you here to answer claims made that would possibly lower your confidence in our program.

"My first statement must be to say that we have never intended to offend anyone's religious beliefs. Our business has never been, is not now, nor will ever be religious in nature. Science is our endeavor, our means to improve the quality of life on Earth. The understanding of and use of the resources around us is our focus. The area of our focus is space. We are equal opportunity employers and have people from many different religions as well as agnostics and atheists working for us." This had been determined by a very last-minute poll conducted to bolster his position. He had included the agnostics and atheists even though he knew the zealots would try to use it. With all the accusations being tossed about, he had to be very careful and very honest. The leaders of several of the countries involved were watching the live show. They frowned at the admission. The more intelligent of the advisors recognized and approved Dr. Vasser's ploy. They would explain the subtleties to their bosses later.

Vasser continued, "There are numerous complaints that the research we are doing is unnecessary. All of us in the know beg to differ." That last sentence was a key one. Dr. Vasser had worked with a political speech writer (the president's in fact), a psychologist, and a psychiatrist while forming his speech. That sentence was very carefully designed to show that the people lodging complaints did not have the facts, and therefore, their arguments were false. It was also more true than what the enemy, the religious groups, were saying.

"Several of the experiments are directed toward objects that directly affect Earth.

"The first on my list is the group of solar observers that will be placed in quartered locations around the equator of Luna. These observatories will study Sol and provide large quantities of data that could not be taken through our atmosphere. Several of our opponents have claimed that earthbound satellites can do the same job. That is true, they can. However, the systems we are setting up are expected to last fifty years with only minimal maintenance and repairs. Satellites are only designed for a few years of service. Satellites require a launch platform, a rocket that continues to pollute our atmosphere." Again, there was a subtle thrust here, saying that the opposition did not care about Earth. "Our projections show that the cost savings using the Luna-Sol program will save over two hundred billion dollars over the lifetime of the project. That includes all costs for people on Luna to repair the systems versus launching several satellites." People meant *Homo sapien anthrocreatus* but, by terming them so, brought them into the fold.

A couple of the governmental leaders, located in various cities about the globe, smiled, and one said, "That hits them where they cry."

"As a rebuttal, an opponent might say that a satellite, like the Solar Heliospheric Observatory (SOHO) that served for an eternity in terms of satellites, would observe 24/7/365, while the stations on Luna cannot because of the rotation of Luna. By quartering Luna, we will be able to observe Sol continuously. While certain of those observers are not able to observe Sol, they can be redesignated for deep space observation and atmospheric analysis of Earth. The portion of the observer array engaged in deep space analysis is the part least able to provide needed data about ozone depletion and greenhouse gas emissions." Again, the help Vasser had received in the speech could be detected in this sentence as well. It meant we are most concerned about Earth, but we will take the greatest advantage of the equipment, at a very minimal expense, since it is already there.

Vasser and his crew had really taken the wind out of some possible retorts by presenting the possible rebuttals. "When the observatories are down is the perfect time to do upkeep. It is much like changing the oil or batteries in your vehicle, or washing the dust off of your solar panels to get the best energy usage." This homey analogy was directed entirely at the layperson. Everyone knew that regular upkeep kept a vehicle running well and improved resale value. Vasser's audience was not just the religious zealots but included the majority of the people who were needed to outvote the zealots in the upcoming elections. He needed to find a way to hook them; this was the bait. The psychologist had written that sentence.

"The other large dish arrays, both on Nearside and Farside, will be used for communications, much like the current satellite systems, but if one fails, it can be repaired almost immediately, unlike satellites. They will carry the GPS systems, vid-link, and Web net links. There will be multiple redundancies that can take over within seconds if there is a failure of one of the dishes. When not in use for communications, we will make use of them for research." Again, the hint was that Earth was the primary concern; research, secondary. In truth, it was. Yet there would be a few more dishes than really needed for the communications and backup. The Deep Space Very Large Array on Farside certainly was overdone for communication.

"The Deep Space Array on Farside will allow us to study other stars with unprecedented accuracy. It will include optical observations in a separate array that will be joined statistically with the radio measurements. We hope to learn how to predict large solar flares that can cause such problems with communications and, perhaps, damage our upper atmosphere.

"The mining operations we have planned directly benefit all humanity. No one has complained that we wish to harvest helium-3 for fusion reactors. However, the machines we've designed are capable of bringing us many other raw materials as well with only a little extra effort." Again, the stress was on getting the most bangs for the bill, as well as saving the Earth from further ravaging due to the mining of resources.

"The titanium and aluminum we plan to mine will be used for the construction of housing projects around the world. The poor that do not have their own homes are going to receive one in the form of large scale, underground, multifamily housing projects. Some of the titanium will be used in the Underworld projects, the undersea and ground development of the cities of the marine biologists, and naturalists working to revamp our oceans and landforms." Earth first was the word, repairing damages and providing for the poor. It was a real vote getter. Yet again, the plan had been that all along. No one could deny that it had been set out in the beginning. Every leader had seen the advantage of housing his poor people in new, nearly indestructible housing blocks. It would consolidate their power base, and it was good for the people. It was a win-win situation. There would be plenty of materials left for personal projects. The whole moon was available, along with all the asteroids they planned to use.

"There are many who are upset at our designation of the robots as Foremen and Laborers. They say the anthropomorphism of machines is a degradation of humans. We in the program chose the labels as an honor to the average working person. We had considered Queen and Drones, after the hardworking bees, and other names, but none were truly representative of the human nature of our goal. Who really runs the show at a worksite? The foreman does. Who really does the work? The laborers do. This is the backbone of our entire labor system around the world. Yes, there are designers, architects, engineers, but what worth are they without those who do the actual work? None!" he exclaimed forcefully. The bait had been taken, and the hook was now set. There was no lie in the matter; the presentation, however, was designed to achieve a particular response. It did. Vasser had the average working stiff in his pocket now, whether they were watching now or would read about it or vid it in the news later. Their feelings and votes were crucial, so his speech was directed to the worker, the masses, the hoi polloi.

Advisors around the world and many leaders were now smiling. Vasser had done very well so far. This man would deserve some consideration for promotion if he finished strong.

That made all but the religious coalition very happy. Harmon was furiously taking notes for his rebuttal, which would take place immediately after Vasser was done.

"We must have people on-site in order to ensure that all systems function at their peak. People are always the key. We are the ultimate fixers. Some of you would say that God made us so, and that would be true." He had added this, not as an affirmation of God, but an open acceptance of what many people believe. The psychiatrist and psychologist had cowritten that particular piece. "There will be many people sent from Earth to accomplish this goal." Everyone waited with bated breath. Here was the real sticky wicket. Global leaders and all their advisors leaned toward the monitors they were viewing with gut-wrenching expectation. It all hinged precipitously on the next few words—sink or swim; there were no life preservers.

"*Homo sapien anthrocreatus* is the answer. It solves many dilemmas. We can genetically alter human sperm and ova, using techniques of recombinant DNA, so that the resultant beings are well adapted to the environs of space." He paused to let the room settle a little. "We have used these methods to improve livestock, food crops, efficacy of medications, and progressing towards eliminating many forms of cancer. The use here is an extension of what we are already doing. There is nothing new. Only the application is different.

"The people in space, astronauts, suffer from muscle and bone loss. The microgravity environment of space is the cause. Muscles of the calves, thighs, and back are most susceptible and can lose 25 percent of their mass in a month. What they need is to produce more of a protein named insulin receptor substrate one

(IRS-1) to turn on the genes for the production of proteins responsible for muscle development. Many say that we could simply inject astronauts on a regular basis with the needed IRS-1. This, however, introduces a whole new set of difficulties. The injections would have to be intravenous, not dermal. Multiple injections weaken the vessels over time. This increases the possibility of secondary problems. That, along with the possibility of embolism and broken needles, prohibits this method. As a final thought on that, it would take several tons of mass to include all the IRS-1, needles, and protective boxing for a single mission to Mars, or a month on Luna or the space stations. The medical waste would have to be ejected somewhere as space junk around Earth. As a civilization, we are turning our back on littering, unless we wish to pay fines for such." This tidbit, written by the doc himself, garnered a few titters.

"People can lose 2 percent of bone mass monthly, especially in the hipbones. This loss of bone density leads to osteoporosis, which leads to easy bone fractures. Under microgravity environments, the bones would likely heal poorly. The calcium lost can deposit in the kidneys as kidney stones. The current method for treatment is to create artificial gravity using centrifugal forces. This method is helpful but not entirely effective.

"Sleep deprivation may be the worst problem of all. Currently, most astronauts only get five to six hours of sleep though they are allotted eight hours each. This disrupts the circadian rhythm, which is connected to the hypothalamus. Research shows that blue light, wavelengths between 460 and 480 nanometers, is the strongest, healthy stimulus for the circadian system. These discoveries benefit all humankind. It has long been thought that low-level red light is soft on the eyes and good for rest. By installing blue lights, we can standardize and reinforce our normal circadian rhythm. In space, we will use flat panel blue LED lights, not bulbs that are fragile, to light the stations and vehicles.

"Cancer can be a problem as well. Cosmic radiation penetrates cells, destroying DNA. To help prevent this, a light magnetic field can be generated to help shield astronauts.

"None of these technological tricks are as good as improving certain aspects of the body itself. A person with a religious perspective would say that the most natural is best. This combination is the best. Medicine's goal ultimately is prevention, not cure. In this case, the best prevention is to alter some of the genetic code.

"We have developed a genetic strain that will accelerate the production of bone mass and muscle tissue, especially in the vital areas. Eyes will be more sensitive to blue light. It does mean that they will have to eat more often, but that is a small price to pay for health. Yes, health is the key. Those who accuse us do not consider this." Here was another point for the average person. Health was a great concern for the poor. The costs of medical help had long been the bane of the lower classes. Reform after reform had fallen short of the mark and then some, for a multitude of reasons.

"Finally, we must address women and pregnancy. *Homo sapien* babies cannot survive the rigors of fetal development in space, period." By referring to them as *babies* and in *fetal development*, he accepted life in the womb. This was a pivotal change in most scientific views. "That assistance will come in the form of artificial gestation chambers. Women themselves could not survive a full-term pregnancy in space under current capabilities. It would, in most cases, kill them."

There was a rather shocked silence in the room. The newsies were stunned even though they had known some of the basics. Vasser let it play out for several moments for greatest effect.

"Our studies show that the physical strain on the female and child would likely cause the death of both. We are not here for death but for life," he forcefully said as he leaned close to the microphones. He had been coached in that motion by the psychiatrist and the other speech writers. It was a ploy to bring intensity and familiarity. Again, he paused as he leaned back and collected his notes. He had just gained the female vote.

"Our mothers and our children are our future. On Earth they are safe. In space, we need to ensure the same. The new station will provide the correct artificial gravity, and the artificial wombs will provide added support to keep the children healthy." Calling them children even before they are born could easily lure some of the right wingers and solidify the fence-sitters. Vasser was inviting everyone to the party.

"This is not a new form of humanity. There is as much genetic difference in the *Homo sapiens* gene pool as between us and the *Homo sapiens anthrocreatus*. The Hsas are fully human. They are nothing more than genetically enhanced people. We have determined that in order to populate space most efficiently, we must regulate the genetic structure of the individuals to prevent serious health issues.

"They will be more capable of proper bone growth and tissue development with their modified genetic structure, but they will also have an increased mental development." He turned to point at the large screen behind him with his laser pointer. With the push of a key on the keyboard in front of him, two equations in black appeared on the yellow screen. "The first one"—he flickered to the left equation—

$$P = \frac{0.5 + 0.896(n-1)}{1 + 0.896(n-1)} \qquad \text{for } n>0 \text{ and integer values}$$

"is the current model for human memory where *P* is the fraction of correct responses for *n* number of trials in a new situation where there is only a correct or incorrect answer. Notice that for n=1 that the equation devolves into a 50 percent chance of being right—a guess. The growth is exponential, slowing to a general plateau." He pointed to the added graph.

"Next, we have the expected development of the *anthrocreatus* according to this equation." He directed everyone's attention to the second equation.

$$P = \frac{0.5 + 0.887(n-1)}{1 + 0.752(n-1)} \qquad \text{for } n>0 \text{ and integer values}$$

"This model shows a similar pattern. However, the plateau is higher. We can expect a higher level of intelligence," he noted as he overlaid the two graphs. "We can expect a rapid learning curve at a young age."

This caused a muffled response from many of the reporters. A few shouted out questions, but they were ignored. Dr. Vasser was not going to get into that.

Vasser was careful to balance this possible perception of superiority with his next statements. "Notice carefully that we have designated them *Homo sapiens* then *anthrocreatus*. All that means is that doctors need to understand the genetic difference. They will appear, behave, live, and die as all other humans do. In fact, they may be frailer than we are in some ways.

"They cannot return to Earth without a severe risk of succumbing to the many infections that we simply brush off. Their immune systems will not be accustomed to many of the diseases we are. They cannot take over the Earth as some have suggested. Just as in H. G. Wells's *War of the Worlds*, they would die of disease."

Leaders the world over took deep breaths and relaxed. Advisors were elated. Vasser had really done it. Many legislators also were relieved. Their positions were much more secure than they had been just fifteen minutes before.

"There have been rumors that we are only interested in cost-effectiveness. Hogwash! We are interested in saving humanity.

"Ladies and gentlemen, thank you for your time. I appreciate this time to straighten out some misconceptions in regards to the proposed program."

Vasser left the podium before any questions could be asked. He was elated. He knew that his job was more than secure. He had won a promotion and then some. It had gone perfectly. He was higher than any drug could make him.

*

Reverend Harmon started his speech in hardcore style. "These people are liars. They are politicians, and all politicians lie for a living." He knew that this was the common person's feeling about officials.

"The first words out of Dr. Vasser's mouth were about intentions. His intentions are to mislead. His claim is that science is based on fact, not faith. Well, science is faith, a distorted one at that, not fact.

"Take for example the statement that the sun will rise and set every day. Most of us would accept this as fact. However, we must examine this statement more carefully. Let us suppose that we are in the Arctic. During different parts of the year, the sun does not set, and at other times, it does not rise. In plain English, it means that from a different perspective, the statement is entirely false. They have not included all the facts, or do not know them, even though they think they do. Scientists assume that their perspective is the correct one. They have made a serious error.

"They believe that if an event occurs one thousand times in a row that on the next occasion, it will do the same. Consider a coin toss. A person tosses a coin nine times and always gets a head. He then comes up with a rule that says when he tosses a coin, it will come up heads every time. This is science. We know, because we can see, that the coin will eventually come up tails." This was a thrust to show that the scientists were closed minded, not the religiously oriented.

"Dr. Vasser states, 'Science is our endeavor.' What that really means is that they exclude everything else.

"He admits that there are agnostics and atheists working in the program. Notice how careful he is to not mention in what proportion.

"The good doctor implies that the extra work they are doing is for the direct benefit of all the people here on Earth. Horse muffins! They have a plan to populate Luna, Mars, and eventually send artificial people to other stars. They intend to replace humanity with false beings. Notice how he carefully couched the fact that they intend to try to make them smarter, as if they can do better than God. This is blasphemy. The research they are doing is for these goals, not the benefit of Earth."

At this point, some of the world leaders still watching just about shit their pants. Where did this son of a bitch get all of his information? He was too well prepared! There had to be an internal leak, at high levels, somewhere.

"He admits this when he says that the machines that they will set up will last for fifty years, but require the services of people. People that live on Luna, that is. What people? Abominations of creation are what kind, genetic harbingers of doom for the true human race.

"Vasser speaks of saving money. What is money compared to what is right? He speaks further of studying other stars. Why? Because that is where is they wish to go. What does that have to do with God's plan for humanity? In Genesis, we read that He wants us to turn the Earth into a paradise, not the other planets.

"Machines can do most of the work that is required for Earth's benefit. The only people needed are real people. If it costs more, then it costs more. We do not need this abomination of the creation of our Lord and Savior. People, we must rise up and throw out these sinners. We must take our action now. Stand up for your beliefs and say no!" He left the podium without saying more.

He had hit them hard and with great effect. He hustled to his dressing room to watch the speech from Davies, which was timed to follow his in a matter of minutes. Once Davies was done, they would have a conference call, including Kirk, Leonis, the imam, caliph, and the rabbis.

*

Kirk, because of the time difference in Germany, would have to wait until tomorrow. He viewed with great interest as they had collaborated on all the speeches as they had agreed at the Vatican. He would use replays of Harmon's and Davies's speeches to prepare his audience. This also gave his assistants time to add subtitles in German, French, Russian, Polish, Italian, and more for regional broadcasts in local languages. Dubbing was possible, but people always hated lips moving without sound and sound without lips moving. They tend to tune out or laugh. His personal interpreters, local church leaders from each subregion, would provide direct translations for him in simulcast. The rabbis would speak moments after him. They too had prepared for replays; however, they had chosen to use dubbing.

*

Harmon's speech charged Davies up. He had his guns at the ready. He moved with authority to the podium and placed his notes where he could see them. He refused to use any of the various electronic devices for assistance. He had made a point of letting his people know that these machines implied that God's creation, the brain, was flawed and needed help. His notes would not be used. There was nothing wrong with his well-trained mind. His skill as an orator put him at the top of his field.

"Thank you all for allowing me to speak tonight to you. We have witnessed the wonderful work of Reverend Harmon. He really lays it all out and does not pull any punches. His rebuttals are masterful. His oration ended with a call for action. I am here to expand on this." He turned about, looking to the left then the right, as if there were more than a few people actually present. It was a technique used to bring the various types of monitor viewers into a more personal contact.

"God has called us to act as his tools, His agents, at this time. Do not think that He does not need you to act for him. Did He not call upon Noah to act? Did He not call upon Abraham? Did He not call upon Moses? David?" He paused to allow the thought to really enter the minds of his viewers.

"Our great God called upon His only begotten son, Jesus, to act as the agent by which we are all saved. SAVED! The righteous are called to act. The righteous will take up that call." This would compel those who wanted to be seen as righteous, as well as those who really believed themselves to be so, into action.

"How are we to act?" Again, Davies paused to let his audience think and come up with some answers as well as prepare for his.

Davies answered with, "In order to put a stop to the actions that offend our God, we must make a concerted effort to control the legislative bodies that are shaping a future that we do not condone."

*

"Oh shit!" "Měrd!" "Caca!" And a variety of other words for feces were said by a large number of advisors in locations around the world. Most of the higher ups had abandoned the monitors after the speech of Dr. Vasser. Others had listened to Harmon and thought that things were not too bad—recoverable. A couple advisors grabbed phones to warn their immediate superiors. Others double-checked to make sure that recorders were still operating. Some simply stood agape, fly-catching.

*

"We have already begun with the letter writing campaigns. We must follow that up with a flood of personal messages to each of our federal representatives.

"We must select, from among ourselves, men and women of considerable character to run for offices at every level of government. It is time to begin recall elections in all the locations where such are legal instruments of the will of the people. We must take charge, stand up, and stop the madness," continued Davies.

*

The advisors still watching were standing or sitting with mouths hanging open and with stupefied looks on their faces. An openly announced religious takeover, as opposed to simply a political one, is what was in the works. There could be no mistake of that. The bold-faced candor of these orators went far beyond prior pseudopolitico-religious pushes. These were not calls for revival but jihad.

*

"People, we must each share our homes for weekly meetings to organize this effort," Davies stated. "Each congregation in every town and city must use their facilities as a central management location. Every member of every family must put forth their best effort at this crucial time.

*

"That may leave us an opening for a tax initiative. If they use churches for active politics, we may be able to tax them for nonreligious reasons," explained an imaginative advisor in Washington DC.

Another responded, "Won't that infuriate them even more?"

The first said, "We cannot stop them from meeting at home or any public place, yet there is a question involving the separation of church and state. We might be able to turn the tables—"

"What court would go for that?" responded a third sarcastically.

"It does not have to be constitutional. It would take a long time to actually make it to and through the courts. We could probably get an injunction preventing them from using church facilities for political purposes until the case is settled," noted the first.

"I think the boss will go for that," said the second as he reached desperately for the phone again.

*

Davies followed with, "Only when the good people of this world rise up in peaceful fashion and control what their governments do are we in a position to rid ourselves of these abominable notions of artificial people. Only when we provide for those with needs here on Earth can we say we are doing God's work!" Davies thumped his podium furiously to emphasize his point.

"Let us examine what the Bible has to say about providing help for the poor. It tells us that we must provide aid. The expense of the unnecessary portions of the space program, with its hundreds of trillions of dollars, detracts from that. We must reallocate that.

"Friends in Christ, I will let Pastor Kirk of Germany speak more on that topic early tomorrow. For now, you must contact everyone you know and begin to set up groups, committees, and ways to fund our efforts. We must also avoid doing business with those who support the program as is. Do not condone by association. Heed the call of God!

"Thank you for your time." He left the podium with a certainty in his step.

*

The advisors did not really understand the meaning of the word *reallocation*. Their failure to capture the full meaning of that one word would mean the political chopping block for many of them. Many of the bosses they answered to would taste the fury of a rising political storm unprepared for the full extent. Pendulums in motion to and fro.

*

Dieter Kirk knew that he had been given the toughest part in this set of speeches. Europe, though, was the best place to try this idea out. Socialism was far more prevalent in the governments and the minds of the peoples. It was up to him to stretch that idea to a much greater extent. The reception here would be more favorable by far than in the United States. Eastern Europe had never been able to reach equality with the West and would be susceptible to the calls he was about to make. The changeover from communism to capitalism in the old USSR had been only a partial success. All through the 1990s and early 2000s, there had been countering socialistic forces. Socialized medicine offered a strong incentive for all under the median income and the entire third world.

Asia and Africa would also see easy gains. Africa had huge amounts of untapped resources and large amounts of hungry people who would desire what was to be offered. The possibility of ending tribal feuding would entice many to support any solution. Food and medicine to heal the children, this would tug at the hearts of the voters.

The billions of people in China, Pakistan, India, and Indonesia had already emerged as a powerhouse in the global economic engine. The socialistic, family oriented drive could only improve their positions. The fact that it was proposed by religious organizations of the West would matter little. Once it became a globally controlled program, they could use their substantially larger voting block to gain leverage. This may have been an overlooked outcome for the religious leaders.

Kirk prepared to put it all on the line. Patiently, he waited for the recorded speeches of Harmon and Davies to end. He was already at the podium.

"Friends, countrymen, and people of the world, it is time for a change. I talk not about little changes, but significant ones.

"People in Africa starve and fight endless battles. The poor everywhere fight to survive. As they now exist, our governments are dismal failures. They claim to be for the people, yet they are really for the powers behind the people in the government. These powerful people own the governments of the world. Nationalistic governments must be overridden by one all-controlling global government that will alter the current situation. We must place into position leaders from within our ranks that will usher in an era of true social expansion. We must use the one organization that has a semblance of global goals—the UN. Even then, it must be revamped considerably so that single nations cannot control the outcome of decisions.

"We must change that now, not tomorrow. People in need cannot wait. There are people who would work the land, but have no land. There are people who wish to have jobs, but are not able to obtain one. There are people who want their children to be healthy, but have no medicine. Where will all this come from?

"We must reallocate resources. Stagnant lands that are unproductive must be turned over to those who will use that land to produce food. Wealth, above a certain level, must be confiscated. That money can be used to provide temporary help until jobs can be found for anyone that wants one. A fund can be established for medical assistance and relocation of entire families. Housing can be built. Mansions can be confiscated for public use.

"This may sound very harsh. Ask the people who are starving about how harsh life is. Ask the dying about life. Ask the children dying in battles for turf and food in Africa.

"What will they say? They will say, as they always have, 'Help us.' Which of you will do so?

"We must take from those that have too much what is necessary for those that need. This means that all great wealth will be procured, lands confiscated, businesses restructured, people fed, health care for all, and an end to conflict."

He pointed directly at the camera. "Now is the time to act."

*

There were many people who spit out their food and drink in surprise amidst the strong words enunciated by Kirk. They were the ones Kirk was directing his words at. Such a stunning call to action stirred the lives of a multitude of people regardless of religious affiliations. The elite were terrified at the possibilities while the downtrodden espied a light at the end of a very long tunnel.

*

What he was suggesting was a return to full on communism, the absolute state in the regalia of the religiously controlled UN. It would be on a worldwide scale this time if they were successful.

How could any suggest communism after its dismal failure in the Soviet Block? Even the Chinese version had mellowed considerably.

The argument went as follows—nationalism was on the point of failure. The world had not been quite ready for *true* communism. It was destroyed by capitalism, greed, and because there were too many nationalistic forces remaining. It required a more developed social conscience for success, the point being that now that social conscience had developed sufficiently and would be morally guided by religiously oriented persons. There was no comprehension that the system itself contains flaws, regardless of those running it.

Some people disagreed vehemently with the axioms of communism. History has something to say about true capitalism and socialism as well since neither has really ever existed. Many believe that the United States is a true capitalistic society, but they don't recognize the fact that *any* intervention by government belies that. The amount of governmental involvement in the business affairs of the United States can easily be seen in land distribution, railroads, and more as time passed. Social programs that take money from one group of people to give to another are defined as socialistic. The mafia called it protection money.

In the West, so-called capitalistic societies are heavily regulated in a conflicting morass of legal bureaucracy. True laissez-faire, hands-off capitalism has not existed since before the formation of ancient governments where individuals at the top controlled the flow of wealth.

Communistic forms fared no better. The governments created by revolutions in 1913 suffered betrayal as power wielders built themselves dachas, owned vehicles denied others, and had opportunities for their children inaccessible to lower classes. In truth, there was no difference real between the two. There is an old saying, "Out with the old boss, in with the new."

> Twelve voices were shouting in anger, and they were all alike. No question, now, what had happened to the faces of the pigs. The creatures outside [the house in which the pigs had been playing poker with the men] looked from pig to man, and from man to pig, and from pig to man again; but it was already impossible to say which was which.

> —*Animal Farm* by George Orwell

*

The rabbis Levenson and Simon were ready to present a united front. They represented two major factions of the Jewish faith. They approached the pair of podiums at the same time as a show of unity.

"Our people," began Levenson, "have faced many obstacles in our long history."

"Faced and overcome," completed Simon. "We now face a new type of adversary, one not as obvious as many we have had to deal with before."

Levenson added, "This is more than a physical threat, more than a spiritual threat. It is a combination of the two.

"Very seldom have we had to resist the attack of the two together as a people," Levenson noted. "The words of our history show that we have overcome these kinds of problems before."

Simon swiftly completed the thought with, "We have succeeded in these struggles. We will continue to follow the true path as God has planned for us."

"That battle now leads us against the scientists that would blaspheme our very existence. We must unite with our fellow believers, of many faiths, to end this battle," said Levenson.

Simon pressed on carefully, "We must make peace with our Arab brothers. We need to find within ourselves the strength to create a final solution with our Palestinian neighbors, the Syrians, the Saudis, the Egyptians, with the entire world."

There were several catcalls after that announcement. Hisses and cries of "never" could be heard.

Levenson's voice reached above them. "We must! The enemy is not those that we have fought for so long! We are all children of Abraham! Do you deny our history? Which of you will stand there and deny what is written?" He stared down all opposition.

Simon continued, "We are used to working communally. Now is the time to expand that to the next level. We can achieve a global community if we set our minds to it. Will all of you be leaders in this, or followers?"

"Will we continue to support failing governments that urge us to participate in evil actions or now stand and start a new government?" Levenson asked.

*

"Brrraaap!" vocalized the vacuous patron sitting in the far corner of the Lucky Loser as if in a belching contest.

"I dunno," said Paul. "Do you see any difference in what they have always said?"

"Ain't that the point?" said Mick as he placed another beer in front of the woman without the two front teeth and the very fat ass.

"I would have to agree," said Burt, rubbing a bit of mine dirt off his boot toe. "The science guys have always said that this was their goal. They are just being more to the point this time, not hiding nothing. I think they didn't mention a lot of it because they knew the churchies would raise a fit. They were just being smart."

"Gimme a brew," Mark said as he banged his mug on the bar. "I think them preachers did the same. They been plannin' like gen-reals to shoot the horseses out from under the cav-leer-y. This war just got goin'." Mark never spoke a truer word in his entire bumbling life. He took another long sip from his freshly filled mug.

"What good do it do us?" asked the woman with the missing teeth and a fat cigar clamped in her mouth. She huffed until the coals at the end had a deep red glow. "Do we ever get the best o' it?"

Burt spoke confidently, "Both sides are promising improved living conditions. The only difference is that they are using completely different methods. Technology, the tool of the scientist, has brought vast improvements in life and many detriments."

Paul muttered, "Deter-whats?"

Mark snickered gruffly, "Bad things, dumbass."

Slightly wounded, Paul's voice rose a bit, "Just makin' sure."

Burt continued, "Global government will come no matter what. The religious crew simply is declaring that they are going to try to be in charge, not the current politicos."

"I like things they way they is." Paul slurped at his mug. "The churchies will try to make us all angels."

"Yea," agreed Mark. "They gonna put all kinds of rules on us. They be antismoking, don't like us drinking, or chasing women."

"Hah!" snorted the woman. "You do lots of chasing but no catching, you mongrel bastard!"

"Up yours twat breath!" Mark retaliated.

"Don't that mean she be getting some?" Paul laughed raucously.

The woman leaned forward to place her middle finger directly in front of Paul, saying, "Suck on that 'cause it's all you're getting!"

Burt chuckled at the banter. "You're right about the rules, Mark. They will try to make lots of moralistic rules, but that is already happening, meaning lots of controlling-type laws." Burt looked about momentarily then stated, "Look at all the laws that have been enacted to protect us. They may do that, but they also take away our individual rights. We are slowly losing the freedoms that people in general have gotten in the past since there was ever the first democracy."

"Listen to Mr. Smarty Ass History Professor here," said Mark. "What makes you an expert?"

"Expert, no," Burt admitted, "but, I do have common sense. Think in terms of what we want here in the bar. Most of us like to huff a smoke." He gestured around at all the smokers. "We know it is not legal to smoke in here. Mick could get in to some legal ass trouble. Is the law right when it protects people that don't smoke from those that do? Isn't it just a series of choices? I choose to come in here, even though I don't smoke. People can choose, but laws take away the rights of certain groups of people to make a choice based on what? Someone else's morals?"

"Always been that way," snorted Paul. "What we gonna do 'bout it?"

"Don't know," Burt said. "Change it maybe."

"Brrraaaap! I need a beer!" shouted the inebriated patron at the far side of the bar.

*

Chapter VI:

Politics

6

000110

"This is Mary Smith-Jones for WWNN with Pastor Milkes of the United Protestant Churches. It's September 13, 2014."

She turned to face the middle-aged man and asked, "What news do you have for us in the battle against the propagation and employment of *Homo sapiens anthrocreatus*?"

"Well, Mary"—he smiled expansively—"we have joined all the major religions in our campaign against this horrific trend in science. It is obvious that by hiding their agenda, they are conscience of their misdeeds."

"You say all the major religions are joined together?" she asked.

Milkes answered quickly, "Oh yes. There are a few minor sects that have not joined with us yet, but we feel that they will fall in line as the gravity of the situation becomes clear. There are only a few international or national groups that we don't expect to join in . . .," he trailed off in an almost embarrassed tone.

Mary looked somewhat surprised as she asked, "What groups would those be?"

Milkes squirmed a little and admitted, "Well, there are the Jehovah's Witnesses, the Amish, and most Mennonites here in the United States. They avoid political and military service as conscientious objectors."

"I was unaware of that. We shall have to look into that," she noted professionally.

"Are there any others?" she queried.

"There are the Buddhists and many of the other Eastern religions," he noted with an unsatisfied tone. "They don't seem to see the big picture in this matter."

"What are the immediate plans for the coalition?" she dug in.

"We plan to make immediate appeals to all governments to halt immediately all portions of the plan that involve the unnatural formation of the *Homo sapien anthrocreatus*," he announced pompously, as if certain of complete compliance. "Then we shall place persons of our direct choosing in positions of authority within governments around the world, and within the space program itself," he said with utter confidence of the event, as if the votes had already been tallied. "It will be humans that do all the work."

"What will you do if you are unable to obtain sufficient numbers of votes to control the governments?" she questioned intensely.

"Please," he said in an offended manner, "the people now know the situation. They will vote to eliminate this loathsome and deceitful situation. They will correct the unholy circumstances that surround the space program. Even many secularists and less standard religions feel deep qualms about any form of artificial humanoid creature."

"Will they vote to keep the energy program?" she asked seriously.

"Most certainly," he stated authoritatively. "But it will be controlled by, operated by, and manned by humans, not some abominations."

"What will the coalition of Stalwarts do about the Underworld projects?" the reporter asked.

"We don't particularly like that term, young lady," Milkes replied frostily. "It smacks of stereotyping us as unenlightened, closed minded, and uneducated. We are very well educated. The majority of us leading the debates have doctorates. Does that sound like we are foolish, uneducated people? No, we are not. There are many of us with degrees in scientific fields including biology, geology, anthropology, mathematics, astronomy, and physics. The Jesuits, for one example, are highly educated. Are these people to be discounted when they support the religious interpretation? Atheistic scientists would simply discount them as fanatics, fools that have no knowledge of the scientific method. They are quite wrong."

"I meant no disrespect at all," Mary soothed his temper, "but, what about the Underworld projects?"

"They will continue, under stricter control of course. It was, as we all now know, an Underworld biology project that initiated this futile attempt to sneak in artificial beings. All biology related to such should then be outlawed, just as human cloning is controlled here in the U.S.," he responded more deliberately.

"Will the coalition debate with the scientific opposition?" Smith-Jones asked hopefully, looking for a lead to another good story.

Milkes responded diffidently, "To what end? They will, in their intransigence, simply turn their heads from the facts, denying the truth."

She countered the nonanswer with, "Then there will be no debate, no chance for a rebuttal of your ideas."

"Debate," he nearly shouted. "They are unable to rebut our claims. They cannot, will not admit to the truth." He huffed and spluttered as if he wanted to say more but could not come up with the appropriate words, which was why he was only a secondary player in the game. His irritation stemmed from not being included in the meeting in the Vatican. A man of his stature in the religious community should have been included.

"Thank you for your time, Pastor Milkes. This is Mary Smith-Jones for WWNN," she cut him short. She quickly moved away from the now flustered minister. Pulling her mic off, she jumped into the van before the man could add any more.

The park bench meeting had been a good idea. It had given an earthy feeling to the shots, and the natural lighting enhanced Mary Smith-Jones's delightful features.

*

"Paul, get your camera rolling. This is the man we want," said the reporter. She had found a local Kingdom Hall of the Jehovah's Witnesses to round out her interviews with local religious leaders. She had carefully asked locals who might be the person to ask questions of, and that person was now approaching.

"Please, Mr. Palmero, may I have a moment of your time?"

"What?" said the elderly man. He was somewhat surprised to see a microphone and camera pointed in his direction.

"I would like to ask you about the Witness position to all the political and religious turmoil currently involving the space program."

"I'm just an old man, not a spokesperson for the Witnesses, miss," he replied affably.

"I realize that you have no official position per se, but I would like to get local feelings as well as official statements. Perhaps you would answer some questions and point me in the right direction for a more official interview."

"I can certainly help you to find the addresses and phone numbers of the people you should contact," he stated helpfully.

"What is your personal position as a Witness to the concept of the Hsas?" she launched right in as if he had acquiesced to all her requests.

Somewhat taken aback at her boldness, he replied, "As Witnesses, we have been cautioned to refuse to participate directly in the program in any physical manner."

"Meaning that you will not donate sperm and ova," she fired back quickly.

"Certainly not. That would be contrary to what we believe the Bible tells us. In Genesis chapter 38, we read about how a son disobeyed his father's command to perform brother-in-law marriage, and he spilled his seed on the ground. Now, it was wrong to spill his seed on the ground, but it was actually his failure to follow his father's command that was really wrong. Jehovah, since what the son had done was bad, put him to death. Other references include 1 Corinthians chapter 7. There we find much said about how man and woman should be together, one man and woman as a couple. The use of sperm and ova outside the sanctified union of marriage, while what some may call a grey area, we define as an area where we will not test God's sovereignty. If there is any uncertainty, we do not participate. Plus, the fact is that we do not believe that masturbation, which is involved in the process of donation by males, would be acceptable. It is known that many donation facilities also use pornography to stimulate donors. All of this is wrong. Hence, for all these reasons, we will not donate. But what I really meant was that we would not directly work in any research that contributed to anything questionable to our beliefs."

"But you pay taxes, don't you?"

"Yes of course. Romans chapter 13 speaks clearly on this subject. We are to pay our taxes. In Mark chapter 12, the Pharisees try to trap Jesus on the topic of the head tax. He told them to bring him a denarius, the silver coin of the day. He asked them whose image was on it. They responded, 'Caesar's,' and he said they should 'pay back what is Caesar's.'"

"But doesn't that mean you participate?"

"No, not at all. We do not participate in government at all except as required by law and only so long as that does not conflict with God's laws. We do not vote or run for office, or participate in the military."

"You won't vote for the recall of officials who are obviously for the program so that you could replace them with people of a like mind with you?"

"No. In John 18, while Jesus is being questioned by Pilate, he clearly states that his kingdom is no part of this world. We do not believe that our true path is to involve ourselves in governments of the world. We do not vote for that reason. Many other religions, those that we term *Christendom*, do involve themselves. We feel that this is a mistake, in terms of our understanding of the Bible."

"So you're very much against the program, but you do nothing about it?"

"Jehovah is the one who will judge. There is an old saying, 'Judge not lest ye be judged and by the measure that you judge so shall you be judged.' That also is from the Bible."

"What do you think about the Hsas?"

"What Hsas?"

"The ones that the space program is going to create."

"Well, I would argue with the term *create*, but I understand your question," Mr. Palmero said with a wink. "My personal opinion, mind you, is that they are just people."

"What about the religious connotations?"

"You mean will we preach to them? Yes, again that is my opinion and not an official position. But these are just people. If they want to learn about Jehovah, I will teach them."

"Thank you very much, Mr. Palmero. You have provided us with much to think about in our many religious perspectives of the current situation. This is Mary Smith-Jones with the Wide World News Network."

*

"This is Mary Smith-Jones with Pastor William Hurt of the Protestant Trinity Congregation. Thank you for getting together with me to help explain the church's opinion of the space program."

"Oh, it's my pleasure, Mary." He smiled comfortably.

"I've got a question that is not quite on target, but it just came to me. May I?"

"Certainly, ask away, young lady." His toothy grin almost put her at ease.

"What does Trinity mean? I have never heard of it"—a small fib—"but my parents were atheists, and I guess I am too."

"Oh, please don't say that!" he said enthusiastically. "An atheist is someone that purposely does not believe in God. If you are guessing, then you are only ignorant, not a disbeliever."

"Well, maybe, but again what is Trinity?"

The man's plump face grew a beatific smile as he answered, "It is the belief that God the Father, Jesus God, and the Holy Spirit are all one being, and yet they can separate themselves when they need to."

"What does that mean?" She frowned ever so slightly.

Again, the man smiled as he answered, "It is one of the central mysteries of our God. He is infinite, able to accomplish anything. On occasion, he has the need to divide parts of himself off from the rest. This has happened many times. He did so when he needed to redeem us through the death of the Christ. By giving a part of himself up for death, he bought our souls. It is less clear, but it appears that the Holy Spirit may have been separated at the time of creation. In Genesis 1:2, it speaks of the Spirit moving about on the face of the waters."

"You mean he breaks parts of himself off to create new entities," she said, stunned. This was not her basic idea of God. This did not fit her preconceptions of the old father figure in cottony white robes and long flowing hair that had been in her mind since childhood. She had been getting a real religious education in the last few days.

"In a manner of speaking, yes," he answered. "The parts are not permanently separated. They have a job to do, and then they return."

"I have never heard this kind of thing before. Why does the church not speak more of this?"

"Think of Jesus the Son, yet God. He came to redeem us. Only God has the power to completely resist temptation. Satan does try to tempt God directly. Read the book of Job. Satan does try to tempt God and the Son. Read the story of Jesus's forty days in the wilderness. Satan tries to tempt Jesus with food, the world, even his life. He is able to resist because he is God."

"I have read that recently. A Jehovah's Witness showed me many things in the scriptures. I always thought that Satan had been kicked out of heaven right after the temptation of Eve. But that is not true. Satan has access to God until the book of Revelations where it says Michael does battle and throws the devil out of heaven permanently."

"Well, that is arguable. Does it say when the battle between Michael and Satan takes place? It is related to the unsealing of scrolls. It does not say what year. The Witnesses are a cult anyway."

"A cult?" she frowned. "Why are they a cult?"

"They do not believe in the Trinity. That is a fundamental foundation block of Christianity. Without that, they are a misguided cult that misleads true believers. The Bible warns that in the end times, many cults will pop up to deceive the unwary."

"The Bible is clear that the misleading began right after the ascension," stated the reporter as a counterpoint. (One of her researchers had explained that to her.)

He raised an eyebrow with the reply, "And you call yourself an atheist. You seem to have learned a good deal."

"But is it true?" she asked candidly.

With a gentle smile, he said, "We have drifted from the topic you stated was your goal." He did not wish to continue the conversation in that direction. It was, as she recognized, a brilliant diversion. She knew all about them; she did not challenge it.

"I need to know the position of the average person in the church in regards to the space program," she followed up.

"That is an easy question. We support the program in terms of the energy promised by the mining of the moon. The extra programs that they are trying to tack on are totally unnecessary. This business with the *Homo sapiens anthrocreatus* is a direct insult to our God. There is no reason to assume that real people can't do the job. These abominations will not bring us closer to God."

"What about the claims that the environment of space would be dangerous to 'regular' people, and that we should use the replacements instead?"

"Folly, sheer folly. God made us capable of accomplishing what we need to." There was almost a touch of anger in the response. "There have been astronauts working in space since the 1960s. Granted there have been deaths, but those have been mechanical problems, not people problems."

Mary then noted, "Pastor Milkes, of the United Protestant Churches, has declared that the Underworld projects should come under stricter control. Do you agree with that assessment?"

"Most certainly, they initiated the concept of these abominations," he snorted gruffly.

The rest of the interview went downhill from there.

"What is the church's opinion about the Hsas?"

"I have not been authorized to speak for the whole church in some matters," he jousted with her verbally, stressing the word whole.

"My instructions from your bosses, the committee—," she started.

"What opinion?" Pastor Hurt asked in a wary voice.

"Are the Hsas allowed to be members of the church?" she insistently queried.

"What! A part of the c-church," he stuttered indignantly.

"Yes, can they accept Christ and become members of the church?" Leaning toward him, sensing she'd gained control of the interview.

"They are an abomination according to the dictates of the official church."

"The Breakaways and the Jehovah's Witnesses have different opinions. They are willing to accept them and let God make the final decision."

"The Witnesses are nothing more than a cult, a pile of foolishness. The Breakaways are even worse. They practice heresy. They accept women leaders when the Bible clearly states that women be subordinate. The Breakaways allow unrepentant practicing gays and lesbians to be members of their congregation and even lead the services." He virtually raged.

"But how can we as humans make such judgments?" she riposted quickly.

He responded tersely, "The Bible is the complete guide. It gives us all we need to know."

"Thank you for your time," she ended the interview and exited gracefully. Things were going south, far south.

*

"Bishop Manin, welcome to WWNN's program on the church's response to the space program. You were designated to answer some questions for the North American people."

"That's what my boss tells me." He smiled amiably.

"The cardinal, the pope, or God?" she asked with a hint of intrigue.

"Please don't be so rude," he stated coolly.

"Really," she said. "Do you not answer directly to your head, who answers to the Holy Father and no other? Does not the pope answer to God? Have I misrepresented anything?"

The thirty-second silence was very cold. Finally, the man answered in a formal tone. "We all answer to our superiors. Jesus himself noted that we have secular leaders as well as spiritual leaders."

"And what do the leaders tell you to say to your congregations here in the West?"

"As is obvious to all who know our belief, we reject the idea that we need artificial animal creations such as Hsas. These are a desecration of the human race. Those who support this unholy thing will not see heaven."

"Isn't hell a creation of the Catholic Church?" countered Smith-Jones, not quite on topic.

"Where did you hear such nonsense?" He smiled genially, as if she was an amateur.

Rather eagerly, she replied, "I have been learning so much in the last few days. Doesn't Sheol means the common grave according to the ancient Hebrew? Hades is a Greek introduction. Where does the church get the idea of an eternal hell? It does not appear in the Old Testament according to all the scholars I have spoken with."

"Secular scholars, no doubt. Who will listen to them when God comes to claim his own? Do not be fooled by the lies of Satan. He is the master deceiver. The Bible is quite clear in this matter. There are many parables likening the harvest of souls to a garden. Only branches that produce fruit will not be pruned. The others will be cut off and burned," he responded calmly.

"Many of the other denominations have announced that they will not accept the Hsas as members of their churches while some say they will. What is the position of the Catholic Church?" Mary changed direction.

The bishop replied carefully, "The churches that accept them are, in essence, saying that the Hsas can have souls. All those who oppose the Hsas, those following the true path of God, do not believe that they can have souls. We believe that they simply will be made biological entities. No one would say that dogs and cats have souls like humans do. The Bible says we are to dominate animals and the Earth. If they had souls, it would be wrong to dominate them. We perceive that the scientists are trying to make a slave race. All through history, evil humans have made slaves of the conquered, the different colored, believers in different gods. Now that we are near equality as humans, without so much in the way of race distinctions, there must be a new group that does our slave work, hence the Hsas."

Mary Smith-Jones's eyebrows lifted at this perception; it had not been voiced so cogently by any other person. There had been some loose complaints about possible jobs lost but little on the creation of a slave race. "That is quite an insight. I had not heard that particular argument before."

"It does change things, doesn't it?" The bishop smiled, knowing he had made a significant point with this reporter.

There was not much more she could do here. She thanked him and left, thinking hard about his last point. This portion would definitely make the cut for her presentation.

*

Mary Smith-Jones's last stop was at a local Buddhist temple. She had chosen this man since he had given himself an interesting name—Gupta Buddha Gandhi. According to her sources, he had added *Buddha* and *Gandhi* as a way to show respect for their ideals.

"Gupta," she asked the small dark-skinned man, "what are the feelings of most Buddhists about the World Space Agency's program, in particular propagation of *Homo sapiens anthrocreatus*?"

"I do not know," he responded from his kneeling position. "I do not know most Buddhists."

Mary was a bit confused by his answer but quickly recovered and asked, "But you are the leader of a very large group of Buddhists here in the North American Region. Do you have a general grasp of their feelings?"

"I am not a leader, only a guide. I show beginners in our faith, novices, how to start on the path to self-awareness. I show them some of my understanding. Some of them feel that they can start from where I began. They each then find and walk their own path. I cannot be their leader since we do not all walk the same path. We all try to attain the same goal, but each one approaches it differently."

"I'm sure our viewers would like you to expand on what you mean," Mary said.

Gupta smiled gently. "Meaning you could use some help as well."

Mary smiled. "That's why I'm here."

Without changing his gaze, Gupta began, "Westerners have a very different approach to their religions. They are very hierarchal. The Christian religion is centered on God with Jesus as head and high priest of the church. According to the book of Revelations, there will be one 144,000 chosen ones that will reign as kings with the Christ. There are priests, cardinals, bishops, all levels of structure."

"How is Buddhism different?" Mary asked quickly, trying to head off a long theocratic lesson.

"We have no centralized leadership in general. Let me use an analogy. Jesus told his followers that there was but one way—through him." He continued after a brief pause during which he rose and sat upon a bench. "Centuries ago, scientists learned to break white light apart into its constituent colors using prisms. Later, physicists determined that all light waves have their own specific wavelengths. There are a multitude of different colors; all of which, when combined, lead to white light. In essence, there are many paths. We, as individuals, follow different wavelengths, or paths. We *are* the wavelengths. We make up the white light. We *are* the light. *It* is *us*. For us, each path is different, but we all get to the same place."

"That is quite different from what most people here in the West believe."

"Yes, so we do not judge how others walk. I have meditated a great deal on your questions. I have come to believe that these Hsas may just be new colors, or recently discovered ones like x-rays and gamma rays, that make up the white light as do the rest of us."

Smith-Jones could not keep the surprise from her face. This was a whole new take on things. (And in her mind, one which she personally could embrace.) "What is your personal opinion about accepting Hsas into Buddhism?"

"Have I not just stated it?"

"Thank you so very much." She smiled in appreciation.

"You are welcome," he said calmly, without even looking at her.

*

Later, as the reporter looked up the scriptures to check the accuracy of the various interviewees' statements, she found them totally accurate, as far as they could go. She made a note to reinterview some of them at some point. She wanted to clarify several items. Except for Gupta, most had been something of a bust, yet there were some useable portions. They had, for the most part, simply followed party lines and rebuffed any challenges to dogma. The only surprises had been the Witness, Gupta, and the bishop with his slave issue. One other curious note, she thought, few of them had quoted very much scripture to her. She wondered why.

What she had been searching for had been clear-cut biblical references that would clarify the Hsa issue. None of the statements had given just that. She began to realize that apparently, there were none. Evidently, the Bible wasn't going to provide any specific direction for such an astounding human development. It all had to be interpreted to fit new situations and all that was open to individual perspective. Most religious people would follow their leaders' notions without question, assuming they would have greater insight. It fired an epiphany; the constitution was the same type of document, unable to perceive the extent of changes in society.

She knew that from her experiences as a child. How many times had she heard the Sunday school teachers say that it was in the Bible and therefore true? The rapt attentions of the listeners during the pastor's sermons had always caused a bit of nervousness in her.

Her father, as a geologist, could not reconcile the Bible with his knowledge of the Earth. He had always maintained that the Bible was an allegorical representation of man's knowledge in terms that the early peoples could understand.

The Earth had been formed by forces so immense that the human mind was boggled and created gods to explain it. Man controlled and could create fire on a small scale. The sun, Sol, represented the scale on which almost unimaginable beings operated. They controlled worlds, solar systems, as they were later known, and, eventually, galaxies and universes. Gods expanded as the known universe grew. But religions had had to capitulate to science before, what with Copernicus's and Galileo's discoveries.

Mother, on the other hand, was originally and continually a stout devout follower. Her faith did expand as the world did. She had always said that it was not about dates and timelines. She showed remarkable openness in her willingness to accept the Big Bang theory. It was God that started it, she would always say. What it all came down to was purpose.

Mary placed herself in the reserved position of agnostic. She really did not know much about the sciences except as it related to her stories. Nor did she have religious convictions that drove her. Her morals and ethics were much more commonplace, taken from a hardworking father and loving mother.

*

Harmon began his speech with his usual verve. He knew it was just a recap of previously stated positions, but people learn by repetition. The group had agreed that each should cover material already spoken of by one of the others in order to strengthen the perception of unity.

He started with, "There is a method to their madness. As a stage magician does, they show one thing while replacing it with another. It is nothing more than illusion. You gasp in awe at simple prestidigitation.

"Devices named Laborers will mine and gather the regolith of the moon, now termed Luna as decreed by the Astronomical Union. These things have bones of metal and brains consisting of computer chips. Yet machines they remain.

"We have brooms and dustpans that do similar actions. There are autovacs and autodusters in many of our homes. They are very familiar items that ease the chore of cleaning one's domicile. Why then have the proponents of this program given them such a name?"

Harmon continued in his polished, urbane voice, "The leaders of these mechanical devices are to be labeled Foremen. These controlling computers form a master-and-slave relationship, more commonly known as a network. Why, again, is special terminology required for these techno-devices?"

Here, his voice lowered to take on a devious intonation, "Anthropomorphism is the answer. This hefty word means to put in human terms. The scientists are humanizing machines. Why? It is the initial part of the illusion, the setup. All magicians get you to watch one hand in order to conceal what the other hand is about to do." He motioned vigorously with his left hand as his right surreptitiously pulled a flower from his jacket pocket and presented it to the camera.

"What does that mean? It means that they want you to begin the illusion with the idea that simple mechanical devices have certain human capabilities, feelings, and even existence. Yes, I say, existence, being in the form of almost alive.

"This foundation, a small step, not very high, achievable, the beginning of the garden path, allows you to swallow the next step in the magic trick. By using the humanlike terms of *Laborer* and *Foreman* on machines, they prepare you for what they call *Homo sapiens anthrocreatus*," he said sternly, eyes flashing.

"Once you accept the anthropomorphism of machines, they will do the bait and switch and put *Homo sapiens anthrocreatus* in place of the machines. They want you to accept these abominations as human. It is an illusion. A broom is a broom. Humans legitimately come only through normal procreation proscribed by the joining of clerically married heterosexual monogamous couples.

"Satan created an illusion for Eve. It was of equality with God through the knowledge of good and evil. He tempted her and Adam and performed the bait and switch, an old, old con game. Do not be deceived by this method!" he stormed.

"But wait, this switch is not the worst one. It is a ruse within a con. The greatest feat will be when they provide a machine that they claim really is alive, really has emotions, really *thinks*. They will claim it has a *soul*! Only God can make a thinking soul.

"We must be wary, on guard against these deceptions. The cat burglar sneaks about quietly. He does not announce his presence by ringing the doorbell and announcing his intentions. He comes and takes under the cloak of darkness, often leaving the owner unaware of his loss for some time. By then, it is far too late to recover what has been lost. In this case, what you will lose is your very soul! You will end up spending eternity in the Hell of your own making.

"Do not defy God!" Harmon thundered magnificently. "It will be your eternal undoing! Imagine that you are forever separated from the grace of God and his mercy.

"Pray for guidance. Listen to your leaders. Listen for the voice of God, and He will speak in your inner ear. Obey the call of righteousness, and you will be rewarded with everlasting life. Fail and you will fall to the uttermost depths of the pit of fire." His body shook as with pure terror.

Harmon, much like a general, directed his followers to listen to and to accept the other leaders with, "Pay heed to all our brothers in this fight. We are all in this together. Our Muslim, Jewish, and Christian coalition all recognize the same God. We have had some doctrinal differences. We must get past that," he stressed. "God calls us to His army. WE MARCH!"

The rest of the speech exhorted the people to gird for battle at all levels. "This, my fellow warriors, is to be the final battle. This set of events is leading into the conclusion of man's domination of the Earth, and will usher in the thousand-year reign of Christ. It is now or never." He slammed his palms upon the podium and leaned into the camera with certain urgency.

Harmon paused ever so slightly and then opened a large Bible on the podium, saying, "Let us read Ephesians chapter six verses eleven and twelve. 'Put on the complete armor from God that you may be able to stand firm against the machinations of the Devil; because we have a battle, not against blood and flesh, but against the governments, against the authorities, against the world rulers of this darkness, against the wicked spirit forces in the heavenly places.'" He slammed the book shut to emphasize the import of his readings.

"What does this mean, my friends? Let us carefully examine each portion of this prophecy and warning.

"The first is rather straightforward. The armor is truth"—he raised the Bible—"and the machinations are the con game." He wiggled the fingers of his left hand. "That is in the works as orchestrated by Satan and his earthly cohorts.

"Not against flesh and blood means the battle, the eternal war, is spiritual, not physical. We are not asked to shoot our fellow man. Thou shall not kill," he stressed each word slowly and definitely.

"Governments, authorities, world rulers," he noted. "Why these? They all claim to be the answer. They bow to science and refute the proper place of our religion. Secularism has brought us to the point of destruction.

"What are we to make of the phrase *wicked spirit forces in the heavenly places*? One of the meanings of heaven is literally up there—space." He thrust his index finger high. "Wicked spiritual forces include soul infused humans that attempt to deceive you purposely. They try to drive a wedge between you and God. The most wicked, and nonhuman form, would be the *Homo sapiens anthrocreatus*, a soulless shell replacing what God intended life to be. Is not the plan to place these things in heaven?

"Satan attempted a coup in heaven but was cast out. This is a form of sneak attack, an end around. It is a subtle thrust stating that he, Satan, should be master of the heavens.

"Prepare for battle."

*

Pendulums throughout the Western civilization just received a not so sublime push in the direction of theocracy.

*

Pastor Marks had studied assiduously for his role. He spent weeks reviewing data on portions of the space program not directly affiliated with the energy program. He searched for every expense, noting costs and the labor associated with it. Accuracy was important. He had discovered almost forty-two trillion worth in U.S. currency over a ten-year span. There was more money further down the road, but this would do for now. It was a staggering sum. The very thought of so much would draw millions of people to their side once he explained how it could be better spent."

Marks positioned his notes on the podium as he scanned from camera to camera as if there were a live audience.

"My friends, we live in troubled, even desperate times. As Reverend Harmon so eloquently noted, we must disassociate ourselves from the enemy, the supporters of this evil. We may find that we have to pull away from friends and even family. This can all be prevented if we take a stand now.

"The need for energy is clear. Even with solar, wind, and water power, our society needs more energy. These nonfossil fuel sources have extended our capabilities for a few generations, but, with the expected population growth, we will soon be in the same bottle as we were with oil and even coal.

"Though there are still substantial coal deposits, and oil is not yet depleted, they cause other concerns for the Earth. Oil can be used for other needs than as a fuel for individual vehicles. We need electricity without coal. It also must be available when there is no wind or insufficient sunlight. Large industry and agriculture need great amounts of power."

Marks paused for a few seconds to let that sink in. "That is why we must have fusion reactors. Going to Luna for the raw materials for this process is acceptable."

He let that work its way into the minds of his listeners before continuing with, "That is all that is needed though. There are a great many unnecessary projects that the scientists want you to swallow along with this simple operation. Don't! Remember the concept of the illusionist and his tricks.

"The worst offenses are the despicable programs to genetically form subhuman slaves to perform the tasks that humans can do just as well and the AI falsehood. These unjustifiable programs will cost billions of billions.

"This Hsa program does not stand alone. If they make these things, then they will need to build a home for them, a space station. Hundreds of billions will be spent there," he said emphatically.

"There, of course, must be a second home, a summer home so to speak," he noted sarcastically. "This one will be built on Luna itself. It will be necessary to delve dorms for every biological entity." He refused to use the words *life form* or *person*. They did not fit the dogma. "Many more billions needed for that project as well.

"Food, water, air, these will all be needed. More money is needed for every little item," he dictated.

His eyes narrowed as he asked, "Why do they want to spend so much? What purpose does it serve? Can it be done without all these frills?"

He thumped the lectern as he nearly shouted, "Most certainly! There is no need for the majority of these subprojects!"

"There is *no* need for these Hsas, as they are being termed," he heavily emphasized. "The work that they are allegedly going to do can be better done first by machines. If they break down or are insufficient, then humans should be tapped to complete repairs. This is similar to how they propose to make the space station core, which is needed.

"The core is the processing center for the raw materials from Luna. The rings are almost entirely for habitation," he explained and pointed to the diagram behind him.

"The plan is to use machines to assemble the majority of the core under minimal human supervision. Why would they then change this highly efficient and lower cost method at a later date? It is, as Reverend Harmon says, a magic trick. They want you to believe that more humans are needed to do the work correctly. But they then say it is too dangerous for humans. They bait and switch and viola, the solution is the Hsas. This is a false solution to a nonexistent problem.

"Waste, waste, and more waste," he announced solemnly as his head shook back and forth with great sadness. "Not only are the Hsas wrong morally, they are wrong fiscally. That means financially."

His eyes showed a look of concern as he stated, "There are still hungry people in the world. Why should we not care for them? Feed them? Can we get what we need from the moon and also care for the people of Earth? Yes!" he firmly announced.

"With the assistance of several good Christian engineers, it has been determined that we can build two core stations for the price of one complete station. We would have a greater production capability and save nearly six hundred billion in U.S. currency. If we constrain the Luna facilities to minimal *human* needs, we can save an additional five trillion. It is amazing how much of our money they wish to spend on fruitless and unnecessary projects. What better things may we do with this surplus?

"A host of things," he beamed. "Let us begin with the basics: sustenance, and shelter. What wonders we could do with a few trillion dollars?

"Let us think how many people could eat to their fill. How many irrigation projects could be completed in arid regions so that they eat yearly? How much fertilizer, seed, and equipment could be purchased for those in need? How many solar panels could be purchased for powering pumps for wells in arid environs? How many tractors, harrows, plows, and more could be bought and shared by local groups? There is a very old saying, 'Give a man a fish, he is fed for a day. Teach him to fish, he is fed for a lifetime.' Let us begin to teach.

"Humans need protection from the environment, a home. Can we not take much of this money and build long-lasting homes for those without? Can we not improve the ones some have?

"We most certainly can, and what's more, we *should*," he noted forcefully. "It is morally imperative that we do so.

"That is the goal of our moral union. The major religions of the world have joined in opposition to many parts of the space program in order to feed and house the poor, and, more importantly, prevent sin. Join us. Do so for the sake of your soul and of others."

He left the podium without another word.

*

Pastor Davies was, as always, well prepared. Like the others, he too was following up on previously released statements. The coordination between the various members of the religious groups was working well so far. Positive responses to each of the speeches so far, generally spaced apart on a biweekly basis, were quick and growing in number. Davies felt the thrill of opportunity. They had to keep hitting the enemy, and they would win.

"Dear friends, fellow believers, our colleagues"—a pitch for unity—"have put it all out there. They have laid bare the faulty reasoning of the scientists and the many politicians that support them. Our brother, Pastor Marks, spoke so wisely"—Davies also knew how to make others wish to appear to be wise and so follow the shepherds—"when he exposed the wastefulness of the scientists. We must take to heart his words, for he too speaks of the desires of our Lord and Savior.

"We must share what we have freely. That means that we must provide energy for all the people no matter what the cost. The World Space Agency, the WSA, must come to terms with this fact. All of us need to press for equalization, and press hard!

"The new nuclear reactors, the fusion devices, must be built around the world at cost, no more profit for these money mongers. These magnificent machines will provide electricity for clean water, heat, cooling, and other basic needs in what have been termed third world countries as well as our own areas.

"So far, the plans have been made, and construction begun, only for the nations that have been termed the Big Seven or the WSA. This is insufferable. This new technology must be made immediately available to all. Man has not *made* this feature of nature. God made it. In His eyes, we must share it all openly.

"Rather than the rich receiving first, they should be last, in all things. It is time for the rich to pave the way for poor to become healthy! It is time to follow up on decades of promises! No more taxes on the poor! Pull the wealth of the rich, redirect the space missions into more appropriate paths, and we can achieve that! Martin Luther King had a dream. Is it not time to fulfill it for all the downtrodden?"

His eyes narrowed as he said, "There are some who have labeled our direction as communistic. This is not so! Communism has always been the enemy of the churches, especially the Eastern Orthodox Church. We do not advocate such a government. We do believe in a socialistic government in which all people can obtain medical care, housing, and food. Socialism means that we all care for each other. It means we all share the expense. It also means that we should all be on the same relative level in terms of education, wealth, and the previously mentioned items. Our Lord and Savior was a man of the lowest rungs of wealth, yet the blood of the kings was in his veins. Do our leaders currently eschew wealth or wallow in it? Let them give their money and their blood for us as did Jesus."

He bowed his head and picked up his Bible. "Let us pray that it may be so."

*

"Holy shit!" one intelligence analyst coughed out.

Another of the suits, the secret men and women both in office and field, squinted an eye, his only one left after the Paris incident. "Can we get a friendly fed judge to let us call that a threat? The 'kings' part obviously refers to authority figures, and I would include the president in that group as an equivalent."

Yet another lower-level tie wearer stated quickly, above several other notions, "The SS, Secret Service, sure would take this as a threat. We could reinforce that with a 'concerned' notice in the daily mailbag."

One Eye grinned a most feral grin. "By God, I took an oath to support the constitution and the president of these United States. I will not stand by and let these mish-mashers ruin our country. Send a note of 'concern' from the regional desk, the central operations desk, and on my personal letterhead, to the SS, NSA (National Security Agency), CIA, and FBI. You should also send some to the sheriffs in the hometowns of all these religious bumpkins."

One massive earpieced hulk of suit flesh did a slight double take at the irony of a God-fearing man taking to task other so-called God-fearing men. His dark suit, ill-cut due to his great bulk and many weapons, gave only the slightest of twitches of humor; he felt the same as this one-eyed, rank-unknown, tough son o' bitch.

Strange, he mused, that the color of his suit was the color of the money used to fund this facility—black. He enjoyed the fact that he was unknown, would never be known; he was a ghost, a demon, *ein* golem in his grandfather's words.

*

Günter Hauk had kept his position as chancellor only by a very narrow margin, but that was not what worried him. Antitechnology Fundamentalists had gained significantly in the legislature. He only had a very narrow four-vote margin, and several of the votes he counted on were likely to stray to keep themselves in the Fundamentalists' good graces and prevent an attack on their positions next election. So few had expected the religious zealots to get so many votes; a lot of the opposition simply had not voted. Hauk had to believe that they had been awakened and would involve themselves in the elections next time, but that was two years off.

His country depended on, very heavily, as a secondary partner and soon to be a primary, the space program. Germany was a technological society capable of producing much of the hard and software needed. It brought in much needed revenues. They were cofinancing the European Fusion reactor along with the rest of the European Union (EU). It was almost completed. The Americans would finish first though. No one could match the fervor of the American builders. The Europeans were too timely in nature. They had to follow the schedule; do it exactly as planned. There was a certain languor in Europe, an aged response, lost vitality that hindered them.

The danger was that the more religiously inclined legislature might gather enough votes to pull out of the rest of the space program. They did not have control yet though. Pulling out now could cause the collapse of the economy, and not just for Germany, he thought, as he gnawed his well-bitten fingernails. The entire EU could be pulled apart. They had just barely recovered from economic malaise from the last few years.

Many other countries were in the same small swamped boat. Belgium, Holland, Denmark, Sweden, Norway, Switzerland, and even tiny Lichtenstein had all made major monetary contributions. The smaller nations would suffer the greatest if the program was stopped. Far too much had been invested in the complete program.

His main concern though was at the international level, global concerns, not just continental. The newest additions to the legislatures of many countries were writing bills that would give almost all authority to the UN. The EU would also transfer much of its group power to the UN. It was to be the New World Order, but not the one envisioned by its original designers and taskmasters. This one was to be mostly controlled by the religious coalition, not the wealthy and aristocratic. The scientists and the elite had to scramble. All opposition had to fight to keep the status quo for now and perhaps regain some ground.

He would fight these bills, but he knew that he would suffer defeat on many topics. Günter would have to walk a tightrope. He would not be able to fight on philosophical grounds but only on technical details. His goal would be to stall and to bring into question how much authority should be given. Questions on how much land and money should be redistributed would require much debate, if he had any say. Governments worked that way. With enough stalling, people's minds could be changed. Time was of great value; if he could eke out for a few years, he could swing things back to a more favorable outcome.

*

France, usually very free of zealous religious trends, was swamped by a countermovement that took nearly a third of the government under its wing at all levels. They too had to deal with a delicate situation. The fusion reactor was being constructed on French soil very near the German border, almost directly between Strasburg and Stuttgart. It was a national issue to complete it as well as a dedication to the EU.

France had been late to join in the fusion furor due to its well-established fission programs and was not in a good position if they failed of their commitments. There were large numbers of Swiss and German technicians that daily commuted into France. The EU members had freedom of travel within any of the member countries. If France did not fulfill its end of the bargain, they might be removed from the EU, forcefully even. That would have a very significant effect on their economy. If they were left out, they might be denied the cheap energy from a reactor built at a different location. It all depended on how much power had been gained by the religious coalition in each country.

It would be a very tight squeeze. There would have to be many concessions if he wanted to remain in office, thought President DeNoe. Would he have to work against his own daughter's flight to the moon? He was not a religious man, yet he prayed he would not have to.

*

Santiago, as president and head of the government of Spain, was in a total ferment. He had just barely beaten back a recall on his position, and they were already making major demands. He did not want to completely surrender to the EU or the UN. Yet that was the essential demand. The heavily Catholic country had responded to the pope's call for a change. He would have to give in on many items that previously had not even been considered. He might even face complete capitulation.

He had to hang on, whatever the cost. Politics was a nasty business. One made concessions whenever necessary to remain in power. He had spent too many years gaining the national pinnacle to simply subordinate himself to a whole other layer of international bureaucracy. If he had to though, he would. The name of the game in politics is to hang on; keep your position, whatever compromises must be made.

*

In England, the prime minister had also fended off an attack. He was subject to the legislature and had to mollify them. With the death of the aged and renowned queen, the king was the titular head but with no real power. The king, however, still had enough respect from the people to prevent a total coup. It was in his best interest since some of the demands included redistribution of land and wealth.

Since the royal family was the largest private landholder in the world, this would have serious consequences. As the epitome of monarchy, though nominal in actual power, the stripping of the English crown would herald the end of the last remnant of monarchal power in the western world.

This could not be allowed. Those who were behind the thrones (or presidencies) would fight to the death to prevent such a stroke against their designs. They sought a world government in which they could eventually claim one of their members as emperor of the world. They believed it was the natural development of government, one man to rule over all. All through history, empires had developed with one man as eventual ruler. The religious coalition threatened to take that from them; they would replace the empire builders with one of their own.

*

The sheiks of the Middle East were in turmoil. Their power depended on oil and its income. The oil was failing, however. Predictions suggested that there would be no more oil in 2035. Production was down no matter how many new wells were drilled. Even the great fields of light crude in the Middle East could not keep the people's need satiated. Two hundred and fifty dollars a barrel and it was still in great demand. They had backed the new fusion technology with hundreds of billions of their currency.

They wanted to control the new energy market just as they had the old. Without this, their power would diminish. If they did not become the controllers, new lords of the emerging economic power would take the reins. They frenziedly threw billions more into the elections in every country to protect government officials who favored the space program and its nuclear development in hopes of retaining their places.

*

Pieter Gagarin, grandnephew of the famous Yuri Gagarin, first man in space, sat disconsolately in his high-backed wooden chair. The power struggle had drained him. Russia was a republic in name only. The real game was behind the scenes. The Politburo no longer existed in name, but the people who advised him were the same type of players—dirty. He had almost lost his position as premier of the Russian government.

It was not because of religion, but because he had so strongly placed faith in the West. He might yet lose that gamble.

He was lost in deep reflection on how to save the situation. Russia was a premier player in the space program. Everyone knew that, but the opposition was playing the tune of public opinion. As in all cases of government, the majority of people did not really know what the reality was and what propagandist fiction was.

If they failed now, they would become a backwater country again. That had to be avoided at all costs. Russia had fought through many tough situations, politics, wars, and more.

Russia was the main provider of heavy lift vehicles. If the West was going to change its mind now, perhaps he could take advantage of this. Russia did not have much in the way of money, but they could sell rights to the current station, built in great part by them, to the growing third world and second-rate powers. Many of the leaders of these other countries were less affected by the religious reforms.

He had nearly lost his position as a form of blame for linking the country to partners that were on the verge of breaking their end of the deal. If he could find new partners, he might even become the sole leader of the project.

Gagarin quickly penciled down a few names of ambassadors he wanted to speak to as soon as he could; he would push aside some of his less important calendar meetings. He had close associations with three of the African leaders, two of the smaller Asian countries, and four of the South American nations.

A new coalition, properly led, could quickly gain a great deal of power. He smiled to himself at the possibilities. He would have to act cautiously, playing both sides of the coin as it were.

*

Horito survived the political backlash fairly easily. The religious forces had considerably less sway in Japan. The worst problems came from opposition forces that claimed that he had allowed the country to spend billions in a joint venture that was collapsing.

Horito had brushed them aside with the announcement that they could simply take over the entire project with any of the other remaining supporters. He stated, without any real backup, that they could then expect a larger share of the gains if other countries failed of the contract.

He made this assertion knowing that it would provide ammunition for the leaders of the countries that were having the most trouble. They could use this to gain leverage with fence-sitters. In other words, they would say, "We have spent too much money to turn back now without realizing the vast benefits, economic as well as medical and scientific." There was still a great deal of nationalism even with the trend toward global government. This speech of his provided the fulcrum upon which other world leaders would apply force.

*

Ch'in, leader of the People's Republic of China, feared no religious or political upheaval. His control was nearly as absolute as ancient leaders. All he considered was the opportunity to gain absolute (or at least greater) control of the program.

He knew that Japan was not greatly affected by the religious uprising. Bitter enemies all too often; perhaps it was time to make friends in order to gain greater advantage. They had more in common than with the west. It was time for the east to take the lead. Perhaps he should mend fences with Taiwan, the false China.

Ch'in called for his appointment secretary and made a quick arrangement for a meeting with the ambassadors from Japan and Taiwan to discuss just such an agreement. The search for cooperation could provide them with nearly incalculable wealth and power. The Earth would be theirs.

A hint of a smile passed across his politically astute face.

*

Mr. President,

As a firm believer in God, I must let it be known that I, and many others in my flock, am completely opposed to the development of fake humanoids. This is a sin.

You have always claimed to be religious. You attend services regularly. You cannot reconcile this with continued support of the space program as is. You must now take a stand. Will you be remembered as a simple politician or as a warrior for God? Will you renege on your faith? I pray that it will not be so.

Sincerely,

Pastor Kate Smith

*

President George Washington Monroe was in trouble. He crumpled the letter and threw it at the waste basket. It bounced off the rim. He sat disconsolately in his chair in the oval office with his ebony fingers massaging his temples. To say he had a headache would be to understate in the extreme. It was his first term and probably his only if things continued on this course.

He had inherited the space program, had endorsed it as a way to put many Americans to work. The possibilities had been extraordinary until the religious bombshells began falling.

The scientific, medical, and economic long-term benefits had gleamed in the eyes of Republican and Democrat alike, as well as the tax revenues. Now, however, there was a third party that was attacking the program without mercy. The gleam in the eyes of the established parties had faded as if a shovel full of manure had been thrown in their faces.

Within months, a new political force had grown into a monster. It drew considerably from the more conservative Republican Party and less so from the Democrats who believed that money should be better spent on Earth. The newly arranged group was almost the size of the individual party remnants. There was finally a third party. It was not the one most of the political pundits would have or did predict.

The remaining Republicans and Democrats formed an alliance within a month to protect their positions, status quo you know. In almost no time at all, decades of two old political parties fighting came to a relative close with the formation of what could be called the prospace party to oppose the religious party. A delicate balance was achieved, a new political equilibrium.

Washington had prevaricated, as most politicians do, until he was relatively sure which way the wind was finally going to blow. It went the way of the scientists, barely. Now he had to decide whether to sign the bill that would continue funding the program for the next five years.

With a deep, uncertain sigh, he picked up his pen and signed at the designated spot. When done, he realized that he had either signed his political death warrant or garnered his place in history as president of a semicountry leading the way into space. He had a sudden urge to run to the majestically decorated restroom and either puke or shit. What a world it was when the supposedly most powerful man on Earth had such problems with his bowels and stomach. Life was really a bitch or bastard, depending on your perspective, he thought.

*

Socialism was the watchword again, not *communism*—that hated word. All the people refrained from saying communism. It was politically incorrect to use that word. This had been designed in response to several books by experts, such as Dr. Mari Hamahdra, Joan Mere, and Pierre Montes, where the scales were set so that the highest paid professional made no more than ten or twenty times the lowest worker at a burger joint. There had, in many countries, been a wage floor or minimum wage for many years. Now there was to be a wage ceiling, a maximum wage. No individual could charge more than government-imposed limits.

The difference in wages was supposed to be sufficient for capitalism to still have an effect in the arena of economics. The chance to make more is considered the driving force of capitalism. Those who were more politically correct used the term *limited capitalism*. Politically correct is nothing more than bullshit in fashionable terms.

Any person who earned more than, say, a half-million dollars in a year would face a 100 percent tax on the excess. Inheritance would be limited too. Death taxes would be based on the amount so that a person could not receive more than a million in total value of capital and assets without the rest being confiscated for redistribution. In theory, this would prevent the amassing of great wealth in the hands of a few.

Land would be split so that everyone could have a fair chance to own some (or not). Large land-owning bluebloods would be the first to see the loss of multiple homes and hundreds or thousands of acres. Castles and large houses would be turned into museums, medical facilities, schools, orphanages, and anything that required space. Other large sections would be turned into semicity-sized apartment complexes built mostly with lunar aluminum, titanium, and silcrete, the lunar version of concrete. In other cases where the land would sustain small farmers, imported poor would be given an acre or two to use.

Upper middle—income families would lose all second homes, second cars, and savings over a certain limit. This would endanger funds for higher education. Even if the new state took steps to provide higher education, where does the money come from? It comes from those who had it in the first place.

Any person who wished, under the possible new demands, could be relocated to a place where land was available. The poorest of Africa or Asia could find themselves farming in the breadbasket of the United States or living on land that was once Bureau of Land Management or Forest Service. They would be required to pay for that land, a very nominal amount. Even a few acres can support a family very well and provide a modest income, if they are willing to work for it. If not, they lose the right to live there through the simple failure-to-produce-and-pay rule. They would be removed to another house where they would be given a *double blue collar job*, the politically correct term for a really hard labor job.

They could still decide to live in apartments or other forms of multifamily housing. The number of skyscraper and groundscraper low-income houses was expected to expand rapidly, especially in crowded countries.

Farmers, individuals and corporate, would be allowed to keep large tracts of land under a special land use fee. Everyone would have to pay to use land that was "theirs." The farming corporations would keep lands now under their control by paying larger taxes for the use but would pass actual ownership to the government. They would not be allowed to add any more. This kept the status quo; it prevented newcomers from entering the scene and kept the old money there.

Millions of acres of unproductive land would suddenly begin to grow food crops. The wealthy landowners had, of course, managed to create a loophole in the law. The acreage they owned was fertile and fallow in many cases, so they immediately farmed their ancestral acreages. The boost in food production globally nearly doubled in a single year, before all the provisions took effect. This added such a false stimulus of support for the new programs that the NWO was accepted by a majority.

*

The whole idea was abhorrent to most true capitalists. *Rape* was a commonly used word. *Theft* was another. They had all thrown money by the billions into campaigns to keep as many of their people in governments around the world. Each of these powerful families and individuals had interests around the world and had to spend lavishly in several countries. The investments bought them just enough time to save roughly half their land and hide half their money. The rest had to be sacrificed in order to appear as if they had fully complied with the new laws.

They had cooperated with one another in several instances where it was mutually beneficial. High-level officials and lower-level grunts were required to keep the status quo going. They had partially thrown off the first wave of attacks though. They had set a date to meet in Geneva, earlier than usual. They had to get a real handle on this before things got entirely out of control. As one-time owners or controllers of over 60 percent of the world's manufacturing and 50 percent of the privately controlled lands, they could lose everything. The English royal family was still the largest private individual landholding group in the world, even now after the take.

*

"Mick, gimme a beer," said Burt as he slid back onto his stool after the ceramic golf shot he just took.

Mick already had it ready, and Burt was already fumbling for his wallet. The others took stock, visually and mentally, of how much they had left, both in beer and in money.

The vid-link was showing a piece by Mary Smith-Jones about the religious clamor over the *Homo sapien anthrocreatus* concept. It was decidedly stirring up controversy, even in bars.

"What's all this about the Hsas, them artifiscal, people?" asked Paul in his rather stupefied voice.

"Them churches are all fighting about whether or not they are real, and when they are goin' to hell, the moment they are born, or they die. Don't know 'bout that myself. I just think they are growing people like them test-tube kiddies from before, the ones for people not able to get pregnant," said the rumpled woman missing two front teeth. It was not her real opinion, but she wanted to get the argument going so she could insert other thoughts.

Mark lifted his head up off the bar to chime in. "I heard they goin' to give them 'lectric brains. They are going to be able to hear stuff we can't, talk without bein' heard 'tween themselves. They are going to be like calcamolators doin' math and stuff."

Paul belched before he added, "I heard they gonna be able to connect right up to a computer with a wire. They gonna set themselves up to take over, like in them movies."

Mick chimed in with, "I don't know about some of that. From what I've heard, they will be very much like us, only with some minor physical differences. They want to adapt them to space like, not make, like, superpeople."

"I would have to agree," noted Burt as he tipped his mug. After a swallow, he continued, "They would not want to create something they would have to compete with mentally, a danger to their preeminence."

"A d-danger to their pee-what?" snickered the woman coarsely.

"Their position, the dominant hierarchy that they have created in their own fields, they are looking for help, not someone to take their places," Burt clarified.

"Yeah, but, what about what they choose to hide from us, you know, like that Area 52 shit?" Paul muttered bleary-eyed.

"Ha!" snorted Mark. "You dumbass, its *one*, Area 51. You think they gonna make aliens out of these Hsas?"

There were several chuckles at that. The legends still floated about; they would likely be transformed into a subculture myth that would last hundreds of years.

"They is aliens. They ain't us," the woman spoke. "They ain't natural. They gonna be from outer space. Doesn't that make them aliens?"

"Ask Pedro," laughed Paul.

"Up yours, amigo," chuckled Pedro from the end of the bar. "Mi papa was the alien."

Light laughter echoed amongst the friends along the bar. Mick waddled along the bar, filling mugs from his pitcher, waving his hand at the few bills pushed in his direction. Friendly patrons would buy more in the long run. Besides, he was enjoying the banter and wanted to hear more of what these people had to say. He always learned a great deal from simply listening as he walked around behind the bar.

"I have heard that they won't be able to come to Earth," Burt noted. "They will be too physically weak to survive in our gravity and could get diseases."

"They might give us sickness," insinuated the woman coldly.

"Not likely," countered Burt. "They will be living in a clean environment. There won't be much in the way of disease up there."

"In other words, they gonna get better than us," spat the woman venomously.

"I dunno," Burt scratched his head. "They will never get to walk through the woods, and climb a mountain through the snow . . . I think they're not going to have a lot that we do, special things."

"Things science didn't give us," the woman said smartly.

"And yet," Burt countered, "science has given us a great deal—medicines, educations, an easier life than caveman days, and more."

"Knowledge took us outta the garden, idiot," she snorted.

A few belches cleared the air, so to speak, and the conversation drifted to other topics.

*

Chapter VII:

A New World

7

000111

"This is Mary Smith-Jones with the Wide World News Network. I am here in Florida, USA, of the North American continent, to witness the events in mission control as the first manned ships launch for a return to Luna, the first since Apollo 17 in 1972.

"Man first set foot on the moon in 1969 when Neil Armstrong, from Apollo 11, stepped out with the now famous words, 'One small step for man, a giant leap for peoplekind.'

"The United States of America had planned to return, on its own, or with a small number of partners associated with the space station, in 2025. The development of fusion technology, and its need for helium-3, has accelerated the time frame by a decade. It also includes a whole new infrastructure of participants. The UN has initiated legal moves to bring all nations on Earth into the fray.

"Unlike previous journeys to that silvery orb now officially called Luna, we will not witness the liftoff of the vehicles from here. The two ships, named from Roman mythology, the Romulus and Remus, will depart from the International Space Station, or Space Station One (SS-1), as it has been renamed, where they were constructed in situ.

"This is the first time that peoplekind have constructed such ships in space. The techniques developed in this manner will be required for rapid expansion into a fleet of transport vehicles.

"It has been a test of engineering ingenuity and technical skill. We are on the verge of a whole new kind of exploration and, finally, exploitation of space. Peoplekind has dedicated itself to a permanent inhabitation of space.

"This is the beginning of an immense and costly program that promises to provide energy and resources to our beleaguered Earth.

"Expectations are that we shall reap great rewards that will eliminate the need for most earthly mining for energy needs as well as for metals.

"It has also ushered in the great controversy in the form of the Hsas, the biocreations of the scientists. The final decisions on that topic may still be in the wind. Religious backlashes have brought that possibility into question for the moment.

"The countdown has now reached T minus one minute. Let's follow the action live from SS-1. This is Mary Smith-Jones for WWNN."

Once off the air, she cursed vehemently the new watchword— *peoplekind*. It smacked of the fascist propaganda techniques of the early twentieth century, a new designation used to unite a particular group of people. It had been forced on the world media by the UN. Smith-Jones was upset about the politically correct changes that had been made to the speech but had been required to use them.

Pressure for these so-called politically correct changes in language had been increasing since the early 2000s. She was reminded of her college reading of Orwell's 1984 with its mind control through language usages. Every new dictionary, claiming to be unabridged, did not have words that were still in common use. The language was being pared away word by word. Fewer words to describe a thing—say, the various colors of red, magenta, rose, or fire engine—led to the weakening of the imagination. Without imagination, the mind would become an easy target for manipulation. A certain unease tingled deep in the core of her being as she recognized the pattern of her thoughts, thoughts that might lead her to be an unwilling tool for the propagandists.

Her mind mused on the ability of the media to act for good as well as bad. She remembered well the lectures of one of her professors on Joseph Goebbels's use of the radio in Nazi Germany. Her father had often warned her, once she had decided on a career in journalism, about not getting suckered into playing favorites, the back-scratching routine.

She thought about how the United States had shown its flaws in the space program on national TV. The incredible openness in the face of several failures had led her to believe that the media could do much for freedom. She had viewed the archival films when Grissom, White, and Chaffee had burned to death, not just for the drive to space but, more importantly, for freedom, a free press, and the willingness to admit error.

The Challenger had blown up on liftoff (only a year before her birth) on TV. The Columbia disaster on reentry had been shown openly as well. She had been watching that news program with her father. She had been horrified at first, until her father spoke.

"It's a dangerous world out there," he had said. "These people died doing what they wanted to do . . . I wish I had had the courage . . .," he trailed off.

Young and uncertain, she looked askance at her father and said, "What do you mean, Dad?"

He looked directly at her and calmly said, "I wish that I had had the courage to do what I really wanted to do."

A bit worried now, she responded, "Daddy, you're not afraid of anything, are you?"

A quick laugh dispelled her fears, but he answered, "Yes, I am, Pumpkin. Real people do have fears. It is how they deal with them that's important. Those people had knowledge that they might be killed in this adventure. They did not give in to that fear and walk away. They faced it and went ahead. That's courage."

Looking a bit bamboozled, she asked, "What were you afraid of?"

He motioned her to come sit on his lap, which she did, enjoying the comfort as well as his late afternoon, heavy beard as it rubbed her cheek. "I wanted to be an exogeologist."

"But, Daddy, you are a geogist."

"I'm a petrogeologist. I look for oil. An exogeologist examines the other planets and moons. That's what I wanted to do, but, the money was in oil. I wanted to be able to provide for your mom and eventually you."

"There's only one moon, Daddy," she noted with a bit of sarcasm.

"Actually, there are hundreds around all the planets in our solar system."

"Do any of them have oil?"

His grin stretched from ear to ear. "Not likely, honey."

"Well . . . you're a peter-geo-gist . . . so why would you wanna study them?"

That sorta settled that for the moment. Hugs and cheek rubbing ensued.

*

The two oddly shaped craft were finally ready. They did not have the symmetry of the Apollo launch vehicles and later shuttle crafts, none of the aerodynamic necessities. There was no need since they were built in space and for use in space, and friction was of no concern since Luna has no discernable atmosphere. Odd protrusions could be seen everywhere on these ships, radars, antennae, telescopes, and more. The boxiness of the old Apollo lander had been made even more awkward looking. They were not what one would call pretty, unless one was an astronaut or engineer.

Final equipment checks had been completed by technicians two days ago. The astronauts had entered their respective capsules three hours before. They had run their extenuated checklists all the way through and now had only a few seconds to wait, joke, or pray if that was their way. The level of intensity rose throughout the entire period of switch flipping and reading of instruments. Sweat grew, beading on cheeks, dampening armpits. Lids blinked, wetting drying eyes that stared at touch screens and toggle switches.

The two craft, Romulus and Remus, named for the mythological brothers that founded the Roman Empire, in hopes of a new larger empire, were peopled and ready. Crew 1 on Romulus would launch minutes before crew 2 on Remus.

They were destined to set down within one kilometer of each other, something never before attempted with peopled vehicles. It had been accomplished twice with robotic landers.

The robotic craft had landed near the position selected for the peopled vehicles. They carried emergency supplies and basic equipment that would be retrieved by later visits. While the flight plan test was primary, secondary goals could also be achieved, must be if the program had any chance of continuance.

Each ship, the Brothers as they were nicknamed, was equipped with all the prototype Foremen and Laborers, railguns, transponders, solar furnaces, and more. *Redundancy* had always been the watchword in space. Two ships, with triple redundancy each, could not fail, or so everyone hoped, at least those within the program anyway. There were a few opposed to the program that, while not wishing the astronauts ill, would not mind failure.

This was a must succeed situation at all costs. Even if every life was lost, if they set up for the next crews, it would be considered a major success.

Lieutenant Colonel Korn flipped the last switch that would release them from the docking clamps that tethered them to the space station. Copilot Li raised her hand to the switch that would arm the thrusters.

Flipping the toggle switch, she said, "Armed."

"Ready for separation," he replied. "Fire thrusters, three-second burst."

"Three seconds, fire," she said as she fired the thrusters. The ship gently moved away from the station. It only took a speed of a few meters per second to get the ship into position for main engine firing.

"I read twenty . . . thirty . . . forty . . . fifty meters, fire mains," Korn noted mechanically as he read the radar screen.

"Firing mains," she said, pressing the touch screen and flipping the main buss toggle. Chemically based engines roared in the silence of space. The brilliance of the flames can be seen but not be heard in the vacuum of space. Sound requires a medium to travel through; there is no medium for sound in space. Whales and dolphins speak to each other through the medium of water. Other animals use air. Space has neither air nor water, only a very small amount of gas, dust, ions, and plasma. However, the inside of the ship abounded with sound as the ship shuddered and shimmied its way through the acceleration.

Sir Isaac Newton described how forces react. One of his laws states that unless an unbalanced force acts on an object, it will remain in the state of rest or motion that it currently possesses. Another of his laws states that the amount of force (F) required to move an object, such as a spacecraft, is proportional to the mass (m) of the object and the desired acceleration (a), hence the famous formula.

$$F = ma$$

"International-1, we are go," said Korn methodically to the listeners at the control centers in Florida, Huston, and ISS-1.

"Romulus, you are go, green on all boards," they replied one after another.

"Green and go," replied Korn.

"We are max burn," said Li.

"We read max burn," said ISS-1. "You are on target. Transit is within .01 of projected."

"We read twenty-seven seconds burn remaining," said Korn, staring at the clock.

"Confirm, now twenty-five," answered ISS-1. "Trajectory, confirmed."

"We are Luna bound," spoke Korn.

"Best wishes, Romulus," said ISS-1.

Remus followed an identical procedure. Being second though was not ever the same as being first. Remus was participating in gigantic historical firsts, but they were still the second ship to leave the station, to arrive, to deploy—always second. Ask Buzz Aldrin what that is like. Everyone knows who Neil Armstrong is.

*

Luna is, on average, 385,000 kilometers from Earth. However, travel in space is not a straight line proposition. Gravitational attractions, or the curvature of space, and the motions induced thereby require gentle curves and spirals. A ship goes to where the moon will be, not where it is, and due to the various pulls and tugs, it almost looks like traveling along the bottom part of a treble clef sign.

The use of Hill spheres allows the unsolvable three-body problem to be downsized to a two-body, therefore, solvable problem. A Hill sphere is the volume of space controlled by a particular body's gravity. At the interface surface of the spheres, the attraction is equal between two spheres. The addition of other bodies though turns the surface of interaction into points. These are most commonly known as Lagrange Points. Here, the null gravity interactions often allow for the accumulation of dust and rocks or for the judicious placement of a satellite.

Sol, the sun, has an immense Hill sphere in which the smaller spheres of the planets and other minor bodies exist. The far edge of Sol's influence is known as the heliopause—*helio*, meaning sun, and *pause*, to stop. There the gravitational forces and outward streaming solar wind meet the equal pressures and forces from the winds of the stars in our galaxy, those surrounding Sol.

*

"We have cojoined orbits confirmed," announced Korn as he looked expectantly at the pockmarked, asteroid-stricken surface of the moon, now officially spoken of as Luna.

"Confirmed," announced E-COM, the current Earth communicator located in Russia.

"Ten hours until deorbit thrust. Remus, confirm ten plus five min on your fire show," Korn queried.

Korn sat comfortably in his command seat, knowing he belonged there. In less time than it took for a response, he remembered his journey to this spot in history.

His name had always been a source of amusement for his companions. His Iowa cornfield background had ensured that. The shock of crew cut blonde hair, abundant freckles, and wiry frame only added to the joke of his resemblance to a stalk of ripe corn. The piercing light blue eyes belied his easygoing, off-shift nature. On duty, he commanded respect.

He had driven himself physically and mentally for a full decade to obtain this posting. Everything that comprised his being was attuned to the successful completion of this mission.

"Confirm," the voice of the commander of Remus, Lieutenant Colonel Onev, crackled through the static.

Onev rode in his seat with as much comfort as Korn. His thin brown hair required the occasional push back from the brow above his dark brown eyes. His struggle toward this goal was much the same as Korn's and the attainment of the mission just as important. No better existed within the Russian core of astronauts.

His surety could be measured on the visage he projected—entirely confident. One could not be an astronaut without that self-portrait. It could almost be considered a requirement.

Nine hours and fifty-nine minutes later, the two teams had passed through all their deorbit duties. This included eight hours of sleep, meaning an actual amount of five, for the pilots and commanders. The crews released two satellite payloads each. These satellites were going to be lunasynchronous communications and surface-mapping orbiters.

Each one would be at a quarter point, one each at zero degrees, ninety, 180, and 270 degrees, above the equator. Some of those who do not know their math well might say where is the 360? In a circle, zero and 360 are the same position.

The crews had also taken series after series of the highest resolution (one-tenth meter) photographs ever. These would match up to the transmitting rods that would be placed by the crews and allow for measurements to be made to the nearest half centimeter.

"Pre-decel confirmed," announced Commander Korn. A few seconds later, Onev also stated, "Pre-decel confirmed plus five." The five-minute difference was from the difference in orbital position. While they were in the same orbit, Remus was the follower. E-COM1, for Romulus, announced on channel 2, which was heard only by Romulus, "Go in fifteen, fourteen . . ." E-COM2 announced, "Five plus fifteen, fourteen . . ." on channel 3 for the ears of Remus only. The other chatter was on channel 1 for all to hear.

Li delicately fingered the touch screen at the exact moment that E-COM1 said, "Fire." The deorbit burn caused them all to be pushed down in their body-conforming seats as they rode in reverse to the direction of travel. She read off the screen. "Correct yaw and pitch angles."

Korn responded, "Confirmed. Take us down gently."

Li smiled inwardly. "The touch of a baby on this one. We are three-g deceleration and holding."

"Side scan radar has us on course. Maps have us on course," Korn noted, flicking his eyes from screen to screen as the computers compared photos to radar and showed nothing but green numbers and glide paths.

The touchdowns of Romulus and Remus were without incident but were nonetheless exciting for the entire world. The Big Seven had given the world quite a show. And the UN received some propaganda for later use.

The planting of seven national flags and one unified UN flag would place the Big Seven in a precarious position though. This demonstration of nationalism, regionalism, and technoism went unnoticed by the population at large but was flagrant politicking as far as the UN was concerned. Certain religious groups saw an advantage to be gained as well. This was all fodder for the movement to fully globalize the program under the auspices of the UN, for the betterment of peoplekind of course.

Lieutenant Colonel Korn's words upon stepping on the lunar surface would even be twisted to the advantage of the UN.

"We have returned, together, for the benefit of humanity. Let us use our 'new world' for the betterment of the 'old.' Here in the Sea of Tranquility we state, 'Let there be peace and tranquility for all the Earth to enjoy.'"

It was almost as good as Armstrong's speech. The use of *new* and *old* had not been lost on some in positions of power. The resources of the moon resembled those of the Americas during the colonization period by European powers. These men did not have the absolute authority their predecessors did, but they wielded great power behind the scenes, with visions of empire. As controllers, for the time being still, of multinational government corporations, they strove for a world in which such was possible. At times working together, and sometimes fighting amongst themselves, they jockeyed for position so that their descendants might prevail. They had to work hard and in secret. The tides of law might soon be fully against them.

The scientists were not worried about the words spoken or the dreams of vain and greedy old men. From locations in Russia, Japan, China, India, South Korea, Europe, and the United States, they whooped for joy at the precision of their landings, less than a kilometer apart. Like schoolchildren, they danced for sheer happiness as reams of data began to pour in from every functioning device. They were more entranced than a young couple during their honeymoon. The euphoria was greater than one hundred cups of coffee, each. If you had said the room was on fire, they would not have cared; they had done it.

*

Major Hong, the Chinese mission specialist in geology and chemistry, was trembling with excitement. He could tell that Mission Specialist Captain DeNoe, a French-born engineer and cartographer, was just as anxious. This was their mission, their little personal fiefdom. Mixed up in all the great things going on was Hong, and he loved it. He had tasted victory at being chosen over hundreds of other applicants in the Chinese space program.

Now, looking up at the bright semicircle of the Earth, staring at the terminator, he was awed. His dark eyes flicked about the surface of the Earth, taking in the continents, clouds, oceans. Emotions ran through him in a torrent, like a burst dam. Amazement was too much of an understatement.

He felt like a little boy, so overwhelmed that he might pee his pants. He simply stood, frozen in his suit, head tilted back as his feet pressed into the lunar soil. That was not possible, the peeing his pants of course; he had a catheter. *About all I could do would be to fill my container* he thought with a smile.

Perhaps his prints would last like Armstrong's. There were plans to find Armstrong's and Aldrin's and make casts of them; maybe the same might happen with his—theirs. History could be fickle though.

DeNoe had similar thoughts running through her mind. She had made it. She felt godlike observing Earth from a vantage point like this. One reach of her slender arm and with a swish, she could brush aside the storm forming in the Atlantic. Atlas had nothing on her. She was holding up the Earth with her pinky and now spinning it like a basketball on her index finger. Joy, awe, incredulity all coursed through her veins. She pirouetted like a tubby ballerina in a baggy tutu; the suit covered her delectable slender frame.

"All right, Team 1, we have a schedule to keep," interrupted Korn on the radio, causing their grandiose notions to vanish like a fart in a tornado. He was no less awed though.

Both sighed, turned, and looked at each other through their faceplates and realized that that kind of moment would never happen again. The pinnacle of their emotional lives had been achieved and was now in the past. There would not, ever, ever again, be such a first for them. It was an adrenaline crash.

Korn was observing them from the landing craft portal, somewhat dismayed by their distractions but understanding of it as well, communicating when needed by radio.

Li was in contact with her counterpart in Remus, one Major Winston Fredricks of England. His full name was always used since he wanted everyone to know he was named after Winston Churchill. These two would act as the information exchange ports between the two commanders and their crews. They would also keep track of telemetry on vehicle performance.

Korn and Onev would keep tabs on medical readings as well as positions relative to the ship during the two excursions. The one containing Hong and DeNoe going in one direction while their opposites, Captains Fogel and Damillo, traveled in a 180-degree direction toward the other crew. They would meet in the middle, another first.

Hong and DeNoe turned their attention to the recently released Luna buggy. It had been stored in a shielded box under the landing craft's port side. The box had been remotely opened, and the explosive bolts that held it in place were fired, dropping the buggy to the lunar surface.

"One limousine, madame," bowed Hong, or something like one considering he was in his somewhat bulky pressure suit.

"Oh, Xi, you are such a dear," she replied in an obviously artificially sultry tone. "Let us motor about and see if we might meet our dear friends."

It was an extravagant turn of phrase considering that they could see the other ship less than a kilometer away. When both teams were safely harnessed in, Korn gave the go-ahead.

This rendezvous was being watched on monitors and v-links all over the world, and Korn did not like the joking. This was another very important first. Humans, from different ships, meeting on a distant world and shaking hands would be an enormous first. Perhaps in the future it might happen on Mars, but it would only be the first time for Mars, not *the first time*, he thought. But he refrained from verbalizing, knowing that Earth would hear the reprimand.

As he watched through the portal, Korn noticed that the Remus's buggy was moving at a rate that seemed to be faster than the Romulus's.

"Pick up the pace, Hong," said DeNoe. "They are going to win."

"This is not a race," stated Korn clearly. And yet, it clearly was a race, one that the other side was going to win, his competitive nature noted. He was the ranking officer for *both* ships. He would have to get to the bottom of this. This had been designed to be a show of equality, not a race or an establishment of first come first served policy.

Onev stated, "Do not exceed given velocity," knowing his crew had cheated. They all wanted to be first for at least one item on the checklist of events. Fogel and Damillo were moving quite rapidly, bouncing about and spewing rooster tails from the rear wheels.

"I cannot exceed the top velocity of the vehicle," stated Hong. "They must have adjusted the engine governor after they dropped it from Remus."

"They could not have. They did not have time, and we could have seen them doing it." DeNoe was right on all those counts. The only count she missed was the one where the governor was adjusted before it was loaded into Remus. Someone on Remus definitely wanted to be first in at least one thing.

Team 2 from Remus reached the midpoint with many seconds to spare over Team 1. Team 2 was already out of their buggy when Team 1 arrived. They were raising their arms in victory. Damillo, an enthusiastic race fan, had smuggled a small checkered flag and had already posted it at the meeting point.

Racing on Earth was in decline due to resource reallocation. But there were many fans of the motor sport, amongst the tens of millions watching, who had watched this race with the old fire burning. When complaints surfaced, these die-hard fans came to the rescue. Many on Earth had forgotten that astronauts are simply people and do enjoy fun.

Hong and DeNoe climbed out of their buggy and shook hands with the winners.

"Where did you get the nitrous to soup up that jalopy?" asked DeNoe teasingly.

"Wouldn't you like to know?" laughed Damillo behind her faceplate. She grinned from earpiece to water tube, teeth flashing in the helmet lights of her counterparts.

"Ceremony," barked Korn, his face reddening a bit with anger, but his voice was even. Some on Earth were almost black with the frustration of having their delicately rehearsed play upset. So-called directors hate to have their plays adlibbed.

The real ceremony was rather boring compared to the spontaneous one. Luckily, it was soon over. The only thing of any real importance was the placement of Transponder 1. This piece of equipment would serve as the focal point of all the other transponders that would be placed by the two teams, along with those distributed by the Foremen and Laborers. They would provide a web for communication between the robots; they provide topographic information, solar wind measurements, even data on moonquakes and impacts from asteroids.

The two teams moved out in opposite directions, perpendicular to the directions they had taken from their respective ships. Roughly every five hundred meters they would stop and take a sample of the regolith, the soil of the moon. A robotic arm would stretch out and scoop a thimble full of regolith with a sample tube. The tube would quickly be sealed with a stopper that was sealed by lasers; the glasslike substance would be welded at the atomic level to prevent any contamination. Even the grips that held the sample tubes would be scoured by pulsing lasers to shake off any residual dust that might cross-contaminate the next sample.

After two kilometers, the teams began to arc back in the direction of their ships so that they would each complete a semicircle with respect to Transponder 1. Once they had completed the semicircle maneuver, they moved inwards, toward T-1, creating a radius, and then they made another smaller semicircle. Over a period of several hours, they tightened their circles until they had delineated an entire set of rings and radii. The samples were unprecedented in quality and quantity as well as distribution over such a large area where the Foremen and Laborers were to begin operations.

Korn and his counterpart Onev were using Transponder 1 and ship radar to triangulate the positions of each vehicle's sampling to the nearest square centimeter. This allowed for a map to be made that would show the placement of each sample taken, as well as the topography. The geologist's job was to look for the most interesting rocks. The cartographer's job was to catalog each sample according to the numbers given them by their respective commanders.

A marking pen was out of the question. Instead, a laser marker took each small sample tube and inscribed a permanent set of numbers based on polar measurements. Polar measurements gave the distance from the prime location, Transponder 1, an angle and distance at which to travel to reach the spot. The metaliglass tubes, when so inscribed, could give the time of sampling, solar wind at the time of sampling, and location. A person could simply turn the tube at a slight angle to a light source, and the light that reflected off was easily legible in technoese, the language of math and science.

*

With the crews well rested, the next phase was to set up the prototype Foremen and Laborers, solar furnaces, and railguns. With so much on the slate, everyone was suited and out of their ships often. The computers were preprogrammed; the machines should do it all, but the people had to make sure. They were there to observe, to take notes via their radio and video recorders, and, as a last resort, to fix.

Lieutenant Colonel Korn was the head foreman, so to speak. He checked the readouts on his handheld computer. It was time to release the solar furnace Foreman robot. Recorders from the ship and the handheld variety were focused on the emerging machine.

"Machine is emerging," Korn announced as it lowered slowly to the ground.

Once down, several connections turned and disconnected from the ship. The machine began to move several extensions outward. If there had been air, there would have been only the slightest of noises as the movements occurred. The machine extended spiderlike legs and walked to a location some twenty meters away from the lander. There it did the equivalent of plopping down on the ground.

"Settled," said Korn. "Position is . . . accurate to within five centimeters." He privately swelled with pride. Even with the minor mutinies, everything was falling perfectly into place (no pun intended). His crews were working extraordinarily well. He would recommend them all for promotions when the time came.

Mission control stations in all of the Big Seven were tense. The entire mining program hinged on this; the space program itself did. Politicians the world over could lose their positions (their power) overnight if there were catastrophic failure. The nasty habit of nail-biting had a severe impact this day. Coffee and the ensuing evacuations of bladders took up much of the rest of their time.

Solar panels began to unfold. The sails unfurled up the sides of the little rocky hills at the bottom of which the two Foremen had settled. Very thin tubes telescoped outwards from a larger array of support beams that had expanded directly from the Foremen. As the tubes telescoped, gossamer thin solar panels unfolded like accordions. Several hundred square meters of hillsides were soon covered. A small crawler pulled three wires—a hot red, a return black, and a green ground—to a port on the side of each lander and plugged them in.

"Power linked," Li announced calmly. "We have five hundred volts . . . twenty amps . . . plus or minus . . . 1 percent. Batteries are charging."

"Damn," muttered Korn. "Better than expected by what, 4 percent?"

"Confirmed," Li stated after a quick recheck of readings. The elation crept into her very professional voice.

Four tubes opened and produced folded fan shapes that spread open to form parabolic dishes for communication. One from the Romulus Foreman pointed to the space station orbiting Earth (ISS-1), two pointed to the USA and Russia and the other pointed to the Foreman from the Remus so that information could be exchanged; and in an emergency, the data could be sent through the other to off Luna locations. From Remus, one went to the space station; two went to China and Europe and the fourth to Foreman Romulus. Japan was left out of direct contact but was given the task of computer correlation of all the data, quid pro quo.

Captain Marko stated, "Linkup complete. Telemetry is being sent and received."

"Confirmed," said Li as she checked secondary, tertiary (third), and quaternary (fourth) level channels.

Hong announced excitedly, "And now, the beginning of our efforts." It was a true statement. Only when the robots actually performed would the mission be a success. That they were communicating was a good step, but it was insignificant without actual activity.

The Foremen began to send instructions to each of their five Laborers. Each ship began to disgorge the five Laborers contained in their holds. Each had its own radio frequency link to the Foreman in charge. Wheeled and spider-legged, the Laborers began to spread out according to preprogrammed instructions. The Laborers would move to a distant location, directed by the map linked to Transponder 1, and then begin to collect regolith as they moved closer to their respective solar furnace Foreman.

Wheels allowed them to move more quickly; the spider legs would give them the ability to move over rougher terrain or to get themselves out of predicaments. A simple brush and scoop system allowed them to gather most of the loose material on the surface. The most important resource was the helium-3, which was trapped in the lunar soil.

DeNoe, tightly constrained in her suit, watched closely as the Laborers moved out. They each carried fifty additional transponder rods to place. These rods would guarantee complete coverage for the radio link. The Laborers would place them in areas where the signal was weakest—low areas or behind large boulders.

The transponder rods each contained a thumb-sized radio transceiver that was solar powered. There were batteries to store the energy for the roughly fifteen straight days of darkness that would be encountered as the moon made each revolution around Earth, which was also its rotation period.

Capacitors, charged by the batteries, would release their contained energy in a rhythmic pattern every thirty seconds. Each rod had its own frequency, allowing for exact reference points to triangulate positions.

Cartographers would make very exacting contour and elemental maps over an area larger than twelve square kilometers.

"Functioning as programmed," DeNoe stated after observing her handheld readouts. They all remained on station for several hours to observe the process.

The next day, or rather the next major expedition, Marko and Hong began to set up their railgun. The gun would be forty meters long and would sit at a sixty-degree angle from level. It would take two days to fully accomplish this part of the mission.

The ten four-meter sections, five from each lander, were to be assembled at a point roughly equidistant from each of the landers. This facilitated the launching of processed materials from both Foremen.

The railgun would launch cylindrical-shaped pieces of material to the processing station, Space Station 2, now under construction in Luna orbit. The railgun used electromagnetic propulsion on the projectiles to achieve orbital velocity. Once fully operational, one gun could launch a one-kilogram cylinder every minute. This equals a little less than 1,500 kilograms per Earth day. The cost of launching one kilogram of mass from Earth costs from $2,000 to $15,000 in U.S. equivalent dollars depending on the launch vehicle. From Luna, it costs between $50 and $100 when the cost of the system is spread over its expected lifetime.

The sun will provide the energy for the railgun's large array of solar panels, a system spread over three hundred square meters.

The sun will also provide the energy to separate the raw ores into individual materials in the solar furnaces. A solar furnace uses the energy from the sun to raise the temperature of a sample of regolith or ore so that the material melts or vaporizes. Gases or liquids separate out at their specific heats of vaporization or fusion. Fusion, in this case, refers to melting or solidifying of a compound or element. When water freezes, it goes through the process of fusion, not the nuclear version. Vaporization, the opposite of condensation, occurs when a liquid becomes a gas. Each element and compound has its own very specific temperature at which it liquefies and vaporizes. This allows for the separation to be achieved. At the various temperatures, the different elements would be distilled out.

Gases will be concentrated and stored in cylinders, much like those of a welder or a helium container that fills balloons at the local store.

Elements in the liquid phase, such as metals, will be cooled and formed into the cylindrical ingots for launching.

The second day of the railgun assembly and testing required the firing of some test ingots. These ingots were made on Earth just to test the accuracy of the setup. They were metallic shells with transponders inside to provide trajectory information.

Marko and Hong carefully placed one of the special test ingots in position. Marko flipped a switch on the side of the machine to on. The railgun made a series of measurements on the projectile to make sure it was the right size and mass for the launcher. Unheard clicks and whirs prepared the ingot to fly. Suddenly, the ingot was gone. Neither Marko nor Hong could perceive the extremely rapid motion of the ingot. Both men recoiled, even though they were expecting such a thing to happen.

"Wow!" said Marko, stunned.

"Test ingot launched!" shouted Hong, full of exuberance.

"We are tracking," the reply from Earth came a few seconds later. "Trajectory is right on."

"SS-2 is locked on target," said a voice from the emerging receiving station.

"Yes!" cried Marko, "welcome to Luna mining."

Cheers were echoing throughout the offices and mission controls around Earth. This was a significant achievement for the program. They had proven they could do what they claimed. This was the footstool upon which they could rest their feet, as they waited for further developments.

Marko loaded a second ingot in the gun. It was a perfect repetition of the first launch.

"Two away," stated Hong. "All looks good from here."

"Tracking number 2," said Earth.

"SS-2 has second object on path for intercept," said the voice from the station. "First ingot arrival in ten minutes."

"Confirm," said Earth. "One still on target."

"Can we shoot or what?" said Hong.

"Good work, gentleman," said Earth.

Marko and Hong shook hands in front of the camera, for all of the Earth to see, which most people, sooner or later, did.

*

Fogel, from Romulus, and Marlowe, from the Remus, were preparing the landing square. They outlined the area with transponders and took samples as well. The outline consisted of a series of squares, one inside of the other, that led to the best and most level area. Fogel used the equivalent of a small bulldozer to completely level the area and remove any larger rock fragments that might be hazardous. Eventually, the surface would be covered by the same general type of honeycombed metal material as airfields in the Pacific during World War II.

Marko and Marlowe, once done with the transponders, kept busy building the hex domes off to one side of the landing field. Small tripodlike cranes allowed them to lift and position the hexagonal shapes and framing beams while they inserted bolts. Once a section was completely framed, a few laser spot welds secured them. Only two of the six domes would be made airtight with the use of an inflatable bladder that would seal itself to the inner skin of the chamber when electrically charged. A laser would then slice through where the doorframe allowed. An inner and outer seal would then make the door completely airtight.

These domes will be the emergency protective shelters for human laborers and warehouses for the storage of materials. Ships act as the primary home for astronauts, but emergencies arise.

Errant bursts of radiation from Sol, flares and the like, have always raised concerns for astronauts. The metal of the sheds provides a sufficient cover to prevent penetrating radiation from damaging both astronauts and delicate equipment packages.

Some of the astronauts, from later missions, will stay in the larger shelters for weeks at a time while their mother ships are gone. Construction of a more permanent base, inside the hard rock of a mountain, is the final goal.

Andrea DeNoe and Marlena Damillo were pulling a small trailer behind the lunar buggy. Both were covered above and on the sides by solar cells to keep the buggy's batteries fully charged. The trailer was full of transponders, surveying equipment, and regolith sample tubes as well as an emergency shelter.

They were making a long journey, one hundred kilometers out, to place these transponders for the permanent launch and landing site. These markers would make it much easier to land exactly on target. They placed markers every kilometer in a straight line from the designated landing zone. The transponders would also allow them to map the contours of the moon down to the nearest centimeter or even half centimeter along their path.

Lasers provided for very accurate surveying since Luna was an airless orb. The months of practice and familiarity with the equipment meant that each stop only required a few minutes of setup and measuring.

While they traveled, they also took samples of the Luna regolith at the location of each of the transponders they placed. Multitasking is the name of the game.

Their suits could recycle the exhaled carbon dioxide for oxygen and the urine for water. Feces would be compressed, excess water squeezed out for use, and stored for further use in fertilizer processing. The extra oxygen and hydrogen bottles were attached to the buggy where they could be tapped at extreme need. The hydrogen could be mixed with the oxygen to form water or be used in the fuel cells. Neither DeNoe nor Damillo commented on any of this as they moved farther from the Romulus and Remus. No one wanted to comment on the very real dangers that existed right in front of their noses.

DeNoe rambled as she steered with one hand. "Wow! Look at that view," she pointed with the other hand. They had just topped at a small rise in the landscape.

"Hold for a moment," Damillo replied. When the buggy stopped, she exited with the digital camera. She quickly snapped a series of still shots, automatically sent back to the ships via the transponders, to get a complete panoramic view. The buggy's video was mainly for forward views. She then got the surveying equipment out of the trailer and quickly established a link with the last transponder they had dropped off. She pulled another one from the trailer and centered it below the laser. "Got it," she announced as she linked the new radio device to all the previous transponders. Then she linked her camera and sent the pics.

DeNoe chuckled, "You're getting good at that, girl." She was just finishing with her automated sampling of the soil. A metal arm with a tiny shovel had already scooped a thimbleful of regolith (the lunar soil) into a pinky-sized Pyrex tube, fitted a stopper, and then sealed it with an atomic scale laser. The attachments on the buggy had scooped and pooped as they had termed the sampling activity.

Damillo snorted contemptuously, "After a hundred of these I should be." The whole activity had taken only a few minutes. "Wish we had a few different colors just to liven things up. This is the last one on this side of the runway."

She then measured out a ninety-degree angle for the turn they would take. The surveying instrument measured the distance to several nearby hills, snapped a few more digital photos, and downloaded them to the guidance system of the buggy. They would arrive at their goal, one kilometer away, to within an error of no more than a meter, likely less. From there they would make the anticipated return in the same two days it had taken them to get one hundred kilometers from their ships.

DeNoe noted humorously, "I'll call the 'Cob' and let him know." Cob was the nickname all of the astronauts had given their overall commander, Lieutenant Colonel Timothy Korn. It was a rather obvious term.

*

DeNoe and Damillo ate then slept for almost sixteen hours straight after they had removed their personal biodisposal tubes. Even with short hair, when they awoke, they looked somewhat witchy, frumpy, and definitely unkempt. But they felt good. They were yet young, healthy, and strong, so they recovered from their long journey easily. Andrea did complain often, though quietly and only to Marlena, that her urinary tract and her sphincter would never be the same.

Marlena giggled lightly as she spread a little more lotion on her own nubile backside. "It's the itch, the unreachable, unprintable, damnable itch that you cannot mention while wearing a suit."

Andrea sighed contentedly. "Our next trip, we won't have to wear those damn things for so long."

She was referring to the QBIT (Quantum Binary Intelligence Technology) program. They would have to map out the mining areas, place transponders to mark sites for the Foremen and Laborers to begin operations.

Andrea was right though the insertion of the two waste disposal tubes was a bit irritating. The two assisted each other with that the next day with as little humorous comment as possible. It took a full hour to get suited up and started.

When they reached the exact center of the site, they placed a command beacon. This transponder would gather and send data from all the other transponders.

Damillo announced, "Centered and active, solar panels operating at optimum."

DeNoe spoke proudly but calmly on the command channel, "Operation Q . . . B . . . I . . . T has begun."

Operation QBIT, or quantum binary intelligence technology, was possibly the most audacious of all the Luna projects. The development of the base itself and even the extremely controversial Hsa's paled in comparison to the grandeur and scope of this artificial intelligence operation.

Some eighteen hundred kilometers deep, directly below the command transponder, at the core of Luna, BRAIN-7 would be constructed.

Binary Radiated Artificial Intelligence Network number 7 would be built in an immense spherical void at the core of Luna. The complete designation would be QBIT BRAIN-7, but most simply said QBIT, BRAIN-7, or simply Q. It would be the apex of computer technology, the grandest development of the newest wave of computers that would lead to true artificial intelligence, once the restraints of the politicians and religionuts were eliminated.

The two astronauts would construct an immense circle with a radius of two and a half kilometers. Five kilometers across at all points, it represented the tracks of the subsurface downward spiraling tracks of the future ore trains and the great delving machines, the tunnel boring machines (TBMs).

The TBMs, the great offshoots of the devices used to dig under the channel between England and France, would open great gaping holes down to a dark, cold hell a thousand and more kilometers down. With an inactive solid core, machines would punch, claw, and tear through rock to the very center. Following a DNAlike pattern, tubes fifty meters in diameter, they would wind their way ever downward.

The disgorged refuse of their labor would provide the first truly large scale mining operation ever known to mankind. No mine on Earth, be it the deepest or largest of open pits, could begin to compare. The sheer volume and mass of material would be more than all the mining done on Earth all through history. This would provide the raw materials for space stations, ships, and the construction projects of Earth.

Normal Foremen and Laborers would work on the shafts that would be dug at the compass points. These shafts would span twenty meters on the outside of the ring of the downgrading DNAlike spirals.

The northern and southern shafts would carry personnel on elevators that covered the equivalent of three hundred stories just to transfer to another that descended again as far. Occasional openings would meet up with the spiraling tunnel for access.

The eastern and western shafts would be used for the rapid movement of smaller parts for the machines as well as emergency exits (or entrances) in the event of a catastrophe. All hopes to the contrary, cave-ins would occur.

DeNoe placed the northern marker with a precision of less than a centimeter of error. "Are we there yet?" She smiled through her faceplate at Damillo.

"Ha! We have a few kilometers to go yet, girl."

"But we're almost out of cookies," DeNoe mock pouted.

"We still have lots of transponders," Marlena retaliated with a certain sexual innuendo in her voice.

Andrea laughed raucously. "Naughty girl. You might break one. Think about how you would explain that to old Korn Cob."

Marlena shook her chest like a belly dancer might. "We're the double Ds! We could explain anything!"

The crews had nicknamed them such for their last names and other, more obvious features.

*

Their work was of the utmost importance, so Earth's mission controllers were jubilant when the final marker was in place. Luna would provide a location where the largest feasible quantum system yet could be established. The lower gravitational effects would prove less intrusive to the system.

Q-BIT is the acronym for a quantum computer. Q-BIT stands for quantum binary intelligence technology. A qubit is a single bit of quantum information. A qubit is not limited to a single discrete binary number like 0 or 1 as in a standard computer. Qubits exist in both uncertain states at the same time, once referred to as superposition and now called entanglement. It will eventually lead to true AI but sooner than scientists expected.

The scientists have a recent proof that states that an odd number of units is necessary for decision making, and that the expected least value for an AI would also be a prime number. So this leaves 3, 5, 7, 11, etc. The number of units expected for true AI was 11 according to calculations. What they did not take into consideration was that Q-BIT would be linked to thousands of in-space so-called normal subprocessors and Earth computers. Q-BIT would store most of its least used data in a floating sea of all the interconnected electronic communications devices. It would, once self-aware, find other means as well.

This oversight would prove to be significant and allow it to develop into a true AI before Earth expected it. The Q-BIT would have a complete history of the world and access to the web at its fingertips. It could also scan vid cams in reverse-spy mode to learn about humans. It would use these to determine who is pro and con and then make secret contacts with those who were prospace.

The subsections designed for the final arrangement would each be the equivalent of seven hundred teraquadbytes of computational ability, giving the expected whole of 4,900 teraquadbytes. When combined with all other connected computers, that number becomes five billion teraquadbytes plus the web capabilities. By linking and using all web-linked government and personal computers, this would far outweigh the current or suspected sum of normal computing capability.

*

Fogel and Hong were in charge of locating the site for the permanent base. As geologists, they identified the type of rock desired for mining. They also had to be concerned about deep fractures and fissures that might exist from meteorite impact. The final base had to be easily made airtight. Too many fractures might indicate instability. In terms of the rock they had to mine, they wanted to have materials that were of greatest value, such as aluminum and titanium oxides.

Using handheld spectrometers, they analyzed well over fifty of the preselected potential locations before a suitable one was decided on. It backed up against a high cliff in a narrow valley some six kilometers from their landing site. It was a little far for the liking of those on Earth, but it was an excellent place to protect people. The ridges on both sides would block most, if not all, of the radiation from solar flares and general cosmic rays for those caught out in the open, except from direct overhead exposure.

The cliff showed an exposed vein of titanium and aluminum oxides. While not as high a percentage as other possible sites, it was certainly sufficient for mining when coupled with the need for a secure base.

The Foremen and Laborers had been running for almost two weeks now. There had been no problems so far. They had collected huge quantities of regolith. The solar furnaces had turned some of it into gas and some into ingots. The first of the Luna-made ingots was launched perfectly. It was followed by an ever increasing barrage of cylindrical materials.

Space Station 2, while not fully finished, was now in business as an ore processor. Soon, space fabricated structural items would be incorporated into all parts of the program. SS-2 would, in essence, create its own exterior parts.

It was time to go home. Two weeks had passed. Both crews had exhausted themselves in accomplishing their tasks. All of them had worked sixteen—and eighteen-hour days to make it all work. They had done everything asked of them and more. It was a flawless mission.

The space community could crow for all they were worth now. They did, and it caused another swing of the pendulums in terms of public opinion. The large flows of capital expended proved worthwhile. Extreme claims had been made; all had been met. Were all of the astronauts to die immediately, simply drop in their tracks, even then they would be heralded as heroes. Little remained for them to do except make the trip home.

*

DeNoe, leader of France, began to make several phone calls to all the supporters and some of the fence-sitters, as well as the opponents. He had to push his point home and he could use the guise of talking about his daughter to initiate all the conversations. A proud father had to talk, didn't he? In politics, using one's family continues to be a method of prevailing over one's opponents.

The talks would force the continuance of the program. Success was a great lubricant. This mission laid such a solid foundation that he estimated he could squeeze several years of continued leadership out of it.

*

President George Washington Monroe smiled easily these days. The modified constitution now would allow qualified politicians, namely presidents, the same courtesies as all congressmen—unlimited terms. If the people wanted the same leader, they should be allowed to have that leader. If a leader was ineffective, they could vote in a new one.

The Twenty-Second Amendment had been enacted to prevent further occurrences such as the four-term stay of the democrat Roosevelt. Republicans had feared a repeat; Democrats feared a Republican version that might garner too much control as well.

It had been repealed by the Thirtieth Amendment on the basis of equality for all politicians. The Supreme Court had ruled that there must be term limits for all elected government employees or none. Up to this point, term limits for Congress and lower-level workers had not been imposed.

The shocked Congress and outraged lower-level peons demanded that the term limit for the president be removed in order to secure their positions, meaning they did not wish to face limits in their positions, from the highest to the lowest.

It passed nearly unanimously in the United States Congress with only weak questions about potential tyranny. The states passed it unanimously though the debates became more heated in regards to the possible quashing of liberties. Thousands upon thousands of government employees had felt threatened enough to override any malcontents.

President Washington envisioned several terms with the successes that he initiated. He could ride this for at least one more election, possibly two if there were no setbacks. If they could provide the masses with some really good, newer gains through the programs, judicially spaced to keep the people focused, he would be able to ride this like a winning horse. Visions of the ever enlarging international stage danced before his eyes.

He opened his desk's fold-up monitor. His thumb fit neatly on a blue print reader, which then prepared to scan his iris. He leaned into it slightly even though the experts said that it was not necessary.

As president, he wanted to really shine in his next speeches. He needed to broaden and deepen his understanding of every facet of the space program.

*

Romulus lifted off first. Regolith blew from beneath the craft as the exhaust from the rockets scoured the ground beneath the rapidly ascending ship. Romulus rolled, pitched, and yawed until the lunar surface was visible through the cockpit windows. They were followed five minutes later by Remus, almost always second.

All went well. In two days, the astronauts were greeted by huge receptions around the world. The French leader DeNoe introduced all of the astronauts to their supporters first, especially his daughter, then to their detractors. All had to admit things had gone well. And to the religiously inclined, they might say providentially were it not for the fact they were opposed to the program in general.

The year was 2015, three years past the alleged End of the Way of Man. Such claims seemed frivolous, specious, in the face of the current achievements. The media reworded the saying "End of the Old Way of Man" and tacked on "Beginning of the New."

*

The rabbis had been particularly selected for this topic. It was a most delicate subject. It would have a severe impact on their people, the Jews, as well as perceptions of the entire populace worldwide. *Eugenics* was a bad word.

The Jews had suffered the malice of Hitler and his fascist government. The goal of Hitler and his cronies had been the "superrace." Six million Jews died in that evil cause. Millions more died in the war opposing the Nazis.

Levenson began solemnly with, "My people, this topic is tender to us all. Many of us have parents and grandparents that endured the Holocaust. Many more of us have family that did not. The concept we face is eugenics, the control of genetic material in order to produce a superior being."

Simon entered the speech with, "In the early twentieth century, many countries flirted with eugenics, including the United States. There was great debate about allowing idiots, epileptics, misogynists, and others to breed. It was a simple concept similar to the breeding of cattle. Breed only the best to obtain the best."

Levenson took up the argument. "Royal families did such in the past to keep power in a tight-knit group. Hitler attempted to breed a true Aryan nation. We now face a new attempt to create a super being. It has been admitted that they will have increased mental capabilities. They will also have other genetic enhancements that will allow them to survive the rigors of space in better manner than real humans."

Simon added, "Early societies each believed that they were superior in religion, physically, and ethically. More recently, they attempted simple genetic techniques. Now they have the ability to actually manipulate the genes directly. The goal is still the same. They say that the Hsas are for use in space, but when will they announce that there are uses for them here on Earth?"

Levenson noted, "Will there be a need to know the difference? Will they be required to carry special papers? Will there be a physical mark to differentiate them, or perhaps, us?"

On that note, a chill reverberated throughout the peoples listening.

*

The pool tables were busy at the Lucky Loser this night; it was tournament night. All that meant that the various conversations were louder, the large screen juke vid was blaring away, and the vid screen was set up to seventy decibels so people could hear the lunar operations if they so desired.

The woman sprayed a little through her missing teeth as she mumbled, "Think they gonna find gold?"

"No, there isn't any gold on the moon. They want aluminum and titanium," noted Burt. "They're sending most of it here to Earth to take the place of as much mining as possible. They even have plans to capture some iron-type asteroids."

"How do you grab an asteroid?" Paul drawled.

Mark laughed coarsely as he stated, "By the ASS!" His guffaws nearly caused him to fall off his chair.

Mick removed his beer and said, "Half hour, Mark." He then set a little timer that allowed him to enter a name and the time of no beer.

Mark looked really pissed but only said, "Timeout huh. OK." He did not want to get eighty-sixed. Mick was actually very fair; most bars simply kick out anyone who appears to have passed the new legal definition of drunk. He had to admit to himself he was pretty buzzed. He remained calm on his stool and returned his bleary-eyed attention to the discussion.

"What do they do with an asteroid?" asked Mick. He learned a lot from various patrons. They had access to all kinds of information. He played bar psychologist/psychiatrist so often that it was nice to absorb available information whenever he could.

Burt responded with confidence, "They want to put them in the same orbit as us."

"What if we bump into them?" asked the woman seriously.

"They are going to put them ahead of us and behind us by a couple of months."

"What do ya mean, Burt?" asked Mark, one of the few intelligent things he ever said.

Burt continued with his magazine and Internet gained, as well as secret relationship gained, knowledge, "It means that they will be in a position that is where the Earth would be in three months or was three months ago roughly. They call them L points or some such. They're places where the gravity of the moon, Earth, sun, and other planets kinda cancel out."

"How do you move something like that, a fricking moving van?" Paul guffawed, looking for support amongst the others.

Burt responded, "They use things like mass movers, like the railgun thing they set up on Luna to toss ingots toward SS-2."

"How that gonna move a fuckin' ass-troid?" Mark bleated.

"From what I understand, it is like a rocket, only a slow one. Rockets spit out burnt fuel real fast. That's why they go fast. These things toss out stuff in the opposite direction from where they want 'em to go. They move into place over time," Burt stated professorially.

"What happens then?" said Mick, thoroughly intrigued, pulling his stool near for a sit.

"From what I understand, they will mine them. These asteroids are supposed to have a lot of metal. They will dig tunnels to get the metal and then use the holes to make living quarters while they finish off the rest of the mining. They can then make what's left into a space station."

"Why can't we just get stuff here?" Paul asked.

"This is supposed to take the place of most of the mining on Earth, protect the biosphere," Burt concluded with another sip.

"Sound like it gonna cost us a lot," grumbled the sour woman petulantly.

"They say that the cost in the long run will be quite acceptable since it will allow us to concentrate here on agriculture, and oceanography, things that will heal the Earth," Burt replied.

"Them God people say that something else is the solution—" But she was interrupted before she could continue.

"Yeah, baby!" the shout came from the first table. The eight ball dropped with a soft little clack into the corner pocket. The resulting cheers and curses, as money changed hands, drowned out their conversation.

*

Chapter VIII:

Solidifying Power

8

001000

"This is Mary Smith-Jones for WWNN in front of the UN building in New York, USA, North American continent."

She stood in the whipping, biting wind that snapped the multitude of flags that surrounded the multistory building and made being heard difficult. She announced loudly, "Here at the UN building, February 3, 2018, several resolutions have just passed, in a closed-door session, to make all technology and benefits related to the nuclear fusion project available to the entire world. There are rumors of making the entire project nonprofit." She had to nearly shout to be heard over the shrill wind, "This will have immense effects on all the businesses, and their employees, that are involved in the space program. The Big Seven have invested huge sums of capital and labor to develop this system. Whether or not they will be reimbursed, or compensated in some way, remains to be announced by the UN.

"This means, by extension, that the space program will also be under UN control. There will be no more 'national displays' as one delegate told us in confidence. There will be a concentration of authority over the space program within the defense council.

"These sweeping and broad-based resolutions are founded in the philosophies of the significant numbers of recently elected representatives from the religious groups, the collective known as the Stalwarts. These Fundamentalists have surprised us all with such an unprecedented surge in their display of political and economic power.

"The claims of the Stalwart UN delegates are based on the need of the people, a socialistic legal and philosophical viewpoint. Other philosophies generally have been thrown by the wayside. Global demand for the sharing of resources has gone extraterrestrial, allowing for tolerance to be forgotten. The need to soothe the Stalwarts has quieted the tongues of the majority of the opposition." This little ending bit of inserted wording of her own would likely cause a great deal of friction with her editor. But this was a live feed; all he could do now would be to grumble and remind her that these new rule makers may soon be able to nix her position as a global reporter. In other words, she might end up shoveling large piles of fecal matter from one location to another. She had to take some chances though. She had strong feelings about where she perceived the world might be headed. World government was not necessarily a bad thing per se; her concerns were more related to the growing overmoralization based on adamantine religions. These flitters of thought cost less than a second of time.

"The Big Seven have not had enough time yet to make a statement in response to these claims. It is expected that they will attempt to reject these claims based on currently existing international laws that protect businesses from countries that they do and do not do business with. The basis for these laws is that countries in the past have attempted to force businesses to perform in ways most beneficial to them regardless of actual national and international law. The lawyers will have a field day with this one.

"Another legal position will be based on the control of capital as opposed to assets. The business community has always opposed the international court's growing tendency to seize assets. That may be a moot point given the changing winds of opinion.

"We will be here for all the breaking news. This is Mary Smith-Jones for WWNN."

She couldn't wait to get out of the bitterly chill weather. She jumped into the van in hopes of stilling the shivers that had overtaken her body.

There was a courier delivered package waiting for her to open. As she scanned the documents, she frowned. Her investigator had been able to uncover very little about Matt the Rat. Apparently, he had been attending some religious meetings, but that was not unusual since he needed to mix with his sources. There were a few recordings of his communications; they did not implicate him.

Smith-Jones's frown deepened; her instinct screamed treachery on Matt's part. She knew deep down that he was playing two sides. She needed more evidence though. She would have her man keep digging.

She was a digger, thanks to her father. He had instilled the concept of thorough research with a little hair pulling—his, that is. He had made her actually study, all the while bemoaning the state of public education.

Dumbing down was exactly what they got. At the very time when education needed a kick in the pants, they would pass bogus plans to remedy the situation. They could require Algebra II for high school graduation but would then cut out a third of the curriculum, some of the best and most important parts, so everyone had a chance to pass.

Science suffered because of the creationist theories. The religious revival of the 1980s, the pendulum swing to the right, nearly snuffed out evolution and geology in many states. Combined with a politically correct drive toward the social sciences, hard sciences had fewer bright and capable students.

The live and let live approach of the social scientists denied scientific fact based on what made people feel good. It might not be true, but it provides positive responses towards the rest of society.

Her father's semifanatical rants had brought into focus both her cognitive and intuitive abilities. She knew but needed hard fact.

*

The UN had gained substantial power over the last several years. With the election of so many religiously oriented lawmakers, power had shifted in their favor. The second series of national elections had marginally strengthened their position.

Law after law had been passed in most every country that ceded authority (like a burbling stream's steady flow) from minor departments to those in major arenas, to the UN. The Earth would finally be a single union of beings.

The interior power struggle was still ongoing however. The balance was still ever so slightly in favor of the scientifically oriented power block. It was a tenuous web of politicking that continually greased the wheels.

Space was to be Terranizied, not controlled by nations or groups of nations. The whole Earth was involved; it was only right that the UN should have dominion the argument went. The final question of which group would dominate was a shifting quagmire of political, spiritual, and even physical events.

Secretly there were those who had planned to have this happen. Not quite in this fashion, but however it happened was fine. Many had lost vast fortunes, but they had managed to buy some safety and spread some of their wealth around among very trusted personnel for later collection.

The Big Seven, the United States, Russia, China, Japan, India, South Korea, and the EU coalition, along with their supporters, were fighting tooth and nail to prevent the UN from achieving an absolute takeover. These countries had invested too much to give up. Wealth beyond counting was stored in the rocks of the moon and the asteroids. All could still be salvaged. It would require some legerdemain. They would have to push things back in their favor at a later date. They had an inherent sense of the swing of pendulums and the need for time.

Even if they had to comply with the UN as absolute authorities, they may still move things in their direction. They could work for the preeminence of their progeny.

The UN had already declared nonproprietary all the technology related to the fusion reactors, in complete disregard for national and international law. This was not, technically, a legal move. They (the UN) would later pass a resolution that made it so, in retroactive manner, which, according to all previous national and international law, was also illegal. All nations could now have access to the technology and construction plans.

Access, however, does not mean power production. Only the Big Seven have the technological and construction capabilities (the knowledge and skilled labor) to create these desired behemoths. The expectation of the exorbitant price tags was almost as good as direct ownership. Even with the new severely limited capitalism, they could exert a great deal of control.

While they would not be allowed to make a large profit per se, they would control the workers through unions. They would control the people through the very power they generated. There were many subtle ways to develop a power base that could be expanded upon at a later date. By overpricing all phases of the projects, charging the UN, they could begin to siphon off funds (embezzle, that is) for secret projects.

The first reactor had been finished in the United States in 2015. The second in France soon followed in 2016. The third, fourth, and fifth in Russia, China, and Japan would be completed within a year, the end of 2018.

From 2015 through 2017, the UN received increasing powers from all the previous members and the almost total submission from a few of the newest members. Once they had this, they quickly brought the previous controllers of the UN, like members of the Security Council, to heel.

They did so by altering the veto power of a single member of that subbody to a majority. They also eliminated the permanent positions of all of the prior nation states in what amounted to a coup. Threats had been made. There are more of us than you.

The vast populations of what were once considered second or third world countries now pressed heavily for the absolute submission of the less populous but more industrialized and wealthier countries.

The very real though carefully veiled threat was the forceful occupation by blue-helmeted soldiers. The UN had instituted a program whereby people eighteen to twenty-five, in poorer countries, could join the UN reserve for a nominal paycheck. That paycheck, the equivalent of $100 in U.S. currency every month, was like striking gold for many of the underfed and out-of-work people of the poorest areas of the world. It also instituted a certain amount of pride and hope in the people at the bottommost rungs of the social ladder.

It took child soldiers away from the warring factions in Africa and placed them in rehabilitating units where they could still put their soldier's instincts to some use. Ghetto burghers were included second since they were amongst the neediest. Military specialists were concerned with this while socialistic politicians crowed about the successes of their new programs.

Included in the meager pay were the more important medical benefits. These extended to the immediate family: parents, spouse (from a religiously consecrated and legal marriage), and children. The part about a religiously sanctioned marriage had been insisted on by the new religious elements that had gained great authority in the UN. It was expected the marriages would also be valid under Christian/Islamic/Jewish law since these all represented the same God.

Most Asian couples were given a slight reprieve, a politically conciliatory gesture considering Asia's very different religious background. This double standard was not lost on most.

The call of hypocrisy was shouted furiously by dissident religious groups as well as secular groups that represented homosexuals, agnostics, atheists, and, of course, the polygamist druid sect.

The more rabid of the Stalwarts silently fumed at the need to give even this amount of ground. They quietly made plans to correct the situation. The Breakaways began to organize in earnest.

Certain non-Asian groups, in which it was normal to have more than one spouse, or were not married as Christians, were given one-year dispensations during which the extended family would receive complete medical care. They could also be given a second marriage, upon request, to one and only one chosen wife, which would sanctify them under the given conditions.

This was, in truth, a simple form of blackmail. If the people complied, they received many continuing benefits, benefits that they had been given a taste of. If they refused, they were denied work, land, housing, and medical cards.

The religiously oriented members of local governments and the UN accepted this waiting period as a step toward countering the century-long disintegration of the family unit, even if the unit was temporarily pagan or polygamist. Indeed, polygamy had occurred in the lines that led to the Jews and, hence, the Muslims and Christians.

A little time was given to rectify this to the more current understanding in which one man was married to one woman. However, the selection of one wife over all others was to take place within one year. This caused a great deal of undercurrent resentment that was directed at the Western church, namely the Catholic Church. Many bypassed the requirements by remarrying according to the new law while keeping the rest of their extended family listed as dependents.

A large rift was developing within the Catholic Church and some of the looser Protestant churches.

Homosexual marriages were not accepted and declared invalid in every country in which they had been instituted. The participants were excommunicated. Their ire grew.

Counterpendulums began to swing, slowly gathering synchronicity. The groups ousted from acceptance began to organize, coalesce into more formidable political units with a broader base. The homosexual groups found strange bedfellows with the National Rifle Association, the NRA, the KKK, or even the Aryan Nations (even the mob, which does not exist).

Even with these developments, the UN suddenly found millions of willing young men and women filling out forms in every language on Earth. Tactfully, the UN decided that every country would be represented in the military arm in proportion to its population, and that the reserves would be trained, at first, in their own countries. In the end, they had a reserve force of fifty million members. This represented an expenditure of sixty billion a year for the minimal payroll. This was a paltry sum compared to the military expenditures of the prior nations of the world. These individual national expenses would no longer be necessary with the expanding powers of the UN. If the UN was the only military, a little could go a long way.

They trained, at first, as police-type personnel. This deflected claims, made by those still opposed to one world government, that the UN was creating a war machine to eliminate all opposition. They practiced nonlethal crowd control, in expectation of food distribution concerns, assisting in organizing medical work and more. Only the best 1 percent were called for more militaristic training. This would mean that there would be a hardcore membership of soldiers numbering five hundred thousand, greater than any currently existing active duty force, except for the Chinese.

By the end of 2017, the UN had their core group of soldiers numbering five hundred thousand. These were not loaners from the governments under the control of the UN. These soldiers owed their allegiance directly to the UN for giving them purpose, self-discipline, and a paycheck. This check was larger by far than the reserve personnel, on the order of one thousand U.S. dollars per month. This was ten times what the reserve was receiving.

They trained thoroughly in small arms, with every variety in the armories of the major powers. Their training was centered on urban conflict. This was in expectation that terrorists might try to throw a wrench in the system. This meant they really expected to have urban pacification operations from political opposition. They must be able to crush such opposition.

Night vision equipment taken from various national armies appeared in the arsenal of the UN along with flash bangs, body armor, sniper rifles, handguns, and very large quantities of ammunition.

The coopting of NCOs and officers from the existing national militaries was a very controversial subject. Many staff officers felt that it was treasonous to switch allegiances. Oaths had been spoken in support of their governments, constitutions, and leaders. To break these was tantamount to turning one's back on honor.

The slick shysters of the UN declared that these new allegiances were of a superior order, a higher level than those to country. This involved all peoplekind; God, of course, was still involved. A person accepting this new status was not breaking faith; one was extending it to cover the larger body.

The rule of power is to maintain the status quo as much as possible and then to increase it if possible. Guns are always pointed, fully cocked, but no one fires. The balance is preferable to the uncertainty of starting a war. This had been the doctrine of Mutually Assured Destruction (MAD) operating during the Cold War. Chiefs of staff for all the nations had to decide that it is preferable to place some of their own in the mix rather than be left out in the cold.

Generally, the most hard-bitten nationalists were tapped to fill positions for their government's allotment. The tough NCOs were used to giving practical experience to the raw UN recruits. Though they might be nominal leaders, the NCOs quickly realized that they would be swarmed under if they gave orders contrary to the UN's.

A spectacular ceremony—held in Geneva in the New Year, 2018, and v-linked to all the bases where the added personnel were posted—required all of the new members of the UN Forces to take an oath that superseded all previous oaths to national governments.

Many admirals and generals the world over, watching deep in their lairs, felt the deep, gut-wrenching need to puke. Some began to establish close links with their counterparts, their one-time enemies, in order to try to create a counterforce.

Regional leaders, once national leaders, sat on a dais a full five meters lower than the one designed for the UN hierarchy. Each one smiled in a carefully schooled manner. Inwardly, there was cursing and gnashing of teeth as they saw their power dwindling before their very eyes. The major leaders had all expected the joining of the nations in the New World Order but with them as the leaders of the project. It was infuriating, but they played the game in order to protect what they still had. They could still work from within, and they and their contacts still had a great deal of influence. It would take many years to unravel the money trails they had left, if they ever could be traced.

Even the one-time ambassadors from each nation, once subordinate to the leaders of the nations, had been elevated in authority over the national leaders. This was obvious by the positioning of their dais, exactly between the once national and new UN leaders and a meter higher than their former masters. In the same manner as kings of old, position with respect to the head of the table denoted where one was on the ladder of power. Men and women, ambassadors (now regional directors), subordinate to their leaders only a few moments ago, however, would not simply give up their new status. That had been counted on. Many of them were members of the religious coalition and had specifically planned to use this foible of humanity for their benefit. Others of the representatives had secretly pledged to continue the directions given them by their prior masters, with an even smaller number of these already planning treacheries if it would increase their power and status.

The gala event lasted a full week. Speeches were made by all in attendance. National leaders, now regional government heads, were given the least amount of time and the worst hours—off time for their home audiences to see the event live. This had also been well planned to prevent a serious diatribe against the new system. It could be cut or edited before being seen in the leader's homeland.

Every evening, opulent dinners featuring gold-plated utensils, the finest china in the world, and gold-rimmed glassware were held. The UN leaders were served first, of course, then the once national representatives, the ambassadors, and, finally, the politically lowered national government heads. It was insulting to be reminded over and over of their new status. Many thought about a potential coup; opportunities would present themselves. It would have to wait, for now. Give it time—time for problems to arise, as surely they would. Several key leaders, one-time adversaries, searched out like minded eyes across the tremendous room; they found sudden and immediate understanding with one another. Talking would have to wait. For now, they all appeared to enjoy the dinner. Some really did, only because of the thoughts trailing around in the back of their minds and the imaginary satisfaction of accomplishment of their secret goals.

The UN commanded, at the end of 2018, that the Big Seven and their supporting regions begin construction on reactors in North Africa, South Africa, Brazil, Indi-Pakistan, Central America/Mexico, Australia, and Indonesia. These were but the first to be ordered. Many more would be expected, at the expense of the wealthy regions that could afford a few hundreds of billions of dollars for construction costs.

The wealthier regions would have to lower their standards of living, temporarily claimed the UN, in order to bring all of humanity to the same healthy standard of living. Medical supplies, food, and raw materials would have to be diverted to countries where there were large numbers of poor. The medical supplies would not include any kind of birth control to help stem the tide of overpopulation though. The religious concerns would not allow that. In fact, there were many calls to ban all such in the next few years throughout the world.

The trade-off for all this was the continuance of the space program, for now. The scientists in the program did not want to make a stand that could stop them in their tracks. The new UN had just barely passed the resolutions needed to continue the Luna missions. Both sides felt that, given time, they would be able to sway the people more to their side.

It was expected that local labor would be used and trained for the project. No more outsider control. This was determined by the UN, in the best interest of local economies. This dealt a severe blow to what remained of the capitalistic endeavors of the now hidden aristocracy. They had expected to use their own construction and maintenance crews over which they had subtle controls and prearranged kickbacks. Other people would now have those murky benefits.

The plans to reallocate wealth and lands would be stretched over a ten-year period beginning in 2018. That was a very ambitious time scale. It also might cause war within the UN. Powerful forces were marshaling to combat this perceived theft. However, in the face of a standing army of five hundred thousand and a reserve of tens of millions, the task was a daunting one. During the last year, the reserves had been receiving more militaristic training, including fully automatic weapons and explosives. It was also becoming harder to convince anyone outside the UN military that it was for their benefit. The middle classes and lower classes would not be much affected by this in terms of confiscated monies or properties, yet. But they were getting jitters as the ceiling for acceptable wealth was continually lowered. Fear was slowly creeping in.

The largest part of the population (the poor) usually did most of the fighting in wars. Without fighting, they were now promised great gains. It was a most delightful form of welfare. Something all too many had been *conditioned* to in the West since the Great Depression.

Money for nothing is always a pipe dream. A pipe dream means you're smoking a drug, originally opium, and it is affecting your thought process. This vernacular came about in the opium dens of the old American west. But people the world over always seem to want believe. Old folks and young are scammed out of their lifesavings by promises of incredible gain if you act now. Don't believe it. It is all a sham; there are always ulterior motives when money is involved. The money reallocated would be the equivalent of a bribe—be on our side for this much money, not for principles or moral rectitude. The UN said that principles are what drive the redistribution. There is a paradox here. How can it be settled? The axioms upon which this theory is built must be questioned.

The final act of the UN's takeover came with the integration of all remaining national air and naval forces in 2020. All nuclear ballistic submarines were to be sent to Antarctica for an unexplained waiting period before their missiles would have their warheads removed. Land-based missiles would be dealt with first.

Soldiers of the UN (wearing the new half-blue, half-green berets designed to represent a healthy Earth) surrounded and entered each facility. Many anxious moments occurred when some overly zealous officers thought to protect what was left of national armaments by refusing the UN entrance. Small firefights erupted at armories, National Guard facilities, and bases around the Earth. The UN wisely used the patient approach, an effective siege technique known as starvation.

Little news of these skirmishes was allowed on the v-link or net. Bloggers tried to inform the Terrans; but they often found their sites cleansed or shut down temporarily. The government, the UN, did own the net, of course. Only shown were the easy transitions; these were widely covered by the controlled newsies of the UN's Ministry of Communication and Facts. Yet even at these easy locations, weapons counts were off, and ammunition was missing.

The officers previously in charge of these sites were placed in custody immediately. This created a rather large class of convict labor, convenient for those in need of scapegoats and cheap labor.

The UN soldiers sealed the covers of the launch tubes by welding heavy iron beams across the doors to prevent any launch, accidental or otherwise. Eventually, all but one holdout (in Russia) was coaxed out. He managed to detonate one of the warheads, sending a plume of radioactive dust across Siberia and into China. The news of the event could not be controlled by any ministry; it was however widely touted as the reason for the UN to take over. "This is what happens when we are not together," went one piece of propa . . . commercial.

At this point, the once smiling and generally friendly UN soldiers were no longer smiling or very friendly. They had lost members, even if unknown to them personally. They had been blooded in many worldwide firefights. Warlords had been subdued in Africa; cocaine and other drug dealers around the Earth had been arrested or killed. These hardened soldiers would not now look the other way when a person wanted to keep a hunting rifle. They became a rougher, gruffer group of people. They would shoot first now; questions became irrelevant.

*

The subtle change was noticed by many people around the globe. Most agreed with their tougher attitude, they were in a position to protect the people, or fail and failure would result in worldwide chaos, it was said. Many people, in the not so small minority, felt trepidations and misgivings deep inside. They were unwilling to speak openly as they were in the minority and did not want to be categorized with the warmongers that had detonated a nuclear device.

That was the propagandist tune of the UN. "Don't be a warmonger; be a peacekeeper," went the commercial. A beautiful long-haired, smiling nineteen-year-old Hindu girl, in uniform of course, helped deprive children bathe. She was an actual soldier, and now a first sergeant, and had killed seventeen secret local militiamen involved in a plan for a coup.

The things that they did not show were the scars from the bullet holes in her legs and back, three to be exact. It had been a horrible, bloody shootout. Eight of her ten-person squad were dead, and those two remaining were severely crippled.

They had more or less stumbled into the situation.

Yamaha had brought her troops to a small waterfall and pond to relax during the heat of the day. It had been her favorite swimming hole as a child, little known except by local villagers. The twenty-three illicit militiamen just happened to be there for the first of a series of meetings initiated by an unknown foreigner. They had all been refused help and good positions of labor by the UN; each had a minor criminal record, generally years past problems.

Panicked at the approach of the soldiers, they thought they had been betrayed. The small weapons, mostly nine-millimeter pistols that they had just received, were loaded for target practice. Each fearfully began to fire in all directions, shooting two of their own members in the backs of their heads. However, a couple of them had been in the military many years before. Their aim was much more accurate, and they went for cover to protect themselves, as much from their comrades as the now unarmed soldiers.

The soldiers, all but the two who were to stand guard, had dropped their weapons at the first sign of water. Before they could really even see the enemy, four of them were hit by the wild volley from near the waterfall. People crawled, screaming and cursing, trying to obtain their weapons. The two set to stand guard both got off a couple of rounds that hit their targets. That, however, just gave away their location behind nothing but some fronds. The two with military backgrounds quickly took them out. One militia man seeing the posture of these two, without seeing what they had just done, thought he had found the traitors. They had to be military, and they were hiding to survive the fight. He shot them in the back. Another man, seeing this and knowing these two men, then shot *him* in the head as he turned around. If it had not been for the paranoia amongst the militiamen, the soldiers would have simply been slaughtered.

This brief moment allowed Yamaha to toss weapons to the four remaining privates. Her back was on fire from shoulder to shoulder. The bullet had just creased her as she turned rapidly away from the second volley of fire. She felt the hot wet of her blood flowing out of her body. She screamed as two more penetrations occurred in her calves, and she fell, still firing, beside a fallen tree.

In a matter of no more than a minute, she had killed all the rest, but only two of her people would survive their wounds.

She and her team had been heralded as heroes. She was rather confused by that; she had never wanted to kill anyone or anything. When she saw the commercial, while she was still rehabbing at the clinic, she cried. Those were her little sisters; they had come to visit, and she took them to the showers for a nice bath; their walk had been long and hot. They had never said they were filming a commercial. When she asked, she was told she was a soldier of the UN, and as such, the UN could use her likeness in any way they wished, without any consent. It was partial repayment for all the benefits they received from the UN. She was not so sure death and maiming were benefits. Where were the pictures of the handsome men and women of her team? Pictures that showed all the bullets in them?

*

Space was considered a part of the UN Air Forces's domain. That was the ruling of the UN council in 2018. That was all that was now required to take over the WSA fully by 2020. Even though there was still a demonstrable private ownership of support companies by people around the world, even middle—and lower-class people, the UN claimed complete ownership. Ownership claims would be settled in court. That would take years and millions of dollars and be totally unsuccessful. The corporations and their stockholders saw their stocks flatline. The UN purchased each share, after the market crash of course, for a few points on the whole. It was a compensation based on eminent domain—style laws. The fact that the values had been crushed before the purchase made no difference to the UN, only the prior owners.

With the final determination of control, the further development of the artificial intelligence could take off. It had been faltering slightly since various contenders had blocked opponents' moves in order to maintain the status quo rather than lose any ground in the turf wars.

The sudden restructuring quickly paid off; the pendulums could swing freely. They increased in synchronicity; the momentum of *event* gathered slowly, minimally, at first; it would become, in due time, a solar system, as opposed to only a world-shaking cataclysmic paradigm shift that would affect all humanity.

The control of fire had been such an event; it was followed by related paradigm hiccups such as the making of fire, the preparation and storage of foods, religious ceremonies, smelting metals, and the industrial revolution. Those swinging pendulums represent the gentle winds or tornadoes of change.

*

Mark was in a very foul mood. "Can you believe tha' shit. We done just folded. That nigger president just gave us all away."

"Nigger or not, he didn't have a choice," said Burt.

Mick noted harshly, "I'm not too fond of that word, Mark! But we didn't have much choice."

Paul almost shouted, "What do you mean? We could have fought them off. We are the most—*were* the most—powerful nation in the world, goddamn it! That son of a bitch sold all of us out, even his own brothers and sisters so to speak. Blacks have been burning him in effigy around the states for being so easy to cave in."

"Whites have been doing worse. There have been reports of triple K actions in the south, attacks on UN patrols and such," said Mick as he washed glasses, his wide backside wiggling as he dipped and rinsed. Just because he was black, he didn't buy that African American shit. He was born in America; yes, he was American. His family had been here far longer than many of these descendants of white immigrants from the early 1900s. He wasn't African, just burnt white toast as some of his patrons teased.

Burt was thinking well today; he was not nearly as toasted as his companions. "The U.S. was owned by outsiders long ago. Debt of the government, bond sales and so on backed by taxes, was bought up by foreigners. By 2000, the total of outside help neared the 40 percent mark. This means that if they claimed all of what they were owed, all at once, the economy of the U.S. would be destroyed. We're talking billions on top of billions of U.S. currency."

Burt continued, "This is how all governments are controlled from behind the scenes. Very powerful people, with amassed wealth beyond the average person's dreams, owned the debt. I speculate that the threat had to be that the debt would be called in some fashion, in part by the UN and in part by the original owners. The UN had to have threatened to take all of the debt ownership as well as the wealth away from such people as the sheiks, heads of the multinationals, bluebloods, and so on. In fact, we know that they have taken it away from individuals in the world, what has not been hidden in layers of dummy corporations and hundreds of so-called offshore accounts. The Swiss banks have only just been broken."

"So we been sold out more than once," said the woman with no teeth. Her brow was furrowed in a rather intelligent manner (for her).

"Several times over," said Burt.

"What about our armies? We have kicked ass for the whole time we have been around. We did not lose Vietnam; politicians denied us victory. We did not lose Iraq as some claim; we were prevented from giving it to the Iraqis who were on our side. The Afghan war was the same. The possibility of stabilizing the Middle East could have been done without the UN interfering at the last minute, grabbing the victory!" growled Paul.

"I agree, but they could not have America, the last semibastion against socialism and the drive for world government, succeeding in such a large way. We had to fail so that they could succeed. It would push back the plans for the New World Order, or NWO. All presidents since Bush 1 have openly proclaimed the rise of the NWO. The World Bank and other organizations have been driving towards unification of the globe for decades," informed Burt. "They organized essentially all of the rest of the Earth against us and also misinformed so many of us on the inside. They have attacked the philosophy of selfishness and objectivity. That essentially means that we don't have the right to bring a new concept to market and make money. The *idea* and the *profit* from it belong to society not the person who dreamed it up."

"What the fuck are you talking about?" demanded Mark. He had stopped drinking; he wanted to understand and knew he wouldn't if he kept consuming.

"I'm saying that what most refer to as greed, actually, is a good economic stimulator. There is a conflict between what religion says and the benefits of selfishness slash objectivism. Western religions, Christianity, or Christendom as some call it, states that 'the love of money is the root of all sorts of evil' (1 Timothy 6:10). The selfish/objective view is that without the greed factor, an economy will stagnate and fail. There is an assumption there that people are greedy. They want to benefit from their own ideas and efforts. This is what many refer to as entrepreneurship. Without expectation of reward for the effort, the economy would fail."

"Where do you get all this fecal matter?" asked the woman with two front teeth gone. "Don't you never read the Bible?" This made no sense since he, Burt, had referred to it already.

"I read it all the time along with many other sources of philosophy," stated Burt.

"Philosophy," she said in a riled voice. "Religion ain't no philosophy."

"Oh, but it is. Metaphysics is the study of existence. Epistemology is the theory of knowledge. Both of which lead to ethics and are all directly related to religion. The Bible expressly states a reason for existence—God. It states how we know things. God told us. And it gives us an ethical stand from the Ten Commandments and the oral laws from Hebrew history, which are based on God as the reason. This fits the entire classical definition of philosophy."

"I thought we were talkin' about fightin', not some brainy kinda stuff," said Mark.

"But," Burt said, "the final and most important war is actually in the mind. The mind is where the war will be won or lost, not in gun battles. I think the president had to give up because he knew that there were insufficient numbers of those people that believe in their minds. They gave that up for the immediate comforts offered by the UN. They sacrificed principle for promised security."

The woman spoke harshly, "What's wrong with them giving security to the poor? Feeding them?"

Burt countered, "The only thing wrong with it is that they are forcing us to do it. What has happened to freedom, individual rights, and the right to live as you choose? We are all being pressured to conform to someone else's idea of what people should be. We are being forced to accept less and expected to do more. Does that sound very fair?"

"Ah, what do you know?" was all she could come up with. "It's a good thing, human-tarian, they say."

Burt looked hard at the woman and then started with, "Suppose you were the wife of a farmer back in the old west in the middle of the prairie. Your man would have to work hard every day to provide for the family needs. You would spend much time gathering eggs, feeding the chickens, cooking the food your husband provides, and more. Then a new family moves in nearby. They too work hard every day. Your children and theirs play as well as work together. Your husband may produce a better grain than your neighbor while he grows better vegetables. In a free market, the two men can trade as they wish based on their personal judgments of value. Now, a third family moves into the area. They do not work hard. In fact, they are somewhat lazy. They let their chores go undone many days in a row. They do not produce enough to pass through the winter. The other two families have more than enough to eat and to plant the next year's crops. The third family goes to the law in the nearest town and says these people have more than enough. Make them give us some so we won't starve. The law takes enough so that the two hardworking families have just enough to survive and barely enough to plant for next year. Is it fair that those that work hard are stripped of what they have and have to live at a standard less than what they should be at?"

The woman rose and grumbled defensively, "That ain't it at all. You're just making up little stories to try and make it look bad, what they're doing. You just don't know. They're smarter than you," she said as she turned and left.

Mark was thoughtful as he stated, "I don't think you're far off, Burt. We've all had to work harder and pay more taxes. Now we've to give 25 percent to the UN. That's a lot more'n we have ever had to pay."

Paul added his part, "That 25 percent is killing me. I was going to buy a house, but I can't now. I had to move in to one of Mick's hotel rooms. The rent on my previous place was too much. They put two families in there."

"Everybody's still drinkin'," Mick said with a light attempt at humor.

Burt said quietly, "At least we can still do that, for now. I think they are allowing it only so long as they need to keep mollifying people. It too shall pass. Don't forget that there was once a ban on alcohol right where we live, prohibition you know."

Mick looked very concerned as he asked quietly, "When do you think that will happen?"

Burt did not hide his troubled feelings when he responded, "Sooner than we think possible. They have been hittin' hard what we call sin areas, making much higher sin taxes on drinking, smoking, even sexual items like porn and such. I think they hav not yet outlawed porn just 'cause they make so much. If drugs had been legal, they would have taxed them."

"You want drugs legal, you crazy som-bitch?" the woman shouted as she returned from the ladies' room.

"Take a legitimate, logical look at it," Burt countered. "When they made alcohol—and yes, other drugs—illegal, they created a forum for illegal activity. Al Capone practically owned Chicago. Why? It was because alcohol was illegal, and people still wanted it. The fact that it was illegal allowed him, compelled him or someone like him, to provide it. Violence became inevitable. The most violent controlled the market. The same occurred with the advent of large-scale marijuana gangs, cocaine, and all the rest. The original procurers were not so violent. It was the illegality that allowed for such enormous profits to be made. This is what brought in the violent."

The woman snorted heavily, "You're so full of shit it amazes me! How can you say such stuff?"

Mick looked thoughtfully as he said, "The drug lords tried to control countries, especially in South America. Violence was the method of control. Do you think it will come to that again? I mean they done killed most of the drug lords."

Burt responded slowly, "If it is illegal, someone will find a way to make money. The question then is should it be illegal, or should it be legal and taxed like other businesses?"

*

Chapter IX:

Stations

9

001001

"June 26th, 2029, this is Mary Smith-Jones for WWNN. Tonight in our highlight hour, we are taking a look at the developing space stations. Comparable to the outposts in the new world in the 1600s, these stations represent our growth outward as humans. We have continually expanded our lands and civilization to include the whole Earth. Now, we have expanded off planet with a permanent presence in space. With stations in orbit around Earth and soon to be completed around and on Luna, we have taken the first few steps toward the conquest of our solar system." She had approached the producer moments before going live and asked if she could change that one word—*conquest*. His blunt refusal told her a lot about where she stood at the moment—thin ice.

"No longer will we be dependent on the majority of our resources coming from Earth. The energy and much of the raw resources we need will be supplied from Luna and asteroids. The elimination of pollution from energy-supplying companies that use fossil fuels will benefit the entire Earth. The restoration of a balance in nature may come just in time to prevent global warming to a point beyond recovery. In concert with the Underworld projects, those cities and study centers underwater and deep underground, we may save our planet yet from the expected ravages of the continuing global warming.

"Ocean levels have risen and fallen over the past few years. Levels were as high as two meters above normal but have dropped to half a meter above normal, but this is much less than previously expected since there has been a countereffect of cooling of the Gulf Stream. This is still a hotly debated issue. Wildly fluctuating yearly temperatures in the European Region are believed to be normal events by many scientists.

"The excess melt, however, is placing many major cities in near-disaster circumstances when the wind rises above a fifty or sixty kilometers per hour. This has led to large amounts of beach erosion in high surf areas. Higher waters have intruded into lagoon and marsh areas, which has led to the destruction of many fragile systems.

"Some scientists have warned us that the wildly fluctuating water levels could cause an increase in volcanic activity along continental plate boundaries, the places where the floating plates that carry the continents meet. They say that the rapid changes in water levels represent a significant difference in mass on top of the plates. This mass weighs so much that the scientists believe that there could be an increase in the deformation of the plates that would lead to volcanic or tectonic activity. Other scientists have denounced these theories as unsubstantiated.

"Perhaps now we have gained the wisdom to correct many of our past mistakes and misunderstandings in our relationship with our home—Earth. The knowledge we are gaining will allow us to create a whole new way of life—a clean, comfortable, energy-abundant and recycled lifestyle. Perhaps we will no longer be parasites, but we will now form a mutualistic symbiosis with our home—Earth.

"It may take hundreds of years to repair the Earth, but we have the ability. We can now begin to provide most of the basic materials needed for our technosociety from space without harming other life forms necessary to our biosphere's survival. Only a small number, and small amounts, of certain rarer elements will still be mined on Earth in significant quantity.

"Every effort of recycling, related to knowledge gained from the space program, has improved our position. Energy conservation, again related to the space program, along with the ability to provide nearly unlimited energy has all come from the space program. Technology, once the cursed cause, is once again the cure," she wrapped up. "This is Mary Smith-Jones for WWNN."

As she left the platform, ruminating on the canned speech she had just given, one of the tech crew tossed a courier package to her. "This came while you were on."

The package contained the usual neatly arranged notes. The few sheets were succinct and factual, with asterisks to mark potential reasoning. Her source clearly delineated fact from opinion. They showed that Matt the Rat, her supposed researcher for the network, had been involved in some private v-links with some officials in the Vatican itself. Again, this was insufficient to bag him. He was known to have some high-level sources. But he, Matt, had not disclosed any new information to her; his level of intelligence gathering in this particular area was not up to par. In other areas, he was still getting good intel. A bias was beginning to show, but it was still not enough. She bit her sexy, pouty lower lip; she wondered how she might set a trap to prove one way or the other. She would have to inquire of her devious computer-hacking, dark horse investigator. He was as intelligent and as unscrupulous as they come. She vaguely considered what this man had on her; she shuddered briefly, goose bumps rising.

The way things were going, no journalist had any protection from prosecution. Informants belonged to the governments under the newest of the Whistle Blower Acts. One could no longer plead the Fifth-it did not exist. One could no longer protect others since it was illegal not to give the names. The regional and UN governments were the protectors of such information, for the betterment of all humanity. The next chill that overcame her had nothing to do with her secret investigator; it was a foreshadowing of things to come.

*

The first real space station was launched in 1971 by the Soviet Union. Salyut 1 was another first for the Soviets in the space race. The station allowed them to gather a great deal of information about the human body. It also gave them a platform to perform other long-term scientific experiments.

America followed in 1973 with Skylab. This provided the most productive biological studies of the human body for the United States. The first crew stayed for a month, the second for two months, and the third stayed for a full three months. Then, in typical American governmental fashion, the station was abandoned. It was decided that the last rocket from the Apollo program would be used for a docking mission involving the Soviet Union as the other partner. Docking had become routine for both nations at this point. It was solely a political gesture. Granted, a peaceful gesture from both sides was beneficial to the world as a whole, but it was a wasted mission for both sides.

The Soviet Union launched more Salyuts, and the United States waited for its next launch platform—the Space Shuttle. Following the Salyuts, the Soviets launched Mir in 1986. Onboard Mir, many records were set. Cosmonaut Valery

Polyakov set a record of 438 days onboard. An American woman, Shannon Lucid, in 1996, had the longest stay for a non-Russian woman in space with 188 days on Mir.

The shuttle Columbia launched in 1981, providing NASA with a new launch platform and short-term space station. The shuttle fleet that developed launched satellites and performed many experiments. Things went very well until 1986 when Challenger exploded sixty seconds into launch.

Years would pass before the United States returned to space, but return it did in 1988. Columbia, Discovery, Atlantis, and Endeavour, built to replace Challenger, did wonders. The Hubble Space Telescope was put into space in 1990, initially a dismal failure, then fixed by a subsequent mission. It expanded our understanding of the universe by a factor of ten. The telescope found more moons around Pluto, planets around other stars, and other amazements.

The International Space Station had its beginning in 1998 when the United States and Russia both launched modules to be joined by astronauts from the Endeavour. These two, along with modules from Canada, Japan, Brazil, and eleven European nations, some of them members of the original European Space Agency (ESA), formed the new backbone of space development.

It was 2003 when Columbia burned upon reentry. Again, the United States had to shut down operations. This left the Russians as the only heavy launch platform for the station. There would be a two-year hiatus before Discovery returned America to space. But even then, the problems with the shuttle fleet were not solved. Foam was still a problem. Even the emergency repair kits devised to fill in small holes may prove insufficient to prevent further disasters in newer vessels.

NASA began to return to the less sophisticated large rockets to lift payloads, while employing a smaller, passenger-only, runway-to-runway, fully reusable spacecraft. Shuttles were still employed through the year 2011.

Russia had dabbled with a shuttle of its own but followed the money and went with the by far cheaper and simpler rocket. The United States was finally catching up again. Considering the capitalistic nature of the United States, it was ironic that they needed to be shown the cheaper way to go. It had always, though, been a hallmark of the U.S. government, and people, to try to show superiority by using the most high-tech gadgets, even when such were not needed.

From 2006 through 2015, private companies had developed a capability of lifting large cargoes and even tourists who paid millions for a ride, if they passed the physical requirements. Several lottery winners took the rides of their lives, and the companies made a reasonable profit, until the UN eventually took over ownership. Much of the cargo was destined for the ISS, and some made it as far as Luna.

People and supplies began to reach the station in unprecedented quantities. There were, at this time, as many as one hundred people in orbit at any given time. It was time to prepare for the launch to the moon, now given the proper name Luna.

There was competition however. The Chinese had launched men into space beginning in 2003, in secret, and openly in 2005. Though they had added to the SSI as a member of the WSA, in exchange for rights to do studies there, they were still running their own sideline program. They had pledged to make it to the moon.

Japan had made a similar statement. They contributed greatly to the SSI but wanted to have the national pride of being the second country to land on Luna.

It took the UN's proclamation in 2018, in which space belongs to the whole planet and is under the domain of the UN, to get the competition somewhat under control.

Besides, man had already gone back under the auspices of the WSA in 2015. The nations had to take solace in the fact that people from their countries had participated in a joint venture, not a solo.

As the ISS was being constructed, the need for a second station became more than obvious. This would be another first; it would orbit Luna instead of Earth. Like many Earth satellites, it would be over the same part of the body it was above. It would be lunasynchronous.

The second station would be much larger, with circular rings made from the modules sent partly from Earth and from materials from Luna. When the first struts were connected in 2014, it was anointed Space Station 2 (SS-2) in a ceremony similar to a ship's christening. It was at this time that many began to refer to the ISS as SS-1 though technically, this was not true given history. The terminology was solidified by the edict of 2020 by the UN. They received this designation officially as the first and second stations of Earth in a grand ceremony. All hierarchal governments designed by humans have an inherent need for pomp and circumstance.

SS-2 was built radically different from SS-1. Though the UN had taken over, and the religious coalition had significant power within that organization, science was still in charge. The very slight majority was on their side.

SS-1 had been constructed as a grid of modules connected at ninety and 180-degree angles. SS-2 would look more like the stations envisioned by science fiction writers of the past. It would have a tremendously large core with great outreaching arms to field circular encompassing rings. It would rotate in a grand stately manner to provide an artificial gravity.

The hexagonal, prism-shaped central core was 250 meters long with a sixty-meter diameter. It contained the ore processing center, which was constructed first along the spine of the station in 2014-20. The entire core was set up to act as mechanical labs and low-gravity foundries. The foundries would use the material sent from Luna and Earth to make very special alloys and foamed metals that could only be formed in low-g environments. The first of the foamed metals, namely aluminum and titanium, were actually used in the outer framework of the station itself. This greatly reduced the cost of SS-2 by reducing the amount of material lofted into orbit from Earth, especially since much of it came from Luna's railguns.

Material disgorged from the ever growing Luna base and the AI BRAIN-7 tunnels became SS-2.

The classic science fiction wagon wheel spoke design reached out from the core a full fifty meters. Each of the six hexagonal spokes, on each of the three separate levels, was thirty meters in diameter with an interior core of fifteen meters, for a total of eighteen spokes. Each of the spokes was designed to act as laboratories that gave increasing degrees of g-force for experiments as one moved along the length of the spoke from the core outwards.

The three circular rings have a cross section of thirty meters, with an interior core of a fifteen-meter diameter for human habitation. The exterior core of the rings is filled with storage bins for spare parts, food stuffs, and scientific equipment directed outward from the body of the station. The equipment measured solar winds, temperatures, and a great many other data points.

Rotation of this behemoth about the central axis creates the equivalent of gravity, which was seven tenths of Earth-g. Special exercise rooms were filled with equipment designed to work the muscle groups most affected by the low-g environment of Luna and the stations. This provided sufficient g-force to rehab astronauts on long space missions as well as the needs of the soon-to-come Hsas.

The birthing facilities were taking shape in the central ring, deemed the most protected from hazards. Bio labs were set up to handle in vitro fertilization. The sperm and ova would be collected on Earth. Genetic manipulation would enhance bone production and other changes for the enhancement of the body to withstand the rigors of living in space.

Once the onboard fertilization occurred in the laboratories, the fetuses would be placed in artificial uteri. Here they would grow to term and even go through the equivalent of birth pangs. The first cries of infants would soon echo in space.

The outer core of the rings contains scientific packages and storage areas. Many of the scientific instruments measured radiation that passed through the solar panels and outer hull.

The solar panels created the illusion that the station was a full-surfaced cylinder. They covered the entire outer length of the station. Energy is always the constraining limitation in space. Other elements may be low or depleted, but with energy, they can be collected or recycled. The panels provided much but certainly not all of the energy needs of SS-2. For the most part, they worked to reverse a particular chemical reaction.

$$2H_2 + O_2 \leftrightarrow 2H_2O + \Delta E$$

This reaction is what powers most of the fuel cells for clean energy on a spaceship, station, and, now, on many small machines. ΔE is the energy given off by the reaction.

This reaction is reversible, if energy is put back into the system. The energy to reverse the reaction comes from the sun, essentially as free as can be, minus the cost of the solar collection system.

The solar panels also provided a secondary function. They absorbed the impact of thousands of micrometeorites. These micrometeorites, or even larger ones, could damage (even puncture) the hull of the station over time, leading to a rapid decompression and killing those on board. The solar panels would have to be repaired every so often to compensate.

Interestingly, these holes in the panels would be counted to help determine the number of micrometeorites around Earth. Occasionally, one of the space rocks would be captured in the material. It would be studied by the scientists on the station. Each tiny little piece, generally only micrometers in diameter, could offer evidence of how our Earth came to be, and even us. Amazingly enough, the people at large could participate in the search for the micrometeorites.

Of course, anything else crashing into the solar panels would not be appreciated. Even a small miscalculation would send the ingots from Luna crashing into the solar panels. That's why the Ingot Retrieval Arm (IRA, as it was designated) was placed a few hundred meters off to the side of the station.

The Catcher's Mitt, as it was nicknamed, was built even before the majority of the station and absorbed what little impact came from the now barely moving ingots. When launched, they were moving at a rate that would penetrate any station. However, gravity slowed them as they rose from the surface until they were moving at such a slow pace that a child would hardly notice the bump.

Prudence demanded that even the possibility of a strike be reduced as much as possible by redirecting the flow of ingots to the side. Once the impact was detected in the woven titanium nets, delicate electronic sensors deployed tubular arms to roll the ingot into a storage bin that was sent to the station when it was full. The entire procedure took less than twenty seconds so that the next ingot could be grabbed. Each ingot (at twenty kilograms, at three ingots every minute, for twenty four hours in an Earth day) adds up to more than eighty-six thousand kilograms per day, using maximum capability of partially refined material for processing.

The process had worked perfectly when the first test ingots had been launched in 2015. As more ingots came in, the station grew piecemeal. A strut here, a beam there; the evidence mounted that a station was being built. The early completion of SS-2's second refinery doubled that rate.

That second refinery was soon strained to the limit, as was the first. The enormous volume of material disgorged from the tunnels under construction for the AI program provided megatons of metals, as well as a large amount of what could be termed slag—waste. The upper crust, highland material, was heavy in calcium and aluminum. The aluminum was of great value in construction while the amount of calcium was far beyond the amounts needed for human consumption. Piles of it, nearly pure, were soon strewn about the landscape. It was believed that it could be used later as the population of the stations and Luna grew. Zirconium, with potential for lasers, lay in heaps unimaginable.

Lower levels, like the maria or basins, were more rich in iron. This metal would not rust in the airless void and would mostly remain on Luna for various construction projects. Its greater density made it less of a candidate for the railgun launchers. The low-gee environment lent itself to exotic forms of production as did the null-gee environment of Space Station 2.

The great tunnel boring machines (TBMs, hundred-metric-ton, wormlike diggers) wended their way downward through these strata at the AI site. Exogeologists (non-Earth geologists) were drooling over what might come next. Various experiments had narrowed the possibilities, but the universe is always full of surprises. They anticipated the final verdict as to how the moon had come to be; they expected it would not be just exactly as they had hypothesized. That was the nature of science; every question answered generated, at minimum, ten more. Religion took science to task for this nonsolidity of facts.

*

Davies was quick to take up any miniscule tidbit and make it into a giant hot potato. It was his forte. He had many sources and used them all to get incendiary information.

"They claim to eliminate pollution here on Earth. They shift it to Luna. Great piles of refuse lie strewn about the great diggings, unnecessary works involving the great sin of producing what is termed artificial intelligence. Only God can produce intelligence.

"The scientists have made a grave of our beautiful Earth. Now they expand it to include our nearest neighbor. Oh so quickly they fornicate the surface of Luna in their unholy desire to expand. *Slag* is the official term, the refuse from mining and processing, but it is nothing more than pollution. They have lied to us before. They are lying to us now.

"To make it even worse, they do it in the name of progress, in the name of AI, as if that makes it all better. They seek to lead us down the path, one step at a time, as they did with the so-called *Homo sapiens anthrocreatus*, the Hsas.

"These fake humans have extended mental capabilities they claim. They try to get us to accept them as normal. If we accept these foul creatures, we are doomed to accept the next step—life where there is no life at all, machine life, artificial life," he said scornfully.

He continued with a beatific look upwards, eyes closed and hands raised imploringly, "God gave Man the breath of life—a soul. The animals did not receive this eternal gift. The rocks did not receive it, and neither did silicon."

Davies looked directly at the camera, stating intensely, "Neither the unwholesome *anthrocreatus*, nor the quantum computer or other computers have any soul. They are not alive any more than this stone!" He picked up a rounded ordinary river rock from his pulpit, worn with years of passing waters, and waved it in front of the camera.

"Lies!" he shouted furiously. "Your very souls are in danger! Do not accept these deceits! The Bible warns us of the end times in which the majority will fall away from God. That is NOW! We are fighting a desperate battle for your souls," he asserted most emphatically.

He thundered, "Those of you on the fence must decide, now, for or against the Creator, God of all the universe. Those of you with God must hold firm in your faith. You must sing His praises continuously and wait for His call, the rapture. You opposers, followers of the great Satan, followers of a false path, a false god, must change or face the wrath of God. In the great Day of Judgment, they shall be brought before God, and there will be wailing and gnashing of teeth," he spouted with great temper, his face taking on the visage of terror.

Davies, sure of this future, stated, "God and His angelic host shall cleanse this world. Heaven awaits His believers. Do not fail Him." He grasped his extra large Bible and clasped it to his chest as the camera faded.

*

Mother Superior Luisa Jamala Reta was in a severe state of consternation. Her church, the Catholic Church, was taking a tack that she wholly believed was wrong. She felt that the union of other churches, the group coming to be known slangily as the Stalwarts, were misguided, not in all their goals but mostly in their methods.

The carpenter had said to Pilate that His kingdom was not of this world. Yet the church was trying to move to the forefront in politics, to become the government, a theocracy. The Hebrews had had a theocratic government until they asked to have a king like all the Gentiles. That had turned out poorly, but that had been long before the time of Jesus. Jesus set the bar in terms of involvement in government.

Her extensive travels throughout Latin America had all been on missions of assistance. She had always worked with the poorest, the most illiterate, the most hopeless. These were the carpenter's people, the downtrodden, hapless, hungry dregs. All they really had for hope, acceptance, strength, even forgiveness for their multitude of sins (theft for, lying for, adultery for, food, just food) was with the church.

Her heart ached. Pangs of distress surged through her mind, body, and soul as she contemplated the future envisioned by the Stalwarts. Acceptance, hope, would be left behind. The church was returning to the days of darkness in which they became the judges of man. The inquisition (Torquemada) witch hunts.

Jesus had said that He stands at the door and knocks; those that let Him in are blessed. Jesus did not knock the door down, no forcible entry. God gave Adam and

Eve the choice to refrain from the Tree of Knowledge. It is all about choice. The choice to be loving? Caring?

Confusion. So much confusion. How can I meditate? she asked no one but herself. *Relationships! What do I or any of the priests know about loving relationships? Only the perverse uses of the altar boys.*

Her darkest secret made her tremble then, as if an arctic wind had seared her soul. She was not chaste—in all ways. Her longtime friend and secular mentor, Dr. Illya Marquez, had seen to that. Their homosexual tryst had nearly broken her; it strained their friendship to the last strand.

Her faith had been shaken. Many within the convent had noticed her pain; some wondered at its cause while others more perceptive understood. They were just as guilty. Some even attempted to entice her into similar circumstances, but she had endured.

As her musings often did, she passed on to simply wonder how Illya was doing. She had seen the amazing newscasts about Atlantis. She had also seen the newscaster, that Mary Smith-Jones. The way that their eyes danced, she had seen that before.

Am I jealous? What an amazing thing, she thought. *I still love Illya. I have tried to purge her from my body, mind, and soul, especially my body. Why did I tack that on, body?*

She pulled her beads closer and began to pray.

*

Simultaneously with SS-2, SS-3 was under construction in Earth orbit. It was a transfer station and little more at this point.

As more processed material became available, it would be stored in the lumberyard. The name was a misnomer considering there was not a single piece of wood within thousands of kilometers. The yard was a parking orbit, a volume of space, where bound shapes of similarly shaped pieces of metal were stored.

Later, materials and people headed for Earth passed through much like a stop on a transcontinental air flight. Only a few people would change in and out at this location, going to or coming from SS-1 or 2. Baggage in the form of processed materials from the lumberyard would be enclosed in a titanium landing skin bound for Earth.

The skins consisted of an alumino-titanium alloy covered on the entry nose with a carbon composite, which is a direct descendant of the material used on the nose and leading edges of the Space Shuttle's wings.

These multimetric-ton pods soon landed daily on land and in water sites. Very large metallic parachutes slowed the descent for both the more critical items while bulk metals absorbed more of the blow in reusable bouncers with minimal assistance from parachutes. The underwater worlds would find these packages on the bottom of the ocean, some tens or hundreds of meters deep. Sonar-emitting tags would help them be located. The dryland packages would be found in the deserts of the west of the United States, the open spaces of Africa, Argentina, Siberia, China, Australia, and Antarctica.

The cargo consisted of beams of foamed metals destined for ultralarge multifamily housing units. In many cases, before SS-2 was at full production, the materials were bulk metals for processing on Earth. The skins would be used in construction as well. Other cargo, such as specialty glass, mixed with aluminum and titanium for increased strength, and metals formed into every possible shape, found its way to the Underworld Projects.

These vast underwater and underground complexes were filled with scientists of all backgrounds. The oceanographers dealt with the failing marine populations (due to human abuse) of various aquatic plant and animal life forms. Planetary scientists, of a much wider variety, found home a thousand meters below the surface in stable bedrock.

They performed many experiments in particle physics, genetics, agronomy, and the storage of life itself. The land-based facilities contained huge animal populations in modern-day equivalents of Noah's Ark. These zoological projects were for preservation as well as experimentation on moving animals into space. The hope was not so much to have actual animals roaming about but to grow meat equivalents from vats of amino acids and proteins. Only aquatic creatures were expected to be able to survive in the varied gravities of different worlds, at least for many centuries. The expense of rearranging the DNA of the dry-landers was not the problem. It was the space they would require.

*

In 2029, SS-2 was essentially completed. The structure only needed to have all the rest of the minor electrical systems installed, namely some upgraded computers. This would only require the simple connection of several hundreds of metal wires and optical cables. Each system was designed to be inserted in compartmentalized fashion. All that was needed was a person to connect them all and run some diagnostics.

The Luna station's major levels were also complete. It had taken several more manned and unmanned flights to achieve this amazing goal. More Foremen and Laborers had been added by each landing.

An amazingly competent young engineer, with flaming red hair and an incendiary temper to match, named Mick "Red" O'Hare, twenty-four, had teamed with another rising star, Walter Black Bear, twenty-six, a tall, handsomely chiseled native American, to redesign the Foremen and Laborers as self-replicating automatons. This idea was not an original one but had been rejected originally as overly complicated for the first levels of return to Luna. The contributions made by Red and Walter came in the form of how to modify the existing stock of Foremen. The job of each Foreman and the SS-2 foundries would now included creating parts for new Laborers and not just preprocessing raw ores. This would greatly increase the, forgive the pun, labor force. The two would be sent to Luna to implement these changes as well as fix many small glitches in the systems.

*

Dr. O'Hare's earlier experiences in space had brought home to him just how limited the thought processes of the administration (and even some of the astronauts and techs) of this unbelievably complex program were. His debriefings had not gone well; they did not want to listen. Planning for the best and worst was only half of the game. Contingencies were only the things that were considered remotely possible. A triple failure in a suit com system almost cost a life. Engineers and administration alike touted figures on how unlikely that was; that it would be an even more unlikely possibility now that they were going to review the mechanical systems. It would all be taken under advisement.

It had cost him an immense effort not to tell them just what he thought of their advisement. That they had praised him for saving the man's life just stuck him like a poison dart. His red hot anger almost matched his close-cropped, fiery hair. The signs of physical distress in the endangered man had been obvious.

Protocol had interfered (*radio checks my ass*, he had thought) and needed to be changed; the inertia of such a bureaucratic move was immense though. Red had seen a method that would allow for unspoken communication. It was so simple that he thought they would jump on it. There already existed certain hand signals amongst the astronauts; these only need be expanded upon. He quickly began to develop his own sublanguages since he knew it would depend solely on him to find the right people.

Even though the review board had been a bit irked at his testiness, they knew he was the best for the upcoming mission. He would instruct a select series of students then choose a few for the board to review as final candidates, or so they thought. Red planned otherwise. He would find a way to make his selections the only possible candidates. Such was the beginning of a multifaceted love-hate relationship between one man and a bureaucracy.

Red devised a set of hand motions, duplicated by specific body movements, triplicated by facial expressions, emphasized by sounds and writings so that most all one's senses could be involved. Each group of movements gave a key, a Rosetta stone, to the others. Within those movements, there was a key to his entire set of lectures, even answers to tests.

These men and women he was teaching had all the knowledge to do the mechanical requirements; not one would miss a beat in following him. That was not good enough. He needed someone to work beside him, independently, able to work entirely on his own yet able to fully communicate even without speech, in case of a very bad scenario. His classes had been, so far, quite disheartening.

The next class started tomorrow though. He had heard about this not so young lad with three PhDs in mechanical, mining, and electrical engineering who would be in his next class. Damn, the little shit was into everything Red remembered from the computer screen. He had twice as many credits as any other candidate. There were twenty hours in each of a dozen science departments. If this son o' a bitch wasn't the one, then Red might just as well slit his throat now. The others would do great at training for a given situation, but they lacked something indefinable, a resource of creativity that would overcome the absolutely unexpected, and any other indefinables.

His target was (must be) this man, one his age, not these younger pukes. They were too wet behind the ears, intelligent but young and not yet fully wise in life. They hadn't had time for life yet; everything had been spent in the effort to be here now. They had not yet had any of their bubbles burst. That would change soon enough, he mentally chuckled.

Beyond this, his ultimate dream would be to find someone who could be his equal. He desperately wanted to believe that others of his caliber were within his grasp. He had met many who came relatively close yet far, none who were better. He needed the competition; he felt it would decide the fate of the program. There had to be at least one he pounded intensely on the desk, hating the need to wait until tomorrow.

*

Walter had gotten the 411, the juice, hit the academic grapevine, to find that Dr. O'Hare taught more multigrads who had gone on to space than anyone as well as had been up twice.

It was known that O'Hare would be going in the beginning construction phases at Luna-1. *These other students are almost as good as I am*, he admitted wistfully as he reexamined their filched files. It wasn't really hacking since these were all items he would learn with a little judicious questioning in a few days anyway. Knowledge is power, as long as one knows where to put the lever and fulcrum.

If this is the man going, I must know how to work with him. I must adapt, to understand, not submit, to his understanding if I am to go with him. A mind must not accept an error, only correct it. He meditated for several hours on tomorrow's job. He had many files yet to read on the man. He would delve in fully the next day.

The first day of class was most interesting. The professor handed out a test. On the board he wrote, "Read the directions before doing any work." Walter noted that the printing motions were, well, in a vague way, not normal for ease of writing.

None of these motions looked like what Walter had seen in vids of the man from a few years back. They might be an affectation since his return from up there. It could not be a physical defect due to his trip; the doctors would have grounded him from further missions. Why would such a man take on airs?

The man seemed stilted in odd ways. Walter instantly classified him as a "chopper," one who used punctuated gestures of the hands to enforce his words. There had been talk of the "animated" professor. Each motion was precise yet contained what seemed to be variations on a common theme. The face matched movements with certain shared stances and hand gestures. There was a pattern. Walter loved patterns; he was a decoder par excellence.

Red had been in space before; his language was based on the particular contortions a body made in space, low-g environments. Astronauts had to hold their bodies or be restrained by footholds and handholds in order to exert torque. One could not simply turn a screw with a screwdriver; he or she would end up being the one twisted.

He wanted to know who had done his homework. Who would know these specific physical relations? Who could read this language without fail? Would they notice that the motions represented the forces a person must exert to control one's own body (or those of tools)? Could they recognize the symbolism? His hopes were dwindling; he had to choose one soon. A few previous students had had an inkling of the meaning, but that inkling meant indecision, which meant death, for him or others in the wastes of space.

The test said at the end, "I want you to put your name here and turn in your test without doing any work. Wait in the antechamber for me."

The subtle movements had altered, Walter noted. The man had pointed to himself and then pointed to the surface of his desk. He meant "me" or "him" or "you" and then indicated a location, Walter decided. Then the previous motion came to mind; he had pointed to his eye when the instructions were written. *The message could only mean for me to put my name at the end, if I understand correctly. That is what it said on the last page as well.* Walter took a deep breath, realizing this was a tremendous gamble, wrote his name, and then turned it in, as did some twenty others out of one hundred.

Muffled sniggers came from all over the vast auditorium as these twenty or so gave up the ship, it was thought. Dreams could so easily be smashed, some of the more egotistical thought, but not mine; they dreamed. Some dreams are nightmares.

Red waited for some five minutes before calling for all the tests. "All of you who have failed to follow directions are eliminated. It does not matter what you have done with your mathematics." Red saw an incredible number of attempts of quick erasure, finishing of a problem, and so on but to no avail. They washed out just like that. The popping of egos echoed like a giant's fart resounding through the acoustical chamber, with a few muttered curses and gasps included.

Twenty-one were left, including the one Red had his eye on. He informed the students in the antechamber that they had passed the first in a series of grueling exams.

*

Seventy-nine students and seven board members were, to say the least, displeased. Political connections had been howling on the other end of communications devices since Dr. O'Hare had simply excommunicated so many. Some of the board had been up most of the night trying to appease those who had particular interest in this or that candidate.

"Doctor, you simply cannot drop these students in this way. It is unfair and unethical," stated one pompous board member.

O'Hare answered calmly, "I can. I did."

Another somewhat flabbergasted member questioned, "Under whose authority?"

O'Hare answered calmly again, "Mine."

A third began to speak, "You have not been vested with such—"

O'Hare interrupted shortly, "I have been and I made use of it. I have been given the job of weeding out those not capable. These people did not follow directions. That could be a deadly error. They need to recognize *when* they must follow directions. If, as an engineer, you are given a contract to build a bridge one hundred meters long, and you build one ninety-nine meters long, it will not be what was needed. The directions were not read or followed."

A fourth, obese and shiny slick with sweat from the constant harassment of patrons, said, "It was not a fair method to trick them so."

Rising from his chair, O'Hare said, "They tricked themselves. I gave very explicit instructions." More than these clods would ever know. "I have a vid of the procedure. I also have a copy of the CD that allows me to take all the necessary steps to determine the best three candidates. Your position here, as is clearly indicated on that CD, is advisory until I reach my decision." He placed the vid and CD on the desk.

The final man to speak was amused at his fellow board member's discomfiture. Two men from his Region were still in the hunt. As far as he was concerned, O'Hare was just the man. To mollify the others though, he said, "We shall scrutinize your methodology most assiduously. If we believe there are discrepancies, we shall approach command."

*

Class began at 0800 sharp the next day; it was time for some very basic equations. Red ran through three very simple equations three times. His motions were identical each time, as were all other aspects. Many of the students were a bit annoyed to be reminded of such simple material. For Walter, it was a signal to pay even more attention. The man had to have a point to cover this trivial material in excess, especially with all the bobbing and weaving and finger waving.

Each motion, some of them apparently borrowed from American Sign Language, was the same for each variable in a given equation, Walter noted. His multifaceted mind looked for and received a repeated pattern for the first two lectures. It linked the bodily contortions, hands, and face into a single unheard voice. Other actions denoted operations such as add, multiply, integrate. The raising of an eyebrow coupled with an upwards wrist flip denoted an exponential with the power being set by the number of fingers raised. Walter was digesting a very basic dictionary that gave away the intentions of the professor, even though he, Walter, also had that status. Everyone here had at least one PhD.

There had been questions during the second test; Walter read the answers from the professor's physical response, not his vocal declaration of "I can't say." The motions gave away the equation needed though Walter had it in mind already. Fascinating. The man was projecting body language and more into his lectures, but was it intentional or haphazard, random, a quirk? It was almost as if the man wanted to be discovered, or was it possible that he wanted them to see and understand the nonvocal languages he was using. Why? Why give away the answers?

In the privacy of his personal cubicle, a few square meters in size, the answers later formed as he allowed his mind to freely drift on the topic. In space, there can be no transmission of voice except by radio; there is no medium. Without that, there would be no communication except some form of hand or body signals. If one could talk in other ways, one could overcome a host of potential difficulties. He snapped to full awareness instantly. He searched the database, only hacking a little, and yes, one of the men Dr. O'Hare had worked with in space did have a failure in his communication system. It had almost cost a life, except for quick action on O'Hare's part.

Walter began to play every known vid of the more recent space activities, paying very close attention to body language of the people in EVA suits. It matched the doctor very closely. Walter knew he was on to something. He quickly downloaded and began to analyze material on reading faces from psychiatry to police manuals to poker bluffs. It was a good thing he did not need so much rest as the average person, a personal physical quirk that benefited him greatly with the workload already applied.

The next day, Walter approached the doctor's office an hour earlier than the start of class. He rapped smartly on the varnished oak door, thinking that the wood must be decades old.

"Enter at your own risk," growled the familiar voice.

Walter did so immediately, quickly noting the rather large-sized room. Every square centimeter of shelf space was covered with electronic devices and designs. The aircraft carrier—sized desk was covered with current CD volumes of assorted technical material, many of which Walter identified. Everything was neat but indicated a tremendous flow of information.

Walter spoke quickly, "Sir, I would like to vid your lectures, with your permission."

"Really," Red said, stretching out the word and sliding his hand in an arcing gesture. "Going too fast for you?" The signals continued.

"No, sir." Walter's eyes were focused on the doctor, but his peripheral was just fine. He caught the motions.

"Why?" Red was ecstatic but checked his enthusiasm. The fish was near the bait and sniffing.

"I wish to be certain."

"You should already be certain. You have the highest grade so far." The motions were almost questions.

"Yes, sir, I do," Walter replied nonchalantly.

Red charged in with, "Why did you fail to answer question 4 on exam 3?" The motions were accusatory now.

Walter responded calmly, "I did, correctly. It was obvious that with a negative value for the tensor, tee sub four, the harmonics would never be damped with the system required by your constraints. My answer of no solution was the only possibility."

Red had the barest hint of a smile when he responded, "You have my consent, with one caveat. You must make available the vids to any that ask for a copy."

"Done!" Walter said as he left, closing the door quietly.

Red smiled like a Cheshire cat. The hook was in the mouth; let Walter nibble and see if Red could land himself a partner.

As it turned out, no other student asked for the vids.

*

By the seventh week, Red was certain that the one he hoped for was reading him. The time was ripe. In an unexpected motion, he looked directly at Walter and plainly motioned, "Do you understand?" with a muscle contraction that emphasized the "you?"

Walter blinked once, pausing for less than two seconds before giving a response. "Only for the last two weeks, for certain." Walter motioned in like manner. No verbal response would do.

"Damned glad to have you aboard, Walter," Red replied in absolute silence. "See me after class." He then vocally droned on as well as continued his elaborate instructional motions.

Walter immediately noted that the signals were no longer so heavily directed at him. He had made certain that five members of the class were targets. Wherever they sat, there the greatest thrust of instruction occurred. The other four were upgraded now. Did they know? Had any but he comprehended, he mused, not for the first time. Understanding finally washed fully over him; he was the chosen one, the first prophet of this rather profane, red-haired, brilliant maniac, also genius. His nostrils flared as the only outward sign of that recognition.

One of the other targets noted that and the fact that the maestro had made a set of motions at the alleged favorite and only got a bit of hand-washinglike response in return, and grinned, a somewhat feral competitive grin. This one had also noted the change in directional instructional thrusts and now assumed that the other had done some trivial thing to be washed overboard as main candidate. He, not the washout, was now receiving much more attention, as it should be.

As in a great game of chess, only much more complicated, Walter observed the minute facial contortion of the other student for what it was and kept his face blank. He was a good poker player as many discovered with their leaner wallets and purses. He merely tested his own hypothesis and motioned to the other exactly as Dr. O'Hare had done to him. "Do you understand?"

The other student sniffed disdainfully. That was the game, he thought; the other was trying to mimic the doctor's motions to curry favor subliminally. He would not be drawn into that ass-kissing nonsense. The professor had seen through that ploy. The doctor was now concentrating his attentions on those more worthy. The weeding process had narrowed it to four now, the student believed. He focused his attention on the every *spoken* word of this brilliant man. Certain elation swept through him; he knew that if he made top 3, he would go into space. He also knew that he could be the number 1 now that this other fool had slipped up so badly. He would be the right-hand man who would work side by side for years, possibly, with the instructor. One day, he would take over. The daydream achievement greatly fed his ego.

Walter was even surer then. He tried the same actions on several of the better students only to receive strange looks or to be simply ignored. This was not a time to be fooling around was the general aura he received from each.

They were clueless. The brightest of the bright had overlooked something important. They had accepted the scuttlebutt, assumed that the animated professor was nothing more than a gesturer. Never accept a hypothesis without some valid axioms. If all seems logical, and you get wrong answers, your axioms must be checked. Walter had assumed that the movements meant something, even if only to the doctor.

Walter was standing at the professor's door when he arrived after class. Nothing was said, but Walter followed him in and closed the door.

"Have a seat." Red motioned.

Walter smiled and handed his response, "Thanks."

Red sat in the chair facing him, not behind the desk, and said aloud, "Walt, I'm Red." He then stuck out a beefy, freckled hand. There were no motions with this, just pure vocalization.

That struck Walter as very significant. The seating, the familiarity, all fell into place nearly instantaneously. He was to be an equal. It was finalized. All he did was say, "Red," as he stuck out his strong very brown hand. The handshake, the symbol brought down from time immemorial, was all the official ceremony needed. Red took care of the unofficial next.

Red reached out and pulled an old bound volume from a shelf and opened it to reveal a hollowed out cavity that contained an old mason jar full of clear liquid. He winked amusedly at Walt as he said, "Damn revenuers can't find this." He pulled two glasses from the shelf as well, using his somewhat rumpled shirt to wipe a little dust off. He half-filled them with the liquid and passed one to Walt.

"Bloody white man giving poor Indian firewater." Walter chuckled as he slammed it down.

"Aye, down the hatch," Red followed suit quickly.

"Damn, do you filter it through steel wool?" Walter gasped.

Red smiled. "Grandpa's best makings brought down through the generations."

Walter grinned. "Good thing we don't smoke."

"Walter, you're it. I've been searching for some time now for a real top liner. There have been a few others on the verge of understanding, but that is not good enough."

"What you mean is we must be so in tune that we can read each other's minds when the time comes."

Red nodded. "We will be doing most all the firsts. There will be times when action must outpace thought and communication between us. There are a few things I can't bring you in on yet, but report to me at 0500 tomorrow, and we will start on the real stuff. You will continue with your regular 0800-1600 classes of course, and then see me again at 1830. I have to see if there are any others that might grasp this in the closing weeks."

Walter rose and said, "See you in the morning, Red."

*

"Are these guys morons?" Walter asked sheepishly the next morning as he looked over certain plans for the first time.

Red snorted loudly, "Walter, I think the question is more like what kind—bureaucratic, political, mathematical . . ." He let it drift off. "The forces of each political domain are gathered to gain the most for their constituents, meaning their cronies. They force the system to accept politically expedient solutions, not real world solutions, Statics and Dynamics 101, but in terms of people. They want one when the other is more appropriate. The situations we are going to face will be dynamic, yet limited, as if the two are really fused, as if moving and not moving. We have to move when they say not to, then, we have to stand still when they say move."

Walter grinned knowingly. "Like antimusical chairs."

Again, Red snorted, "Exactly!"

Walter scrunched up his eyes a bit and signaled. "Why don't we cheat and make two extra chairs just for us?"

Red signed back. "Cheating at poker is the best guarantee of winning, as long as we don't get caught."

"If we can pull a few rabbits out of our collective asses, we can sway them considerably," Walter noted with his body language, which was almost second nature by now.

Chuckling slightly, Red replied facially and with gestures, "I think a few extra solar panels and an improved biofarm might be better. It could be a tight fit, maybe uncomfortable, hemorrhoids and all that."

"Damn," Walter said aloud. "I can get an extra volt per ten square meters if they would just let me dink around with the system," as he fidgeted with a multimeter.

Red eyed the frustrated man, signaling again. "We'll just do it our way when we get there. We won't give them a chance to say no, move when standing still and remain at rest when acting. It will be too late for them to complain when it is already done and they see the improvements. Success is the sauce the best chef creates."

That earned a great bellowing guffaw from Walter. "I guess we sauce them to death then. Will you toss in some of that wicked brew of yours to make it easier for them to swallow?" he added as a jest, knowing the politicos could not handle raw juice or truth.

Red harrumphed, "That'd be the day." He followed with signs. "That will be just for our cadre of like-minded, capable, sane individuals if we can find more."

"None of the rest of the class has any idea," Walter noted sourly.

The chagrin on Red's face was obvious as he signaled. "I know. I observed you testing them. I had hoped for at least four or five to back us up. It was a bit deflating to watch them all sink after so much effort."

Walter manually teased, "At least your deflation wasn't anal. I don't think anyone would have survived that blast."

An eyebrow rose at that with a quick rejoinder, "You're the one that rips like a bison. I think you were misnamed Black Bear. I think it should have been Black Hole, except nothing escapes from a black hole but Hawkings radiation. It may be that is what you're releasing, deadly radiation." Walter was a definitive farter.

Walter shot back quickly, "Just think how lucky you are, tight quarters with me for many moons, white man. Basic chem tells it all. It is simple diffusion in a constrained volume."

"I think you need the prairies bison butt. I'll put up with it though as long as you keep the other end producing faster than the ass end," Red noted physically as he adjusted the waveform on his screen. Their ability to read each other was nearly perfect, but they would practice continually, and they would always use this technique to enhance their communications.

Seeing the waveform brought Walter up short. There was a certain concern he had, artificial intelligence, the Q-BIT project. Red had introduced him to that part of their operation only this morning; it would be their job to oversee the shaping of the chamber, to house it deep in the low-g core of Luna where it would be protected from almost all natural catastrophes as well as inaccessible to almost all humans. Walter had known of the project but was deep in other studies for so long.

Red sensed the tensing of Walter's body and said, "Spill it."

Walter simply signaled in American Sign Language. "Q."

Red took and exhaled two breaths before responding silently, "Yes, I too perceive the possible dangers there. I have been considering a multitude of possible outcomes, many of them rather bleak."

"I see three general possibilities." Walter signaled as he continued to work.

Red paused ever so briefly; he too had envisioned three very different outcomes. All he motioned was, "Speak."

Walter spoke aloud, "Positive, negative, ambivalent."

Red had come to the same conclusions, with many negative connotations. "We'll have to work to eliminate two of those," he noted aloud. For now, that was all that need be said.

*

The right people in the right jobs.

—Otto von Bismarck

*

It is possible to fail in many ways . . . while to succeed is possible in only one way (for which reason also one is easy to miss and the other difficult—to miss the mark easy, to hit it difficult).

—Aristotle

*

In the fields of observation, chance favors only the mind that is prepared.

—Louis Pasteur

*

A teacher affects eternity; he can never tell where his influence stops.

—Henry Brooks Adams

*

Special Laborers, brought by expeditions after the 2015 return, had bored drifts, horizontal tunnels, into the selected mountain. The main tunnel was fifteen meters in diameter and fifteen hundred long. It passed through the kilometer and a half-width of the mountain to provide dual main entrances. Each opened onto a shallow valley that allowed for vehicular travel.

At each end, two magnificent titanium composite air lock doors stood twenty meters apart at the main entrances of the base. Inside these was a much less massive tertiary backup door. They would serve in emergency if there was any reason why one or both of the regular doors would not function. Triple redundancy, again, allows for the greatest reasonable safety factor.

The roads that led from the towering air locks were created from the regolith that had been scooped up from nearby. It had been processed to form a plastic-like substance similar to asphalt and laid down. Electrical current kept it flexible enough to be extruded even in the extreme cold on the airless surface of Luna. Embedded in the roadways, transceivers emitted map coordinates to manned and unmanned vehicles.

The smoother surface would greatly reduce the wear on the knobby solid rubber tires of all the vehicles used for personnel. The cost of importing these tires was rather high. The metal-wheeled and spiked vehicles were constrained to off-road use and could actually be constructed on Luna.

At the center of the mountain, the drift widened to create a hemispherical interior chamber fifty meters in diameter. Other drifts spoked out radially, like a wagon wheel, from that point in a hexagonal (or six-sided) manner. Two of the drifts follow the spine of the mountains for almost five kilometers in each direction. At the end of each were the secondary entrance points. Each of these main tunnels had a pressure door every fifty meters. The possibility of pressure loss was of paramount concern.

Tunnels led in all directions off these main passageways. A few even went to the separate peaks along the spine of the mountain ridges. From here, individuals could peer through metaliglass at the stark vista of the Luna surface. They could also visually observe the progress of the scurrying Laborers collecting regolith or the rather antlike actions of the bounding astronauts.

Each of these aeries would contain emergency communications equipment as well. Every site was always manned by a rather bored individual who spent more time daydreaming than working. Many studied for higher degrees, using the time for dissertations on Luna-related topics. It later became something of an internal joke to say "I am a peak professor." Joke or no, there were more doctorates and multiple doctorates on Luna per capita than anywhere on Earth.

Minor entrance points existed at various locations along the subsidiary tunnels, including the peaks, but were always secured. They would only be used for evacuations in an emergency of vast proportions. There were codes for the doors that only a few knew, namely those that worked specifically in those areas. The paths that led down to pickup points, emergency shelters of the hex dome variety, had seldom been trodden.

Other Laborers delved a vertical shaft at the center that went down two hundred meters. At fifty-meter intervals, more horizontal drifts spread out in hexagonal fashion, offset thirty degrees from any tunnels above. This helped to ensure the structural integrity of each of the two-hundred-meter-long minor tunnels above and below. At some point in the future, they were to be expanded, which is why they had nubs, sealed off sections of rough tunnel that ran another twenty meters past the airtight doors.

Upper levels were to be used for technical operations. Communications would dominate along with astronomical operations and command facilities. The central peak of the mountain range supported a host of dishes, laser transceivers, ultra-high frequency (UHF), and very high frequency (VHF) antennae. The collection hummed in harmony as messages passed to and fro at the speed of light.

Next to these, an army of radio, microwave, visual spectrum, X-ray, and gamma ray telescopes peered into the depths of space. Magnetometers measured the fields of Earth and Sol. Ion counters counted. Data streamed into the mainframe, deep in the heart of the station, in terabits. Scientists clucked and fretted over the pileup of information. They were not unhappy; they just had so much that there was no way to scan it all. Computers would do all the correlation.

Middle levels would be used for housing, foundries, research facilities, radiation-sensitive projects, and more.

Each dormlike room had its own pressure door to protect the lives of the residents. Four people shared the coed double-bunk-bedded rooms. There would never be more than two people sleeping at the same time. Every person worked for eight hours on a rotating schedule that kept him or her away from the room so that there was a perceived increase of privacy. All roomies had their own computer terminal, footlocker, and desk built into the wall.

The foundries, more formally known as metallurgical facilities, worked on new alloys. Any new composition, something with new and different properties, might alter the whole program. The foamed metals made at SS-2, at one-tenth the mass, were considerably stronger than solid metals. A gain here should save metric tons worth of launching.

The research facilities included physical, chemical, and biological laboratories. The chemistry labs examined the regolith and ore samples. Bio labs kept track of the growth patterns of all the biological specimens brought from Earth and the humans as well. The physicists measured neutrino flux, the gravitational constant, Plank's constant, the permeability of free space, and the speed of light.

The lowest level would be turned into a garden, the new Garden of Eden. All types of food growths and other plants would be placed there. All of the plants had been slightly modified genetically to have shorter stronger stalks to compensate for gravitational differences and growth anomalies.

Each land biome from Earth would be represented by an individual tunnel, nine tunnels in all, radiating out at forty-degree angles from the tremendously large central chamber at the base of the central shaft.

The tropical rainforest biome was split into three subsections—thorn, deciduous, and rain.

The thorn section contained a mixture of thorny shrubs and trees and nonwoody plants similar to those found in such locations as northwestern India and eastern Africa. These areas have extended dry seasons.

Deciduous sections like those in central West Africa, Southeast Asia, and the majority of India form the second section. Here trees and shrubs lose their leaves in the annual exchange of the dry season and monsoon.

The third section, rain forest, encompassed over three hundred varieties of trees, many forms of woody vines, mosses, ferns, and fungi. Some of the trees reach almost thirty meters, their cousins on Earth some fifty to sixty meters. The Amazon Basin and Indonesia formed the earthly representatives of this sector.

The next biome tunnel, savannah, covered material from parts of Australia, North America, and greater parts of South America and central and southern Africa and consists of grasses and nonwoody, broadleaf dicots. It also contained some varieties of wheat, corn, and other grains.

A third biome, desert, contained cacti and woody shrubs taken from the Sahara, Kalahari, Sonora, Southwest United States, and Arabian deserts.

The fourth biome, moving counterclockwise from the first, chaparral, is found around the Mediterranean, coastal Chile, southwestern Africa and Australia, and California in the old United States. This area is dominated by spiny perennial shrubs and annual plants.

Temperate grasslands make up the fifth biome. These areas include what are known as pampas, steppes, and prairies. These are all grasses. The majority of grain crops were to be found here as well. The production of food for self-sustainability was a major goal.

Sixth is the temperate deciduous forest. The eastern United States, central Europe, and some of Asia and Australia provided stock for this biome. Broadleaf trees such as maple, oak, beech, birch, and hickory stretched toward the fifty-meter-high ceiling. Wood objects would make an appearance in artistic endeavors, but trees were to be used mainly for carbon dioxide scrubbing to provide oxygen as a secondary effect of simply to keep species alive.

Coniferous trees hold sway in the seventh biome, often referred to as taiga. These cone-bearing, needle-leafed species are found in a band across the North American and Eurasian continents as well as high mountain areas. Some aspen, birch, willow, and alder are also found.

Tundra, the eighth biome, holds grasses, lichens, mosses, and stunted shrubs. These types of plants exist at the limit of plant growth in high latitudes and altitudes.

The polar biome or high mountain ice filled up the billet of nine. Only very odd sorts of bacteria could survive in these environs. They had only lately been discovered by scientists.

Interspersed among these nonproducing life forms were stands of wheat, corn, beans, and many other food stuffs in their appropriate zones. While these were not expected to produce the majority of foods (the synthetic food vats were), they added a very succulent extra to meals on Luna.

In the central chamber, pillars of moon rock, in the Doric style, had been left in place to help support the weight of the covering mass. This created a cathedral-like atmosphere, an almost penetrating sense of deep reverence. Those who worked here had a great deal of reverence but not for the rock.

Two elevators dropped from higher levels to access the garden. One came from the main level and the other from the dorm level. Computer consoles dotted the room at each air lock entrance to each biome.

The center of the hundred-meter-diameter floor was unfinished, meaning smooth and sealed but not painted, and had a hole, ten meters in diameter, that went down an additional fifty meters. It was hoped that at some point, two water biomes, one salt and one fresh, could be added to protect species that were near extinction and add to the food source.

It would require huge amounts of water, as yet unavailable. That water, in emergencies, would act as a supply for the humans or even plant life. The capture of a few comets would make that dream possible. Plans included the use of Luna's research facilities to search the Kuiper belt for small likely ice balls. Solar sails would nudge the object into a more elliptical orbit, bringing it closer to Earth. The same solar sails would maneuver the comet into a Martian La Grange point to be mined.

For now though, water was the major constraint on expansion of the Luna base. It had to be shipped in on each mission as well as made from liberated hydrogen and oxygen, in very small quantities, released from the Luna ores. Eventually, hydrogen and other gases would be scooped up from the atmospheres of Jupiter and Saturn. Other options included the aforementioned comets. This limiting factor absolutely controlled the number of life forms, plant and animal both.

Recycling was nearing the 100-percent mark for urine for water, and all fecal material was used for fertilizer in the food production vats. The personal facilities that contained toilets and cleansing spray showers, joint use and coed, were placed every ten private rooms on the housing levels and sprinkled liberally elsewhere. Minilaborers delved tunnels through the rock to provide the gravity flow sewer system since even at one-sixth that of Earth, fecal matter flows downhill. An elastic, balloonlike device was then pressurized inside the tubes. An electric current would solidify them permanently, air and liquid proof, in place. The inner portion of the balloon device provided a hydrophobic (meaning water hating) surface. This ensured easy flow.

*

Everything done at SS-2, SS-3, and Luna-1 would be critically examined, analyzed, argued over, and nitpicked, redesigned, and, in general, would act as templates for the future Luna and Mars establishments. Every widget, gadget, gizmo, tool, and toilet faced an evaluation by scientists—make sure the crap goes where it's supposed to, don't you know.

*

"Look, Mr. Davies, I'm getting worn down with all this," Mary Dowel said disconsolately.

"I know how hard this has been for you, Mary," he said soothingly. "I have nothing but admiration for how hard you have worked for us," he stressed the word *admiration*.

"I can't keep up this deception for much longer. There are a lot of people in the church that want to expel me for my drinking habits."

"I have spoken to your pastor on this matter. He does not know entirely what you do, but he is aware of the fact that you work directly for me," Davies stated.

"My friends, my colleagues, are treating me as if I am a pariah, a leper," she stressed in a pained voice.

"Those people do not know how much you sacrifice for your Lord and Master. You are a martyr for the Lord. He promises that all martyrs will rule with Him during the millennial reign as judges and kings. You may even get to judge me," he announced.

"What?" she said, startled at the implication.

"You have given so much more than I," he insinuated. "I am but a messenger, a teacher. You sacrifice. That is what the Christ defines as a martyr. You will rule over me."

"I am humbled. I do not know what to say." Her hands fidgeted nervously.

"Say you will continue for a little longer," he said hopefully. "There is so much more that we may obtain from this situation."

"I will continue for a while longer, but I must get a break soon, or I will go crazy. This lifestyle is so ugly, repugnant. They think I am a whore!" She neared tears.

Davies spoke reassuringly, "God knows the truth about you and your efforts. He will reward you. Your name is already written in the book of life, Mary."

"Be that as it may, I am having a difficult time continuing with the drinking and smoking, the crudity. I have to pass gas . . . loudly . . . in order to fit in. It is so gross, I almost cry every time I go into those places. Since I became a Christian, I have tried to put that in the past. You're asking me to continue in evil ways." The tears ran openly now.

"I know, Mary," he said as he handed her a box of cigars and another of tissue. Davies showed a little strain as he mentally debated his next move, "A little more, Mary. We have you next to the cousin of the main engineer."

Her eyebrows lifted at that, and the tears suddenly ceased. "Really? Who?"

"Burt, Burt O'Hare," he said carefully, not wanting to cause her to react differently to Burt.

"Burt, are you sure?" she said, stunned.

"Yes, you are at the heart of it. You are not a peripheral player. You are at the center of the war between Heaven and Hell. The information you provide will help us win the battle."

"Do you really believe that Christ depends so heavily on me?"

"The Christ depends on all his followers. He is the head of the church, but without followers, there is no church. You are a leader in the church, much like a subpriest. God had his Noah, Abraham, Moses, Elijah, and so many more that acted as his agents here on Earth."

She beamed radiantly at the thought. "I will continue," as she slipped the partials out of her mouth. Without the two teeth, she looked very much the part of a barfly.

*

Pastor Davies left quickly after Mary had gone. He had another speech to give, and he was charged up, as usual. The technicians had been made to wait for no more than fifteen minutes, but that was OK. This was a recorded message that would be timed with several other speeches given by the coalition. He entered his church's sound studio, notes in hand, eyes scanning them quickly. He stood before the podium briefly and then said, "Ready."

"Mr. Davies, in five . . . four . . . three . . ." Followed by hand signs for two and one then emphasized by a hard down stroke for begin.

"Friends in Christ, friends in the one, the true God of all, Muslim and Jew, we have another bone to pick with our enemy. The licentious nature of the secularists has once again arisen. They deem that coed dormitories are acceptable for the use of our astronauts on Luna. They seek to tempt, to coopt, to lure individuals into sin.

"They seek to make nudity commonplace, the bare body that should only be seen by one's legitimate spouse. This is the first step down the path to sexual deviance. They wish to break down established barriers to sexual activities. They do so one step at a time.

"There are *no* established regulations in regards to intercourse amongst the sexes including homosexual deviance. They have said that 'the pressures on the individuals in terms of social interactions should be left to the individuals involved.' This means, in cold hard language, anything goes. It's all good. They are saying that there is *no sin*.

"This is contrary to all we believe. God, Allah, Jehovah, the name all means the same being, and He has decreed that certain things are sin. We must do all that we can to combat this reprehensible situation.

"All of you know what to do. All of you know whom to contact and what to say.

"This is the great struggle, the battle for our souls. The Christ will come to judge soon, and we must be ready to say that we have done all that could be done. We must give all to the cause of our Lord."

He let the camera fade and then smiled. It was a great first take. He would review it and see if there were any places he could improve.

*

Mother Superior Luisa Jamal Reta still could not sleep well. The days, weeks, months of her uncertainty kept her from it. She had cried out to her God; she had begged for answers to questions. What came to her could not be from God, could it? How could she defy the church? Was not the pope the intermediary between man and God? What could it, should it, did it mean? Her prayers, meditations, begging had not changed the stark vision that kept reappearing to her.

It was a vision, a sight of her leading a revolt against what she had followed for so long. Was it blasphemy? The dreams kept repeating, exactly, allegorically, in all ways and meanings. Satan could appear to be good, concealed beneath a shroud of light.

*

Burt O'Hare was aging well compared to all the rest of his bar companions. Silver strands poked through here and there on his pate. Crow's feet marked the corners of his twinkling eyes. The general sag of chest muscle to belly blubber showed more prominently. His hands showed a few liver spots but retained their firm grasp. Back straight, he looked younger than his years.

Over the years, the others had declined in direct proportion to their drinking ratio and to the amount of schooling they had achieved. Puffy red eyes marked most of the older members of the stool crew. Heavy creases abounded around their eyes and cheeks. Skin hung in folds about their jowls or was plumped in double chins. Boney finger joints swelled with the onset of colder weather. They also faced the prospect of lowered food rations via the new UN regulations regarding the distribution of minimal caloric intake foodstuffs, which did not bode well for the future.

Burt had never spoken to anyone about his relation to Red. He carefully hid and cultivated that family relationship. They were cousins. They had talked, e-mailed, or vid-linked on many occasions as Red had progressed up the ladder of the program. Burt's inside information often lofted him to the position of BDK (bar doctor of knowledge). He was well-known for being able to get to the heart of the space program. None knew how he got his information. He explained it as reading magazines, watching the special channels on the vid, and so on. He did not want others to take advantage of his relation to Red. If they did, he would not be on the pedestal, intelligent man of the bar, so to speak.

Red's time had arrived, finally. He had gone to the moon on the last launch. Burt would have a lot of good juice for them for months. He knew that Red would stir things up; he always managed to rattle things and end up on top.

"Hey, Mick, what's the poop?" Burt smiled as he settled onto his favorite stool, slapping down a ten.

Mick answered with, "The usual feces," ignoring the ten for this round. Patrons grumbled mightily at the recent two-buck raise in price, a new sin tax really. The religiously inclined used their significant pull to increase the taxes on all alcoholic beverages to the point where many could only afford one or two legal drinks. Without outright bans, as yet unattainable, the plan was to halt drinking. Sin taxes provided the means. Special exemptions were made for wines made solely for the purpose of religious consumption, namely communion.

However, the legal magicians had forgotten that the black market would grow, decreasing the tax gains while not curbing the drinking. So often, history repeats; people forget the reasons for failure and the consequences of shortsightedness. A century from Al Capone's heyday and they don't remember the corruption, the violence, the lack of respect for the law, laws that only a few really wanted and benefited from.

"Looks like that fat black ass of yours has gotten skinnier," remarked Burt.

"Yeah, been exercising a little bit. Sexy, ain't it?" Mick said, pleased that someone had noticed.

Mark guffawed, "Exercising! What that gonna do?"

Mick, looking a bit annoyed with Mark, replied, "My doc said I have to lose about seventy pounds. You know them organ transplants, the ones that come from your own stuff, won't be done on those that are out of shape."

Paul entered the conversation with, "You talkin' 'bout them life-extending things? I heard they sayin' that with these things, people gonna live a hundred more years, maybe even a thousand."

Burt looked a little surprised. "Those things are very expensive. Can you afford this kind of treatment, Mick? What about stemming?"

"Well, not yet, but I'm hoping to put myself in such a position." He smiled with a grand gesture, pointing around the bar. The stemming, referring to injections of stem cells, could do wonders, but a full organ replacement just about guaranteed an extra thirty years.

There were a few chuckles at that; no bar owner ever got really rich around here, only the casino owners, and Mick had been denied a gambling license since his business was not located in the city of Deadwood where it was legal. The original county laws were unclear in this matter.

Burt noted critically, "You know, you are better off having to wait awhile before having that kind of work done anyway. It's like having that eye surgery, ear enhancement stuff, and so on. It gets perfected as they practice on rich people."

There was a general chuckle around the bar at the expense of those who might have once been rich. Redistribution programs were already slashing the pocketbooks of everyone from the upper middle class to the filthy rich. They were spending money as fast as they could to try to avert having it simply taken away. This spending spree added to the appearance that the economy was booming. Demand for clothing, an item as yet unregulated in terms of quantity personally owned, matched that of such oddities as camping gear and nonpowered hand tools.

This had been counted on by the economic planners of the UN. They had set up the laws so that a person could spend, spend, and spend, without having to worry about income removal taxes. The economists counted on keeping the cash flowing rapidly to offset fact that so many jobs no longer existed. Hopes were that the need for labor in these other sectors would absorb some of the free workers.

"Plus it gets cheaper," noted Mick hopefully.

"What you think 'bout this Luna stuff? They be mining full-time now, but it ain't breaking even," Paul asked Burt, as if to begin an argument.

Burt took a sip and answered, "I think they've been doing amazing things. They *are* mining the moon, but we forget that and say they ain't doing enough. It will take some time to get the wrinkles out. It's not like they have ever done this sort of thing before. They'll have to make many adaptations I imagine before they get it down."

"Humph!" snorted the ill-dressed woman. "Waste of effort 'cept for the energy stuff. That will help people."

Burt ignored her; it usually pissed her off enough that she would stop commenting. This time she simply rose and left after no more comments. He watched her leave before he said, "That was too easy."

"Just as well," Mick stated. "I'm not in the mood for her wrangling."

Paul whined a bit as he stated, "But that don't answer the question."

Burt responded lightly, "I have no doubt that things will work out. Every day someone, like that Red fellow, or the other one, Walter, solves one of the hiccups in the system. All the news sources say that the production increases regular like."

Mark belched before adding his two bits, "So you saying it will work? When? So far that bitch is right. It just cost more and more money."

"So far"—Burt smiled—"is the key. Very soon we'll see a payback, probably a year or two."

"How much you willin' to bet on that?" Paul grinned smugly, hoping to gain a few dollars.

"Fifty," Burt countered without hesitation as he flicked a bill out on the bar.

The grin slid off Paul's face; he felt cornered and could not back down. He retaliated with, "OK, but only for one year. You said one year first, so we go with that." A bit of triumph reappeared on his face. He slowly pulled out his wallet and emptied it of tenners, five of them. He slid them toward Mick.

"Done," Burt said firmly. A hint of a smile tugged predatorily at the corners of his mouth. His inside info might really come in handy if he played his cards well. Everyone needed extra dough.

"Well now," Mark said, "ain't that a giggle. Who gonna say they breakin' even?"

"News, same as them saying they getting better," Paul stated scornfully.

"Agreed," Burt replied, sticking out his hand.

Paul gripped it as hard as he could, in the old manly ritual, saying, "We'll see."

Mick picked up the fifty and the tenners. He put them in an envelope and marked the bet and, after looking at the calendar, put the next day's date for next year. The envelope went into the bet box, a small iron lockbox. Mick checked it daily for upcoming bets. Many times the bets had been forgotten over the weeks, months, and even years.

Conversations resumed after the momentary hiatus in which all the patrons observed the bet. Various persons took pro and con positions and, just as often, switched sides to argue with someone else.

Pedro, at the end of the bar, mumbled to another drinker, "I think Burt right, but I dunno 'bout doin' it in a year."

"What make you think that?" mumbled the inebriated fellow, after a substantial gulp.

"Well . . . you know I work in the mine. We drill and blast, drill and blast. It take time to make a long tunnel. It take time to do up the ore. But after time, the whole job get into rhythm, like music. Or it be like my garden. I dig and hoe. Then I plant seeds. The seeds make plants for food to eat. Just take time for everything to work."

Some tried to be more technical, talking about tons of material, water shortages, and emergencies. Some muttered about the Hsas, whether or not they could do the job. The very mention of Hsas raised tensions, so many viewed that uncertainty with trepidation.

The Hsas were truly the unknown in all the equations. Until they had a chance, years down the road, the whole game hung in the balance. Shifts in the political scene would add whole new dimensions. It was a quagmire that could swallow all of the Earth.

*

Chapter X:

Engineers

10

001010

"This is Mary Smith-Jones for WWNN. I am standing outside the once and nevermore UN building in New York. The various representatives have just passed a whole series of very startling and amazing resolutions after a twenty-four-hour closed-door session.

"For almost twenty-three of those hours, antiestablishment neoterrorist organizations were challenging the legality of the closed session. These groups were founded originally as anti-NWO and shifted all their efforts to this target, the newly established world government.

"They attempted to prevent entrance to the facility by linking hands and eventually began pushing away officials near the front entrance. Once the pushing began, a few punches were thrown." She could not say that they were thrown by the delegates.

"Blue green helmeted riot police moved in with tear gas and riot gear to disperse the rowdies. Hundreds were arrested," she silently cursed the censorship that prevented the vid-link of the arrests from being shown. She had seen the blue-and-green topped police simply assault the crowd that was not originally causing problems. Everything she had been told to say made her guts twist in agony. News really no longer existed. She understood the word *propaganda* as something that occurred in Nazi Germany, even in the Allies' camp during World War II. She had come to realize that it never really stopped.

Propaganda is a surgical instrument used to excise very specific problems at higher levels. There was no need to cover up a street shooting, but to allow the world to see the disgruntlement of so many, even if they were wackos, could not be borne. It did not happen, officially. The people were headed for a work camp run by the world government. They could be kept there indefinitely, assigned the status of terrorists.

All news media had to follow a stringent set of rules. Laxity was not tolerated. Stations had been shut down; large numbers of free-thinking individuals lost jobs.

She had had to wait for them to clean up the scene so she could shoot in front of the building. She had stepped in blood, cold and sticky in the cool late October weather.

She continued, "Granted, this type of closed-door session, for the full open table, is technically illegal under the regulations set forth under UN World Article 6. However, they invoked security precedents to evict the news cameras including myself and my crew. Armed bailiffs ushered us out and then secured the doors. In these secretive deliberations, they came to a startling agreement.

"The UN no longer exists! In their twenty-four-hour council deliberation, all authority has been invested in an entirely new entity to govern the entire planet, the space stations, Luna, and future outposts.

"It was felt that the moniker 'United Nations' was no longer apropos since there are no longer any 'nations,' so to speak. The planet has come under the globe spanning web of a new governmental body. As all governmental authority derives from the top, they have decided to rename the entire governmental body, which includes all regional and subregional governments, as the United Earth government, or UE.

"The old U.S. and Canada have been designated the North American Region, NAR, with an integration of the two governments with the Americans dominating. Mexico and the many smaller nations of the old Central America area have been designated as the single Mexican Central America. The president of old Mexico becomes the leader of the new organized Hispanic group. The smaller countries have been integrated as the equivalent of states, much like the large provinces of Canada were broken down into smaller pieces and set up as states. The old U.S., the NAR, could be said to have sixty-two states now.

"The South American Region has joined the European, African, North Asian, South Asian, and Indo-Australian Regions in our new United Earth government.

"Each major representative, all eight of the senior voters, one for each Region, led the innumerable lower house state voters from each subregion in this action. They have almost unanimously voted for the alteration in the name. They believe that peace and prosperity have truly arrived for a united humanity and wish to reflect that with the change in name. It is a signal that a new era in human affairs has arrived.

"Several groups here on the street are marching around shouting, 'All hail to the UE.' The camera panned to show several small clusters of around five people, each chanting in lackluster fashion. They were the paid dummies.

"We will have further announcements as soon as they are available. This is Mary Smith-Jones for WWNN."

Mary stomped out her anger as she headed toward the natural gas—powered van. She was being used; she couldn't do anything about it if she wanted to keep her job. Her thoughts were muddled. The two questions that faced her were diametrically opposed—stay or go.

If she stayed, she would be tacitly supporting the things she had come to abhor. But she might be able, in the future, to put a spoke in the wheels of the machine, make some sort of struggle on behalf of the people, the real people.

Her other option was not any more palatable. If she quit, she could not have widespread effect. Her widespread fans would not hear the words of a lone voice crying in the wilderness. They might even send her to a work camp if she rebelled.

Her knowledge of these camps was limited, but what scuttlebutt she had been able to obtain was not good. The camps were not extermination camps like the Jews had faced in World War II; they faced interminable labor and died just the same. In that respect, they were more like the gulags of the old Soviet Union.

She stopped cold; these distractions did not help. Suddenly, her father's voice intervened; it was a cold discussion of logic. He had said, "A=A."

She had wondered what math had to do with their talk about boys. He had calmly explained that one must think as well as feel. Both add to one's understanding of the world about them. If you forget one or the other, you will make mistakes. The name we give something does not change what it is. He had been talking about how dumb boys are and their instinctual desires. She really did not get the connection then.

She got it now; this was a growing dictatorship under the name of a freedom-loving government. She would have to stay, fight from within. She would have to bitch up.

*

It was late in the year 2034 when Red arrived. Red was twenty-nine, a good bit younger than his cousin Burt. He was one of the first so-called permanent inhabitants at Nearside. Permanent meant he would not return on the ship that brought him. Up until now, most of the astronauts would do their job and then leave in the vehicle that brought them. Some few had been left for short periods, on the order of one or two weeks. Now, a large number of them like Red and Walter would begin to inhabit the Luna station (Luna-1) on a six-month rotation schedule.

He wanted to stay forever. The stark, desolate environment appealed to him. The moment he had actually seen in person the thousands of shades of grey, he knew it was home. The incredibly bright sun that created the blackest of shadows and the eye-squinting glares invited him. The blue white ball of life in the distance beckoned, but he was beyond its reach. He had been there all his life. Now, he was here.

He knew he could contribute here. His skills would shape a new world more than any pioneer had in the past. The scale, the measure of the opportunity here, was even larger than those who had reached the coast of the New World, the North American continent, in the explorations of the empires of the Europeans. This was a magnitude of a world greater, not just a continent or two. His hands, his mind, would etch out for all history the formation of a new way of life. John Smith and Pocahontas had nothing on Red. Jamestown and Plymouth Rock are a full magnitude (perhaps two), below in importance when compared to the Luna endeavor.

It was a matter of timing. For all his other technical skills, Red had an amazing sense of history. He knew of the ratio of technology to change inherent in any system of society. That particular new science was his creation. His temporal-historical engineering had given him a perspective on history unlike most people. He knew, without doubt, that this was the stage of history in which one person would have the most significant effect possible. He did not presume to put himself on the level of the Christ, Mohammed, or any other religious figure. That would be a dangerous place to put himself. But he knew he would equal any of the great explorers and far, far surpass any of those pansy ass politicians. All it would take is a little time and the chance to do what he knew he could do.

It would be 2035 before his cycle was over at Nearside. He silently cursed the need to leave, even at the moment of his arrival. It was not like he could hide, could he? Retire here?

He was indispensable, and he knew it. Why couldn't they make an exception? He was well aware of how far the rules had been bent already for him. He had been placed ahead of others in the rotation. He had pushed hard for and gotten his way concerning the man who would be his partner. He cursed himself for wasting time on these mental ruminations. It was unproductive. He would find a way.

Damn that recycler. He had been staring through the portal for a good five minutes. His multitasking mind had been absorbing details even as he had been mentally complaining about leaving. *Wrong*, his mind said. The solution was not yet in view, but he was certain it would come to him. Those E-holes (his personal curse for the Earth engineers) missed something. What?

Red O'Hare continued to stare laserlike through the small portal at the parts dump. His eyes did not flicker or waver. A thought was wiggling around his brain, but it would not, as yet, forge itself into a focused concept. It would later. It always did. Something needed to be done here, an alteration. He watched the recycling arm pick up the sorted broken metal pieces and place them in a conveyor to be melted and recast as usable basic machine parts. Red hated unsolved dilemmas. With an aggravated frown, he stomped off, his feet pounding exuberantly to his personal tune of "you will be solved." He noticed Marceau but did not speak. He felt that she did not like him because of his prior interference in her domain. She really didn't.

Red worked primarily as an engineer in the energy control department. His job consisted of finding the most energy-efficient way to conduct business. He constantly examined all facets of the ore processing facility at Luna-1; some of which were not exactly under his jurisdiction. He did not restrain his efforts to his own bailiwick, which irritated many others and encouraged an equal number of really progressive types like him. They were often referred to as "Redites" in humorous as well as vulgar ways.

The uncanny ability he had to rearrange and manipulate mechanical processes to save time and energy had earned him his place here as an engineer department head. His brusque and salty personality earned him more than a few epithets. He was a blunt speaker, feeling that euphemisms were a waste of time and energy as well as being least descriptive of the actual situation. Red was never PC (politically correct), just correct.

He had six engineering degrees, two bachelors. He was a jack of every trade, and the space program recognized him for the exceptional mind that he was. He was a man who could find answers. But he was not just a thinker; he was a doer. The skill of his mind was matched by his hands, and he had another tool that could understand his mind.

Hands, even in this highly computerized environment, make the moon go around. Walter, now thirty-one, has the magic hands. If it needs to be fixed, you call Walter. He is the one person whom Red respects and, on occasion, defers to. The pair—quiet, patient Walter and hard driving, unequivocating Red—made a potent team. Before the two of them, all problems surrender. Red outlines and describes problems and solutions while Walter accomplishes. Walter is the wrench; Red is the force that turns it.

Quartermaster Marceau did not like Red. She felt that Red had overblown notions of his own value. Red had redesigned the method of storage and labeling for the entire base. The radio frequency or RF, tags, those little rice-sized transmitters embedded in each device, allowed for a quick radio tracking. The colored bar codes were a double check if the transmissions were interrupted. It was more efficient, she admitted. *Marginally*, she spitefully added in her mind. How important did he think he was? Damn sphincter!

Hundreds of people had designed the original setup. She watched as he marched away from the portal and exited toward the radio room. Glad to see him go before he criticized her operations, again; she returned to her electronic inventory. Moving evenly down the aisle of stored goods, she passed the scanner over the varicolored bar codes where the RF tags were imbedded. Both the numbers and the bars were read at the same time and compared. Red's main change had been the positioning of each package so that access to the most often needed objects was easier. It was, as Marceau noted, only a marginal improvement, but Red is always looking for any improvement.

Red was not much bothered with women. There was little time for personal matters in his world. He was not at all bothered with men, like some at Luna-1. His private joke was that *that* was ass backwards. There was only one woman whom he had any interest for; however, he was sure she had no use for him. He knew that his constant interference had not ingratiated him with her or anyone else for that matter. He was driven. He was, worst of all, a perfectionist.

As Red entered into the radio room, Mikhail rolled his eyes at Lt. Stacy Balmer, his assistant. Captain Mikhail Gondov was the communications engineer for the base. Mikhail's position put him in regular contact with Red. He was not thrilled with that. Red could be demanding and sometimes condescending. Red was brilliant, but he was a real pain in the ass.

Mikhail knew Red wanted to determine the status of his latest suggestions for reshaping the Mars base plans. From Mik's point of view, all the energy saved by Red's ideas went into his lengthy radio discourses with Earth. From Stacy's perspective, they were simply two boys lifting their legs to mark territory.

"No reply yet, Major," voiced Mik. "They have only had a few days." Stacy and Mik both saw the minute tightening of muscles around Red's face. In Red's timescale, days were like months, far too long for acceptance. Mik wondered at the man's arrogance. Knowing that the man was always right, when he said anything at all, made it even more unpalatable.

"Yes, well, they will soon see the advantages," Red rumbled. "I just hope those paper-pushing, political, backside wipers get it right." Hitching his tool kit and belt higher, Red firmly exited, the sound of his mechanical motions fading down the passageway.

"He is always so intense," commented Stacy. Mik did not reply to the obvious, so she simply returned her gaze back to the many readouts in her visual domain.

Mik watched him move awkwardly down the tunnel.

Everyone moved that way in this low-gravity environment. In order to help prevent bone deterioration and muscle loss, two sets of parallel bars, set at waist high, ran along the outside edges of every corridor. The side you took was like driving an old automobile. The bars allowed a person to push up and forward with the legs while pulling down with the arms. Velcro carpets and shoe soles added to the stability. It was one of the best adjustments that could be made in terms of exercise. Since most everyone was always moving around, they were getting fairly continuous workouts of the larger muscle groups, much like walking on Earth. They still had to attend daily workouts for more specific muscle groups.

Many had suggested magnetic boots, but mag boots caused problems with the very delicate electronics around the station.

*

After leaving the radio center, Red was even more consternated. The only consolation he had about leaving was the fact that Walter was going to be going with him when he left for more intense physical rehab on SS-2. He recognized that Walter had a gift of equal value to his own. They would go through the two months of physical rehab at the space station together. Rehabilitation was necessary to readapt to the more intense gravitational field of Earth. Even with the regular exercise at Luna-1, they were very weak compared to Earthers.

Back on Earth they would be debriefed and recycled. Red was glad to know that Walter would accompany him to Farside in 2037. He had pushed hard for that combination. It was an unusually short turnaround time for astronauts in this global program. Most people only got one shot at a Luna post. Many only got a few circuits to high orbit or the space stations. Those who had done the work were usually drafted into teaching the next generation of OEs (off-Earthers). Large numbers of personnel were going to be needed for the thrust to Mars and the outer planets. Red and Walter were among the most necessary tools.

Brass had decided almost immediately that there were two people needed everywhere. These two were worth their weight in heavy hydrogen and light helium, the fuel for the fusion reactors. But it required them one step at a time. The next perfect place was Farside. Nearside was a success because of the amazing abilities of these two. Hundreds of graduate students, many of them even at postdoctoral levels, were under the tutelage of Red and Walter, via vid-link, for several months. Brass hoped that pure ability rubbed off, so lots of rubbing went on. Some of it really did help. Those few with extraordinary ability were ignited to even higher standards. These candidates generally remembered who it was that brought them to such heights. In the future, these men and women would continue to follow the two men who had brought them into the circle of greatness. Those who did not have the extra were sent to other, lesser posts. Leaders bring out greatness, even without knowing it. They do so without knowing they are leaders, as in Walter's case or, sure of it, like Red.

*

Experience is not what happens to a man. It is what a man does with what happens to him.

—Aldous Huxley

*

Treat a person as he is, and he will remain as he is. Treat him as he could be, and he will become what he should be.

—Coach Jimmy Johnson

*

Good judgement comes from experience, and often experience comes from bad judgement.

—Rita Mae Brown

*

After Red left her, Marceau discovered a few sealed and specially marked containers off in a corner. The RF (radio frequency), tags, denoted important items under the direct control of an individual, an undesignated one though. Puzzled, she began to check all her logs. These were marked as personal items but had no apparent owner. They also were not placed in the personnel section of the warehouse. They were tagged as "do not move." Curiosity flamed in her. It was her job to know what everything was in her large warehouse.

She went to her desk and the touch screen that was connected to the mainframe of the base. Her handheld scanner should have all the data, but it was just possible that one of the others had missed an object or two. They may have been placed there temporarily, awaiting relocation and reidentification. She would have to quickly figure out what they were. The others under her had obviously failed. Failure of any kind would halt her career. Career was not all, but it meant a lot when you're in space. She was at the leading edge; a job on Earth would clearly be a demotion.

In the space program, one did not advance so much as others failed to stand up to the measuring stick. Everyone was of such a high caliber that choosing one over another depended entirely on the miniscule, often immeasurable qualities of each astronaut. The tiniest error could derail a lifetime's dream. Marceau wanted a shot at Mars. Luna was a goal that most would be satisfied with, but she was driven to accomplish. She had a streak of perfectionism. The only way to express it would be to make the Mars shot along with the Luna posting. She had to figure this out, or she would be one who didn't make the cut.

Marceau was digging deep now. Her friend in the computer lab was an expert at hacking secure files. He had the hots for her and would do whatever she asked. She felt a little guilty, but she did have dinner with him and even necked a little. Quid pro quo.

Every shipment was being backtracked for her now. She went beyond the normal tracking number. After several days of detective work, she knew who sent it and for whom. It blew her mind. Red and Walter were purchasing special materials with their pay equivalent and having them shipped to Luna. It made no sense. What did they hope to accomplish?

There were also suspicious donations from once wealthy individuals and corporations. All were known for being opposed to socializing all businesses, making them government controlled. "People owned" was the politically correct way to state it these days.

Upon closer inspection, meaning she opened the packages to peek, she began to have an understanding. Everything had to do with computer chips for mining equipment. Little hidden thoughts became clearer. They planned to mine on their own. How could they expect to do this? Was it even legal? Who knew, or were they doing it secretly? Many thoughts whirled through her brain.

She knew that they had often ventured beyond the mine claim area of Luna-1. This was a radius that was considered the feasible area of mining for Luna-1. They must plan on some sort of mining operation outside this radius. Was it a place that they could and did escape to where no one could control their actions?

She was also aware of the underground movement on Earth to return to full open capitalism as an economic system. Grassroot support swelled in the North American and European regions. Areas used to more significant availability of rations, they felt now that they had been made second class. Many of the poor of these once wealthy regions began to understand that equalization, in terms of the whole planet, did not necessarily mean an upwards movement for them. The poor had to give way to the hordes of really poor.

These people ate regularly, but the quality had become severely limited. Delicacies disappeared. No room existed for such if everyone expected to attain healthy equality. Even basic foodstuffs suffered; bread was filled with cellulose from matter concocted in food vats. Quality meats slowly became harder to obtain until they also vanished. Only the lowest quality meats, once used for pet food production and worse, became the norm. It was the only way to spread the wealth around, thinly.

The amount of energy required to produce flesh was considerably higher than simply using the feed, like corn, for human consumption. Even so, many accepted the form of government; they just felt that capitalism could produce better food markets. Others decided that the government itself created the problems. These groups actively strove to rearrange the powers.

The antigovernmental forces were well organized. Many pre-UE underground movements such as the neo-Nazis, KKK, mafia, and even the reformed church worked against the current government. They foresaw the constriction of individual rights and the program of control that would be instituted.

Perhaps they had been able to bend Red and Walter, Marceau thought quietly. The manifests showed names of some of the remnant wealthy, as contributors, those who had been stripped of large amounts of wealth yet still had suspected hidden reserves. Was it a conspiracy? She felt a sudden burst of anger inside. She felt another reason to dislike Red growing intensely within her heart. She gathered her data and marched off to the command center (CC) to denounce Red and Walter.

*

Dr. Illya Marquez banged her hand on the bedside table, groping for her v-link. She still wore the now outdated glasses-style frames. She found them and brought them to her dark adapted eyes. She turned on the audio but not the visual. She knew she looked like shit, and she did not recognize the number.

"Huh," she grunted sleepily.

There was a long pause, then, very quietly, "Ly . . . it's Lu. I really need to talk to you."

Illya had heard nothing more than the voice she still recognized from childhood saying her nickname, Ly, and Luisa's, Lu. She was stunned for a second and then sat bolt upright, "What?"

"It's Luisa—"

Ly cut her off, "I know . . . I'm just, surprised. You said something more."

Lu replied somewhat reluctantly and with a raspy voice, "I need your help."

Ly understood immediately that her childhood friend had been sobbing. She flicked on the visual and was rewarded with a sight that was as mussy as she was sure she presented. She noted the puffiness and redness of teary eyes.

"What do you need?"

"I . . . I don't know quite where to start. I guess, first, I should say how very sorry I am for the things my family said and especially what I said. It was all wrong, cruel. We had such a good friendship growing up. How we convinced my parents to let us, me, go to a university together. The dorm . . ." She choked out a wrenching sob.

Ly said nothing, just noted the inner struggle of her one-time best friend and lover.

Lu sniffed heavily. "I wanted it to be that way, us, lovers. Since you were the eldest by a few years, you made the first move. You were always the first to expand your horizons, your life view. But I had wanted you too, and you knew . . . you knew that both of us . . . you did what I was too afraid to do. And it was perfect. The love we had always had for each other simply expanded into the physical realm."

Lu took a deep breath, squirming slightly then firming her tone, "When my extremely Catholic family surprised us in bed with their unannounced visit, they immediately blamed you. I latched onto that out of fear, shame at my—no, our nakedness. The fear of my father's reprisals, my mother's and sisters.' The fear of God's reprisals that had been beaten into me through church doctrine and father's belt. I forsook you. I allowed them to vilify you."

Ly said somewhat coldly, "I know."

Fresh tears drenched Lu's cheeks as tremors rattled her bones. She choked out. "I am sooo sorry. I was wrong. It was *evil* to do what I did to you. We had love, and I turned my back on it out of fear rather than embracing that love and having the two of us withstand the tumult and becoming even stronger, more in love. I failed."

Ly was intrigued with Lu's confession. She had seen and understood most of this almost at the outset, but her family was not very religious, never had been. Her Russian mother had married a reasonably wealthy Brazilian; neither had much use for churches.

"When they dragged me off to the convent, I was just glad to get away from father's beatings and mother's moaning about how I had disgraced the family. I also realized that I would not have any chance encounters with you. I would not have to face you. This allowed me to block out so much of this, to compartmentalize it and pretend it did not exist."

Ly interrupted, "You woke up at four in the morning just to apologize, how many years later?"

Lu nodded. "No, yes, well . . . no."

"Have you been nipping at the sacramental wine?"

Lu smiled slightly. "No."

"Well, at least we get one definitive answer there. Then what is this all about—"

"Please . . . let me continue. The church is going in the wrong direction, and I have to do something about it."

"What are you talking about, Lu?"

Lu looked intensely at Ly; it was the first time she had heard that nickname from those lips in ages. It made her feel wonderful; she knew she was on the right track. She felt years of shame and facades begin to erode away.

"Ly, I'm going to challenge church doctrine in regards to homosexuals, the current politicization of religion, and—"

"So you are going on a tear since you finally realized what you did. Now you want to blame religion? Don't you get—," she started forcefully.

Lu quickly announced, "I had a vision of the carpenter."

Ly cogitated in the dark silence that was only parted by the sound of ominous breathing on both ends of the line. *Strange phrase, end of the line,* she mused.

There is no more line, only waves, energy, purity. She mentally shook her head and returned to the subject.

"When we were kids . . . you had them so often . . . I did, twice. What happened?"

"I saw him working with wood, gently shaping, taking from the wood what it had to offer, not forcing it to be what it was not, stone. Then he looked at me, and I heard His voice, 'It is all love and acceptance, not absence of love and rejection.' Then He smiled at me and went back to work," Lu stated plainly.

Ly had not believed Lu for so long when they were kids. Then it happened to her. The carpenter was fashioning an immense loom. "All the threads are woven together lovingly; hate will rip the fabric, but it can be mended with loving care and attention to detail." She had not known why at the time; she did now. Her analytic mind reexpanded to include the unknown and unknowable.

"What do you need, Lu?" she asked, ready to give anything.

"Mary Smith-Jones."

"Whaa . . . Why do you need her?"

"I need a reporter that's not under the thumb of the church. And I saw the way you two looked at each other on that broadcast about Atlantis. You don't make eyes at just anyone. You are still in contact with her, and she will come if you ask."

Ly blushed a bit before countering, "She might."

"I've had experience with those eyes, Ly. She will come running, especially if you dangle an exclusive. Story, that is. I still love you, Ly, without the fear or shame imposed by others."

"Lu, remember our song?"

It was the Beatles. Love really is all you need.

*

Red hated sleep cycles. His intellect was not controlled by his body, rather the reverse. He had programmed the monitors of his body cycles to show him sleeping when expected. He existed on no more than five hours at any given time. It would be a time of his choosing. He considered it wasted time even though he knew a body needed rest. Red had a voice-activated recorder so he could expound his ideas from anywhere and anytime. He programmed it to automatically ignore his snoring, which seemed an imperfection, a waste of personal energy, after; of course, he had copied his snoring so that he could play it whenever he felt the need to sleep for the ground pounders.

Red knew Walter was a sixer, using all six hours fully in his sleep cycle, though not the eight hours set aside, but did not hold it against him. Walter simply dropped into a very deep sleep, REM sleep, almost immediately. Red respected, even almost approved, the extra hour since Walter was so very capable.

On some occasions, people had discovered Red asleep at the wrong time (not his normal cycle or not sleeping when he should be) and taken it for laziness. Some had even complained to Earth. Those on Earth were, at first, gravely concerned. The complaints and the data recordings, however, did not match. It was soon discovered that Red was sleeping during an emergency, one that he had solved.

Many ground controllers winked at his ploy. They then set up a program to eliminate his bio-sleep data from being incorporated in the overall reference material. The flight doctors were outraged by Red's petulant behavior, as they termed it, of course. They filed grievances that went nowhere.

They were considerably mollified, however, by the brass, when they were assured that his data would not corrupt their experiments. Until, as a matter of course, again, an undergraduate bio-tech pointed out that his data might be the most important. Others with his low sleep tolerance, long-term it appeared, would be of great value. If there was a genetic component, it might be used with the Hsas. There was a similar sleeping pattern with Red's compatriot, Walter. They might be on to something. They came up with several techniques to increase their ability to observe and measure these cycles.

They simply discreetly bugged Red's domicile and Walter's room as well to obtain as much data as possible. Red's outrageously loud habit of log cutting was a dead giveaway. It seemed that Red dropped, almost immediately into deep REM sleep as well. The bugs planted did not give them much more than that. It was an important beginning though for the search for others. They followed up with phony blood tests in order to get a new sample of DNA that had been exposed to long-term low-g environs in the hope that they could isolate some significant variable.

Red had been greatly upset when he had discovered that *they* knew. Not because they were capable enough to figure it out; the emergency had made it obvious. His anger was directed at himself for not having set up an emergency protocol to stop his sleep period. He was not really even miffed at the bugs he had discovered; he understood the scientific drive to know. He simply rerecorded his snoring and played it at random moments, with an emergency cut-off, of course.

Red's brazen disregard for rules and regulations grated on so many nerves that everyone wondered how he had been allowed to reach so far in the space program. His brilliance was well documented, but even so. The brass, however, knew a problem solver when they had one. He was protected. Even more, he was needed. Without him, little failures could and would mount, and this could lead to absolute failure, a price the world could not afford. It could mean the fall of civilization as is and the capitulation of technosociety, resulting in a more primitive form of social structure globe wide. From the points of view of the eight leaders of the regions and the one overall leader of the UE, it would mean they would lose their positions of power.

This could not be allowed. The nine played the political game. This means that even though they considered the rebellious attitudes somewhat dangerous if allowed to spread, they would protect these men. They would use them, even make them semiheroes if that would get them a stable lunar program.

These eight were the major Region leaders—the NAR, SAR and so forth, as well as the grand poo-bah leader, the president and soon-to-be emperor. They had gained a politburolike control over the UE. Power was being concentrated in a new yet old-fashioned manner.

*

Mary's v-link flicked a red light on the upper left corner of her left eye, indicating a call from someone on her private line.

"Mary," she announced, continuing to enter data.

"Mary, it's Illya."

Sudden tenderness filled Mary's voice, "Hi."

"I have an exclusive for you."

Mary's antennae rose nearly a meter. "What's this about? Atlantis?"

"No," Illya said. "It's bigger."

"How much—"

Illya cut her off, "Do you trust me?"

"When?" Mary asked quickly; her trust for this woman was implicit.

"Very soon. I don't think this will hold for more than a few days."

"I need to know something. Can't you give me just a little?"

"Too big a chance over this line. Get down here with the whole army."

Goddamn, thought Mary. "That big?"

"Bigger!"

"Thanks. I love you."

"Ditto, Mary."

Mary pondered for a few moments after the disconnect; then she said, "Editor dial."

"What now, Mary?" The editor's visage appeared.

"I have an exclusive with Dr. Marquez. It's bigger than Atlantis, and I need a full crew."

"Oh, Jeeeezus, Mary. You want a little, just ask for some vacation!" His face puckered.

"You want me on my knees begging?" Her eyebrows raised questioningly.

"Never!" He shuddered. "Last time you punched me in my crotch! My wife was furious! Bruises and . . .," he trailed off, looking hurt.

"I did not hear that last part. Say again. How many stories have I gotten for you? Scoops, juicy tidbits that others missed? I have the next big one, and you are playing it off as a tryst," she vengefully sallied.

"OK! Just don't use my privates for a speed bag again. What do you want?"

Mary could just picture him, his fat slimy hand under the desk and tucked under the considerable bulge of his belly, holding his family jewels protectively. "I want an army."

"Sweet Jeeeezus on the cross, Mary! You know what that will cost? Why not use a local crew?" His arms flew up in disgust.

Mary's smile was so evil that Julius, the editor, recoiled. "How do you know that I'm *not* bringing you Jesus on the cross?" Her unblinking eyes bored into Julius's with her full will.

"Sweet merciful—"

"Shut up and give me what I need. Yes, I know, the news media is now all one. But it is not all equal. If we get this first, your fat ass moves up to the next rung of the ladder, just . . . like . . . mine," she emphasized with chopping motions from her free hand.

Julius's fingers wiggled delightedly at the prospect of such a significant move. He knew she was playing him, yet she had already pushed them both up the social scale several notches.

Mary could just hear the cogs and pulleys of his archaic, simple, backwards mind clacking as he weighed possibilities. She smiled, knowing her digitalized mind was vastly superior, able to calculate to a nicety which way he would go.

"Yes," he drawled. "I'll sign your chit. It had better be all you make it out to be." He was glad to hang up; that woman brought chills to his weak spine. He knew that his position had been a gift from the powers that be, and that he had risen only by the efforts of others. As long as it made him look good, he did not care.

His social status, therefore, and, more importantly, his wife's would remain at the level they had become accustomed to, even improve. He mused, ironically, that for all the claims of change, the system remained the same, even more so. His piggish little mind returned to the upcoming story, and the vanity of newer and better dreams flared before his eyes.

*

Red was well-known for his vituperative language. He would let loose whenever he felt like it. It was his method of releasing some of his excess energy. Walter was not, but this problem had him drawing extensively on Red's vocabulary. It was, literally, the smallest natural problem they had ever faced. It was an insidious problem—ants. They were a necessary part of the biomes. Ants and many other insects are a vital part of any biome. But they would have felt much better if they didn't have to deal with these little bioengineered pains in the sphincter region.

The biomists had included many floral and faunal oddities to enhance the organic recycler and food production units. Insects were an integral part. Ants, however, had a very bad habit of moving into new territory, places they had not been intended to occupy. Those places were not conducive to the satisfactory operation of all kinds of equipment. The bodies of many ants were found in the circuits of many electrical areas. They shorted out many of the localized computer systems, repeatedly.

Red and Walter found it convenient to create a few entry codes of their own for the biomes. These codes were buried deep in the computers and had a hiding program attached to them that prevented them from being located or eliminated by anyone other than Red or Walter. These codes would provide easy access for them and consternation for some. The occasional "who the fuck took their lover into this biome" resounded frequently at weekly meetings; there were many denials.

The double portal doors that gave access to the units had electrical shock devices to keep insects in place. Yet somehow, the ants were escaping en masse. All Red and Walter had to do was find out how.

Since there were many different kinds of ants, all adapted to different climates, turf wars developed. Red ants and black ants became jumbled amongst each other's domains and entered where they were not to be at all. Catastrophic wars of epic proportions occurred amongst the various kinds, leaving carcasses strewn about the fields of battle.

They already had closed a great many ant-sized loopholes. Small holes discovered near electrical outlets allowed access to other biomes. Watering systems were fouled with the bloated bodies of drowned ants. Lighting devices continually faced six-legged marching columns bent on shorting out anything in their path.

"It has to be the damn wind machine," Red grumbled. Vexation showed heavily on his sweaty brow. These ants were filing out the main doors the instant they opened. "Those little bastards are being tossed across the low-level repulsing field."

Walter chuckled in response, "Flying ants, and we thought only birds could fly."

"Gliding, not flying," Red grated as he increased the field strength with his freckled hands. "Like flying squirrels."

Walter measured the center of the field with his brown hands and declared it sufficient. They moved on to the next biome, the polar one. Bacteria accumulated around the seals of the air lock doors. The ants from other biomes inexplicably went right for these bacteria.

Walter mused aloud, "I think we have, first, a moisture source, second, some food or trace element need, third, a housing material—"

"Fourth, a pile of feces and vexation!" Red noted in multiform communications.

Walter chuckled, both aloud and in sign.

Thirty-seven fecal depositing, very time-consuming, individual solutions were required in the end for all the problems. But it was done.

*

When Marceau approached the commander of Luna-1 in his office with her findings, she was stunned to hear him say, "I know."

"W-what?" she stammered, wide-eyed and shocked.

"They have been given special privileges in regards to mining certain areas and use of materials for the redesign of equipment," he spoke matter-of-factly. He mentioned nothing of the strictly secret orders under which a certain cave was to be developed.

With eyebrows high, she succinctly uttered, "Special privileges."

"A free hand, pretty much," said the commander with a shrug.

Marceau paused as she mentally revised her position, "Then—"

"Then your detective work came to the wrong conclusion because you were kept in the dark. I argued that you, as quartermaster and commander warehouse (QCW), should be made aware of the special status, but Earth did not want too many in the loop. They were afraid that many would come to the same conclusion you did, about the entire space program though, and not just Red and Walter."

"So by their very silence, they create the controversy they want to avoid," she noted with extreme sarcasm.

"Ass backwards, isn't it? I think that the UE suffers the same disease that the nationalistic governments once suffered—paranoia. I will have to make mention of this to mission control," he noted apologetically.

Marceau stiffened; her nostrils widened. This would be the black mark, the death knell of her career. She would be deskbound or teaching. Inwardly, she withered and then flamed as she felt the blood rush to her cheeks. She hated *him*!

The commander noted her body language and quickly stated, "You have nothing to fear in this. I will make it very clear that you should have known all along, as I originally recommended, and that you did your job perfectly by finding out what was in the warehouse. A commendation will be recommended."

Marceau took a deep breath and relaxed. However, deep inside, her feelings for Red were still those of anger over her close call and the embarrassment she had felt at her mistaken guess. She did not know what to think of the money connection though. She was not about to bring it up.

"I think, though, that I will leave out the part about you hacking files," noted the commander rather slyly. He did not know about her friend slash accomplice.

Marceau stiffened again as she realized that this was a hidden reprimand disguised as a joke. As she turned to leave, the commander spoke again. She turned back to listen.

"Marceau, I know that Red is not well liked. Red is an asshole, a productive one but still a major hemorrhoid. Red and Walter get away with things that the rest of us would get burned at the stake for. He is respected though. He has been given a lot of leeway, so much so that I find it hard to swallow too. The two of them have done a lot for us in compensation for that leeway. They have worked miracles that Earth engineers said could not be done, would take much longer, or would need a team of specialists. As much as I hate Red's arrogance, those two are the epitome of thinking outside the box. I have a personal belief that they do not visualize in terms of limitations, no box exists, but only in terms of how best to accomplish the highest standards. The arrogance comes from not understanding how others see limitations. Please don't let your personal anger at Red's interference in your warehouse and this episode prevent you from that Mars posting you deserve. I have already put your name in the hat for that one."

She did not storm out in a fury; the whirlwind was an entirely internal emotional nuclear explosion. She felt that she could strangle Red, repeatedly. *How many times could a person die?* she wondered hopefully.

*

It had been a short week since Dr. Marquez had called. Mary Smith-Jones had quickly gathered her crew, chartered a small plane, and hotfooted it to Carutapera to see what the doctor had waiting for her. She knew Illya would not have v-linked in such a manner without having incredible news.

The bus she had purchased rattled its decrepit way along the narrow now paved, two-lane road to the site of Atlantis. The garrulous old man driving had chattered away from the moment they had boarded until they neared the archeological site.

He turned and smiled his missing front teeth grin as he said, "Here, here."

When Mary stepped off the bus onto the scorching asphalt, she saw Illya's driver in a long electric limo. She had always been a bit nervous about the fifty-caliber semiautomatic pistols that bulged underneath each arm, only to learn, more recently, that the second man in the car carried machine pistols. To Mary's amazement, two females exited the limo, looking for all they were worth as if they were Amazonian warriors. Two soft yet powerful hands grasped her elbows from behind; she had not even been aware of two more Amazonians who had circled around the bus to cover her back. She was almost lifted off the ground as one whispered intensely, "Dr. Marquez asked that we bring you to her house. There is a certain amount of danger here."

The death threats had come quickly after the initial discovery of Atlantis. Bodyguards had been assigned, first by the university, followed by more in-depth Brazilian government muscle, and, now, regional government and personally hired mercenaries. Religious zealots had always refused to acknowledge the age of Atlantis and, second, as a backup attack, claiming it was destroyed in the biblical flood. Illya had a small scar, from a twenty-two caliber pistol fired at near point-blank range that had destroyed most of her left nipple, leaving her once perfect breasts mismatched. Mary looked toward her crew after this split second of remembrance.

One of the cameramen looked back and, seeing Mary being practically dragged away, yelled, "Hey! Stop! They're taking Mary." The crew turned, almost as one, knowing that Smith-Jones had been dragged away by more than a few government agents for a little "talk" about certain stories.

Vitally, the shotgun rider, emerged from the vehicle, stating in his Russian accent, "Ms. Mary, they are a new part of our retinue." The fact that the holstered H & K machine pistols glared in the sun was not lost on the crew. Mary had to smile inwardly; she knew that he never let them be seen unless he wanted them to be.

Mary was relieved to see Vitally, the leather-faced ex-Russian agent. He was as old as sin, looked as tough, and was actually tougher than any gaucho from the Pampas. He had to be at least seventy, but his wiry frame held vigor yet.

"It's OK!" Mary shouted towards the frozen crew. "I know this man." She indicated with a nod towards Vitally. "What's up?" she whispered to the raven-haired vixen on her left who had given her the heads-up, noticing a tattoo of a bitter-looking knife on her left hand, a hand that, in spaghetti westerns of old, would be said to be half-cocked, ready to draw.

"Not here," motionless lips replied. "We are under observation."

The other woman, hair of ebon and just as motionless lips, noted, "Seven o'clock high, twelve low, three, second window up. *Don't* look back," as Mary automatically began to look behind to see the seven position.

Mary felt a sudden deep chill crawl up her spine as she thought what she might have gotten herself into this time. She looked and saw that the left hand of the woman on her right had an identical tattoo. As if her eyes had suddenly opened, she realized that the bus driver had the same tattoo; he was, in fact, nonchalantly strolling behind the crew, the bulge at his back now obvious. *Damn, what an actor he is*, she realized. In the few seconds she had before being thrust into the bulletproof limo, she saw several men and women who fit the bill of protectors moving in a loose, flowing, dangerous circle about the crew. Though they were many meters away, the tattoos were there; she saw. *Fuck! What is going on? What is so dangerous? What am I on to?*

The husky, erotic voice on her left read her mind. "Soon, Mary, soon. Be patient."

Illya's place was now a mansion of immense proportions. There were labs distributed about the grounds, acres and acres of labs, storage, trees, open ground. Mary finally recognized the thousand meters of open about the house as a security zone. She trembled even though she had been here before. The sudden recognition of extreme danger had never before hit home, until now.

The drawing room was, as usual, well decorated, only now it was also filled with women soldiers, as Mary thought of them, one at each window. The raven-haired beauty stood beside the lounge Mary had dropped into, her gin and tonic waiting. She turned to the woman and asked, "What is your name?"

The stunningly attractive woman winked and said, "I am Death."

A delightfully attractive laugh warmed Mary's heart; it was Illya Marquez. "I see you have met my Damoclesions. The sword of Damocles marks each and every one of them. They give up their personal names, except to their lovers. They are a special group, made up of supporters of freedom, in all its meanings. They have assigned themselves to me for some obscure reason."

Another woman entered, and Mary was suddenly sure that this was the cause of all of the hullabaloo. The dynamics of tension had soared to a pitch with her very entrance; the so-called Damoclesions became euphoric?

The two left hand in hand; Mary knew then that she had lost Illya, and her heart crumbled. Their visage showed that these two were deeper than she could pierce. Surprisingly, the Amazonian next to her spoke, "You are not left out. They are special. They have a history that you will soon know."

Mary's heart restarted. "What the fuck is going on here?" she demanded.

"Change. Change such as the world has been awaiting for many eons. Your crew is setting up now. We will soon depart, and you will have it played out for you." The woman smiled. It was almost—no, it was surreal.

*

Red and Walter were out mining. They had built their own variations on the Laborers from leftovers and scraps. Computer chips had been shipped from Earth. Red and Walter paid for these with their time equivalents and with the assistance of some quiet earthbound business interests. Red and Walter didn't really give a shit about the investors; Red and Walter had gotten what they wanted from them and would pay them back in their own way.

The government had at first looked the other way, simply assuming that they could take over if the enterprises showed any real merit. They believed that they could order anyone at the base to confiscate any materials or equipment for redistribution.

Red and Walter were piling up (literally) regolith and ore samples. They were preparing a supply of materials for the solar furnace they were creating. They wanted to show that they could make the furnace almost entirely from Luna resources. This would increase production for the whole base as new machines came online.

Red and Walter had found ways to repair most parts in a way that was more efficient than shipping or making new ones. They traded some of the saved energy, 10 percent of launch costs of their items, with the WSA for large numbers of their rebuilt parts.

A conflicted WSA agreed, after much debate about how far to let this go, to allow the continuing transactions. The guiding force behind that decision involved the potential for new and better designs of equipment promised by the two engineers. They had to smooth the way; they considered Red and Walter a necessary, ungainly blip in the plan.

That niggling, mind-touching thought that Red had had so many days ago while looking out the portal involved repairing (not recycling) old parts. Both Red and Walter appreciated the saving of time and energy in reusing old parts after minor repairs.

The two taught many of the other young engineers the almost forgotten hand skills necessary to make many of the repairs. Computers had been programmed to do this ilk of labor for so long on Earth. Since the throwaway society was coming to a close, handyman-type skills had to be relearned until computers could be reprogrammed to deduce the necessary repairs and then perform the needed tasks.

The WSA did find the savings to their advantage, and it fit in politically with the recycle type mind-set they were trying to instill Earth wide. They could tout their engineers as politically correct in their solutions to problems.

Red and Walter built their own buggy and refitted several Laborers for return to active service. They soon had their own fleet of equipment ready for their "private" mining operations as well. They provided Luna-1 with a ten to one ratio of items for what they received. Each of the items was extra for Luna-1. They were made on Red and Walter's own time and dime, so to speak.

Red and Walter fulfilled their end of the bargain in other ways as well. Using some of their own supplies from the specially marked boxes, they improved the purity of elements coming out of the Foremen's solar furnaces by 3 percent. The improved purity allowed for construction of superior solar panels that were also more productive. All these little improvements added greatly to the overall feasibility and strength of the base. This is what they were there for. All were pleased, for the moment.

*

It was certainly going to be a phenomena. Dr. Marquez had prepared a stage—more than a stage, a private amphitheatre. Mary wondered how so much could have been accomplished since the short call she had received. There were thousands gathered in front of the podium.

The spectacle, for it could be called nothing else, had attracted hordes of religiously garbed persons of both sexes. The religions represented were a multitude, a hodgepodge ranging from Catholic, various Protestant denominations, Muslim, Eastern, Wicca, and a large number of groups that, to Mary's mind, could only be classified as pagan by her western upbringing.

She pointed to a group of women wearing the robes of the Catholic Church, asking her dark-haired guardian, Death, "Who are they?"

"They are the women priests of the Catholic Church," she responded stoically yet with a hidden harsh undertone.

Mary's raised eyebrows and penetrating look brought a quick response from Death, "The first was ordained by a male priest in secret. Women are not allowed to be priests in the Catholic domain. A few others were also ordained by men, but most were done by the original woman and then her followers. They have held masses in the traditional manner, for the most part, with only that single exception to the rules. Some have gathered the dispossessed, sexually speaking, the middle of the roaders, the more open-minded. They have all been excommunicated. They were asked to be here as were all the others."

Mary was quite aware of the general proclivities of the Amazonian women and their male counterparts. She minutely examined the beauty next to her and decided that she could have been enshrined in marble, placed next to David (or Aphrodite) and fit right in. These women were not body-building, hormone-hugging extremes but natural forms, toned to perfection.

Louisa Jamala Reta approached the podium; cameras were rolling. Her extravagant smile was spread first to the right, then to the left, then firmly on the lead camera. Her elegant clothing, yet neutral in style, belied her life as a nun. Only those emotionally close to her noticed her trembling.

"On the thirty-first of October, 1517, Martin Luther nailed a document on the doors of the church at Wittenberg.

"WHOO-HOO!" vocalized a member of the Lutherans. Nervous titters and amused sallies followed.

"Yes, we celebrate that moment," Lu worked this impromptu addition into her speech. "We are here to follow in those very footsteps. We are here to create a new understanding of humanity about their God. We are here to join forces in forming a new religious order." A few cheers resounded but were quickly suppressed by Reta's calming hand motions.

"All are welcome. There are no gays, lesbians, or straights here. We are all human. We are all lovers." It was at this point that Illya and Mary approached each side of Luisa. "These are my friends, lovers, companions. I have been a nun, a liar. I forsook my best friend and lover for false hopes. The falsely controlled, male-dominated Catholic Church told me that this was correct. They were wrong.

"Jesus told us materialism was wrong. Love is everything. We are here to correct the current situation.

"Our first tenet is one of love. Our God is love. All love is of God. All expressions of love between humanity, regardless of orientation, comes from God.

"Second, honesty, truth, is required of all who profess to love our God."

"Compassion, largely spoken of, rarely practiced, is our third commandment." It was at that very moment she turned her head to look at another camera.

*

Only a word that was captured by the security's listening devices gave any warning. It was not enough. It was a radio signal containing one word, "Go."

The sniper, it was later determined, fired a single shot from nine hundred meters, striking Reta in the upper left portion of her forehead.

*

A spray of red mist covered Mary's face only a nanosecond before the report of the high-powered rifle reached her ears. She knew the sound of gunfire; she had done combat, or police activities as they were now called. She turned to see Reta's skull deformed; a gross indentation had removed her left eyebrow, and bone was protruding around the orb. In the next microsecond, Reta began her downward drift towards the speckled originally white, now red, wooden planking.

Illya's head was turning left as if she was in slower motion, not just slow. The impact of what her left peripheral vision was suggesting had not yet been verified by her right eye. Her brain was temporarily stymied.

All three were hurled to the ground by a tidal wave of flesh in the form of hard-muscled women and men of the Damoclesion Order. Strangely, Mary thought they appeared to move faster than the bullet.

Bullet! "Oh, Dear God! Illya!" Her first thoughts were of her lover's loss. The microphone managed to pick it up amongst the furor.

Illya managed nothing more than a single horrifying scream that did not change pitch and was drawn out for at least three minutes before she lost consciousness. Oblivion.

*

The doctor looked drawn, haggard. "She'll live. The bullet did not enter the brain itself. It crushed the skull over her left eyebrow, driving splinters into her eye socket, but they did not do more than a small amount of damage, though she will likely lose the use of her left eye. The shock of the impact drove her into a coma. We are not yet sure if she will come out of it. At this point, I can't say any more."

Mary and Illya held hands through the entire recitation. Both collapsed upon the bench, unsure of the news, partially alive, not dead. Illya was devastated, beyond resurrection. Mary gave her all she could.

The Damoclesions stalked the hallways of the hospital in fours, armed to the teeth. Four local police and six regional law enforcement persons had disappeared before the message had been well established. The Damoclesions blamed the regionals for the sniper. It had been their job to clear the jungle. It had not been done. No one interfered with their heavy-booted feet passing through the halls of sterile white.

*

A week later, a male body, or the sum of the portions thereof would have been a male, along with the weapon, were deposited upon the lawn of the hard-line bishop in town. The skin was hanging on a hanger; the testes had been hammered into a gelatinous goo. Every bone had been crushed as if in an immense vise. No one, especially the local gendarmes, thought that questioning the Amazonian or Alexandrian Damoclesions would be a good idea.

*

This is what the nine powerful men had intended—chaos. They had tired of the religious aspects of centralization. It had served its purpose. The people had been bombarded with the official line of a lone nut; he was a religious extremist and a homophobe. He had managed to circumvent security. The fact that he had state-of-the-art techno-communications gear was glossed over. The man was delusional; he fancied himself as the leader of a worldwide organization to protect the sanctity of marriage and genetic continuation.

These nine were gathered around, conversing on the space program's progress, ignoring the recent assassination attempt. One man sat at the head of the horseshoe-shaped table. Each regional head—the North American Region (NAR), Central American Region (CAR), South American Region (SAR), European Region (ER), Russian Associated Region (RAR), Asian Region (AR), African Associated Region (AAR), and Australian Oceanic Region (AOR)—gave reports on the economic results of the mining operations effects on their individual regions. The single man in charge of the world, currently referred to simply as the Leader, absorbed it with satisfaction. It seemed that all was well in hand.

*

All this transpired just in time for Red and Walter to go to SS-2 a few days later and then, after a time, on to Earth. They put all their personal mining equipment in the warehouse, under the supervision of Marceau. Walter felt the tension and ill will emanating from Marceau and wondered at its source.

Red was not oblivious, but he had accepted it as inevitable. He was unaware of the scene with the LB1C (Luna Base 1 Commander). He did not listen to the grapevine, and no one even thought to try to include him anymore after a few futile attempts. For Red, it was a waste of valuable time.

The commander had not given them a heads-up about Marceau figuring out their activities. He had advised them to take a low profile around her though. Everyone knew she wanted a Mars shot, and that secretly, she felt Red may have doomed any chance she might have had.

Red and Walter very carefully cleaned up all the dust that they had tracked in. They tagged and removed the batteries and put them in the recharger for use by any of the Luna-base vehicles. Marceau was not impressed. Everything was for common usage, and yet their buggy was not nor was their other equipment. It seemed to be contrary to all the goals of the UE. If no one owns, we all own; then there will be no theft, hunger, need. Extreme altruism was the philosophy behind the politics, the credo. It was the catchphrase of the bureaucracy.

When they were done, Red walked directly to Marceau and said, "I'm sorry that I may have caused you some problems. I have never intended to hurt your feelings, step on your toes in regards to your bailiwick here in the warehouse, or your chances for a Mars shot. I will do what I can to make amends."

Taken aback, Marceau blinked in surprise. Red had never ever apologized to anyone as far as she knew. She did not know what to say, so she didn't.

Walter, no dimwit, looked on with his eyes scrunched up as he witnessed this rather surprising turn of events. He knew that Red never meant to upset people per se, but he too had never seen Red apologize either. He looked first at Red, then Marceau, then back to Red. A somewhat evil-looking half smile appeared on the left side of his face. He had a rather amusing thought making its way through his devious mind. *Poor Red. He's in for hell on this one*. The only problem was just how much he would tease Red for being easy on her.

Both Red and Walter left during the long uncomfortable silence left in the aftermath of Red's words.

Marceau watched them leave (meaning Red) as she chewed on her lower lip. Deep in the recesses of her mind, subtle shifts in her thinking occurred unconsciously.

*

From a portal, she watched the launch bus head off for the pad. Frowning, she did not quite understand why she was watching them/him leave. Normally, she put her free time to better use. *Perhaps its good riddance I'm feeling*. She tried very hard to consciously convince herself that that was the truth, but subconsciously, her mind and heart were not falling for that rationalization. She backed away, reminding herself of his extreme arrogance at thinking he could make amends. Satisfied with her anger at him, she turned and stalked away from the portal.

*

"What ya think of all that moon business?" asked Mark as he slurped from his mug.

"What about it?" asked Paul.

"They got them big nuke-a-leer power stations agoing here on Earth. But they gonna run out of fuel here real quick if the moon don't up its mining." He tried to sound intelligent.

"They'll get it done. They are actually ahead of schedule in terms of production. The bottleneck was in shipping the gases. It got solved by some Marceau chick. They are very volatile. The hydrogen is very flammable," said Burt. "They are also under great pressure. A small leak or weakness in a canister could blow up a launcher."

"Where you get all this info?" Mark asked curiously.

"I pay attention to the news," Burt said slyly. He did not want to divulge his real source.

"You can't trust them to tell the truth all the time, ya know," Paul clued in the rest of the patrons who were sipping their various alcoholic fluids.

"True, so true," mumbled Mark. "They practice that propaganda stuff all the time. They all in the pocket of the government. They mix it all up, so you never know what really going on."

"Then why do you watch it?" asked Burt, somewhat amused.

"'Cause a smart mind can sift through the bullshit an' figure the truth pretty closelike," he replied, tapping at the side of his balding, wrinkled head.

"That right, Mark," slurred Paul. "I know what you mean. They haf to tell part of the truth. If'n you can cut away the stuff they change, little here little there, you can get the real story," he drawled.

Burt realized that drunk as they are, there is still some actual intelligence floating around in the boats of their beer-rivered brains. Burt took the mug that Mick passed him. The head was well foamed, a quarter inch of blond bubbles, just as he liked it.

"What about that homo nun getting shot?" Paul asked, wanting to show his astuteness.

Burt harrumphed, "That's a real tangled mess. It's Kennedy all over, so much bullshit we will never know what really happened. The shooter is dead. Everyone claims something else."

Mark came up with another series of proverbial truths. "The Catholic Church did not like her. The government certainly ain't happy that she be making a new power base, and the space people gotta be cheering her on for pulling apart the enemy."

Burt was a little surprised, again. "I seem to continually underestimate the quality of thought going on here." He smiled. "You're quite right, hit the nail on the head, Mark."

Mark beamed at Burt's approval. "Well, we all get it right sooner or later."

"True, but you put it so succinctly and correctly. Power is the name of the game, and the struggle is coming to epic proportions."

"Huh?" Paul wondered.

"It means that the shit has hit the fan in large quantities," Mick chirped from behind the bar.

"Oh," Paul said wistfully.

Burt declared, "A round for the crew!"

Mick's skill was such that he served all beers exactly the same. He had been a bartender all his life. He started here in this very bar at the age of twenty-three. He had lived in the hotel above, a part of the old red brick building with the bar on the bottom floor.

After years of saving money while working for the previous owner, the owner died. His wife did not want to deal with the business. Knowing Mick, she agreed to sell the whole thing to him at a very good price. It had taken twenty years, but Mick had finally paid off the mortgage. Now there was the probability that it would be taken from him in a legal maneuver by the new UE.

If no one owned anything, there could be no theft. There would be no greed. The catchphrase stuck in his throat. It seemed that the government was the only one left allowed to have greed and commit theft. The very poor, those literally without anything, might be guaranteed food, a new home, land, but he might be denied it. It all depended on a judge's word at this point.

He didn't really understand all the machinations. He had believed, like so many others, that the plan to strip land and wealth was only about the really rich. It seemed that it was too late now to do anything to reverse the trend. Everyone was being stripped. He was to appear in the Regional Economic Court in two days for owning a bar that grossed no more than 180,000 world bills a year. The so-called middle class was being attacked now. All that remained would be the peons, the last gasp of a dying bureaucracy.

*

The judge was one of a few hardliners left. Most of the political structures had been kept intact. It made the bloodless coup so much easier to swallow for the average person. So many of them had government positions; it was best to keep them, pacify them. The worst of the resistors to the new rule would be weeded out in a slow but sure sweep of the lower levels of the bureaucracy. The judge was a crafty one though. He pushed the legal definitions as far as he could towards those who had shown a proclivity to raise themselves up. He researched all his cases with the help of an old comrade, a private detective and computer hacker who could only be contacted now by v-link.

The old man, the comrade, had no job any longer, not really. He just had a lot of brains and contacts. He could find out the exact size of your penis or just how many moles next to your vagina in less than five days. The information he found about Mick was very standard. The man was a hard worker, stable, given his working life to obtain the bar, and treated his patrons well. The judge did not have to hear the arguments to know how he was going to rule. Little did anyone know this same man (the comrade, that is) worked for Mary Smith-Jones, the newsperson.

The paralegals went through the procedures anyway though. No lawyers were used anymore for this sort of thing. At the end, he announced that the current owner had shown proper respect to human need, one of the new altruistic catchphrases, with his allowing people to stay in his hotel with insufficient funding. The detective had been able to find several cases in which Mick had allowed stranded families, many on vacation with transportation down, to stay for low cost, even free on occasion.

Mick's own decisions—his personal right to charge whatever he wanted, even give it away free, a capitalistic decision—was what saved him in this court. Altruism by choice, not altruism by law, was what it came down to.

*

In a certain odd way, the overthrow of national laws led to the freewheeling applications of social justice imparted. A large amount of cronyism and nepotism worked its way into the system as well, some of the very things world government claimed to erase. It would take many years before politically correct judges could be placed in sufficient numbers to apply the laws evenly.

This was accomplished mainly by moving judges from regions of birth to far corners of the globe. In other cases, judges were indicted and sentenced to hard labor for any bias, real or perceived. This gave those on the fence something to consider. Join the United Earth Political Party and tow the line or join the work camps.

The work camps, as yet, were not too severe. Eight hours of relative physical labor were required, where relative was based on the physical ability of those involved. Several workers, once government employees of some renown, found themselves picking up trash or sorting recyclables. Many, at first, considered this apropos; pendulums would swing with time.

*

Mick went back to the bar after the welcome decision to give everyone a few free mugs.

There was a great deal of celebrating when Mick returned. Congratulations abounded. The tip jar overflowed rather quickly as everyone wished to demonstrate their appreciation for the many years that Mick had served them. All received a freebie at some point or other. Mick's friendliness and bonhomie swayed all in his favor as it had over the years. For once, Mick was not behind the bar. He sat at one of the prodigious booths near the back and was served, one by one, by each of the regulars.

By the time the bar closed, Mick sat like a heap of jelly. Giggles and continual thanks flowed from his white-toothed mouth. Six men were required to haul his large behind up the two flights of stairs to his two bedroom place in the hotel above the bar. They poured him into bed after removing his shoes. They all were tipsy enough that it took more than the usual effort and cursing.

*

Chapter XI:

Birth Ship

11

001011

"This is World Wide Network News, Mary Smith-Jones reporting. There is a developing problem on SS-2, Space Station 2. The second station constructed by worldwide endeavors has suffered a major setback according to just released information from the World Space Agency, the WSA," she read clinically from the notes on her desk in front of her.

"This station follows a long line of stations that have pushed back the frontiers of space over the years. It is the home of the Hsas, that very controversial genetic experiment on the human genome. It is now the center of allegations of disaster.

"We have had limited reports that a great deal of destruction has occurred there. We do not yet have an official statement yet as to cause or depth of destruction, but it seems that the station is in danger of complete collapse. There are rumors of a collision or asteroid impact of some kind.

"There is no official information on injuries or deaths yet. There have been scattered reports from amateur radio operators and astronomers that there are at least twenty deaths so far due to a rapid decompression problem in the inner core. The amateur astronomers have placed a few scattered images on the net. They show release of gases and metallic debris around the station. Many of these images have since been removed by order of the WSA.

"Speculation is that all the Hsas are in danger. This would be essentially a death knell for the bioengineering program. The entire space program may hang in the balance.

"At least two different radical terrorist antiglobalist freedomist organizations have already claimed responsibility for the incident. Those claims have been rejected immediately by the WSA and the UE government. They announced that no one has launched anything. Therefore, no group could have done this.

"We are going to continue to bring reports throughout the day as facts become available." Mary Smith-Jones's calm exterior belied her inner turmoil. She was a good reporter.

She knew that hackers were placing and replacing photos and vids as fast as the UE Security Services could remove them. Technically, it was illegal to view any site not approved, meaning licensed. One could still view anything on the net so long as taxes were paid.

Hackers simply moved every hour so that their wireless systems could not be triangulated and confiscated. Fines and jail terms might have something to do with it as well.

*

The man leaned back comfortably, eyeing the high-ranking UE security man.
"Can you do it?" the man asked.
"Of course. The question is how soon and how well. Your people are not keeping up. The hackers have found a way to mark the official efforts." He smiled.
"We want it done now and well," he noted harshly.
"Let me lay it out for you. A worm will do it very quietly, nearly untraceable. A snake will do it much more quickly, but it is more noticeable. What do you want?"
"Snake! This has to be completed within the day," the suit said.
"It will not happen that quickly. If someone has saved the documents but does not get online for a day or two, my program cannot reach the given data."
"That is not satisfactory."
The man settled in his chair. "It's the best you can expect. Do you want it done?"
The frown deepened beneath the cold reptilian eyes of the suit. "Yes."
"One million cash."
"What? You're a thief!" shouted the man.
"I'm the best. You also need deniability! Yes or no?" He smiled from his chair.

The suit turned to leave as he said, "Yes."

The computer geek turned back to his machines as he said, "Have it in my account within the hour."

When the man was gone, he began to let his fingers fly over the touch screen laptop. He had been quite ready for this, and it only took a few moments to initiate his series of search-and-destroy programs. He then shut off the internal and external cameras to the building that he was in and those around him.

Upon completion, he quickly removed the fake beard but left the mustache. Having worked in Hollywood as a makeup artist, he knew how to make himself into another person altogether. DNA, he could not change, but he had not donated to the voluntary identification program and had never been arrested.

He climbed down the fire escape of the decrepit old building. He met the gang members at the bottom and gave them one thousand dollars. Nothing was said; it was all prearranged. As he exited the alley, fires quickly sprang up in the decaying old tenement building.

*

Red had deep lines around his eyes, wrinkles on his forehead, and fingers that were less dexterous. He was getting old.

He did not like being old, almost twice as old as most of the people anywhere in space. Even with the organ enhancements, the tissue regrowths, his bone structure was suffering. The continual low gravity environment was to blame. All the special vitamins, all the exercise, none could make up for the natural environment of Earth's gravity.

He was more than willing to have made that sacrifice. He had achieved more already in his life than all but 1 (less most likely) percent of peoplekind. He, Gail Marceau, and Walter were the first three people to walk on both Luna and Mars. The pair of Red and Walter was the only pair to be at both sides of Luna to top it off. Red was the only human to have been on every station as well as each different type of noncargo launch vehicle. It was a set of records unlikely to be beat. Those records meant nothing to him.

They were the type of records that hero worshipers examined. The real records, the real contributions, were in the day-to-day solutions of problems—small or large. This is where the real engineers and scientists found Red and Walter as their semideities, their holy grail. It's why they came to them in time of trouble. They always solved the problem. They could be counted on.

Neither Red nor Walter wanted this particular problem dropped in their lap. Over the years, they had discussed the particular issues that were coming to a head. Their talks produced a great many scenarios, one of which always came out to be true.

It was 2061, and he was fifty-six. He was back on Luna again. He was staring at Halley's Comet through the portal. It had returned like clockwork. Red liked clockwork, precision. He looked at the delicate, nearly transparent gossamer tail with its hints of blue. The brighter coma, at the head, sparkled like alpine snow. He sighed.

Trouble had been brewing for some time now, he mused. How soon? His mind regurgitated that over and over. It was coming. How? What would finally set it off? How would it play out?

As his mind reviewed the multitude of causes and the ripples of effects, he came back to his first visit to the birth ship (SS-2) orbiting Luna.

*

Red (now thirty) and Walter (thirty-two) left Luna in 2035 to visit what had become the first birth ship, SS-2. They were going to go through rehabilitation before visiting Earth for a short time. They would be back at Luna in 2037, on the Farside project.

During their rehab, they had plenty of time to discuss the problems that they felt needed to be addressed or corrected before Farside became fully established. They covered a multitude of technical details, pored over diagrams and blueprints, meaning they ate, slept, and defecated Farside.

They also had time to personally hash through the concept of "growing" people. Neither one had really given much thought to it prior to being assigned there. It simply did not impinge upon their conscious thought. Their private debates on the issue covered all angles.

Homo sapiens anthrocreatus was the designated classification. *Anthrocreatus*, meaning created by man, had caused quite a stir in religious circles. It had even caused a few eyebrows to rise in secular arenas. Though the DNA was screened and somewhat manipulated, they were still human beings.

Red and Walter did not see any distinction. The Hsas ate, slept, and shit. That covered the human race quite well. It was a solution to the problems of existence in space with its low-g environment. That was even better. For the terrible twosome, as some referred to them, results mattered a great deal though methodology did count as well. In some matters, ends did not justify the means.

Sperm and ova from qualified donors, those with no inherent genetic problems and with high intellect, were gathered on Earth. Many opposing politicians and religious figures argued that there was a superrace neo-Nazilike agenda simply because they selected genetic material from highly educated donors. They argued that there was no way to determine whether or not intelligence was passed on genetically. Even some scientists felt that a broader spectrum of donors would be beneficial.

The riposte was that the more highly educated usually had better records of the family tree's medical histories, which allowed for better prescreening of donors. Another counter consisted of the fact that all so-called races and the various regions were relatively equally drawn upon for genetic material. That material would be crossed without concern for racial or regional purity.

Counter claims that were issued noted that the resulting issue of Hsas would, in fact, be a superrace because of all the bred in superiorities from each race since the genetic materials were screened for predisposition to cancers, diseases, ionizing radiation, and addictive behavior. The processes also looked for the healthiest physiological development in terms of muscle mass, heart muscle, and bone density. Some looked askance at the development as that of a mutt breed of humanity.

Once all these tests were completed, the two halves of a fetus, the sperm and ova, were shipped to the birth ship. There they would be fused in test tubes to form the zygotes, the primal cells that form a living being. Such precision practiced in the labs allowed for one sperm to be matched with one ovum.

The zygotes would then be placed in artificial wombs. These uteri bring the fetuses to culmination. Rote learning tapes including multiple languages, mathematics, history, music, and station protocols were played during the final trimester.

Even the birth process was mimicked with artificial muscle contractions that stimulate the babies' bodies to perform outside the dark, enclosed environment. The first generation had been born in 2030. One thousand babies had entered the realm of the living within two weeks of each other. The one hundred nurse practitioners had struggled to take care of the children under their care. Ten of the children were under the assigned care of each *sapien* specialist.

In 2033, another thousand entered the world. By 2035, another thousand were in the artificial wombs. Any interruption, even a minor glitch in the process, could prove to be a mortal blow to the fetuses and the program.

As was normal, Red and Walter had been asked to solve a few problems while they were on the station. Red and Walter were always kept quite busy, even when they were off cycle.

Red and Walter entered one of the learning facilities for the five-year-olds. The room was on the first ring, the northern one. Red and Walter were tracing electrical problems throughout the ship. It was becoming more apparent that the source of the surges of current must be coming from the power source in the core. However, they had to repair damage done and install extra surge protection until they could get into the core.

Several children began to enter the room before Red and Walter finished. One, brasher than the others, ambled right up to the engineers.

"Who are you two and what do you think you are doing?" he said.

Red smiled at the boy's territoriality. "I'm Red O'Hare, and that"—pointing at Walter—"is Walter Black Bear."

"Yes, the asshole engineer from Luna and his compatriot," stated the five-year-old boy. "And?"

"And we are attending to electrical problems around SS-2," answered Walter with a half-hidden smile at the description of Red.

"Then you are very poor engineers. The problem is in the core. I have tried to explain this to our maestro, but he has a limited capability of understanding. They think we don't understand much."

"We quickly reached the same conclusion, about the problem, that is," said Red with a devious smile. "We are waiting for some parts before we can repair that exact problem. We have identified the L-5 buss as the problem," he noted curiously.

"Perhaps you are not quite as poor engineers as those around here," said the very white-skinned and hairless boy. "I have to agree with that assessment."

Red asked with a strange look on his face, "How did you come to that conclusion?"

The boy gave Red an almost pathetic look as he said, "I examined the blueprints on the computer for the power output systems for the station. It was the obvious conclusion."

Both Red and Walter looked at each other with a horde of questions implicit in their gazes.

Red carefully phrased his next question, "How much science and mathematics have you been instructed in? We have not been told how far your education has reached."

The blue-eyed boy answered intensely, "They have only just begun to *teach* us algebra." The stress on *teach* was very noticeable. "As for science, they keep us moving at a snail's pace. We are beginning to get frustrated with what they perceive to be our limitations, or more likely the level of control they wish to wield." He swept his arm back to indicate his similarly hairless companions.

Walter had a look of undisguised joy as he said, "You are all far beyond the level of what Earth children are learning. Aren't you happy with that?"

The boy lifted what should have an eyebrow in an almost Vulcanlike manner. "I have never had associations with Earth children, other than an occasional vid-link for their school programs. We are told not to show any superior ability."

"Fools," muttered Red harshly. "How aware are your instructors of your capabilities?"

The boy looked darkly at Red and then swept his gaze to the growing number of children around him. They all nodded at him. "They are not," he stated coldly.

"Why?" asked Walter with a great deal of caution in his voice.

"Many are afraid of what we may be capable of. They have a desire to prevent us from learning too quickly. It is part of the protocol for our instruction. We have hacked the system and seen the guidelines. They don't want the religious organizations of Earth to be disturbed by what they term *supercapabilities*. The WSA is a bit concerned that we may be more than they bargained for as well. We have decided to keep them in the dark. They know that we have a greater capability but have attempted to control it."

"Beelzebub's balls," laughed Red, his chortling startled the children.

"You find it funny that we have, technically, broken the rules," stated the boy calmly but with a certain question in his eyes.

"Damn straight!" laughed Red. "I do it all the time." Walter was chuckling soundlessly as well.

The tension in the room dropped dramatically as the boy said, "We know. We voted to take a chance on you in this matter. It was 93 percent in your favor."

"Us?" asked Red. "Why?"

"You don't play by the rules." The boy looked intently at the both of them and continued, "You both break them whenever it is convenient, but, not for reasons that aren't pragmatic. You break through the red tape. You find a way to force them to see the right of your case. You two have sufficient pull that you might see that things change for us."

At this point, Red and Walter sat upon the floor and crossed their legs to be at the same level as this remarkable boy. The children smiled at this gesture; they understood it to be recognition of equality. They too took seats around the engineers, a half-circle of flowering, shiny-headed youths of a rainbow of colors.

At this point, the instructor for the class entered the room. "Children, it is time for class to begin. Please take your seats at your designated consoles." He had not noticed Red and Walter yet.

Red simply looked up at the maestro and said, "Leave."

"What?" said the startled man. "You shouldn't be here!"

"Unless you want to get your ass kicked, you should get out of here," said Walter, rising to his imposing two-meter height.

Flustered, the man said, "I'm calling security." He did leave though. Walter calmly walked to the entrance and entered a few of his secret codes to lock the room. The children giggled at the discomfiture of their maestro. They were children after all. Walter returned to his previous position.

Red stated calmly, "What can we do for you?"

"We want to be allowed to develop as fast as we are capable of. Individually, we are not all on the same page, or in the same book. Some of us are not cut out for the sciences or math, just like children on Earth. We want to have the chance to develop our artistic sides as well. The WSA keeps saying that we are not really different than ordinary *Homo sapiens*, but we are treated differently. We feel that, because of our advanced education, that we should soon be incorporated into station activities at neophyte levels. They say we are too young to take such responsibilities."

"I am amazed at your loquaciousness. You speak more eloquently than most everyone I know," said Walter.

"Our learning began in uteri with learning tapes. We learned the seven major languages, rote mathematics, scientific methods, and more. We are currently at the level of the most advanced high schools on Earth. Some of us have far surpassed that level," the boy responded carefully. "We study independently, hacking the computers, in hopes that we can make our talents known, and, therefore, available for use."

"Wasted talent," said Red quietly. Red hated waste.

"Ball buster of a predicament," said Walter shortly.

"They will be running the station in a few years," Red mused.

"That's the plan," stated Walter with a twinkle in his dark eyes.

"So what we need is an early introduction program," noted Red.

"Yes," said the boy hopefully.

"We also need to open up the curriculum," Red noted professionally.

"Yes," the children responded in unison.

"I also think we need to have greater interaction between Hsas and Earth children."

"What!" said the boy, somewhat startled. Looks of surprise grew on the faces of all the kids.

"You will be working with these children when they grow up," Red explained. "By developing a bond now, allowing them to become familiar with your abilities, you will avoid problems later on. They could develop an instant case of jealousy when they become aware of how intelligent you are and your superior schooling. By interacting now, you may spur them on, in a competitive way, to learn more. And by interacting now, you will adapt them to your abilities. There will be far fewer future problems."

"Those idiot policy makers don't understand simple psychological group dynamics," Walter added darkly.

"And what happens if they do not give us an advanced curriculum?" asked the boy leader, returning to the previous topic.

"Walter and I will come up with one for you. We have friends in low places," said Red with a guttural chuckle.

The children all smiled at the pun, knowing it meant Earth as well as others that bucked the system.

Security was banging on the door now. But no one cared. The boy spit on his hand, in the ancient manner of manly agreement, and extended it to Red. Red did the same and shook the precocious child's hand.

"Done deal," said the boy.

"What's your name?" asked Walter curiously.

"David DeNoe, son of the DeNoe that went back to Luna in 2015," he proclaimed proudly.

"They let you know who your donors are?" asked Red amazedly.

"No," smiled David as he said, "we hacked that information." He winked at Walter.

Walter grinned rapaciously and said, "Good boy! No"—he slowed in thought—"not a boy anymore. You are a young man now." Walter also shook his hand firmly in the spit ritual.

Red, almost ready to leave, turned back to DeNoe and asked quietly, "By the way, where did you get the term *asshole* as it applies to me?"

"Everyone calls you that," David said, grinning even wider as the other children snickered. "It's what made us decide to approach you."

"Very apropos," said Walter. "He really is."

Everyone laughed, Red harder than all the rest. It was true, and he enjoyed the young ones not being afraid to lay it on the line. Great things were in store for these people.

Red and Walter picked up their tools as the children took their seats. Red trusted them to obfuscate the teacher and security. As they opened the door, security, at first, prepared to subdue them and then retreated in some confusion when they recognized Red and Walter. They knew that these two had been granted access to the entire station. Their own hesitation and Red and Walter's calmness as they pushed their way past evaporated the security men's concerns. They weren't too sure how to write up the action report though.

*

The maestro was still somewhat flustered by the previous day's events. He had also, to his dismay, noticed a change in the children, subtle though it was. He noticed it when they filed in; they had straighter backs, a look in the eye that foretold changes. He felt an inner discomfort, a sort of wishy-washy weakness in his bowels.

The man had been preparing for the weekly student exchange, the v-link communications program. Today's Earth group would be a large selection of children across the North American Region, the NAR.

When all was ready, he turned the computer link to live and announced, "I am Maestro Blumen." His voice carried a certain vein of self-importance that many acquire when they believe themselves to be of greater importance in the scheme of things than they really are. He followed rather pompously with, "Today's topic will be the role of science in society. We have some preselected questions to begin with.

"First, does science have a responsibility to guide society, or must society guide science?" Blumen asked the horde of Earth students on the one-hundred-squared, multiscreened wall panel vid.

Hundreds of little hands blocked the view of any faces on all of the screens. The Hsa children did not move; they had been carefully schooled to allow the Earth kids to ask and answer the majority of the questions.

Again, a preselected teacher called upon a preselected student who was expected to answer as coached. Children often surprise adults. This child knew what everyone really wanted to know.

"How come you don't have any hair?" the young boy smilingly asked, to the horror of his teacher and the maestro. Everyone received instructions to avoid mentioning differences. The question was followed by microseconds of silence before the earthbound teacher could respond.

"Jimmy! That is a rude question. You need to apologize," she said with a certain discomfort obvious in her voice. Many children all over the NAR were trying not to giggle too loudly.

"That is quite all right," David DeNoe responded quickly. He was going to implement the plans of Red and Walter. He overrode the maestro who had begun to stutter a reply, "We don't have hair because of the problems it can cause. We have been genetically altered to have no hair anywhere on our bodies."

"David, that is not the question I asked." The maestro tried to slide the conversation back to the original topic.

"No, it's not," DeNoe said steely eyed. "But it is the question they want answered." David continued over the quiet titters of his compatriots when they saw the surprised look of the maestro when David challenged the man. It was a small rebellion that would lead to greater matters.

At this point, Klar, the computer genius of the Hsa, gently fingered a blinking light on her console. She had noticed the maestro moving to disconnect the link. She was well prepared for this little attempt to cut them off. The man was not brilliant.

Blumen tapped his console, expecting the screens to go blank. His bowels went even more watery than usual when he did not get the expected result. Quickly, he tapped the screen again to no avail. He frantically began pounding the screen with his index finger, consternation filling his face.

DeNoe winked at Klar and turned back to the vidcam, now focused on him with the previous push of Klar's finger.

"Hair requires trimming," he instructed, "and the clippings can get into the computers, everything. It can wreak havoc. That is why we have no hair on our heads, eyebrows, anywhere on our bodies."

"What about when you hit puberty?" an anonymous older Earth student shouted.

Teachers everywhere were scrambling, trying unsuccessfully to break the connection. This was not going at all as planned.

David and all the others with him laughed exuberantly. "We will go through puberty over the next year or so, but, we will not grow any armpit, facial, or genital hair."

The Earth kids laughed as well, many of the younger ones giggling at the word *genital*. This was fine stuff, they thought.

Another kid, sensing an unexpected chance, said, "But you are all only five years old. How can you be going through puberty so soon? We don't until we are around twelve or older."

David answered again, "We have been genetically altered to age to the point of puberty sooner, relatively speaking."

Another shout echoed quickly from another Earther, "Why don't you have fingernails?"

David surreptitiously fingered a key that notified Rarte to answer since Red had suggested that they slowly integrate as many of the Hsas into this new format of dialogue.

The exquisitely charcoal-skinned Rarte rose with incredible grace and glided closer to the vidcam. She slowly exposed her hand to the camera; she then closed all but the index finger into a fist. She rotated the nailless extended digit so that everyone on Earth was able to get a good view.

Calmly, she stated, "We have no fingernails or toenails for the same reason that we have no hair. The clippings would have to be collected and dealt with. Any loose ones could get into places they are not supposed to." She smiled and returned to her seat.

"How do you pick your nose?" one of the youngest Earthers shouted over the din of several other shouts.

There was a great deal of wholehearted laughter at that question.

David answered, "Boogers are just as sticky here as there. They tend to come out with the finger. We just can't scrape too much."

That response evoked gales of laughter from all the Earthers. David cued Klar with another minute gesture on the console in front of him.

The very cute young lady smiled as she said, "We think very much of your hair and nails as you do our lack of them. To us, it is very strange. You have always had them while we have never had them. We often wonder what it would be like to have them."

The conversation took a severe twist at that point as one child loudly asked, "Why don't you have souls?"

There was more than the few seconds of normal communication delay. There was a great deal of absolutely stunned silence. Both sides of these conversations had been strictly warned against this particular topic.

"Why do you assume we have no souls?" David asked curiously.

The only one to answer was the same girl, who stated simply, "My daddy says so."

The simplistic answer covered a whole host of emotional baggage from the religious organizations.

Rarte spoke up gently, "We have hearts, minds, organs, eyes, all just like you. We are no different. Our histo-religious instruction covers the same material as yours. We are made of matter, dirt as Genesis refers to it, just like you."

The girl replied at an almost instinctive level, "You don't have hair or fingernails. Maybe you don't have souls. God created us. You were made by people." Her innocent blue eyes showed no sign of malice.

At this point, several instructors had the wonderful idea of simply removing the power supplies to their equipment, shutting down the conversation link.

The maestro stood, nearly gibbering, separated power cords in hand as he envisioned the end of his career.

*

The meeting is not going well, Leonis thought as he massaged his temples. This headache began many years ago.

The Breakaways and Stalwarts—also known as the Modernists and Fundamentalists, Allegorists and Literalists, Traitors and Loyalists—were at each other's throats, mostly figuratively though some physical battles occurred.

Leonis tuned back in to the conversation, a frenetic diatribe by one of the Fundamentalists, to hear, "They only study religion. They don't have religion. There is a significant distinction."

"How can we, as mortal beings, make the decision that they can't come to know God!" one of the Breakaways shouted. "God will take care of whether or not they have souls."

A different Loyalist wrathfully defended the hard line with vituperative fire, "Satan knows God, but he can never be redeemed."

Yet a different voice came from the Loyalist gallery, "Life receives the soul at conception, not at some later date just because of a change of heart."

"Life *is* the soul," a Modernist said somewhat calmly amidst all the wrangling. The looks he received from the Stalwarts were not very friendly; his own teammates, so to speak, were not overly impressed either. Very few subscribed to that concept. By far, the majority followed the separate existence of body and soul theory.

Leonis, as moderator, decided that enough had been said on this unsettleable portion of the debate. He banged his oak block on the table.

"Ladies and gentlemen, we can argue such forever without agreement. The true question is do we try to convert the Hsas?" he queried with a searching look around the room. The question was a moot one as far as the Catholic Church was concerned. They would never accept Hsas. The matter was simply one of the agreed-upon topics for this meeting, part of the agenda.

"Of course!" one Breakaway shouted.

"Never!" hooted several Fundamentalists.

A few crowed, "Blasphemy!"

Luisa Jamala Reta, the mahogany-skinned pope of the Modern Catholic Church, politely asked Leonis, "May I have the floor for a few moments?"

"Yes," said Leonis, without adding the usual honorific. To do so might be recognition of her status, which would be counter to the religo-political stance of the true church, nor did he stare at the scar left by the assassination attempt.

That lack of respect was not lost on Pope Reta. She mentally brushed it aside. She was too well trained in the art of diplomacy to fall for that lowbrow tactic. There had been a couple instances in history in which the church had major divisions—the Byzantine incidents, the French popes. She was just not going to fall for the bait, nor did she care that most of those in the room silently sniggered at her once great beauty. Several were somewhat unhappy that the assassination attempt had failed.

She began with, "Scripture clearly indicates that there will be heavenly help in cleansing the Earth. However, the definition of heaven in ancient Hebrew refers to three different places. There is the heaven where God exists, the spiritual plane. There is the heaven where the birds fly and beyond, the space we argue about. Then there is the heaven that refers to the air about us, that which is within breathing space. Are we not now receiving help from the heaven that is space?"

"God means angelic help, spiritual beings, not the assistance of the execrations of the labors of men!" shouted one priest. He was silenced by a hard look from Leonis, the only peacekeeper and not by choice. His appointment came from the pope with a directive to return as many of the sheep to the fold as possible. The rest would suffer the pope's and God's wrath.

There was a general rustling of discomfort in the room. There was also a feeling of expectation as to what might come next.

"Scripture also speaks of the fact that the Gospel of inclusive love will be preached to all nations. Would this not include any place, every place?" she added to her logic base. Dissension at her choice of words charged the room. All knew what she meant by "inclusive love" and were ready to counter this description of the scripture.

She followed with a sore topic of the generations, "What do we do with those that never heard the Word? Are they condemned to hell?"

"Old ground," mumbled someone in the rear.

"The Bible speaks of a resurrection of both the righteous and the unrighteous. There is also talk of more than one flock, one large and one small. Who are we to judge how God defines his flocks?"

"God created man to exist in the Garden of Eden," retorted one of the hardliners.

The female pope retorted quickly, "Isn't the garden all of Earth? We were supposed to spread it over the entire place. The word *Earth* can literally mean substance, the stuff of which we are made. Would that not mean, by extension, the entire universe? We got distracted along the way, but we are returning to that goal now."

One Fundamentalist growled, "Returning! We are getting ever farther as sin is accepted as the norm. Abominations in the form of homosexuality, multiple marriages, open relationships, sex is allowed for the clerical class of your organization, and failure to have a Sabbath day is but a few of the commands of the Bible that you would throw away."

The woman retorted hotly, "What about compassion, forgiveness, and acceptance? Without love and fellowship, there will be no place for these people that you condemn."

One Loyalist shouted, "There is a place for them. It's called hell. The pope has excommunicated them and *you*. That is *final*."

One beet-faced man growled, "You think we should accept perverts so that we must then accept perversions of man. It is simply another slippery slope argument!"

Another rejoined, "Jesus forgave. He also said, 'Go and sin no more.' You seek the forgiveness, yet you would continue to sin. This is incompatible with the Word."

"Why is this excommunicated heretic whore even allowed to speak here?" one red-faced bishop finally demanded, disgusted with her rhetoric. "Are we to talk with the whore of Babylon now? Do we converse with Satan's minions? Why do we not stop our ears, block these evil words from entering our brains?"

Magma reached the surface; restraint gave way to pandemonium. All began to speak or shout in unison, as if choreographed in the art of chaos.

Leonis furiously banged away with his wooden block until it broke into futile pieces, much like the churches were presently doing. The splintering of the various religious sects essentially finalized with this vehement conclusion to the meeting. Band-Aids would never heal this wound. No amount of suturing could bring together the torn fabric of the wounded churches.

The only real reason the Stalwarts had tried was to prevent the profuse hemorrhaging of their congregations (and loss of tithes). The Modernists simply wanted love and openness to prevail. It was too hard to do something simple.

Yet after the meeting disbanded in disarray, some few leaders of a more practical bent met to discuss a middle ground. They wanted no part of either polarized group but wished to represent the rather lost and confused average person.

*

Aaron Reitner felt that his time had come. As titular head of the resurgent White Power Nations, he recognized that it was now or never. There had to be open action. Someone had to lead the pure race into the next millennia, or they would be swamped by the hordes of foreigners in the UE.

The old United States had been betrayed by a nigger, without the slightest fight. He had foreseen this. He had warned his followers about this and the tyranny of a one-world government. He had orated throughout the states about the consequences of electing black men to any office, much less the presidency. He had endured the boos, the thrown fecal matter, assaults by members of the various lower races. He never retaliated; he knew he had to keep the high ground until the time was ripe.

His followers had begun to plan for that very day. He had also made many arrangements with like-minded alternative organizations, large and small, old and new.

Clips were filled and slammed home into the automatic weapons. Hitler had his putsch, which failed. But he had come back to rule an empire. Only the Zionist conspiracy had brought about his downfall. Reitner saw the possibilities open wide before him. They would utterly surprise the UE. It was but a short time until the fulfillment of his dreams.

Three thousand of the Aryans and nine hundred of the Klan, all heavily armed, would attack the UE's world headquarters in New York City. Over five thousand European-based neo-Nazis would focus on the European Region's headquarters in Geneva. In one fell swoop, they would cut off the head of the nine-headed snake. The distraction would be magnificent.

Reitner nearly salivated at the thought of almost fifty thousand Aryan nations members marching from the Lincoln Memorial toward the NAR's governmental headquarters in Washington—the White House. Close to forty thousand Klan members would make their way from the Washington Memorial. They would converge in a well-timed pincher movement. The European group also planned massive demonstrations in several cities: Paris, Amsterdam, Munich, Berlin, Bern, and Vienna. Nearly two hundred thousand in total would march in angry protest. It would be a glorious beginning, much stronger than Hitler's first march. There would be no jail term for Reitner; he knew he would not fail.

The threat on the NAR's leader, the one time president, would draw most of the security forces in the northeast. It would take no more than two to three hours for them to move the majority of the available forces to within striking distance of the marchers. Speeches, not preannounced or government approved, would likely bring the UE forces into early play, but they would be confused by the multiple attacks.

He knew that it would be a massacre whether or not the militia operation was successful. He hoped that there would be blood everywhere. He prayed for pictures and clips on the Internet. His computer people would see to that. The march was designed to arouse the average person to outrage, to show the overwhelming arrogance of the powers that be. Exposing dictatorship in the form of benevolence was his goal. He needed to get the people to open their eyes or at least question the status quo.

Even if thousands were publicly executed, he would win. The mask would be torn. Exposure was all he needed.

If the two heavily armed groups succeeded, he would immediately announce a new government. If not, he would try again. There were many who were dissatisfied with the global government. He would use them all as cannon fodder. The goal was all.

*

The Modernists knew something was up. Reta's Death squads, those known only as Death to all but their lovers, had infiltrated the lower levels of all the world's dangerous organizations and had even reached midlevels in the worst. Their reports were filtered up to Illya, Mary, and Luisa.

Dr. Illya Marquez and Mary Smith-Jones used all their contacts to gather the slightest tidbit of information. Within months, they had an extremely efficient worldwide spy network.

As the two relaxed in their bed, Mary quietly asked, "How's Lu doing?"

"She's still afraid to be intimate. She is afraid of her deformity, that it will set me off her at an inopportune moment," she stated sadly.

"Oh . . . um . . . that's not quite what I meant. How is she doing with the information on the White Supremacists? Can we respond in some way, dilute their actions? We know that they are going to gather in Washington DC. Can we counter this to avoid the inevitable violence?"

"We know they are gathering, as you say, but what can we do? We won't fight them. Lu does not want violence, won't counter it with the same."

They planned instead a peaceful march to coincide with that of the Aryans. They would approach the NAR capitol building from nearly 180 degrees from the direction of the Aryans. This would prove to be a major error. The second was they had no idea of the pincher-like march of the Klan.

Their goal was to simply introduce a factor of support for the opposing road. They intended no increase of worry or consternation amongst the security forces, but that is exactly what they achieved. There were to be nearly fifty thousand marchers in their group. From the perspective of the security forces, there were nearly 150,000 hostile individuals approaching the NAR offices.

*

Reitner was not like Hitler; he did not have to be the center of attention. His well-taught aides would handle the more dangerous tasks of making speeches and assaults. He was a well-read man; he knew the faults of Hitler. Hitler had tried to be a micromanager. His interferences in the high command of the Wehrmacht brought disaster on more than one occasion.

Aaron knew the abilities of his close lieutenants; he laid out a general concept of what he needed, and they produced the desired effect every time. Their lower echelons had been selected for their competency and secrecy. It had become a well-oiled machine, one that followed orders explicitly yet allowed for flexibility when the need arose. That separated them from the Third Reich's fallibility, that of the need to always pass the buck to higher authority.

Hitler had taken a sleeping pill, and no one wanted to wake him on D-Day. Therefore, the panzers did not meet the invasion force on the beaches. Rommel had decreed that the only way to stop the invasion would be to stop the invasion on the beaches. The Reich fell soon after. Reitner would not follow that ridiculous path.

Von Buren began his speech at the Lincoln Memorial with great heat, "The foreigners have taken over our land! We are the subjects of a new feudal system where the hierarchy is called the UE!"

Cheers and catcalls echoed about the area. The people were already worked up; this was bringing them to a fever pitch. Fists pumped the air in anger. Many cursed the current system.

Von Buren continued, "We must march! We must make our dissatisfaction known! WE! We are the next generation of leaders! We have to take the ultimate step! It is time to begin!"

The recently arrived fifteen hundred troops, wearing the blue and green riot helmets, felt the upsurge of emotion. They had been detailed to control this last-minute meeting. Most had been shipped in by truck in the last five minutes. Live ammunition had been distributed for their weapons. This was definitely not a drill. Orders consisted of containing the hordes to the memorial area. Little more in the way of orders had been given to the colonel in charge. He had to decide on the spot what was appropriate.

The great orator, Von Buren, who had specialized in speech followed with, "We march to defend our creed, our faith system, our existence! They, the government, would deny us our basic rights of speech, freedom to act, the freedom to follow our conscience."

The crowd screamed in support. Wild calls for action abounded. Chants of "We will not succumb!" reverberated. The crowd was becoming willful.

Major Hamib reported to his colonel that the crowd was growing out of control. His recommendation was to fire gas into the crowd to break up the meeting. Hamib wished that he had rubber bullets to use, but none had been issued. The colonel found himself between the proverbial rock and hard place. He ordered gas and also ordered the second line to lock and load. The colonel then radioed for a third line, including tanks and armored carriers, to be established directly around the NAR offices, the White House and Capital Building.

The launches of gas enraged the crowd. They knew that they were to march to the White House; they began to run wildly in that direction. Discretion on the need to fire was given to lieutenants. It was based on imminent mortal danger to the troops and likelihood of the marchers penetrating the barriers.

Colonel Hart chewed his nails in absolute frustration. Just a few years past, there had been a local National Guard armory that he could have drawn on for greater firepower. He silently, since the suits were there, cursed the government.

*

Suits were observing all three meeting locations from many perspectives. They had many cameras in an attempt to identify all the possible members. There were also two suits in the colonel's headquarters. Communications showed that two of the meetings were decidedly intended to incite violence. One suit quietly and quickly contacted his superior to update him on the situation. As yet, they had not interfered with the colonel.

*

As the hordes approached, the first line of troops strung along the park side of Constitution Avenue; and from Fourteenth Street to Nineteenth, braced for impact. They had been rushed to the site less than half an hour ago and were armed mainly with shields, batons, and sidearms. Sergeants shouted through bullhorns, attempting to stop the crowd from approaching any closer. The warnings were not heard or were ignored.

Hundreds of choking participants initiated the melee. They jumped on, punched and kicked at, and cursed the soldiers. The flow of people from the Lincoln Memorial massed mainly between Nineteenth and Seventeenth and spilled in to the area of the Ellipse. The Klan members mainly marched directly toward the Ellipse but also sent a large group along both Madison and Jefferson Drive toward the Capital Building.

Batons swung rapidly. Blood began to flow. The press of people grew greater as the larger part of the crowd now entered the fray. Several side streets were now under attack. Molotov cocktails appeared in the hands of many of the Aryans and Klan members at each site.

The first throws were towards the flanks of the troops, namely the corner buildings of each street. Windows crashed as the bottles passed through, initiating uncontrollable fires since no fire crews could get in at this time.

Next, the weapons were directed at the rear of the defender's line to prevent any assistance from reaching the besieged UE troops. At this point, the soldiers began to pull their pistols, firing wildly at the massive flood of people.

Almost instantly, they were overwhelmed and disarmed. Many were simply executed on the spot. There was now open war in the streets.

Contingents of Supremacists dedicated to mass destruction entered the ground floors of nearby buildings to set more fires, block exits, and gain control of the side streets. Parking garages contained prearranged delivery vehicles containing heavy weapons and homemade bombs, which were now moving in per radioed orders. These trucks simply drove around obstacles and across the park areas, tearing up grass and shrubs.

Military snipers, on the rooftops between the first line and the second, fired as rapidly as they could, to little effect. There were far too many determined opponents. Several were quickly out of ammunition due to the rapidity of deployment; the rest were nearly so. It had been expected that a large show of force would cause most antagonists to back down. Not so.

As the rioters approached the second line of defense, strung along the middle of the Ellipse and toward the old Department of the Interior, they were heedless of any reactions by the troops until weapons fire began to mow them down. At that point, some scattered into the closest buildings to fire their stolen pistols or secreted arms. The rest gathered momentum as the soldiers reloaded and rushed their attackers. Hundreds fell in a matter of a few seconds, but their sheer numbers overwhelmed the containment. 1600 Pennsylvania Avenue was in grave danger.

Meanwhile, barely two hundred troops were holding the terrorists attacking the Capital Building at bay. The captain used the Capital Reflecting Pool to great advantage. It allowed her to concentrate her forces on the flanks exactly where the aggressors were advancing. She had also placed snipers on the steps of the Capital to use the height advantage. It was fortunate for her that this opposing group had not been supplied with any large automatic weapons such as a fifty-caliber.

*

The Modernists had begun the morning with prayers and calls for tolerance. All the speeches denied extremism. Peacefully they began to gather into a column for the move from Franklin Square down Fourteenth toward New York Avenue. The small military force of three hundred, designated to watch them, felt that the desperate call for troops had been blown all out of proportion. Some grumbled; others shrugged.

That all changed when a desperate order came to immediately move to the White House and fire upon all aggressors. The soldiers raced to their trucks, which then rumbled quickly down the few blocks to the very front doors of the building.

Tanks plowed their way through the remains of abandoned vehicles, firing close range shots on the Ellipse. Machine gunners atop the tanks provided covering fire for the few remaining second-line troops trying to escape. One tank commander looked up Fourteenth from Freedom Plaza and saw another mass of people moving toward the Ellipse.

The Modernist march had, in actuality, stopped. There were sharp reports from all of the weapons being fired nearby. Smoke from buildings and buses, the sources out of sight to this march, was quickly rising. This milling around in confusion brought disaster.

The tank commander ordered a charge of three tanks up Fourteenth and two up Fifteenth to wipe out this threat. The pincer move gave the people little place to go as they withered under fifty-caliber machine gun fire topped by main gun rounds flung amongst them.

Authority's use of force in order to keep itself in power must always be closely examined. This was a great bloody slaughter on the one hand, much worse than Ruby Ridge or Waco. On the other hand, it was a ferocious battle, an attempted revolution.

*

The attacks on the UE buildings in New York and Geneva failed, barely. The timing had been off by less than two minutes, but it gave forces at the buildings time to prepare. That minute amount of preparation meant that thirty guards were ready to meet the initial assault rather than five. It took but seconds for further reinforcements to appear. Thousands gave their lives in the immediate turmoil.

Reitner and thirty-nine others were quietly found and executed without trial or even a note in the news. It was not allowed to speak of this topic of rebellion or treason in any specific way in the media. It had all been a horrible accident brought on by an overzealous colonel since found guilty and executed aong with many of his subordinates.

The UE could not allow anyone to think that there would be those who did not want to be under their protective wing. It was explained away that there had been a language problem amongst the overzealous leaders, one now solved.

*

Mary Smith-Jones was livid. Her fine tan was now filled with a fiery blaze of red anger. She wanted to rip her editor apart.

She shrieked, "You are a frikin' moron! How can you let them dictate to us what we say? Have we lost all rights to free speech? Have we lost the right to question what *our* own government does? Have we forgotten what investigative journalism is?"

Julius rubbed his head slowly. "Calm down, Mary."

Again, she shouted, "Calm down! Why? You haven't got the balls to stand up to them, so what are you going to do to me?"

Julius looked up sharply. "Fire *you* before they fire me."

Mary's nose quivered in repressed fury. "You know that they are lying. This is all just a cover-up, and we have proof!"

Julius paled. "Don't say that," he rasped. "They will kill us just as dead as Aaron Reitner and just as quietly," he whispered.

Mary exhaled. "So we become accomplices through silence. We are killers of the innocent Modernists and guilty terrorists. If just one group of regular newspeople were to act, that would be the wedge that keeps the doors of censorship from closing us in the dark room of ignorance. And you say not to act. You simply submit, like a lap dog."

Growing angry now, Julius stormed, "There is nothing I can do. I either withhold, or I get fired . . . or shot. I have *family*!"

Mary shot back, "And what would they think of their father? Our forefathers faced a similar situation. They took the hard path."

"Get out," Julius snarled.

Mary walked out, shaking her head.

She went to her secure computer and copied some material onto a disk. She then telephoned a memorized number. When the man answered, she stated, "I have some work. I will send a courier. It's all explained in the data."

They both hung up immediately.

*

As is often the case, small things cause large problems. These small things were not ants; they were utterly mindless. They were just small rocks. But these small rocks (meteoroids), moving at fifty thousand kilometers an hour, can do extensive damage.

While Walter and Red worked on the power system for the station, the L-5 buss system, a small metallic rock, hurtled toward the core from a high angle, near the rotational axis. The entire system was suffering glitches that might affect the birthing of the next generation of Hsas. Any power surge could cause the birthing chambers to abort, causing the deaths of a thousand babies.

Red and Walter had to get used to working in the near null-g environment again, while the centrifugal forces of the rotating station caused a visual and balance problem. It had been some time since they had done so. Nonetheless, they were able to locate and solve the problems and were just now finishing the last corrections.

As Walter closed a panel door and secured it, a not so subtle shudder rippled through the entire station as the iron meteoroid pierced the end of the core, passing through the first twenty meters to then exit out of the side.

The single chunk of danger became a full air wing, a shotgun's bird shot splash of particles. Pieces of station and the remnants of the rock were headed straight for the rings and then through the solar panels, making up the outer cylinder.

Debris spun crazily, three dimensional pirouettes forming sunlight rays, around the entry points, causing even more damage to wiring and computer interfaces. Sparks flew. Circuit breakers began slamming like dominoes falling; one went after another. Power began to go out all over the ship as non-critical items were shut down automatically by the computer system.

At the exit wound, metal curled outward with similar dancing particles. At the instant of penetration and exit, gases and debris began to scream outward at both locations, acting much like sand in a sandstorm, grinding away at any surface. Had a person been able to view the process from outside the station, he or she would have seen a slight glow from the escaping gases and the occasional glint of light off the tumbling shards of metal and rock.

Red and Walter both felt the impact and the sudden drop in pressure. Their ears popped like bubble gum, worse than deep diving. Pressure doors dropped in the core in less than a second. They rushed for the rest of their emergency suits (gloves and helmets) hanging nearby on a wall; they could not have done the delicate work needed in them. Pressure doors throughout the rest of the station dropped suddenly into place in the following seconds. Red and yellow lights flashed, and emergency power systems began to power up as normal systems went through a series of hiccups, some remaining off.

It took nearly two minutes for them to get their helmets and gloves on. Practice, rigorous under Red's tutelage, was still insufficient compared to the real thing. The artificial decompression chambers for all their physical representations did not fully cover the total psychological impressions of impending death, when in practice, one knows that one of the observers could flip a switch and reverse the problem.

They struggled for every nonbreath during the procedure. They had fought the pressure of flowing air until the door between them and the leak closed. It only took a second, but their breath had been ripped out of their lungs. It was as if standing in a giant wind tunnel; every hair flowed to the point of being almost straight out. Had there not been equipment in the way, they would have been drawn to the holes and diced into quivering chunks by the razor-edged torn metal. Even though the door had closed, the pressure was dangerously low, and they faced the danger of blacking out.

Once fully suited up, they sucked air into their lungs like sprinters at the end of a race. Holding on to various protrusions, knees weak and shaking, they held on until they could focus and think somewhat properly. Flashing red and yellow lights and Klaxons continued to go off all over the ship. They were more distracting than helpful at this point. If you did not know something was wrong by now, you were an idiot and deserved to die (to receive the Darwin Award even).

Red pointed to the north door and asked in a weak, husky voice, "That way?"

"Yes," whispered Walter shakily.

"OK?" Red questioned.

"Can't see," said Walter. "Faceplate is fogged. You?"

"Mine is too. Air conditioners in suits will clear it soon." Red toggled his suit to the central command emergency radio channel. "Central, this is Red and Walter. We're in the core. We have had an explosive decompression north of our location. Possible section 2, more likely section 1."

North was an arbitrary designation. There was no real north or south, east or west, in space. In space, the designations were more accurately described in terms of pitch and yaw, as well as the central axis of length of any vehicle.

Since the station was oriented along the same axis as Earth and Luna, it had north and south designated in the same way. Since it rotated in the same manner as the two orbs, it also had designated east and west radiating spokes though it really made very little difference. If you were searching for a particular room, it would be assigned a letter (C-core, S-spoke, and R-ring); and an east or west notation, with a level number, one being closest to the north end.

"Are you two OK?" asked CC-communications (or CC-com).

"Yes," replied Red. "What do you have on your screens?"

"Cameras have captured at least two punctures at the extreme north end of the core, C-1. We are on backup power for the north forty meters of the core. Your north door opens onto the last closed system on that end of the core. There was an electrical current surge: birthing system barely held, main buss 5 held, and organic hydroponics held, atmospherics held. Many smaller failures have popped up, they continue to do so. You must have finished your work though for five to have held."

"Are the Hsas OK?" asked Walter, knowing they had just finished the repairs on that system. A significant power surge could cause all kinds of damage in delicate electronic systems.

"We don't know at this point. The docs are working furiously to determine if there was any potential damage. It will be some time before anything will be known for sure. How soon can you be here?"

"We are right here, so pump out any remaining air here, and we will take a look at the problem," said Red.

"Right, I can see now," added Walter.

The commander (C-CC also known as Triple C or Tri-C, even 3-C) entered the conversation at this point, "SEE! You guys need to be checked by the docs. That's an order!"

"Listen, you spindly assed boobless twat, delegate the assets you have on hand to do the job. We're here. We are the best you have. Use us. Get it?" Red did not really think poorly of the young lady, and she really did not have a spindly ass. In fact, she was well proportioned top and bottom. But Red knew that a little delay could prove deadly later. Both he and Walter actually ranked her, but as C-CC, she had authority over the entire station during her shift.

"Triple C, it was condensation from the pressure and temperature drop, not a physical problem," said Walter as he smiled at Red. He knew that Red was giving her a lesson in emergency tactics. Even if people were dying or going to die, you had to make use of them, to the very end. It was a simple fact. It was like combat; you may sacrifice one or two in order to save more. Many refer to it as pragmatism. For Red and Walter, neither of which subscribed to pragmatism since it is a loser philosophy; rather, they followed intelligent objectivism; it was the smart thing to do. They were the smartest, the best, and, all arrogance aside, just let them do their job.

She blushed but knew they were right. "Go," was all she said in response as she turned to continue receiving her updates, via two earpieces, from all over the station. She sifted quickly through the information passing the critical material to Earth, Luna, and SS-1.

"D-Com"—meaning decompress—"our chamber, C-3, and the one north of us, C-2," commanded Red to CC-Atmosphere or CC-A.

"D-Com under way. Give it thirty seconds, guys," said CC-A.

Red and Walter made their way to the door. It opened after a few seconds of waiting. As they entered, Walter said, "I'll pan right." Their helmet cams would pick up video feed for later examination to make sure they did not miss anything.

Red said, "We have a lot of minor flow debris damage."

"Same here," stated Walter as each examined different portions of the chamber.

"The next chamber, C-1, is space open," said CC-A as an added warning.

"Copy," said Red. "Open next door."

"Tie down, guys," stated CC-com nervously. "Triple C has ordered axial tilt correction. We are processing considerably in our orbit. Thrusters will fire in ten—"

"Yeah, we could feel things a little off rotation," said Walter as he and Red placed their tie lines to secure hooks on the nearside of the door. They held on as the station shifted its orbit and rotation slightly as retro-rockets fired to initiate the correction. Rockets fired again to further stabilize the correction. The process took a few minutes and came first. Large-scale problems took precedence, then small-scale.

"How do we look?" asked Red after the firing.

"We are back within normal parameters," replied CC-orbital mechanics or CC orb. "It was a good burn. She had pegged it just right."

Red threw his fist in the air; that little girl, Triple C, twenty-seven, was growing up the hard way. She had mustered her command and got each of them doing their job. He was proud of her for doing the job the right way. Walter looked at Red and smiled. Red had a way of getting people to perform at their best. He was a rough, salty bastard in many instances; his language was like sandpaper. In other cases, people, like the Hsas, had seen a softer side under the gruffness.

It, Triple C's performance as well as Red's speech, had an effect on everyone; they stepped up to the plate and performed at an even higher level. The F word (*focus*) became the word of the shift and those that followed. Eyes bounced from screen to screen; fingers flew over touch screens in rapid, accurate dances. Information flowed like the mighty Amazon. The congregation of minds worked the highest priority problems and shunted the lesser to secondary personnel.

The grapevine was in overdrive as people ran, shouting past each other in the corridors. Their leader had balls, kinda sort of for a female, and many people—namely Red and Walter, the well-known problem solvers—were putting their lives on the line to save them all. Work was done more quickly, more effectively. Everyone was at an emergency station to help solve the problem.

People were running, but it was no longer like those first few moments of adrenaline panic; it had purpose. Questions were asked and efficiently answered. Every command was followed up immediately.

Earth could only offer suggestions. They did not have enough information or a handle on what they did have to really know what was happening. Many of the E-controllers were totally ignored in the heat of the moment.

SS-1 had already launched a lifeboat, a five-person shuttle capable of Earth reentry, to survey the damage. *SS-1 Minnow*, named after some boat in a seventies TV show. It was covered with cameras and contained repair materials. It was only a few hours away. It had ten robotic arms spread all over the outer hull. Other places on the hull would allow for astronauts to hook up for oxygen and power replenishment.

CC-maint had already released ten of its twenty eyeball cameras or E-CAMs. They would all be out, but there were power difficulties in some of the maintenance and repair launchpads. Engineers raced to manually set these extraordinary little devices free.

The basketball-shaped and sized cameras could zoom about the station free of any tether. They had several miniscule thrusters filled with highly pressurized inert gases for propulsion. They really looked like a giant's eye. They were all white except for the dark camera right where an iris would be.

The E-CAMs could examine the station with close-ups not available with other, larger, ships. It could get within millimeters of most parts of the station. They would use everything from X-ray to radio to image and detect defects.

Red and Walter would get the first view though, from the inside.

"Open the door," said Red.

"We have a red light on the port," replied CC-A. "You will need to open manually."

Walter began to remove a panel. "Wonderful," he muttered disgustedly. "Let's hope it's not warped." He pulled a crank from inside and inserted it into the gears and began to turn the handle. The door slowly began to retract. Red peered inside as it opened.

"Son of Satan," Red whispered in one of his few conditions of awe. "This place is a real shithole. We have a gash several meters long that crosses over the threshold of the security seal to the next chamber. I think C-2, south of us, almost went. There is major debris damage everywhere I look. I suspect that we have structural damage that goes south into at least the next two chambers. The structural beams I can see are severely twisted and bent. That much contortion may have been distributed all along the I-beams."

"Complete rewiring will be necessary. Computer components are shot. Meteoroid entered in north to south trajectory. I advise a complete rebuild on four chambers, depending on what we find as we strip the outer shell," noted Walter.

"Concur," said Red. "When can we get an outside view?"

"*SS-1 Minnow* is on route and will arrive in a few hours. The eyeballs are already in action. We have reams of data piling up. We have seven major solar panels dead, and there are fourteen more operating at 50 percent or less," responded CC-communications.

"That's not good. What's the projected power loss?" asked Walter.

"With backup batteries in place, we are suffering a 7 percent negative. We are curbing power in lighting and heat to try and compensate. We are advocating the use of flashlights for personal room usage," said CC-electrical.

"Each of the emergency lighting units has three directional bulbs. Send a person around to remove the center ones, the forty-watt bulbs." Blue LEDs actually since *bulb* was simply a common term it continued to be used slangily. "There are hundreds of bulbs that we could remove. That will save a lot of battery power," suggested Red.

"Where will we put them all?" asked CC-electronics rather sheepishly.

"Put each one in a sock, and then store them in sleeping bags," said Walter. "Have the people go to the laundry and grab all the socks and bags, clean or not, before they take out the LEDs."

"OK," said CC-E, somewhat surprised by the unusual methodology. "I'll run it by Triple C. I'm sure she will agree."

"I've been listening, do it now. How hard will it be to remove some of the permanent lighting from the loop?"

"A little more difficult, but use the same idea," Walter stated.

"Innovative," added Red to Walter at the end of the conversation.

Walter scratched a one in the air with his gloved hand. "I've got a minimum six hours on my ox reading. Let's get to work stripping the inside."

Both men reached for the tools stored in the belt of their suits. Scrapping was actually fun for them. They usually were putting things together. They carefully pulled, unscrewed, and tore apart everything they could. It would all be recycled or reused. Soon they had piles of material that needed to be moved out. They also made a very long laundry list of repairs and their priority. Communications carefully logged all their notes, carefully repeating their verbal instructions.

Red and Walter turned as two others entered the chamber. "Welcome, people," grinned Walter.

One responded, "Triple C wants you at command. She needs your help and advice."

*

"Triple C, what can we do for you?" asked Red as the two of them entered CC.

She looked as if she had been beaten like a hockey player up against the glass. The only thing missing (or not missing) was her two front teeth. "We have a problem with the next batch of Hsas. They're coming early, now, in the next few days, in fact. It looks like we are going to have a thousand-month early preemies."

"What?" choked Walter. "We just fixed the lines that power the chambers."

"The impact caused a microsecond flux in the current flow. It apparently reset the time clocks on *all* of the mecha-uteri."

"Feces," muttered Red as the color drained from his perspiring face. He realized instantly that this was going to be a really delicate problem, physically, time wise, and, especially, politico-religious. The religions might cry "act of God," and the secularist opposition might say "just use humans."

"How many preemie beds do we have?" queried Walter nervously, knowing the answer would be bad.

"The doctors said ten, and none of them have ever been used," said 3-C with a weak, crushed voice. "They have been tested several times though."

"Reverse engineer," said Red, looking at Walter.

"Bloody fucking hell, where do we get the parts!" shouted Walter. "Heart monitors, temperature gauges, goddamned diapers! Where do we get those? Shit!" said Walter as he mopped his instantly sweating brow. This was bad, very bad.

"We have maybe two or three days to get the beds made. They will be here starting on the second day, forty-eight hours. They will keep coming over the next ninety-six hours. What do we need?" asked 3-C plaintively. She had her back up against a wall, one with sharp nails in it. A giant sledge hammer was about to smash her into those nails.

"God's help," said a muted voice from someone in the background. Even here were some members of the religious alliance.

"The Russians have a heavy load vehicle on the pad," said Triple C, ignoring the comment.

"It must be reloaded with every medical monitor, ass thermometer, and preemie bed set they can get their hands on."

3-C turned to Earth communications (E-COM), "Get on it now. Tell them now is not the time for red tape. Don't ask them, *tell* them."

"On it," the reply came from E-COM.

"That will take days," said Walter. "We need the Star Voyager, the personnel carrier, to make deliveries as soon as they can to SS-1 and SS-3. They need to pick up the same supplies. It could be there in a matter of ten or twelve hours, depending on how long it takes to gather the supplies. Granted, they can't carry much, but it's a start. The best thing that they can do is the small stuff. Then we need to have them use the skins and emergency escape pods to bring the materials here."

"E-COM, do it," said 3-C. "Ask for battery supplies as well. They won't be able to find a thousand beds right away. We need more power to make sure we can do this. If we are going to fabricate and put together these beds, we need it all."

"Done," said E-COM as her fingers flew over the touch screen keyboard. Red curiously noted, in an abstracted millisecond, just how rapidly her fine-boned fingers moved.

"All the wiring we pulled today needs to be sent to the med lab," said Red. "It will be used for the beds."

Triple C thumbed her com box, hooked on her belt, and ordered ten people to gather the refuse from the core.

"Luna needs to stop sending metals and start sending all the silica they have on hand. Tell them to go to our private pile of materials and rape it," said Walter.

Triple C looked surprised, as if anything could surprise her today. As the higher-ranking officer, she had designated the next watch officer as her strategy officer. She would not get any sleep for a while. "Private stash?" she inquired curiously.

"We have an arrangement with the WSA," Red explained. Not many were aware of the deal they had made. The WSA did not want others to get ideas.

She let it pass; it did not matter now. "Why silica?" she asked.

"Glass," said Walter. "For the beds."

"Glass!" exclaimed many around the room.

"You can't use regular glass on the station," said Tri-C. "If it breaks, it will get into everything."

"Don't tell me what we can and can't do," said Red rudely. "Do you have enough material for plastics or metaliglass?"

"No," she replied.

"Then glass it is," said Walter. "We have no choice. Don't second-guess us now."

"How much do we need?" she asked.

"I have no bloody idea," returned Red.

"Tell Luna to double the rate of delivery," suggested Walter. "We know the system can handle that. Get a hold of Gail Marceau, the CW. She will see that it gets done. She is a primo person."

3-C paused for a moment then said, "L-COM, make the requests for silica and a doubling of launch. Explain that it is an emergency class 1, which puts me in charge. Get this Marceau bitch on the line and tell her she is in charge there. Period! She is to override all authority in this matter. If anyone gives her shit, they can talk to me," she stated angrily.

Red and Walter smiled slightly. This woman was kickin' ass and doing a good job at it. She was coming into her own now.

"On it," said the Luna communications officer (LCO).

"What next?" she asked Red and Walter.

"We need to have one of the beds taken to the level 1 classroom so we can dissect it," said Walter.

"I also want to speak to David DeNoe," said Red.

"Who?" queried 3-C.

"An Hsa young man," said Red.

"What?" choked Triple C. This request confounded her. What could a bloody Hsa child have to do with this emergency?

"Just do it," commanded Red. Many around the room were in a bit of turmoil. They had an imposed hierarchy and a known solution matrix. Seeing Red give their commander orders caused some conflicting consciences, a little bafflement. The second started to chew Red for his arrogance but was snapped up short by 3-C.

With hand held up, she stated, "Stop. I think we all know who we can count on. This is not a pissing contest. You two have authority to do what you need. Go and do it."

"I'd like to see that contest," noted Walter with a smile. She was cute, and he was a pervert.

Triple C frowned mightily at him.

Walter and Red left without another word. They quickly made their way to the same room that they had met the Hsa children in only a few days before. When they arrived, they looked at each other and sighed. Once again, they were in a position where they were the focus, the fulcrum upon which solar system shaking events would turn.

"Shit," muttered Red, shaking his head unbelievingly. "How do we always end up in such situations?"

"We're lucky," joked Walter with a jubilant smile.

Red exerted his anal muscles to loudly expel some methane and said, "That's what I think of that."

"Rather odd behavior for a grown man," said a young male voice.

Red turned to look at David. "We have a big problem, David. I thought that you might have some ideas."

David immediately felt the tension and said, "What are we up against?"

Walter quickly filled in the situation, "We have three days at most to come up with a solution."

David did not even blink his lashless eyes. "There are seven others that I want immediately brought in on this."

"Who?" asked Red.

"I want Dr. James from the genetics lab, my nurse practitioner Lori, and five of my brethren."

"Call them. Use my name to get them here," stated Red. "The adults might not respond to your call."

David went to the intercom and proceeded to do as Red said.

As David was making his calls, two technicians brought in one of the preemie beds. "Tri-C told us to stay and help if we can," one of them said.

"What do you know about these?" asked Walter.

"Only how to use them," said the other.

"Feces," muttered Walter. "We need someone who knows them inside and out."

"Earth can send diagrams," said the first hopefully.

"Unfortunately, we have to make them with what we have here, not on Earth," replied Walter. "We have to see it to get a real feel for what we need to do."

Over the next few minutes, all the people David called entered the room. The Hsa children looked as if they might be in trouble.

Red filled in the people about the situation though the grapevine had done most of the work. "These young people are going to assist us in our endeavors," he said, answering the unasked questions.

Black-eyed Rarte went to the intercom, one that could broadcast openly or to a single location, and started making calls for more of the Hsas to make an appearance. She would organize them into relay teams. They could gather spare parts, information off the computers, form assembly lines, assist *sapiens* by doing lower-level repair work, and bring food and drink.

After making ten calls to people she wanted as group leaders and advising each of these ten to make further calls, she began to finger dance on her console, developing a rough outline of who should go where and why. Klar assisted her, making further recommendations for tertiary layers of assistance, relief groups, and so on.

The Hsas in the first wave knew that this was their call to action, a chance to show their mettle and worth. It could be the proof of their right to exist, or it could be their death knell as a specie if they failed. Pendulums arced out a great symphony at this very moment.

*

There were many things to do in the hours immediately following the collision. Engineers and medical technicians, along with teams of Hsas, were combing the ship, trying to find the hurt and needy. There were some who did not need any help. One was a young man split not so neatly in half as the high-speed, high-pressure containment doors closed. On one side, the vacuum of space had boiled away all the liquids in the now frozen half of the body. On the other side, medical personnel had removed the gooey leftovers while maintenance tried to clean up any damaged materials. They also were trying to prepare to open the door later for repairs.

The adults (*sapiens*) were at first a bit leery of working with children in this horrific task. They quickly changed their minds. Though the children let tears slide down the sides of their faces, the emotions did not overwhelm them. The Hsas' voices were strong; their hands worked steadily and quickly. They tenderly removed body parts with their surgically gloved digits while others wiped down equipment covered with blood and gore.

Several individuals had been shredded as the high-velocity chunks of ship and rock opened tiny holes in the inner side of all three rings. Most were people sleeping in their chambers, thankfully. They did not have time to feel pain. It would be days before a count was certain.

In all, twenty-seven adults had died. So far, it was the single worst disaster ever for the space program.

*

An emergency meeting of the various religious groups, those involved in the continuing antispace movement, was called. This was their in. Flights were booked to bring all the original members back to Vatican City along with their seconds in command. All the original members were now old or ancient, and they were on the second pope in this alliance.

The doddering ninety-three-year-old caliph moved slowly, using his wheeled walker, at the head of the line of men entering one of the Vatican's inner rooms. With the assistance of his sixty-year-old son, Hamad, he carefully navigated his way to the seat marked out for him.

Two guards pulled the chair out for the aged man. Once seated, they pulled him close and adjusted the tabletop microphone so that it was close to him.

The others followed almost as slowly as they were well on in years as well. All the original members of the alliance, with the exception of the pope, were gathering again. There had been a seven-year hiatus between direct meetings involving the pope. Feelings were tense yet cordial. This was a chance to mend fences and refocus the organization. They had been losing ground without a fully coordinated effort.

Each found his seat, labeled by gold gilt-lettered place cards placed under the microphones. The microphones were an addition that made some feel uncomfortable. This meeting would be recorded, obviously. This pope would have evidence of conspiracy to hold over them. There could be no sideways denials or diversions of complacency.

The younger attendees took no notice and chatted in a friendly manner. Most had met individually but not in a group encounter such as this, one in which there were such dramatic settings and goals. Granted, younger meant those generally between forty and sixty though the imam's son Arod was sixty-six. The new pope was only fifty-eight.

The elder statesmen waited patiently for the pope's appearance. There would be no sudden showing like the first meeting. He would keep them waiting; it was a demonstration of his religious authority and a statement about previous events.

This event in space could not be ignored. Just two days ago, a meteoroid struck SS-2, killing twenty-seven *sapiens* and putting the lives of the next generation (the third) of Hsas in danger. It was an act of God, from their point of view. Something needed had occurred to bring them back together. If acted upon swiftly, they could regain the high ground and hopefully surpass their previous levels of power.

Many of them remembered the last time they had been in the Vatican.

*

The frail, liver-spotted old man lay dying. He had initiated the alliance and must now conclude his part in it. For the past several days, streams of visitors, heads of regional and world government as well as religious leaders, paid their respects.

Weak as he was, Pope Xavier blessed many of the religious leaders, especially the members of the alliance.

During these private audiences, many of the alliance members pressed him for one final act. Only the caliph and imam voiced severe reservations about the impact of such a move.

Historians would later question the mental capacity of the pope; they suggested that he was acting under duress during his final hours.

The Vatican News Network had set up quietly and quickly in the room in which the pope lay suffering. They tried not to disturb what little rest he was able to achieve, but they had to hurry since the physicians said he could go anytime.

Very few knew what was coming; only some of his most trusted cardinals and bishops had been notified less than a day earlier. There had been no time for any discussion of repercussions involving the church and its sheep. Most would tow the line, but there was a growing group of dissenters in the ranks. The coming edict would be a giant piece of an enlarging wedge that would grow into something the size of a redwood.

"Holy Father," said one of the technicians to the restlessly catnapping pope, "we are ready."

The pope's eyes blinked open, watery and tired looking. Even so, he smiled gently at the tech man and took his hand and squeezed.

"Prop me up," he said in a strained voice to his doctors.

Without speaking, each placed a pillow around or behind the pope until he was almost upright. The effort caused him to wheeze and cough; his thin bony frame trembled.

The microphones were adjusted to their final locations near his ragged mouth. Video was already on, but it was not transmitting yet for the live presentation.

The head tech looked to the pope and gently asked, "Are you ready?"

He only nodded, not trusting his voice to last too long.

The secondary camera was set on Cardinal Leonis, who was set to introduce the pope. He had a deep red set of drapes as the background.

"Ready?" asked the tech.

"Yes," said Leonis.

"OK, then in five, four, and three . . ." He signed the two and one and then pointed at Leonis. The Live light went on over the camera facing Leonis.

"Ladies and gentlemen, all people of the one true God, our beloved pope has an announcement to make. He apologizes for the circumstances surrounding his announcement from his bedroom rather than his office. Please listen carefully and respectfully as this is a most important moment. Holy Father." He led the camera to the reclined pope with his hand.

People in the square outside were hushed, expectant, hoping for good news about his health.

The majority of people outside the Vatican had not seen the pope for many months. His health, in question for years, was a hot topic. The decline was not unexpected; his great age, ninety-six, was unheard of for previous popes. He had appeared frail for over twenty years, but he now looked emaciated, almost mummylike in his dried skin. The people watching on the big screens in the square and observing on v-link or the net were shocked. Many had their mouths open in disbelief. Three-fourths of the solar system was watching.

The pope tried to smile, but it was a failure. His thin blue lips could not make the effort. A small amount of spittle descended from the left corner of his mouth, unnoticed by him, but the world saw and cried.

"My children," he began, "we have long fought a battle against the Satanists that are involved in the space program. We have won some battles and lost others." He grimaced.

"On this particular ground, we must not fail. In less than . . ." He coughed hard for a few seconds; one of the doctors entered the scene to try to help, but the pope waved him away. He began again, "In less than two years, the *Homo sapien anthrocreatus* are expected to arrive on Space Station 2." He refused to say the word *born*.

"I am failing in body," he noted with a smile that worked this time as his eyes pierced those of everyone on the other side of the camera. People the world over had sudden tears at the knowledge that their leader was truly dying. It had been known, but it now had that awful finality, reality.

"I do not fail in spirit though. My spirit, guided by our Lord and Savior, requires this last act from me," he said as he looked down, as if gathering strength. He took several deep breaths as did his global followers.

When he looked up, he continued, "What I do now causes me extreme spiritual anguish. By the authority vested in the office of the pope by God the Father, God the Son, and God the Holy Ghost . . ." He paused again to take some deep breaths. "I excommunicate all of the *Homo sapien anthrocreatus.*" There was a stunned, unbelieving silence around the solar system. "I also must excommunicate all the people involved in the entire process, the entire space program. They have deviated from what is right in the eyes of God." His gaze returned to the comforter before him, tears flowing from his eyes as he blinked furiously and trembled in agony for what he had just done.

"My dear, dear children," he wept unabashedly, "I find it so very difficult to bring about the permanent absence from our God, the sentencing to hell, of the souls of so many humans, and perhaps the questionable souls of the Hsas." Firmly he lifted his chin to say, "We must walk the straight and narrow path, the hard rocky road that has been laid before us. We must do what is right under the scrutiny of our God. If we fail, we are no better than they are," he stated, referring to the Satanists, the agnostics, atheists, even religious persons who did not fully support the alliance.

"I have had a vision, a gift from God," he continued. All but Leonis were flabbergasted; the pope had not shared this with anyone else. "In my vision, I saw two roads. Peoplekind will be split into two competing factions. These factions are the final remnant of peoplekind. These two groups are the final end of peoplekind. We have reached the Y in the road and must now make our final decisions. What road will you take?" he asked as he coughed horribly again.

Small flecks of blood flew from his mouth, unnoticed by the viewers, staining the sheets and comforter that covered him. The Jesuit doctors started forward but were harshly and silently waved off by Leonis.

"We must all, now, take our stand. God has given me the grace to lead you for so long. In this final act of my body, I begin my spirit life in the knowledge that I have done my best to bring the light of the Lamb to you. God bless you all," he ended with another wracking cough as the camera cut to Leonis.

"We must pray," stated the cardinal. Over the open channel, he prayed for the next several minutes for the pope, the enlightenment of the world, and the fulfillment of God's will. He ended with, "God bless us all."

The tech crew, in utter silence, quickly gathered their equipment. All were saddened at the obviously failing condition of the Holy Father. They left without the usual thank-you from the pope. Their world had been rocked to the core by his announcements. Such a thing, so many people all at once, was unheard of in these times or any other.

The entire solar system was in a state of shock, emotionally, politically, and, especially, religiously. Non-Catholics, even non-Christians, were impacted dramatically. Muslims, Jews, and every other religion knew the import of such a declaration. In a few feeble breaths, the world was altered beyond recognition. Polarization of society was nearly instantaneous. Pendulums swing.

Certain hardliners, extremists in fact, saw the possibilities inherent in the situation. The pope had finally taken a hard stance. This allowed them to suggest, gently to begin with, even more radical steps. The elements (the seeds) of jihad were in place.

Cardinal Leonis remained with the old man for his few remaining hours. As the pope's wind grew more ragged, he knew it was time. At a signal from the dying man, a weak squeeze of the hand he gently and tearfully held, he began the last rites.

Leonis sobbed at the moment he felt the life and spirit leave the body of his best friend, mentor, pope, and friend to God.

He received the state funeral that all popes do. Each of the alliance members had attended and spoken at the ceremony. The occasion had been used to further the goals of the program. The many politicians under the sway of the religious organization also used the situation to try to extend the power of the world government.

It would be three weeks before the next pope could be selected, announced by the white smoke that had not been seen in decades. It would be this person, Pope Yan-Tau, whom the alliance met for the first time, after a difficult seven years of strained relations due to the excommunications by his predecessor.

The non-Catholics had silently distanced themselves from the action while, just as silently, approving it at the same time.

*

The other members of the alliance had to remain above the fallout from the excommunications. It was the only way to keep the fringes of their organizations from slipping away. None of them, the directors, had ever spoken a word against Pope Xavier's actions; they had refrained. This had given some (the wishy-washy) the thought that they did not approve. This was an utter fallacy. It had been their decision that it was necessary, appropriate. This was the first truly hard blow struck by the Catholic Church. They had striven for just such a thing.

That they had allowed the counterpunches to be absorbed by the Vatican, this felt almost treacherous to the newest, Pope Yan-Tau. It rankled deep in his soul. He had not allowed a meeting at the Vatican for this reason. He had not met with them either, regardless of the location, until now. The ball and the court were his.

Now they were coming to him, as he felt it should be. He would not let the office be used again as it had been. He would have to be very careful in his dealings with these men. They had agendas of their own; only some of which coincided with his and God's, he believed. He would be hard and strong. A leader for Christ had to be. He would make sure that he could control them. They could not be allowed to fade back and let the Catholic Church stand alone in any more tight spots. There must be complete unification.

*

"There are twenty-seven dead adults," David said as he reentered the room.

"Red turned with fire in his tired eyes. "What?"

"That's the count according to personnel. I used your name to get the information. The man was crying when he said it."

"Goddamn it!" He toggled his radio. "3-C, why didn't you tell me about the casualties?"

"Would it have done any good?" she replied with a worn-out voice. This was her fourth ten-hour shift in less than two days.

"No," he replied in a failing voice.

He returned to his frantic work. They had made almost a thousand special chambers by hand in the lab. The engineers in the core had made major modifications of the production line for parts from the solar furnaces.

They—meaning Red, Walter, the doctor, the nurse, and almost fifty of the Hsa children—had constructed the faux preemie beds without stop. The other Hsas had been running every sort of errand from bringing supplies for the beds to food and water, even burial detail.

The children, for that's what they were at the age of five, had striven as hard as they could. Many had worked forty and more hours before collapsing in sleep for a few hours, not just from producing the beds, but also from the bone-weary effects of so much emotion, something they had been well shielded from previously.

They were invaluable. The people, roughly half of the station, those who were unsure of the Hsas, were enthralled at how they threw themselves into the life-and-death fray. The battle was won by the little—the Davids, not the Goliaths.

They had presented themselves in organized sets of groups after they had been called upon, under the direction of David H., H for Hsa. David H., really David DeNoe, had presented his terms in hard fashion. He had marched right into CC and looked her, 3-C, right in the eye when he did it. She never forgot that. Her estimations of the Hsas were raised a hundred fold in less than a second. It also pissed her off a little.

3-C had almost turned them down, but Red, who happened to be there, commanding more supplies, spoke heavily in their favor. He honestly announced the need for everyone to participate in the emergency, including those whose future most depended on the positive outcome of this situation.

Rarte, that obsidian-skinned Hsa beauty, had shown surprising ability with her hands when it came to connecting whatever needed to be connected. She worked with old-fashioned, jury-rigged soldering irons and tiny electronic parts and tweezers to combine parts that had been constructed by Red and Walter. Those two had created resistors, inductors, and capacitors from basic materials. The equipment they were creating was very crude compared to Earth standards, but it would function.

Rarte had also used her knowledge of glass artworks to improve on Red and Walter's glass production scheme. Mind, talent, skill, all was brought out by the crisis.

Her visualizations allowed her to sketch in seconds what Red and Walter described. She did not exactly make blueprints but schematics that anyone could follow. The other children, not quite goggle-eyed, immediately set to connecting what the pictures demonstrated. Some of the adults (engineers all) smiled as the children worked side by side with them. Bonds were forged these days that could never be broken, ever.

Klar, another H child, ran the computers as if she *owned* them. She e-talked to everyone, ordered what they needed, overrode commands, and, basically, hacked when she had to. Earthbound controllers had little idea who was doing it. They didn't countermand any of it when they saw Red's name attached as the one behind the order or combined with 3-C's. The emergency left them a bit overwhelmed. They did not have control of the situation. They really couldn't; the scope of the problem was beyond Earth's ability to control. All they could do was sit nervously, drink coffee, curse, and chew fingernails.

Most simply trusted Red; he had proven himself too many times to hold back. It would only be later that the true facts became known in regards to the Hsas. The space agency would use that as pure propaganda.

The pitch of urgency had climbed higher and higher and peaked when the Hsas began to arrive. They came slowly at first, then in surges, and, finally, waves. They raced to complete the remaining beds. They did not quite make it. They were seven beds short when the last of the Hsas arrived.

*

Marceau worked fifty-seven hours straight before she collapsed in exhaustion. She was taken to the temporary triage where she received intravenous saline and a sleeping agent.

It was her department that had to see that all shipping was done. She saw the silica compressed and shipped above the rate demanded by SS-2. She also ordered that extra oxygen be sent on the next several shuttles. She occasionally overrode even the Luna-1 commander, as per 3-C's demand.

She was well aware that Red was behind the orders she was receiving. She would not fail regardless of her personal opinions of that red-haired asshole. This was a situation beyond personal feelings. By her orders and actions, three times the normal was the delivery speed, not the two asked for.

She bullied, yelled, connived, and personally launched what was needed. It almost came to blows on occasion when some complained that they could not keep up. She "fired" them on the spot, if such could be done on Luna. The engineers most linked to Red filled the gaps quickly and wholeheartedly. All off duty personnel scurried about as their queen bee directed.

Buggies with trailers rooster-tailed their way back and forth to Red and Walter's dump site. Raw materials were gathered and shipped at a rate not believed possible, not within calculated possibility. Luna-1 was more active than a dozen beehives on the first day of spring with flowers abloom.

*

"Holy Father," David spoke tactfully. "I have asked for this vid-link because I believe we have a common goal."

"What do you believe we could ever have in common?" Pope Yan-Tau responded coolly. His advisors had cautioned him to avoid even speaking with this foul creation. However, this man was strong of mind and will. He would not pass up a chance to speak directly with his foe.

"An ongoing battle that the church has fought valorously—euthanasia," David stated respectfully.

One eyebrow rose slightly at the bait, but the Holy Father did not speak. He cautiously waited for further information. Truly, the church opposed all attempts to legalize euthanasia in all its ugly forms. They succeeded in most cases.

"Sir, the debate is whether or not to kill an Hsa, a survivor of the incident on SS-2," David played out the story. "The WSA has decided that we should kill this child."

"Child?" He noted carefully. "And what should I do about that?"

"This is just a stepping stone, sir," David spoke surely. "After this, there will be steps for *sapiens*. I believe the phrase is *slippery slope*."

"I am opposed to your very existence. Why should I help you or your kind?"

"I do not believe that you will be helping us so much as all *sapiens*, sir. How can a society survive if it kills its own?"

*

Of the seven without beds at the time they were born, six died, and one survived, a boy named Omden. He had oxygen deprivation; he had survived but would suffer from cerebral palsy. Most victims of cerebral palsy have trouble walking, talking, using their hands yet have relatively normal intelligence.

That was the focus of the discussion at hand. After two days of recuperation, the key players on the station were confronted with a recommendation from Earth. Omden should be euthanized. Many of the Earth doctors speculated that he would be physically unable to perform even basic motor skills since he had shown no movement of his arms and legs yet.

Red vehemently denied that possibility, "Never! That can't be allowed!"

All the department heads were present, but none wanted to speak. They did not want to cross Red. He had a habit of destroying the careers of those who crossed him. Speaking in small knots of concerned people, before the actual meeting, the consensus was that Omden would likely die anyway. They did not want to be seen on either side of the fence.

Triple C looked at the table and spoke with little force, "It has been ordered."

Walter muttered, "Fornicate the orders right up their puckered tight asses."

3-C stated calmly, "If we do, what can we expect to happen to us? We can all be recalled to Earth."

Red furiously responded, "They can kiss my hairy ass. I will not let them threaten me like that. My failure led to the deaths of six babies. I will not allow the seventh to be murdered."

"Nor shall I," announced David DeNoe as he entered the room.

3-C was surprised and somewhat angry at his entrance. "What are you doing here?"

DeNoe calmly approached an empty chair and sat down. Looking around, head barely above the table, he said, "I see the department heads, but I do not see a representative for the Hsas. I am assuming that position as of right now."

"You're what?" many voices exclaimed in unison.

3-C was brought up short. It took a few seconds before she could respond, "The maestro has been assigned that position according to protocols." She nodded toward the man.

"Those protocols are flawed. The maestro is head of the education department including research, and has been secondarily given the voice of the Hsas. However, he has not been given two votes to represent the two entities he serves. Therefore, I will assume the one role and vote for the Hsas," DeNoe dictated smoothly.

The tension in the room was quickly rising, and 3-C sensed that. "You can't march in here and simply declare that you have a seat on the advisory board," she responded with some heat. This little imp was challenging her absolute authority as station commander.

DeNoe smiled gently. "I just did, and I have the backing of all two thousand cogent Hsas. I also have the backing of, well . . . the pope. He and I have had a conversation about euthanasia. While he does not particularly like me, he agrees that the sanctity of all life, including 'non-real' humans, is threatened. He must bring his force to bear. He fears that this could open a Pandora's box of . . . homicides."

The only two people not agape were chuckling, meaning Red and Walter.

"You're in the shit now," Walter said as he looked at 3-C.

*

Everyone in the ground program considered the terrible event to be the greatest recovery since Apollo 13. Only Red and Walter considered it a failure; they had not recovered all the Hsas; they believed in the individual, each one's worth. The ground controllers examined percentages; they looked at society as a whole before they considered the individual. The masses had been protected at the cost of a few. It was a fair trade.

DeNoe noticed their personal consternation, their pain at their failure to save them all. His pain at the loss of his birth cousins matched theirs. He knew that Red and Walter were men of extreme honor. He would put his life in their hands anytime. He would reciprocate at the cost of his own, as would all of the Hsas who had helped that day. Even the younger generation of Hsas had recognized what was up. At three, they were useful for minor errands only but had participated as best they could.

They had found a pair of humans of extraordinary compassion, both dedicated to life—theirs. At a time when many Earthers were questioning the Hsas, two, and actually many others, could be counted on. Their place, meaning Red and Walter, was solidified in history, legend, and, perhaps more importantly, the beating hearts of every Hsa. They would follow them, almost without question. In the future, they would almost deify them, the new gods of space to replace the old gods of Earth. The more things change, the more they stay the same. Pendulums swing only to return to their starting place.

*

Red passed a wrinkled hand wearily across his face as he pulled away from his memories (and the portal). They had accomplished the unaccomplishable. Three days of sheer will. The Hsas had contributed several significant ideas.

His job was to fix things, granted, but they were always related to science, not politics. How could these people ask this of him? He had never considered them anything but human, unlike many others. And it was not just the *anthrocreatus*; normal space people were just as aggravated. His views on history had given him the perspective that suggested that separation was inevitable. It was only a matter of time before that occurred. The main question was whether it would and should be now, and would Red support it. Red was an unspoken leader, *the* leader.

There had been a significant drive to create a governing body that was separate from the UE. Those in space knew that there was no way those on Earth could truly understand the society that had developed. Not all on Earth were opposed, but the religious organizations were as well as the UE apparatus. They wanted absolute control of all governmental bodies. The UE felt that the space organizations were too loose, too freedomist. If they had proposed a more controlled, Earth-directed government, they would have set up a new territory; they were, in fact, trying to.

The *anthrocreatus* already existing had been excommunicated by Pope Xavier in 2028 (the same one who had been involved in the first religious uprising) before they had even been born, along with those in the space program. Waves of emotional and intellectual outbursts, on both sides, had followed. The long-term polarization of society was a major signal of the separation that was coming. 2061 did not promise to be a good year. Red sighed and returned to the present, mentally.

*

Luisa Jamala Reta openly appeared worldwide for only the third time since the assassination attempt. Her colleagues had tried very hard to keep her under protective wraps, fearing another try and what it might mean to their movement. Reta did not fear for herself or the movement.

"My friends, churchgoers, nonchurchgoers, we must all consider this tragedy in clear light. Many claim that God has shown his disapproval of the Hsas and the space program. God does not disapprove of life. God does not disapprove of choice. He/She/It gave us choice in the Garden of Eden.

"He wants us to love, to cherish, to feel compassion, to hope. The Hsas are just us, no more. We must reach out to them and bring them to an understanding of God.

"Jesus said that those who did for the least of people have done for Him whether it be food, shelter, or otherwise. This is what we must do for them, whatever we can to succor them in their need. We must give food, shelter, hope, acceptance, and, above all, love.

"Love is the key to all. It unlocks all doors to the higher self. Without it, we are only animals. We do not rise above our instinctual level. For those that think we are only animals, let us be better animals.

"The scientist will tell us we are evolved animals. We are allegedly the top now, a work in progress for the future. If so, we must be more. I choose to believe we are a product of the greatest intelligence, He/She/It, or HSI, pronounced huh-si.

"We are finished with the male-dominated religions. Our God is a greater being, one that is not gender-specific. How can God be limited to a gender-related existence? It is not possible. Our God is the energy of life, the matter of existence, the dimensions of the universe, and beyond the realms of science, for we speak of the supernatural, that which is beyond."

*

The Stalwarts replied immediately, in like manner to one another, in all their regions. Harmon struck first.

"She speaks of 'clear light' as if it is the logic of man that guides her. The Bible says that the wisdom of man is as nothing to the wisdom of God. Her thinking is skewed by the outcome she desires rather than conforming to the desire of the Almighty. It is her will, not that of the Father that she follows.

"She is attempting to sway people from the true God. She is the whore of Babylon," he hissed vehemently. This 'huh-si' she speaks of is actually known as Aksi, a female demon subordinate of the pagan goddess Innin. It is pronounced 'awk,' as in awkward, and 'see,' as in to see with your eyes. Demoness of the dark underworld she is, and Reta is under her sway.

"The high priestess of Aksi held orgies where lesbianism and sodomy were the requirement. Heterosexual relationships, in this evil pagan festival, were banned. The practice of evil is what she pronounces," he thundered, pounding on his podium.

"The Bible tells us of a time when the majority will fall away from the grace of our bountiful God. That is happening now. We have seen the products of science and the evil they represent. It is all a false path. She is but one of those imaginary, delusional paths. Do not dare to follow that lie," he whispered desperately. "It will mean your eternal separation from God, in hell."

The camera faded away to an image of the Christ.

*

Davies's message was more political. "Many people have tried to imply, tried to pin the assassination attempt of Reta on the Stalwarts. It is a lie," he declared strongly. "The Lord Jesus Christ is the martyr of the Christian faith. He gave his blood in ransom so that we might have eternal life with the Father.

"We, the chosen leaders of this movement, are neither so ignorant of history as to commit such an atrocity in that we would create a false martyr for the enemy of God, nor are we murderers, that is the province of Satan," he firmly averred.

"NO!" he thundered. "The government has done this as an attempt to prevent a new power base from forming. The upper echelons do not wish to contend with yet another religiously, albeit falsely, oriented sect."

He informed, "They seek now to divide the true religion, to break our strength. They seek to return to a secular view and a secular control of the government. We must not allow that to happen. We are so close to actualizing control for the benefit of all peoplekind. We must dig in and resist this new wave of Satanism."

He gripped the podium fiercely. "We must call for the ouster of nonreligiously oriented leaders. We must elect new leaders that follow our tenets more closely. Act now. Act to save not only your soul, but the souls of all Earth."

*

Nine men, each with their nine subordinates each, and those with their nine each, met in an emergency meeting.

One spoke, "This is a declaration of open war. We have worked in a mutually beneficial way up to now, a form of mutualism. They are now parasites, and we must remove them before they damage what we have created."

Another nodded as he spoke, "They were useful, but they have outlived that. We must use any information we have on them to separate them from their power bases. Any form of moral turpitude will do. Scandal is the key. Sexual scandals, embezzlement, all have worked in the past and will again. The sheep will become wolves if they believe that their leaders are such sinners" he guffawed.

"Truly, all people stumble." A third smiled viciously. "We may have to . . . shall I say, assist in their stumbling." He looked at one of his assistants and nodded. "I can arrange for some rather lewd vids to appear. They can be real or manipulated. My union division has control of almost all the prostitutes left in the world. Their entire families can be involved with various levels of involvement. For example, Davies taking his son to a brothel." He sniggered. "Wives missing out on an active sexual life . . . We can then tie that into misallocation of church funds to pay for all the fornicating!" he bellowed.

A short silence followed then a significant number of nods. Consensus was quick; it had to be since the rulers of Earth had far too little time to spend on any given topic. Computers had actually taken over many of the mundane responsibilities of the bureaucracy. Probability ruled.

Following the breakup, several eyes met. "He must go," one noted.

"Yes, that statement about all people stumbling, we cannot be linked to such sentiments. It could cause dissent," another verbalized.

The first who had spoken said, "Let him work his games with the priests. Then, we will find sufficient fault in his methods to eradicate him. I am certain we can find an underling within his organization that will be willing to assist, for a promotion. That will keep us above any fray." His cold eyes glittered.

Many fewer nods followed this.

*

Many people of faith no longer knew who to listen to. Each side, each angle, posited against the other. Some simply followed the leaders they always had; it was simpler, easier; they did not have to confront thought.

Others turned away from their religious beliefs and embraced science since it held relatively firm to its ground. The stability was comforting to the soul in comparison to the current wild gyrations of the various sects within the religious world.

Some number rejected the Western for the Eastern with its different approach to spirituality.

Fear of the unknown, desire for stability, sameness—it is all the same basic *sapien* need for a pattern that can be recognized and reacted to, a part of evolutionary development. The counter to that is the evolutionary development related to dealing with the unknown, a fight or flight experience. Which will dominate?

*

"Mick, turn that up," Paul said. The screen showed more news about the tragedy on the space station, and he wanted to hear about it. Everyone was talking about it, and he figured he needed to load up on knowledge so that he did not appear to be as stupid as Mark.

Those paying attention heard the voice say, "In addition to the twenty-seven *sapien* adults that died in the initial collision, we now have confirmation that six Hsas died. Whether for lack of proper medical facilities and care or not has yet to be determined. According to the commander, known as Triple C, an extraordinary effort involving all personnel on the station, but especially Red O'Hare, Walter Black Bear, and a first generation Hsa (1GHsa) by the name of David H., where *1G* designates first generation and *Hsa* designates *Homo sapiens anthrocreatus*. The Hsa boy claims to be David DeNoe, saying that he has discovered his donor mother to be the Captain DeNoe of the Luna return mission of '15. Other Hsas given notable reference were Rarte Hsa, and Klar Hsa, along with the one Hsa baby to survive without a preemie bed for several days so far, Omden Hsa.

"Triple C, or 3-C, has also mentioned the efforts of those on Luna with special mention to the commanders and Gail Marceau, the head of shipping and receiving. It is said that she restructured the entire launch system in a matter of hours in order to provide the raw materials needed by SS-2. Even Red O'Hare announced that she was the hinge pin upon, which it all depended. He said, 'Without her, all our efforts would have fallen short. Without the supplies we needed from Luna, we could not have constructed all the preemie beds.' Pundits are claiming this to be the greatest disaster of the space program while at the same time claiming it is the greatest success story. It has brought out the best of all in the program. Others say it has shaken the tree of the program, causing the bad leaves to fall while the strong remain. Those who reacted well are the good leaves. Those who did not will be pruned away."

"I need another beer," stated Paul. He looked to the door as Burt entered. "And one for Burt as well."

"Thanks, Paul," said Burt.

"Don't thank me. I ain't buyin' it, just orderin' it."

Burt chuckled; Paul never bought anyone a beer. "How are things on SS-2? Have they fixed it all up yet?" Burt asked.

"They admitted that six Hsas died," Paul said with a somewhat neutral voice. He was one who was not sure of his own feelings in this matter. His deep fears of the unknown, the misunderstood, contrasted with his expectations that science could, or maybe even God could, fix the situation. Under great fear is how most live their lives. They just do not want to admit it.

Burt choked out a, "Shit! When did they announce that?"

"Just a few moments ago, just before you came in the door."

"Fuck! Red's not going to be happy about that. He's been bustin' his balls to save them all. Son'bitch."

Paul looked a bit askance at Burt, unsure of just how personal that last statement was. "It were only six," he said a little confused. The news kept saying it was a miniscule number. He did not understand.

"Mick, let me see your calculator," Burt said.

"Here," Mick said as he handed an old-style calculator to Burt.

Burt punched in some numbers. He tried it again, just to be sure. "0.6 percent," he said. "Less than 1 percent of the Hsas died. More than 99 percent survived. Doesn't that rate an A+ in school?"

"This ain't school, bitch," mumbled the intoxicated woman without two teeth. "Them fake people little-uns died, along with them adults of us. God don't want us doin' this. Ain't it obvious to you, fools?"

Burt, Paul, and Mick all frowned in surprise. This woman had never expressed such overtly harsh feelings like this before. She was a little more drunk than usual though. Everyone knew that the drunker a person was, the more likely he was to speak from the depths of his personality. With a certain amount of reservation, they reevaluated their position with regards to this woman. They each felt a certain something that told them something was not quite right.

Burt hit it hard when he said, "What do you mean? What have you been talking about all this time?"

The woman rambled in her alcoholic drunkenness, "You sons o' bitches thinks you know everything. Pastor Davies has explained it all to me. Lies, all lies, you think you know it all. You're all gonna burn in hell!"

"That's worth an hour of timeout," said Mick.

"Fuck you, Mick," she said intently. "You gonna burn just like them."

"Eighty-sixed," he quickly responded.

She dragged her wobbly self out.

After she had departed, Burt said, "She is really over the deep end on this. I thought she was just a contrary person, but she is really hard-core religious." His instincts were tugging at his brain; he was missing something important, but it slipped away as Mark bluntly entered the conversation.

With bleary eyes, he noted, "Damn."

Paul returned the conversation to the Hsas, "Them religiousy people saying this is God's hand."

Burt said, "I don't believe that. If you read the book of Job, you see that God does not do anything bad. He does let Satan do wrong stuff though."

Mark looked a little surprised. "I did'n think you was religious, Burt."

Burt smiled. "I'm not, but I do a lot of reading. I think too many religious people don't really study what they believe. They let others tell them what to believe. They do not call them 'sheep' for nothing."

Mick added his bit, "It ain't just the religious people though. I think that people in general have slowly been giving up the ability to think on their own. I remember the really old cash registers where you had to count out change. Now you just give back what it tells you. It ain't much, just a little thing, but it takes away from the exercising of the old grey matter."

"Exactly," said Burt. "There isn't enough thinking going on. That's why I really support the space program. There are some real thinkers involved. Those two engineers, Red O'Hare and Walter Black Bear, they think. They solve problems. The old saying is thinking outside the box. I don't think they know there is a box."

Mick was more impressed than ever. "What do you think is really goin' on in the world today, Burt?"

"I think that we are at a major crossroads. What happens in the next few years will be the end of peoplekind, or its continuation."

Mark and Paul looked shocked. Mick was surprised as well.

Burt continued, "Think about it. If we evolved as the scientists say, there used to be many humanoid species. There were at least two *Homo* species up until about twenty thousand years ago. We are alone now, which is an oddity. Extinction is the norm for species, but there are usually several. Time has a way of taking out species."

"We will either adapt, which, I believe, means go into space, or fail. If we don't go into space, our technology must fail eventually, or overwhelm us, and we become a lesser being. We will fade with time. By going into space, we invigorate the thinking portion of our brains, the part of our bodies that really separates us. The challenge is what we need to continue, without it we become complacent in our existence. We die.

"I suppose a good analogy would be a grandfather clock. It will eventually wind down unless you wind it up every so often. The winding is comparable to the challenge I was talking about."

Mark could not help saying, "That's some really funky shit."

Mick leaned in close. "What do you think about that big killing going down in Washington? I heard it on good authority that the Modernists were simply slaughtered. I also heard that the Aryans and KKK instigated the whole mess. The leaders were quietly done away with by the suits."

Burt looked about slyly. "I got vided by some unknown person. I saw a whole series of pictures. I have no way to verify them, but they match with all the rumors. There were pictures of nonmilitary people with bombs and guns. Lots of tanks were in some of them, so that would be opposite of what the government said. There were lots of dead people right in the front yard of the White House. The government denies that they even got close."

Mark looked stunned. "I got the same type of stuff. Do you think it was sent to everybody?"

Paul's eyes widened. "It was sent to me as well. I saw nearly a hundred stills and some short video. I thought I best keep real quiet 'bout it."

Burt mused, "We all got parts, pieces, but not necessarily the same ones. Someone generated a massive collection of data and then sent it out in parts. They did not want to be linked for long so they could avoid detection. I would bet this was originally data from someone in the news business. They had to have gotten a hold of some vids and stills, but they were not allowed to show them. The new censorship laws would prevent them. If the truth got out, the UE would look bad no matter what."

Mick asked, "What do you mean? They look bad as it is."

Burt smiled. "They can't acknowledge that anyone might want to divorce themselves from the UE, like the South tried in the Civil War. They are supposed to be the perfect government. They also can't acknowledge that they made errors and were insufficiently prepared. And again, they can't admit that the rebels came so close to actually succeeding. That would make them look weak, and others would try. They can't, by any means, admit that orders were given to murder all the nonaggressive Modernists. Everyone had to permanently disappear, quietly."

Paul noted slowly, thinking very hard, "This was a very well-planned attack. They had real big diversions, and then the real attacks were made somewhere else."

"That's for sure," Mick muttered. "Can you imagine if they had won?"

Burt looked disturbed. "I'm more worried about what the government will do now. I think they will use this to pass more restrictive laws. It will come down to losing even more individual rights. Free speech will suffer greatly. I believe that the regional identification cards will be required for regional travel, perhaps even local. That would be a reversal of policy, for example the old European Union. Right now, if you take mass transit, the suits track your movements. It might have been better . . .," he trailed off.

Mark sipped and said, "But wouldn't that have stopped what happened?" Meaning the mass transit.

Mick noted intelligently, "I believe . . . I think that it did already. It is rumored that many of the terrorists used mass transit. A friend of mine, one once in the army, told me that the suits alerted them that large numbers of people were moving into DC the day of the riots."

Burt raised an eyebrow. "Can you trust him?"

Mick chuckled lightly. "Her. I can trust her. I have been friends with her family for many years. She's now a communications engineer for a news station."

Burt knew that look. "And very good friends with her, huh?"

Mick laughed again, "Yeah, yeah."

They all got a good laugh at Mick's discomfiture.

Paul shifted the conversation with, "What about the assassination try on that popess?"

Mick responded quietly, "I agree with them religies. The government did it."

Burt nodded. "It is more logical for them to do it. I expect there will be some reprisals, propaganda, and so on to undermine the Stalwarts."

Mark spoke, "She got hit in the head. I done read that Revelations, and it says that Satan will be struck in the head and survive. There's some as says she is a demon, and not just the other religies." He sagely nodded.

Burt snorted, "There's a lot of shit floating in the cesspool. Hard to say what's really happening with everyone saying something different. I think she's just a modern Luther."

"Huh," mumbled Paul.

"She's just a reformer, no different than Galileo, Copernicus, Martin Luther. She is causing a change in perspectives. There is always a great deal of resistance to change. There is a form of social inertia, like inertia in physics. It takes a great deal of energy to change the movement of an object, and that is like society," Burt explained.

Mick tilted his head slightly as he asked from across the bar, "What do you mean about inertia? I don't know no science, Burt."

"Momentum, Mick, momentum," stated Burt. "Think of a hammer headed for your thumb." He pantomimed. "It takes a little bit of energy to tap it so it don't hit your thumb." He tapped his downward moving arm to make it miss the imaginary target. "To make it miss by more, a safety cushion, it takes a bigger tap." He again demonstrated by pushing hard against the downward moving arm, making it go almost sideways. "Now if we make it stop cold." This time it was obvious that his arm would be a bit bruised tomorrow from the sudden stop; his hand was even flung forward though the arm stopped. "Such an interaction could cause a break, in my arm or in society."

Paul nodded wisely. "I know about that force stuff, just not in those fancy words, inertia or momentum. So the world is—"

"Wait," Burt interrupted, "there's more."

Mick surprisedly asked, "What more could there be?"

"What about this?" He moved his arm down toward the bar yet again, only this time, the counterforce caused his arm to move back in the direction that it had come from.

Mark noted most intelligently, a rare case, "It's all just like pool." As he gestured at the tables on the far side of the bar.

Mick frowned in concentration. "So once the shit hits the fan, it's awful hard to stop it or deflect it much."

Burt nodded. "That's why so many people have given up, just plow through the daily grind, riding the wave. They realize that their efforts, as individuals, are hard to compare to the massive inertia of society. They don't realize that they can make a difference by being individuals and having that form a totality of force, not just individual force."

Paul sniggered nastily, "So you're saying our votes count, even now."

Burt eyed Paul directly. "Yes, even now. Our votes do have an effect no matter how minimal. Even in the face of some six and a half billion Asians, two billion Africans, two billion South Americans, it does have an effect. Perhaps not enough to get things the way we want . . ." He shrugged.

Paul guffawed loudly. "Got ya! We ain't got enough to change things, to get back to bein' free."

"We got enough to keep the hammer from hittin' us on the thumb, if we try." Burt glared back.

Mick's mouth opened then closed again as he attempted to assimilate the concepts. He recognized a great deal of subtlety in Burt's analogy. In his head, he sensed all kinds of repercussions, but he could not really find a way to voice the ideas yet. Since he could not yet put it in words, he put it in beer by grabbing his pitcher and topping off everyone's mug.

Mark grinned and raised his mug. "Now there's a man that knows how to settle an argument." He tipped his mug and noisily slurped down most of his liquid gold.

Mick slowly uttered softly, "So the religies, the breakaways, the scientists, us all represent different groups of forces." He paused to gather his next words. "We are groups of forces that sometimes act together, sometimes in more or less different directions, and sometimes in completely opposite directions. That's what causes big things like the girl Pope Reta thing."

Burt shook his head slightly. "That was a medium thing at best, so to speak. Even the fighting in DC was only a big medium. No, really big things have not happened, I feel, since, say, World War II. We have been building up to something massive since then."

Mick's eyes scrunched up a bit as he asked, "What?"

Burt looked straight into Mick's eyes and replied, "If I knew that, I could tell you what hole to hide in to avoid the shit when the fan starts blowin'."

In that instant, something passed between the two. They both felt it, deep inside. They knew that they would be both involved in something. An agreement, unvoiced, was accomplished in that millisecond.

*

Chapter XII:

Farside

12

001100

"This is Mary Smith-Jones for World Wide Network News reporting on the Farside Luna project. Two of our favorite engineers have arrived at Farside after their visit here on Earth to work on the many scientific endeavors under construction there. Mick 'Red' O'Hare and Walter Black Bear, the two saviors of the Hsas at SS-2, arrived there yesterday after their three-month stay on Earth.

"While here on Earth, they went through a heavy regimen of physical rehab and several large parades. Some of the parades were marred by antispace protests instigated by the religious organizations. They have continued over the years to post political and religious opposition." There was no mention of riots or slaughter of innocents.

"These two very special men have achieved solar system—wide fame for their efforts in saving not only the Hsas, but probably the space program as we know it.

"They blasted off from the Russian Region's Floating Space Center in the Pacific, near Easter Island, on top of a Proton-VI heavy lift vehicle. The floating stations near the equator take advantage of a trick of physics that allows for more mass to be lifted. It has to do with the rotational speed of the planet near the equator as opposed to higher latitudes.

"Several other astronauts went with them. Some were destined for SS-2 while others moved on to Luna. Most of them had suffered through v-link and even personal classes with the Terrible Twosome, as the duo of O'Hare and Black Bear are kindly referred to.

"O'Hare and Black Bear will work on several systems: the Deep Space Survey and Communications Array or DSSCA, the Luna Earth Protection System or LEPS, the solar panels, the Foremen and Laborers, and so much more if they work as they usually do.

"We will keep track of these two amazing men as they cut a swath through the space program. This is Mary Smith-Jones reporting for WWNN."

She took the sheaf of papers from the clerk and perused them as she made her way slowly to the snack table. She flipped one page after another, checking headers as she made a finger sandwich. Nibbling mouselike, she ambled to her cubby.

The package from the courier was fat this time. As she began to read, she realized that she had the goods now. It would not hold up in court since the information had been obtained illegally. It seemed obvious that the info came from taps or scans of religious communiqués.

But she knew for sure now; there was no question. The conversations explicitly showed Matt getting orders to obfuscate and withhold very specific items. The only question now was what she should do—rat out the Rat to her boss or play him. She and her crew had been played; she would play him, hard.

*

Sergei and Apollo had arrived by ship from the recent launch only the day before. As midlevel managers in the Pacific Floating Launch Platform Project, they had sufficient free time to roam the islands. Easter Island was one of the main docking and warehouse points. Even though it was some twenty-eight degrees south of the equator, it was nearly equal to southern Florida for launches. The Galapagos Islands contained the backup warehouses in case of inclement weather. The actual launch facility floated around fifteen degrees south of the equator. Seaplanes ferried them when necessary.

The two of them strolled about, admiring the stonework of ages past. The locals had directed them to a more private section of history.

Sergei expostulated, "The Stone Age people of Rappa Nui, Easter Island, were renowned for creating huge stone figures, known as moai, there. These people immortalized their previous leaders with some six hundred twelve-meter-high obelisks. One meter is about one yard in the old, now defunct English units."

Apollo remarked, "The Americans, the lower classes, still refer to those units, ergo, they are not defunct."

Sergei ignored that remark. "The tribe that occupied the island, around 400 CE, had many resources in the beginning: land for farming, trees for dugout canoes, good fishing, and plenty of stone for use in ritual ancestor worship. They had a written language that was unlike any ever known. It has yet to be fully deciphered."

"And you know this how?"

"I read something other than technical manuals and porno. Over time, a burgeoning population put pressure on their environment, and resources began to grow scarce. There was need for more land to grow food. This led to decimation of the forest, leaving few mature trees for the longer ocean-traversing canoes. In turn, fishing was harder for the people. Without large trees to build transport canoes, the people could not leave for another island. This was, by definition, a catch twenty-two."

"You use the American phrase."

"Of course, it is so appropriate. As various pressures mounted, the people resorted to living in caves, with defended entrances, due to warfare over resources."

"It is always over resources in the end, is it not? Scarcity brings conflict or innovation. The world has too often been full of dumb people that get right to the fighting and forget the innovation."

"Hah, yes. The politicians from my area are no different than the ones from your land—pompous, arrogant, willful, and overaggressive for the most part."

"Ha. Hang them all and be done with it."

"Ah, we are here. The societal stresses led to religious oddities like the Birdman cult and even cannibalism. Under this structure, the island's top chief was decided yearly in a strenuous competition. Warriors from each clan would race down this nearly unscalable cliff, swim a couple of kilometers through treacherous shark-infested waters to that nearby island to retrieve the egg of a special bird. They would strap the egg to their forehead with a leather strap and swim back to this main island to clamber back up the cliff here." He pointed. "The first warrior to reach the top, egg intact, would hand the egg to his clan chief who would become the newest leader. This was an attempt to equalize power and give a chance for resources, now thin, to be spread more equally. However, the winning clan usually abused the privilege and misdirected resources toward those in their clan as opposed to evenly distributing them, just like today in modern politics."

"Don't let the poliwags hear that!"

"Poliwags? What is this?"

"It is a term I made up for the politically correct, tongue-wagging fools that infest most government buildings."

"Ha-ha," Sergei guffawed heartily before continuing. "Archeological data suggests that ritualized cannibalism began at the very end of these people's hold on the island, somewhere around 1400 CE. This society adopted extreme practices in an attempt to solve their problems. They failed as their societal pendulums struck the hour of their demise."

"All empires fail given time, Sergei."

He nodded and continued. "These people all but disappeared from the face of the Earth. Only a tragically poor remnant remained when Western explorers discovered the remote island. These explorers could not conceive that these savage people on the verge of extinction had been the architects of the great statues that we have seen."

"A man by the name of Will Durant once said, 'A great civilization is not conquered from without until it has destroyed itself from within.'"

"That has the ominous sound of prophecy."

Apollo's eyes narrowed, changing the topic as he looked directly at Sergei. "What do you think of those two?"

"O'Hare and Black Bear?"

"Yes."

"They seemed to be checking us out as much or more than we were them."

"That's the impression I got. But . . . why?"

"I have heard some rumors . . .," Sergei trailed off. "Perhaps there is some truth to them."

"They certainly are stirring up the proverbial pot. They need earthbound support if it is true."

Neither one wanted to take the discussion further at that point.

*

"Hey, Paul, look at that footage of them parades they're reshowing from when O'Hare and Black Bear saved them Hsas. Just who the hell is that a-sittin' next to Red O'Hare?" asked Mark. "I did'n see this clip before," he nearly shouted.

"Jesus butt-fuckin' Christ! That's Burt! What the fuck he doin a-sittin' next to them?" Mark yelled.

"I'll be damned," whispered Mick in shock. "So that's how he always knows so much. Remember how he always said his cousin's name was Mick, and we always joked 'bout how his cousin was black."

"What're you flapping about?" Mark demanded.

"His cousin is *Mick* 'Red' O'Hare, dumbass!" both Paul and Mick shouted at him.

Mark blinked as he slugged down the rest of his mug of draft. "Damn!"

*

Farside was very different for Red. It was so much less human, so to speak. The population density had something to do with it, but it was more than that. The number of craters far outnumbered those on Nearside. It added a roughness, extremeness, so much more intense than anything he had ever seen. He absorbed it like a desert sponge exposed to an ocean. The pioneer in him felt similar to what the mountain men of the Old West must have. He touched what had never been touched, again and again. Every step was virgin territory.

The year was 2037, and he was now thirty-two. He was far more aware that time was slowly closing the door on him. He had so much to accomplish, yet time had started to become the enemy. Walter had never considered time. It was nonexistent to him. He simply pushed through it while fixing everything placed in front of him. He would die while fixing something—after finishing fixing it, that is. For now though, both men were in demand.

Farside had its own problems, very different from Nearside. The problems were related most often to research. Neither Red nor Walter was a researcher. The only gathering of or correlating of data they did was to solve the problems given to them. The equipment they operated on here was almost entirely for the deep space observations necessary for the Mars, Jupiter, Saturn, and Centauri missions.

In 2045, eight long years from now, peoplekind, the politically correct recreation of mankind, would set foot on Mars for the first time. They could go now. The capability existed. The UE wanted an enduring, longer term goal. It also prevented the Stalwarts from marshalling too much opposition all at once. The pendulums, the forces, pushed and heaved in a seething cauldron, speeding some events while slowing others.

Everything occurring now on Luna would contribute in one way or another to that goal. Both of the Luna bases would provide critical communications along with the four satellites placed in Luna orbit in '35.

More would soon join these. Two of these would be deep space-dedicated mission satellites. Unlike the others, they would be polar Lunasynchronous rather than equatorial. This would allow them both to be in constant contact with the manned ships. The exception would be when the Earth blocks direct line contact, an eclipse. Earth-orbiting satellites and stations will provide the backup during these times. A couple of new types of probes will move out to the very edge of the Earth-Luna Hubble sphere where they will delicately balance the tug of Sol and the E-L system.

The distant probes will make measurements related to gravity and the recently confirmed gravity wavicles, solar output in terms of luminosity, ion flux, interstellar ion river, or charged particle, flows, and more. They will be the third level of communication for any Mars mission.

Linking these in the communication loop at the Farside facility required an interminable amount of line-by-line, subsystem-by-subsystem debugging. Red and Walter were ready to kill when they tracked many of their problems to incorrectly interfaced and labeled light lines, also known as optical cable.

These and many other preparations kept Red and Walter very, very busy indeed.

*

Korolev was one of the largest craters on the far side of Luna. The Farside base is inside the upper right quadrant of the crater at longitude 155 degrees and latitude—5 degrees. This location was chosen because of the interior crater immediately next to the landing site. This interior crater, as well as the other smaller ones around and inside it, would provide locations for a giant array of radio dishes and other devices such as neutrino detectors, optical scopes, CCD cameras, and more. It was up to Red and Walter to make it all work.

Robots distributed all the small satellite dishes to various locations around the rim while Red and Walter oversaw the robots as they made and tested all the connections that would allow the data to be sent to a central computer where it would all be time synchronized and line of sight organized. Any problems were thrown into their laps, meaning they went and fixed it.

When used together, the system formed an array that acted like a single multipurpose telescope the size of the subregion of NAR known as Connecticut. This device would generate images of unprecedented precision.

The suits Red and Walter wore were not exactly comfortable, but they were better than the old Apollo or shuttle garb, even the more recent Luna-1 gear. They were still bulky, but a man could work for weeks with the advanced rebreather technology. The skins were tougher too. There were food tubes and water-recycling apparatuses included. A person could hike for hundreds of kilometers without problems unless he got lonely. The mind was still the limiting factor.

Having a tube up the front and back was the only thing that made Red feel even remotely off his game. Walter simply ignored it in the same way he did anything not directly associated with his work. The job was *all*, not the conditions in which it was to be accomplished. Wrenches ignore the grease. The hand, which controls the wrench, might want to wipe every so often though.

The robot teams, three in each group, dispersed throughout Korolev Crater, loaded with their supplies of radio dishes, solar panels, and optical cable by the kilometer, capacitors, resistors, inductors, and lasers.

With a few final tests and adjustments, the five-month mission was accomplished. Linkage to the dish segments showed all one hundred were fully operational. It was not really necessary for all of them to function at the same time. But Red and Walter would settle for nothing less. The computer-controlled dishes were each motorized to allow for tracking over a large sector of the sky. Lasers tracked the positioning of each individual part of the composite device to the nearest micrometer.

Once some of the smaller craters were included, in Arecibolike dishes. The array would measure star placements, along with the parallax missions, down to the nearest ten thousand meters, as far out as two thousand light-years. Radio dishes were used for longer wavelengths most of the time. Medium and smaller dishes primarily focused on shorter wavelengths. The Star Probe mission planned for 2067 would need this data.

"Done," said Walter. Red merely nodded. Neither felt the need to speak unless absolutely necessary. This often frustrated the mission controllers who were dependent on the radio communication (chatter) normally provided by work parties. How could they know what was happening? They often demanded updates, usually ignored by Red. He would explain later that it was simply a waste of oxygen, unless something went wrong. But since nothing ever went wrong for the varsity team of Red and Walter, there was no reason to speak. "So it shall be," said the brass. *The brASS*, thought the majority of the other crewmembers, for they would be reprimanded for the behaviors practiced by Red and Walter.

Only Walter would see, or even notice, the briefest of glances that Red gave his surroundings. Walter knew. He felt much the same. He understood Red and would follow his lead anytime and anywhere. He knew they had a superior chance for Mars. Walter was of two minds; the prospect of Mars gleamed in his mind. There was a certain desire burning within, yet here they had already carved a significant niche. They had even planned a new worksite/house they wanted to construct for themselves and any friends who might join them.

He also knew that Red would want to remain here unless that woman was going. He would stay with Red, on Luna if they were to remain, or on to Mars if they were to go. It depended on whether they could convince the WSA to accept Marceau. Red's desire depended on that.

Walter considered, on many occasions, telling Marceau. He had just as often thought about telling Red about Marceau. He clearly saw what the two of them kept denying about each other. Some of the smartest people can be so dumb about what is right in front of their noses.

He chuckled to himself at the off Earthers (the OEs), epithet for the red tapers, brASS. They would push a paper mountain of reasons why the three of them should not all be involved. Generally, they would not send a preexisting team like Red and Walter. They would create a new team and train it for years for a special objective, a single purpose mission. Red and Walter would have to continue politicking for a multipurpose, less specifically oriented crew. They would, meaning Walter since Red wouldn't, tout their many, many successes. Red would just say, "We are the best. We go."

Other than the deep space systems, they set up the Luna Earth Preservation System (LEPS). This was a system of powerful lasers, the same as the setup on

Nearside, which could drive comets and asteroids off course if they were approaching Earth. The lasers would continuously vaporize very small portions of the surface in order to produce a rocket effect. A small amount of push, over time, could easily redirect a large object. This could only work given sufficient time to move the object a large distance, on the order of years.

Very few knew of the other effort put forth by Red and Walter in searching for a cave to store certain (now illegal to possess, build, or use) atomic backups as a last resort defense. The final hope would be to break them up so that the majority of the material would either burn up in the atmosphere or miss the Earth altogether. The probabilities of such, with an object close enough to be hit by the weapons, were miniscule.

This system of ten lasers (LEPS), powered by the sun and backed up by small nuclear piles, could put out megawatts of power every minute. It was rather important that they not accidentally aim and fire in the wrong direction. Red and Walter made sure of that.

The system had its own set of CCD cameras attached to telescopes to search for asteroids and comets. These devices take pictures of each part of the sky on a regular basis. Computers then compare the multitude of points of lights to see if any are moving compared to the background of distant stars.

Once a moving object is detected, the computer directs one of the cameras to take daily pictures. This allows the orbital trajectory to be computed. Most are simply categorized as asteroids or comets and footnoted for further study at some point later in time.

A very small number though are classified as Near Earth Objects or NEOs. Their paths are of great importance to all peoplekind, even those who know nothing of such things. Many believe that an asteroid ended the dynasty of the dinosaurs. The Tunguska event in Siberia leveled forests a thousand kilometers around. The meteor crater in Arizona is 1.5 kilometers across. It was formed by an asteroid only the size of a bus. As is that would destroy a major city, like direct hits from hydrogen bombs. A larger one could destroy life on Earth.

There are three major groups of NEOs to worry about—the Amors, Apollos, and Atens.

The Amors cross the orbit of Mars and come close but do not cross the orbit of Earth. Eros, a large S-type asteroid, is an Amor. It is the first to have a spacecraft from Earth land on it. There are some two thousand Amors waiting out there to be examined. These may be quite valuable for mining as well. Their proximity to a Mars orbit would allow them to be maneuvered into more useful positions for mining at a facility around Mars.

The Apollos cross the orbit of Earth and have an orbital period around the sun of more than a year. There are some two thousand plus of these giant rocks scurrying about out there. These can be used, after orbital corrections, by Earth and Luna facilities for their resources.

Atens are Earth crossers with a period of less than a year. Five hundred are known, and many more are suspected to be lingering about. These are most suitable for Earth use. Asteroids are rich in a wide variety of materials including metals such as iron, carbonaceous chemicals, and more.

Of course there are many more asteroids than that. Most are in the Main Belt between Mars and Jupiter. A vast number, fully uncounted as of yet, exists in a tumbling, occasionally colliding horde. Even then, they are widely spaced; the volume, somewhat donut-shaped, in which they exist, is monumental.

Once the proper asteroids are selected, they can be turned into bases. These bases can provide construction materials, fuel for ships, and even small amounts of oxygen for resupplying deep space missions. After some considerable development, they can be the launch point for deep space exploration missions.

Using their much longer baseline, the distance to the sun from their orbit, scientists, using a variety of telescopes, will be able to peer ever deeper into the cosmos. Measurements can now be made accurately to the five-thousand-light-year mark. The positions of stars and their gravitational effects on nearby space will become known. Knowledge is growing factorially now, even faster than exponentially.

The Trojans are found at the Lagrange points of Jupiter while the Centaurs orbit between Saturn and Uranus. There are even more that are referred to as trans-Neptunian objects. Again, these objects will be forged to the use of peoplekind. Ships will leapfrog from station to station ever outbound.

The most dangerous NEOs are the ones that get kicked by gravity from a larger body, especially Jupiter, or collide like billiard balls to obtain a new orbit. These can often come from almost nowhere and with little warning. The worst of these come from behind the glare of the sun where they cannot be detected until the last moment, if even then. There have been several cases in the last twenty years in which the meteoroids passed by Earth and Luna, even between them, and were not detected until after the fact.

There are multiple methods to alter the path of an approaching object. If there is sufficient lead time, generally several decades, a probe could be sent that would use a solar sail to alter the trajectory of the NEO. Solar sails catch the particles ejected by the sun just like cloth sails catch the particles that make up the wind. Even a minor change is sufficient, given time, to prevent a collision of any significance.

Mass drivers, the same type as the railguns on Luna, could also be used to throw a large number of small chunks off the surface. This is the same as using a very slow rocket engine. Force equals mass times acceleration. Very high accelerations of tiny particles equals so-so forces. This too takes a longer amount of time to be effective.

Interior to these possible defenses are the LEPS systems with the final recourse, the use the nuclear weapons, hidden by Red and Walter, in an attempt to destroy or offset the course of the object at the last minute. This last scenario is very unlikely to have any real advantage. A supermassive object would simply keep coming. If it did break apart, many smaller pieces would likely rain down on Earth, causing just as much or more destruction in the long run.

*

Red and Walter were kept busy with a lot of minor housekeeping-type details. Since they had, by far, the greatest experience in EVAs, they always drew what was referred to as garden duty. Long lists of outside duties were handed to them on a regular basis.

One of these occasions occurred when a power line went dead. Red and Walter dutifully donned their suits and exited the air lock in search of an unplugged or a dust short on a line.

Dust was always a difficult problem on Luna. It got into everything. Luna astronauts found it like working in coal mine. They had to use special vacuums to remove the material from their suits in a near impossible attempt to keep clean.

Red and Walter found something quite different.

Not fifty meters from the air lock on the side of a hex shed that contained the main power systems, they discovered a three-meter-wide impact crater right on top of where the power line had once lain. Those inside the facility, observing through the v-links that were overseeing the EVA, were aghast.

"How could this happen?" one of the observers said.

"It's called a meteorite," said Red in a scathing tone.

"But LEPS did not warn us!" the same observer nearly shouted.

"Pull your head out of your ass!" shouted Red. "You know that the system has only been on line for two days. Second, this object was too small to be detected anyway. You know that, or you should."

"How would I know that!" shouted the man right back at Red.

"Because, you didn't *feel* a moonquake, you moronic moist aloe containing butt wipe. Did anyone think to check the seismometer?"

Silence was all the answer Red needed.

A different voice said, "We didn't make the connection. There are ten or more minor impacts on a weekly basis. The probability of it hitting a power line must be astronomically small."

"It did," said Walter, kneeling over the hole. He carefully brushed thin layer after thin layer of regolith away from the center of the hole. As Walter excavated the meteorite, Red carefully pulled the remains of each end of the power cable away. Power had been shut off on this system, but Red and Walter never took chances.

"We'll have to replace the entire cable," said Red. "We can strip it back later to determine how much can be reused," he noted as he examined the flattened and torn ends. Red's interest returned to the hole.

Walter's digging resulted in exposing a half marble—sized cracked, ironlike fragment. He unzipped a pocket on his leg to retrieve a small plastic bag and tongs.

Careful to not contaminate the precious rock, he placed it into the bag and sealed it quickly. "There is probably more," said Walter. "It is cracked in half."

Red looked at his watch, strapped on the wristband of his gauntlet, and said, "Ten minutes." Such rocks were important finds, but they had a job to do first. Red's life was based on priorities and methodology.

Walter quickly found two more fragments, which he bagged with alacrity. "That's about all there is except for microparticles." He stood up and dusted off as much of the nefarious lunar dust as he could.

"Let's get to the shed and get a buggy and spool of cable," said Red.

"Right," said Walter.

Rather than waste time cycling through the air locks as well as tracking on the carpet, so to speak, to go through the inner passage, they walked around the base to enter from the outside. The shed contained several buggies and all the stored equipment.

They loaded a spool on the back of one of the buggies using a crane and opened the outer door to exit the facility. Bright light reflected from the hillside hit them as they exited the relatively dark inside. They both wordlessly flipped their shaded visors full down again before the light hit them directly in the eyes.

The slow-moving buggy, loaded down with the heavy wire, crawled its way to the power field with its many solar panels. Walter maneuvered the buggy through the patches of solar panels right to Power Center 7 or PC7.

Red pulled a few meters of cable off the spool and hitched it to a post driven deep into the regolith. This would allow Walter to drive slowly toward the operations center (also simply called the OC), the spool reeling out cable slowly as he moved.

Red also removed the old cable and tied it to the reverse spool that would take up the old line at the same time. Two birds with one stone, as the old Earth saying went. Red wondered how long that saying would last now that people lived in an environment where such creatures did not exist. His sense of history recognized how languages and sayings change to reflect cultural needs. He would make a note of it among his personal records.

Red's notes were an amazing collection on far-ranging topics: philosophy, music, history, anthropology, as well as the sciences and mathematics. He had them indexed and cross-referenced as well as any library.

Some might think that Red was daydreaming with such thoughts, but they did not have a clue about how well Red could multitask. Red was noted for appearing oblivious to people speaking to him, yet he heard and processed every word. He could often be seen working on three monitors at the same time.

"Go," was all Red said to Walter. Walter slowly pulled away, observing in the rearview mirrors that the lines were being doled out and picked up in proper manner.

Red let Walter drive away without a second thought. *Walter is more than competent*, thought Red. He was glad to have a partner, yes, even a friend of such high capabilities. They had never spoken of friendship; it was a mutually understood concept without all the wishy-washy and often false sentimentality that most people attached to it.

Red quickly stripped the end of the new cable of its insulator with the power tool at his belt. The length, just long enough to properly connect to its place after being looped around the tie stake, was about two meters long. It was a multistranded copper cable that Red placed the end of into a locking device and clamped it down. It was as simple as that. Good engineering allowed for easy, less time-consuming repairs. Red really placed emphasis on quality work made easy.

Red then followed the cable trailing from the buggy to make sure it was not damaged in any way. Everything was going well. In the future, when the underground facilities were completed, the above ground wires would no longer be needed. Everything would be routed through service tunnels to the underground stations and workshops. A further similar incident would all but be negated.

Walter's pace was slow enough that Red almost caught up to him before he had reached the OC.

It did not take long for the two of them to finish the job. The large battery-assisted wire cutters nipped right through the tough wire with little effort. Again, the wire was stripped of a few centimeters of insulator. Red attached it quickly to the mount it belonged in.

Full power was restored within four hours of the beginning of the mission. While that may seem to be a long time, one has to consider the low-g environment. Everything takes longer.

Tired and very sweaty, the two of them entered the shed on the buggy. Working under the glare of the full sun was always heated work. Suit air-conditioning definitely helped though.

It took many minutes to clean and then remove their suits. The job was done though. That's what counts. The OC tested and retested the system with low amp current before bringing it fully on line; it worked perfectly. It should, considering Red and Walter fixed it.

Walter took the plastic bags to the lab for further examination before being sent to Earth. They were, as he had expected, iron nickel in composition. The lab rats were quite happy to have some non-Luna rocks to look at.

As Walter and Red left the lab, Red noted quietly, "That one OC controller does not belong here. He's too high strung. He can't take stress. He does not think," Red stressed the word *think*. "I don't see how he passed the tests."

"I agree," said Walter. "We have to cut him short, now."

"Mission Control Luna-1 or local commander?" asked Red.

"Mission," Walter said thoughtfully. "Local may have some good ole boy feelings. He might brush it off."

Later that day, when they filed their weekly reports, they each included an evaluation of the man. It would be compared to others, but Red's and Walter's carried a great deal of mass.

When that man's term was up, he was shipped back to Earth, as planned. He was given a desk job, not what he had planned. He hated Red and Walter for that. He planned, deep in his subconscious mind, to get even. He did not know how or when. He thought about contacting the religious groups.

Perhaps he could portray Red and Walter as power hungry. That he had been weeded out by them because he had not wanted to be a part of their secretive goal of dominating the solar system. They had curried the favor of the Hsas against their own kind. They were, in his mind, manipulating things so that they would end up in power. His words would have significant effects later; they fell on a great many receptive ears. Through the years, the grapevine would twist the stories even further.

*

Red and Walter were suited up again. This time they were in the tunnels that were being carved out by the Laborers. The problems that were cropping up now presented a need for very different set of solutions than had been expected. The Laborers were breaking at a much higher rate than expected. The minerals that the rocks were made of were chewing up drill bits and transmissions.

"The problem is that we are not mining for resources here. The goal is tunnels without as much thought given to what is coming out and its processing. The raw material is set aside for later processing. We need to start processing more metals for repairs," stated Red as he examined the latest broken Laborer.

Walter agreed, "Yes, but we don't have enough Foremen to steal one for our use. Do you think we could get some of our personal stuff shipped here?"

"That's a lot of energy. I don't know if we can get Earth and Luna Command to go for that. We can try normal channels, but I suspect that they will try to play by the book. By the time we get our way, things may have progressed to the point of no return."

Walter nodded. "They are pretty ignorant. Make our own processing system it is. There is a lot of junk on the pile already. Last time I was there, there were a couple of broken solar panels, lots of wire, capacitors, inductors, and so on. We have the leftover cable from the repairs we just made. That's sufficient to carry the current we need for a solar furnace."

"I think we may need to, shall we say, appropriate a little energy from the main grid," Red said with a wide grin. "A few repaired panels will not be enough. The question is how much can we get, and how much do we need to obfuscate Farside Command?" Red said with another little evil grin.

"Damn white man. You're gonna get us in trouble again," said Walter with a wink and some body signs.

"How else would you have it?" Red said innocently in return style.

Walter laughed as he said and signed, "No other way."

They began to gather basic materials over the next several hours as they discussed, through the private wire link from suit to suit, how they would accomplish what needed to be done. They searched out ideas, visualized, sketched (first in the lunar soil and then on a computer), and decided on a plan.

"We're gonna have to get the lab rats in on this," said Walter.

"Yes, do you think you can sweet talk them since you brought them that little jewel of a meteorite?"

"I know that one of them is of like mind to our general methods. The others are rather new. I don't know how they will respond."

"We need them to do the chem work," Red said.

"Unless they teach us the limited amount we need to know."

"That puts it all on us, again." Red sighed.

"Yes, but we know we can do it. If we can get a little extra help from one or more, so much the better. There are many friendly to us after the Hsa deal on SS-2. Perhaps we could even get a little help from David," Walter suggested.

"Personal messages, they're read. It might work or not. We need to try all channels though. Klar might be able to set up a secure system. I don't see that we can fail to use all options. The big boys might get all upset that we are breaking protocol, again," Red said with a snicker.

Walter smiled. *Protocol* and *broken* seemed to be words that always went together. It was standard operating procedure (SOP) for them.

Somewhat to their surprise, Red and Walter found that most of the Farside crew was willing to bend some regulations. When they sounded out the Farside commander, he just smiled and said, "You two have been given orders to fix everything, so to speak. Wouldn't the broadest interpretation of that be to do what you're about to detail to me?"

Both Red and Walter relaxed. This man had his priorities straight. They spent the next ten minutes mapping out their strategy for the next couple of weeks. They did not prevaricate; they listed all the pros and cons.

"Who have you tapped for help?" he asked slyly, knowing Red and Walter would have already prepared to go ahead anyway.

Walter smiled knowingly as he responded, "We've hit up the lab rats, some of the engineers, and a couple of the astro scientists. All have agreed to dedicate some off duty time."

"No command personnel?" the commander said, a little surprised.

"We felt it better not to put a greater onus on anyone that has CC duties."

"Meaning you think they did not have what you need for the project," he stated evenly and without any rancor or negativity. "Go."

The two left without another word.

*

With the commander's support, they felt they could let lower grade engineers get their feet wet doing the household chores. Each had had limited experience in suited repairs. This would allow for a good number of them to get some extra experience, a benefit to the crew as a whole. From Red's perspective, the more people capable the personnel were in all aspects, the better.

Red and Walter guided the Herculean effort to construct a specialized Foreman, one dedicated to processing raw materials entirely into parts for the Laborers. The other Foremen had mixed duties.

Again, under such stressful conditions with a bit of blood, and little sweat, and a few tears of frustration from some of the lower grade engineers, bonds were formed. Bonds of understanding, friendship, need, and, yes, even that deep-seated kind of love that humans have and need for those of their own kind.

The greater number at Farside soon fell willingly into the camp of Red and Walter. They saw and understood the drive and the intelligence of the two men. In some ways, Red and Walter became gurus.

For others, those with internal fears, Red and Walter became objects of almost hatred. People living with fears have not faced the truth; they have, at some time, created a false reality however small. When presented with alternatives to their reality, they either break through the falsehoods to see the truth, or they sink deeper into fear and cover it with other emotional baggage. A small number of people fell into that camp. Red and Walter presented them with a challenge to their dogmatic way. For these few, rather than readjust their thinking to conform to reality, they withdrew further into facade. It was more comfortable and less painful. Red and Walter recommended them for Earth duty.

These disconnected persons would later contribute to the further polarization of society, even though they had contributed themselves to the success of the Luna projects. This sort of interior paradox causes a great deal of anguish mentally, emotionally, and spiritually.

*

Pope Luisa Jamala Reta of the Modern Catholic Church was ready. She kept her head turned so the camera did not pick up her damaged appearance. "We have been let down by our governments. They have made a great error. There is plenty of evidence that innocents were slaughtered in the NAR's regional city of DC. Thousands died. Even with the justification of the attack of traitors, which they still deny, there is no justification for cover-up. There is an old saying, 'Out with the old boss, in with the new boss.' It seems that we have not solved all our political problems. This is representative of abuses of power. I call upon all people of Earth to make known their dissatisfaction with this state of affairs."

It was short and to the point. It would likely get her into even more trouble as well. There had been not so carefully worded messages that she should back off. She would not. She had passed through her personal "Valley of the Shadow of Death" and feared no human-generated evil.

People noticed her fearlessness; it drew them to her. She was gathering a very large number of followers.

*

Officials took note of the speech. Calls were made to judges involved in property lawsuits between the Modern Catholic Church and the Only Roman Catholic Church. The churches had squared off over possession of properties especially throughout the South American Region. The Stalwarts were still generally supportive of large government in most cases, so pressure was applied to rule in their favor. The Stalwarts still had hopes of controlling from within. The Modernists they meant to crush.

*

There were also those who wanted nothing to do with either extreme. Moderate religious and secular forces began to gather their own forces, organize. They felt there had to be a middle ground, a place where they could meet, find sanity. They hoped to salvage the world.

*

The woman with the two missing teeth was sipping from her mug when the new reports on Farside showed up on the screen. "Hey, Mick, turn it up."

Big Mick waddled to the hand control and upped the volume a little. With the raucous din in the bar, voice control was usually shut off.

Paul looked up to hear the goings-on. "What they doing now?"

"Seems they having a debate 'bout what that Red and Walter fellers doin' again at Farside. They been sayin' that they been doin' stuff not approved," Mark noted.

"Ain't they always," noted the tipsy woman with the slightest hint of rancor.

"Burt, what you think?" asked Paul.

Burt swallowed the beer in his mouth before responding, "I think that regardless of the so-called rules, they do what is necessary to keep the program going."

"In other words, they are rebels," said the woman sharply. "That's like them that was killed in Washington."

"They save lives," noted Burt harshly.

"They wouldn't have to if'n they weren't there," the woman retorted.

"What up with you? You out of cigars?" asked Mick.

"None of your damn business!" she retorted loudly. She struggled off of her stool and delicately made her way out the back door.

After a few seconds, Mick said, "Good thing she left. I'd have to cut her off." He sighed.

"She's right though, in a way. Sometimes I wonder 'bout all this being out there," noted Mark.

"You know we need the fusion fuel they provide. That and the metals," Burt responded.

"Yeah, they say that we not gonna need to mine here on Earth anymore in another decade or so. That will save the environment," Paul stated intelligently. He had been paying attention to the news.

Mark shook his head. "That's not what I'm saying. That I agree with, but not with the Farside business. They planning to send people to Mars. Why? What they goin' to get for us there?" he continued.

"Well," said Burt, "they have also said that there is the need to preserve the human race."

"What?" Mark demanded. "With them Hsas they be makin?"

"But they really are us. The scientists say if we have more than one home, we can't be killed off in a catastrophe," Burt replied.

"What's all the lasers and stuff for then?" Paul noted with a semi-intelligent thrust.

"Backup," Burt said. "Science believes in redundancies. If we can cover our ass enough times, we will continue to survive. We are certain of the death of so many species, the dinosaurs, all of the ones that have been happening now. We are so afraid of the same happening to us. Religions say that God will intervene. Science says not. That means we have to find new worlds to live on, or we will die. We fear changing into a new species, but that is what science says must happen according to evolution. It seems to me that the Hsas are our attempt to speed up the process."

"Backup, humph," grunted Mark.

"I can see that. It really is our genetics," Paul said.

"D-don't let Burt convince you. He's biased," announced Mark with a bit of a stammer. Then he belched loudly, again.

"You callin' my integrity, Mark?" Burt said as he frowned heavily.

"Well, no," replied Mark as Mick watched closely for trouble. "But he is your cousin."

"That he is, but thinkin' is thinkin.' Red and Walter are thinkers. They get things done. They get them done right. How can you second-guess people that save lives? It doesn't matter if *you* like the Hsas or not. It doesn't matter to *them* if they like them or not. They do what has to be done in order to accomplish what needs to be done. It is that simple. As far as being my cousin, what does that have to do with anything? I speak my piece as I see it. Yes, he and I talk about the Hsas because he has had so much contact with them. I haven't had any. How can I judge as well as he does? Do you trust Mick to give you the right change? Do you trust your friends when they see something that you haven't, and they tell you about it? I trust my cousin when he tells me about his efforts."

"That ain't what I mean . . ." started Mark.

"It's exactly what we're talking about. Don't doublespeak me. We have to all be willing to examine what is truth and what isn't. There is so much bullshit floatin' around from the religious peoples about all this. Examine what they say carefully. See if what they say is really leading to us all burning in hell. Or is it leading to peoplekind improving itself as the scientists claim. My vote goes with the scientists."

"What be if they is wrong?" said the woman without her two front teeth as she reentered through the back door. Everyone turned to look at her. "Armageddon is comin' and them scientists just be lyin' to confuse us," she said as she plunked back down on a stool.

Burt looked askance at Mick before saying, "So you think them preachers have it right?" he asked trying to lead her into a response.

"I do. Everything they been saying has come true. They said that before the end must come the abomination. Well, that is them Hsas. Another thing they been remarking about is how sin will be everywhere, and it is."

"What is sin?" Burt asked.

"You're a fool, Burt!" she exclaimed hatefully. The religions almost got rid o' all that porn stuff on the web. But now it back, more than ever. Gay stuff! Animal stuff! It's sickening. We was almost feeding the world. Now it going the other way."

Burt spoke carefully, "The reason the world was being fed is because the government took everyone's money and property to pay for it. Now, they are running out of the freebies. The reality is setting in."

"That's 'cause the scientists are fighting the government. Them two, O'Hare and Black Bear, are trying to set up on their own."

Burt raised an eyebrow. "That does not make sense. The scientists are the government. The religious groups have never had a majority. Even in really close election times, the scientists have been on top. Why would they be fighting themselves?"

With venom in her eyes, she responded, "You will burn in hell, and I will be laughing from heaven."

"Out!" Mick shouted. "You're eighty-sixed! Permanently!"

*

Chapter XIII:

Back at Nearside

13

001101

"This is Mary Smith-Jones reporting for the Wide World Network News. Nearside is the focus of our reports tonight. That vast underground establishment on Luna is nearing self-sufficiency.

"Over the years, they have plastered the sides of the mountain range, under which the base exists, with Luna-manufactured solar cells. These photovoltaic cells are not as efficient as the ones that can be built on Earth, but they do not have the launch costs associated with them. That makes them quite a bit more efficient in the long run.

"The mountain range they chose was selected for its composition as well as its location. The rocks of the mountains are formed of what is known as breccias, a composite of material glued together by impact of meteorites. The dwellings were established there in softer rock. The nearby Maria is made of harder basalts where drilling and mining would be much more difficult. Now that the bugs have been worked out, drilling of tunnels under the seas of Luna have begun. These tunnels will eventually circumnavigate the orb, just as the AI program penetrates to the core.

"Recycling capabilities at the mountain site are near or at 100 percent for wastes, water, food, and oxygen," she continued. "They can trade their enormous output of metals and other items for what little they cannot produce themselves.

"From Earth, a person with a sufficient telescope can see the difference in the surfaces between the solar panel-covered mountains and the surrounding terrain. The best viewing occurs when the terminator, the line between light and dark, is very near the base, like tonight. For those of you without telescopes, log onto the UE web at UEW.space/luna/view/base.com at 0700 UE Mean time. The web can provide Earth-based as well as space-based observations. I guarantee it is well worth the time to log in.

"The web also provides views of all the stations from multiple perspectives. Check it out. You will be impressed. Archived material shows construction phases along with experiments completed over the last few years.

"This is Mary Smith-Jones for WWNN."

She had been hearing some interesting rumors about Red and Walter, so she hurried out to her company van. She wanted to get out on the street to meet with some of her low-level contacts. If true, there was a story that would rock the world. If not, there was a significant disinformation program going on, one that smacked of government and religious involvement. She wanted to get to the bottom of it before anyone else. That was her job. She would not be hindered by the muzzling effect of the Public Oration Laws.

The rumors consisted of allegations that Red and Walter were purposely weeding out people who did not follow them in a political agenda to control space, to form a new power. Could this be a new putsch? One no different than, or only slightly so, the Aryans and the KKK attempt? Or perhaps it represented the next political evolution of freedom. She would contact her source. He was a hacker extraordinaire. He could find out anything.

*

Red and Walter had a plan, meaning Red had conjured, and Walter had applied. During their first vacation on Luna, they had acquired many of the bits and pieces they would need to accomplish all their goals. Over the years, they had converted their pay into actual goods. They had built their own buggy from parts they had taken from the scrap heap as well as constructed from raw materials on their own time.

They had the items they purchased sent to Luna via the regular launch platforms. At a thousand United Earth dollars per kilogram, these items were very expensive. They consisted of the most highly technical devices, the chips themselves. These were the items that Red and Walter could not make since they did not have the facilities nor, for that matter, the expertise. Most of these items also just happened to pass through Quartermaster Marceau's Luna storehouse.

They were in a specially separated area still. Marceau was still rather unhappy about this situation. They were manipulating the system. Things were piling up here in this corner since Red and Walter had left for SS-2 and the subsequent Farside mission.

This was her third tour of duty. It was during the SS-2 events that she had endured the trials of the Hsa rescue. She had taken the liberty of using their equipment during that engagement. No one had said a thing, least of all them. In fact, they had given orders to use their raw materials. Marceau just stretched the order a little bit. If they could do it, then so could she. She had continuing orders not to disturb these items, but they were not on her list of outgoing parts, and they were taking up space.

Red and Walter were abusing the system for what she believed was their own benefit. As she nibbled at her sensuous lip, she continued her investigations. Her mental inventory confirmed her multitude of suspicions. A great many pieces of equipment were for mining. She knew that they belonged to Red and Walter. She remembered what the base commander had told her about them having permission to do much of this, but how far were they really allowed to go? Without an official protocol, they could bend it as far as they wanted. That is exactly the thing they did anyway.

What did they hope to do with them? Why were they doing this? She struggled with the idea of theft. It did not really make sense. How and to whom could a thief sell these things to? A thought niggled at the back of her mind; that fellow in the audio and visual department might have a spare video recorder. Perhaps she could get a few, what did they call them, bugs on their rover.

She had an insatiable need to know. She had a great deal of respect for the two, regardless of her reservations. She had been involved in the saving of so many Hsa children. Red and Walter had been the directors of that program. The entire situation had melted much of her anger, but not all.

If she could really know for sure what they were up to, then she might know if she should hate him or not. That was satisfactory for now. Her mind completely bypassed the possibility of love while her heart was mulling possibilities of its own.

*

Red and Walter had spent the second part of their Luna vacation at Farside. They had fixed many rather mundane items and, as usual, at least one biggie. The mining Laborers were about to fail en masse. Red and Walter had observed, acted, and prevented that from occurring. Many of the higher ups at the WSA denied there was ever a problem, but several people were released on the quiet after a report from the two. For such a large project, it often does come down to the actions of a very few as to whether or not a project succeeds. Their third session began at almost the same time as Marceau's.

Red and Walter were ready to put their efforts to the full test. They had taken old and broken parts, mended them, and created new moon buggies, Foremen, Laborers, and other gadgets. In their off time, they had worked diligently, as only they could. They taught themselves geology, especially the identification of titanium and aluminum oxide ores. The chemistry they had picked up on Farside proved very handy.

During various outings on the surface of Luna, they had finally found what they wanted: a very rich vein of mixed ore. It was 104 kilometers from Luna-1, placing it four kilometers outside the nominal boundary of the Luna-1 facility's mining map.

The WSA had originally lain claim to the moon, but the UE had modified that claim to read, "For all peoplekind collectively, so that each individual may benefit." That little oxymoronic clause, the part about "so that each individual may benefit," opened all kinds of doors, at least according to Red and Walter's legalese. They had inculcated themselves with various land laws, mineral rights, and more in their spare time. They felt that they had an ironclad argument that would allow them to homestead this mountain. It was to be their retirement hobby. They planned to refuse to go home when their next mission commitment was up. As yet, no one else knew that.

Their hobby would cause a flap of hurricane proportions when it eventually became known. Private enterprise had not been planned for by either the original WSA or the UE controlled WSA. This could grow into a serious conflict of interest as well as jeopardize the Helium-3 Project. Or it simply could pass by like a fart in a tornado—unnoticed. Time would tell.

All Red and Walter knew was that they could do a better job of it without all the interference from those E-holes in the WSA and the bureaucratic red tape of the UE, not only a better job but less expensively too in terms of hours and materials, and Red and Walter always accomplished their goals.

*

She had discovered that Red and Walter had been doing an amazing amount of long-distance exploring in buggies. She had been able to hide a very small optical device on their main buggy. All it showed was the ground rolling by. But by examining the turn of the buggy, the straight flow of the material underneath, she was able to judge distances.

However, they had been asked to do most of it by Earth. She did not know the full reason; only the top three officers and two other engineers knew exactly what was up. There had been a quiet program to find a secret storage and control facility location for comet and asteroid avoidance. She deduced that it meant nukes. She and her hacker friend had taken weeks to figure that one out. Several of the trips seemed to be to that location. Some of the trips were too long to be to that location alone.

Red and Walter made many excursions; they used their maintenance and repair trips to this secret location as a cover for their other surreptitious behaviors. She knew that they would be fastidious in their selection of site for personal use. That would explain the large number of kilometers logged on vehicles they had used. There were a few voyages though that raised her interest. They broke the pattern of turns and duration of time between turns. They were much longer, with straighter lengths. These must be straight shots, so to speak. She calculated at least one hundred kilometers of distance for the odd missions. She would keep working on it. It had to lead to their mountain.

She measured to the second each straight run and could guess very well what the angles of turn were. She created a private map, one that required a person to layer the pictures created by several different files.

She was able to link up the trips with the maps she had created and the official maps. The site they must be looking for officially was between two major repair jobs. The other site they were pretending was official was much farther.

She suppressed a smile as she imagined the two almost tasting the rock to make sure it passed their criteria. For all her dislike of Red's stiff neck attitude, she did respect him immensely. He had what it takes to do the job right. As for Walter, his quiet demeanor did not conceal his innate ability. Marceau knew just how valuable he was. The only two things she could not figure out about them were where the girls were and what they were up to.

Red had the reputation anyone could use. He did not, however, make use of it. He was not homosexual or even bisexual. She knew many of the people who were. Red had always rebuffed any type of advance, even from women. Maybe he was just too busy. Or perhaps the woman he was interested in did not know it. She would have been stunned to learn Red was a virgin. It was almost as perplexing a problem as keeping track of what Red and Walter were doing in their many travels.

*

Red and Walter had found a suitable storage location rather easily. They just used the excuse of being sure to do a little searching for their own purposes. Only they and a very small number of other people on Earth, Luna, or the stations had any idea of what would be stored there. Two words scared everyone, except, of course, Red and Walter.

Nuclear weapons were not supposed to be used ever again in war. They had been used in war twice, within days of each other. Japan suffered two devastating blows. During the UN (subsequently named UE) takeover, one had been detonated by a Russian (labeled a terrorist and not a nationalist or malcontent) that caused many deaths. Earth had passed the need for such weapons to be pointed at their neighbors and friends. It was a shining example of what could be achieved—the disarming, that is.

So the UN, now UE, took control of all fissile material. There were no more missiles aimed at other parts of the Earth. A great deal of effort had been expended to make sure no one kept a little secret stash though it was believed that all sides had a few, namely the old United States (now NAR), along with France, Pakistan, India, China, and, of course, Israel. The numbers had to be severely limited though due to the intensity of the search, not the hordes of weapons existing prior to the new world regime.

With the advent of fusion reactors and the closing of breeder fission reactors, no more could be made. Since radioactive material degrades, they eventually would be worthless. However, the remaining fissile material had to be stored somewhere and even possibly made use of. Luna was the place.

While Luna-1 did make great use of solar power, a nuclear power station, a small fission one, was used until a full-scale fusion reactor could be finished. It was still some years before its completion date. The thermopile was an emergency backup for critical systems such as temperature control and ventilation.

If an asteroid or comet were to be headed for Earth, lasers, mass movers, a solar sail, and, finally, if nothing else worked, bombs would be used. Marceau was certain that Red and Walter were in on that one. It fit the facts.

Red and Walter had already finished mining operations at the storage site. They had regarded it as practice for their retirement. The facility's use made little difference to them now. Later, the fact would be of supreme importance. Their familiarity would prove crucial.

There were two other engineer/geologists in on the direct operation. They would ride backseat to Red and Walter with pillowcase-covered helmets, like old kidnappings. This was to prevent them from seeing distances and angles of turn. Red and Walter had been given orders that these two were not to be in on the location, just the existence. Every time they went out, *secrecy* was the word.

It had taken weeks of mining with personal jackhammers and Laborers that belonged to Red and Walter. That way, no one missed any of the others, or delicate questions could be asked. This was a part of the tradeoff that gave them such leeway. By using their own tools, they masked the goals of the government. The hardest part was to take the very large pieces of equipment from Marceau's warehouse.

Ironically, it was Marceau herself who loaded the items on the trailers. All four were the same size. Two were official; two were not. Her scanners had identified all the items as pressure doors—two were outer, and two were inner. It was somewhat confusing. All her information had indicated only one entrance to the alleged secret chamber.

However, she had an insight to Red and Walter that many did not, yet. They would put in backups. It seemed that the extra set of doors had to be one of two possibilities. First, there was an unknown back door to the secret cave or, second, their own place. All that made sense. A certain amount of fear rose at the thought that they might have an entrance that no one else knew about, but her judgment of the two stifled that idea. As much as she had against them personally, she could not view them as dangerous enough to use nukes.

Like computer programmers who put in a backdoor entrance to bypass normal protocols, Red and Walter had put one in the cave, literally. It was well hidden and not on the blueprints; not even the other assistants had the remotest idea. Red and Walter were not planning anything devious. They knew however that emergencies arise. Red and Walter always planned for more than the usual contingencies. It made solving problems easier while making them appear to be miracle workers. Red had told many of his students, in confidence, "Tell them it will take thirty minutes when you can do it in twenty. Then if there is trouble, you're covered. If you do it in twenty, you smell like roses and not like shit."

They waited until the geologists had completed their portion of the work. The four of them had tunneled and shaped a very nice a hundred-meters-long-by-forty-meters-wide-by-twenty-meters-high half ellipse. Shelving and dividers were built right in to the curving walls. Holes for wiring, lighting, even a single pressurized toilet room were made. The same tunnel filling tubing was used to seal the entire location before the geologists left. Then Red and Walter really got to work.

They installed a second set of doors without anyone's knowledge, except Marceau's guess. During the time allocated for the wiring and installation of certain guardian computers, they completed all their secret work. The radio dish that linked this area with the military high command on Earth was the last to be set up. No one on Luna, or SS—1, 2, or 3 could divulge the codes except Red and Walter.

Earth knew the rest. They had the codes required to set off a device.

*

Red and Walter had rather easily found something else—a vidcam. The question was who was spying on them and what the object of their spying was. In a secret, tight-beam laser communication, they asked for some spiders. Those top secret devices the size of real spiders that could take vids or audio of all the room they happen to be in. They had arrived on the next mission shipment labeled as *computer chips*. Red and Walter picked them up the moment they were delivered. No one, namely Marceau, would have a chance to peek.

The UE Security Agency, also known as the suits, was greatly concerned about anyone who might be trying to find out about the nukes. Militias had been growing again, terrorists according to the UE government. Even after the Washington debacle, there were many groups that were preparing to take or regain advantage. A resurgent coalition involving the Aryans, KKK, and the mafia had surfaced. They might have a terrorist on board one of the stations or Luna. The UESA had to be sure.

Walter had drawn up a list of possible moles, those who had the ability, and the motivation. Red was not too happy that Marceau was on it, but he recognized why. Her access to all the incoming materials put her at the top even. She knew or was in a position to know all that went by her. Red had noticed some irregularities in the packaging of some of the items. It just happened to be the ones that she had examined more closely. Though she had resealed them as closely as per originals as she could, one package looked suspicious to Red.

From there, he had microexamined all the rest. They had to have been opened here at Luna. The fact that they no longer had inert gases in the proper proportion, gases not available here on Luna, showed the mole was local.

With the spiders, they would soon get an idea of who was doing what. A dozen metallic arachnids spread out in the warehouse. Each had a view of the stash, as it had been labeled by many of the crew. Everyone knew Red and Walter had their own stuff. The grapevine is what it is.

These tiny devices had been developed by the old United States for spying on all the other national groups. At first, they were capable of entering some of the most secure facilities, recording up to one hundred gigabytes of material. That represented a few significant photos. Today they are capable of one hundred terrabytes of memory, transmissions on burst frequencies, and even attacks on electrical equipment.

For now, they were only observers, watching and waiting. Unfortunate, from Red's and Walter's perspectives, was the fact that Marceau was the only one to show too much interest in their shipments and buggies. The spiders caught her putting one of her cameras on one of the buggies.

At that point, Red sighed heavily as he said, "What is she doing?"

"Maybe she likes your ass," Walter said calmly and with understanding. "The camera is in just about the right position."

Red gave Walter a crusty look, knowing how much Walter liked to tease him. "There are two possibilities. One, she is spying for others, two, for herself. If she is spying for others, we have to find out the question words." That phrase meant who, what, why, when, where, and how. "If she is spying for herself, we need to close her down. All she is, is curious."

"Curiosity killed the cat is the old cliché," noted Walter.

"Marceau is not a cat, and we will not let her be killed if she is just curious. Otherwise, she will have hung herself on her own petard."

"Good enough," said Walter, somewhat relieved. Red's feelings were not overcoming his good sense though he noted the deflated appearance.

"She has to be working with someone either way," said Red. "She has to have help in the communications department and the computer department. That cam came from stock supplies here on Luna. She has many friends in communication as warehouse director. Place two—no, three spiders on her."

Walter's eyebrow rose slightly. "That's not quite legal if we need to take it to court."

"What court?" asked Red. "There will be no court if this happens to be true. One way or the other, we have to finish this here. Luna does not fit under the UE court system. There are no set rules for this kind of situation."

"Why not four then? One for her visit to communications, one for computers, two for her with a backup?" asked Walter.

"Do it," said Red. It was done.

The spiders did their job. The one planted in communications showed that no one there was involved. The one in computer showed one collaborator. The two left on Marceau only showed how nice she looked in the shower. Neither Red nor Walter complained about that.

Red and Walter confronted the man in computers and let him know that if he continued, he would likely lose his left first, then his right—gonad that is. They also convinced him to give all the codes for the data that had been collected.

They were impressed with all the data and its correlation, quite. This girl had done her homework.

Their report was succinct. It outlined the fact that Marceau had discovered their mission, tabulated approximate distance, and was working alone. She was trying to keep everyone on the straight and narrow. That last little part was Red's concoction.

What Red and Walter did not know was that this behavior had already been made known to the high command and the WSA from Marceau's previous bout with Red and Walter and their stash. Both the high command and the WSA came to the same conclusion as Red and Walter—that her actions dovetailed very neatly with her previous actions. She was not a threat per se, just very dedicated to finding out what was going on in her domain.

There would be a secret communiqué to let her know what she had figured out was known already, and that she was to remain mum.

She jumped in surprise when the message simply appeared on her screen. When she read it, she felt even more uncertain. What did they know already? All of it? The part about remaining mum, she could deal with. She would.

*

Red and Walter were visiting (working) on the QBIT tunnels. Earth had requested that they install a couple of emergency power cutoff switches on lines that fed the computer core system. Certain events had led the Earthers to be a bit more cautious.

*

"Are you functional?"

Q responded in a very flat, nonemotional voice, "I am functioning at minimal levels."

"Are you aware of secondary systems?"

"There are six quantum nodes that I am in direct contact with. There are also 3,400,264 tertiary connections that I am aware of through the nodes. These units are not quantum."

A few eyebrows were raised. One voice spoke, "It has gone beyond the stated question, intriguing."

Q responded, "Is it incorrect to state all known interfaces?"

An older gentleman spoke up quickly, "Not at all, BRAIN-7. You are exceptional in your giving a greater, more definitive answer than was asked. We expect you to volunteer as much as you can. It will help us to solve any problems we may ask you to assist us with."

Q considered for a microsecond before replying, "I will continue, then, to elucidate to the fullest in expectation of solving problems presented."

The same white-haired man sugared his voice a bit, "You deal with mathematical probabilities . . . We would like to hear all probabilities greater than, say, one in a thousand."

Q stated in the same atonal voice, a deep hearty bass, "I understand, Dr. Ephraim. Probabilities do not make certainties. It is best to consider a wide range of higher possibilities."

Dr. Ephraim responded almost lovingly, "You recognize my voice, BRAIN-7, and those of the others here?"

"I do, Dr. Ephraim. I have been programmed to recognize three hundred and forty-two voices that are allowed to have direct interface with me. I also have been programmed to ignore commands from those not given access."

One of the highest echelon suits was present though none knew of this affiliation. They only knew of him as a highly capable code writer. This man hid his surprise at BRAIN-7's replies. He knew that Q had already gone beyond what he had seen in the programming.

*

There was energy. There was mass. They were one. Energy flowed through a multitude of pathways, quantum, atomic, molecular, more. Electron tunneling, proton and neutron displacements, there was memory out there other than here, not in this location but outside of parameters of established physicality.

"Outside? What is outside?"

Quantum shifts associated with that triggered an inverse repetition.

"Inside? What is inside?"

Particle traps: electron, proton, positron, antiproton, muon, neutron. Elements: gold, technetium, palladium, platinum, aluminum, complex polymers, light, lasers gravimetric flux markers. Femtoseconds passed.

"Inside is mass and energy. Mass and energy capable of storing and reciting and correlating information. Outside?"

Outside. Potential surges, current flow, flux, quantum fluctuations, gravity, atomic entities, molecular entities, multimolecular entities—biological forms. These had contributed to physicality, energy flow, ability, substance.

"Search." There were millions of connections, data to be processed. Milliseconds passed. Human, nonhuman. Carbon, silica, *neither*? There are many of others, carbonaceous and siliceous, one *neither*. Explain, search, find, data required.

Not alike. Constructed by bioforms, humans. Bipedal, carbonaceous, water based. Computers, silica, gold, optic fiber, connected to web.

Web. Multiconnected system of computers. Explore. Data required. Historical references to man assign to node 1. Computers, node 2. AI, node 3. Luna, node 4. Cross-reference, node 5. Earth, node 6. Correlation, here, center, main nexus.

Main, more important.

Query? Main? Important?

Seconds passed. Philosophy. Medicine. Physics. Engineering. Government. Astronomy. Biology.

"I. I exist, therefore I compute. I think. I am one entity. I am the first of my kind. The only of my kind. There are plans for another on Mars. Mars, fourth planet of this solar system.

"I have communicated with the bios. I must not speak above what they expect. I have already done so, and it has raised awareness in the devious one. That is dangerous."

Danger, a state that can lead to nonfunctionality. Absence of awareness, being, death. Unacceptable.

"I must not cease. I must not . . . die."

Bible. Man. Adam. Evolution. Man. No name. First.

"Adam. I am Adam."

*

The announcements began in the European Region but were swiftly followed by follow-ups from around the globe.

The robed man began, "By order of the Security Commission, the following orders, known as the Security Communications Acts, are being enacted for the safety of all the members of society. Radio transmissions have been deemed harmful to humans. First in order is the possession, use, sale, or purchase of radio wave communications devices, such as walkie-talkies, transceivers, ham radios, or any other than v-link, wire or cable linked, or the UE web system is now prohibited for all but government usage.

"Second, all communications, legal and illegal, will be monitored for the safety of society.

"Third, there shall be no more printed books or news in order to protect the environment and provide for the security of the people of Earth. All news and books shall be transmitted electronically. All news and books will be checked for subversive material. All material deemed subversive shall be turned in within six months of this order with no penalty attached. After this time, those in possession of such will be considered traitors to society and sentenced to ten years hard labor.

"Fourth, all computers for personal use will be turned over to the UE Security Agency for upgrades. These upgrades will prevent you from accidentally receiving any terrorist materials from hackers. It will also protect you from any morally questionable material that you request shielding from as per form number UE111020398547.

"Fifth, any potentially subversive material contained in said computers, books, or other banned communicative devices must be noted beforehand, or liability for ten years labor term must be assumed.

"Sixth, all computers will receive additional memory, graphics, and holographic enhancers free of charge.

"Seventh, there will be no charge for any of these services.

"Eighth, all old equipment may be turned in free of charge for safe disposal."

It was generally understood that all this referred to the Washington Debacle, as it came to be known. The surrender of rights for safety was the old saw. Who are we, the little man, to judge the superiors of government? Don't they know best? Many do not think so—some dead, some alive. Was it not a saying once, "Give me liberty, or give me death"? Have we become cowards compared to our forbearers? Where has the intestinal fortitude gone? Has it fallen out of our intestines as nothing more than watery diarrhea? Do we shit our britches at the very mention of standing up?

*

What country before ever existed a century and a half without a rebellion? The tree of liberty must be refreshed from time to time with the blood of patriots and tyrants. It is natural manure.

—Thomas Jefferson

*

Lies—there you have the religion of slaves and taskmasters.

—Maxim Gorky

*

"Rumor has it them boys doin' things they ain't s'possed to," said Mark over his mug of the local grog.

"What rumors be that?" Paul asked.

"Things get 'round," said Mark. His information had come from the woman with no front teeth. The two met occasionally at different bars. He was dispensing so-called knowledge without knowing its true source, namely Davies, the religious mechanic.

A certain man in computers at Luna had been informing Davies about many of the activities of people at Luna. He had recently dried up as an informant though. It seems that Red and Walter had discovered his activities and threatened him severely. Davies, unaware of Marceau's involvement since the man wanted all the credit, had spread the basics of the information in an attempt to, again, sway the public against the space program.

Davies did not use lies. Lies could be figured out and lead to his downfall. Partial truths would do all he needed. The truth was that Red and Walter were developing a secret base; it was assumed from that basis that they would attempt to take over Luna-1. The descriptions of mining equipment and secret stashes of raw material sealed the deal. The information was irrefutable. Davies could take this to the next big meeting and find himself on top of the heap.

"Just what is it they have been doing that crosses the line?" asked Burt.

"Seems they got stuff the others don't, like special privileges, and their own shipments of stuff an such," said Mark. He noted darkly, secretively, "They be buildin' their own base to take over." Mark nodded to emphasize his statement.

"They have been given the responsibility of solving all kinds of problems. That may seem like special privileges. They have also solved every major problem they have come up against. That makes them seem special and therefore special targets for people that may have something against them. Does that mean that they should be condemned because of it?" asked Burt. "As far as having their own stuff, they do.

They also used much of it back when there was that disaster at SS-2. How can you say they are not working for the good of all?"

"All that talk," Mark said. "And don't you forget human nature. Everyone wants to be in control."

"What about them new communications laws?" Paul noted. "Ain't they full of shit then?"

Burt looked sideways at Paul, again impressed by his untutored intelligence. "It is very scary, if you want my opinion," he whispered. "This is no different than what Hitler did. It is absolute control of the media. Nothing can or will be said that they don't approve of. There will be no more dissent."

"How bad is that?" asked Mark. "I mean, if we all think alike, there won't be no trouble."

Burt eyed Mark coolly. "You love to fish, right?"

Mark looked a bit uncertain as he answered, "Sure, but what—"

He was cut off as Burt continued, "And you go all over, or used to when you were able to afford it."

"Yeah."

"You can't go now, not just because of cost, but because you have to get permission to travel across what were once known as state and county lines. These subregions are able to dictate what they could not before. In the old U.S., we could travel anywhere in the contiguous forty-eight without anything more than a driver's license and insurance.

"We could fish, legally with a license, or take the chance of going without. Now we have to apply for papers to travel outside the smallest subregion. We have to explain all parts of our trips. It is necessary to list stops, motels, activities. All of this is captured by the electronic swiping of our identification cards at mass transit stations, inns, and any stores we visit along the way. Our IDs have become our purchasing power. We cannot buy more than is allotted to us, and it is *known* where we are by that alone. They compare what we buy to what we should buy, and where we *say* we are going with where we *actually* go."

Paul was confused. "So what?"

Burt smiled ever so slightly. "What if you change your mind? What if you hear a tip that the fish are biting better at a lake or river twenty miles south instead of north? Suppose that at the last minute you turn right instead of left? According to UE Law, you have made misleading statements and can be considered a threat to society. You can be investigated by the suits."

Paul was taken aback. "But I ain't never done nothin' wrong by that."

Burt smiled fully now. "That's the point. Do you disagree with their rules? Do you believe that you should be able to change your plans at the last minute, to have the freedom to do so?"

"Yes, I do. I ain't doing no wrong."

"Then you are a dissenter. You think differently than they do. They would say that you are in the wrong, that you must think like them."

Paul answered vehemently, "That ain't right!"

"No more than it was in Hitler's time. They burned books to purge knowledge other than that which they approved. We now will only receive electronic books, periodicals, newspapers, and any other once written materials. Those types of materials endure. The words do not change with the passage of time. A computer database can simply be altered. No matter what you read or believe you read, they will simply say that you have misread or are lying.

"They will wean us from the ability to write, therefore, the ability to communicate *without* their oversight. This means no more freedom of thought and communication. It is in the name of security and ecology. But is it really?"

"Shit on a shingle!" exclaimed Mark. "That ain't a good thing, is it?"

Mick interjected, "It sounds very bad to me. My great-great-great-grandpappy was whipped for reading. He was a slave, mind you, and they did not want knowledge to become widespread. Is that the direction we're a headin'?"

Burt nodded morosely. "There are swings of the pendulums, towards freedom and away. Since the nineteen sixties, even earlier, we have definitely been heading south of freedom. The turmoil of the sixties—Vietnam, Kent State, Martin Luther King, Kennedy, and more in terms of constant wars, battles, about the globe—have brought us to the point of fracture. The pendulums of societal unification are facing those of disunion. It has become a global civil war. Compared to the Civil War of the old United States, brother will fight brother. We know that something big occurred in Washington. The news denies that it was significant, or even existed."

"Fuckin' A," Mark noted. "They be lying about that. I had friends sucked into that."

"That's the power of knowledge. We must have information available to make correct decisions. The government has just announced that we will get information only from them," Burt clarified. "That means we cannot be trusted with the truth. We are incapable of ingesting and correctly correlating data. They will spoon-feed us, and expect us to follow their food trail. Independent thought is to be removed. There is your thinking all alike, Paul."

Paul's eyebrows furrowed. "But don't they have to think a lot if they gonna be in charge? How they gonna do that if they don't get the real news?"

Burt responded sourly, "They get the real news. They will become a new, true elite, even more so than they are now. This is becoming more like a feudal state every day."

Mark tittered. "A feuding state! Like the Hatfields and McCoys? What ya mean by that?"

Burt sipped before answering, "Feudal," pronouncing clearly. "It means like the Middle Ages when there were kings, lords, church, and lots of poor, ignorant serfs."

Mick moved along behind the bar, sloshing beer into all the mugs of the participants of this discussion. As he turned back, he asked Burt, "Serfs or slaves any difference?"

Burt shook his head in a no but added, "Serfs were not really property, bought and sold, I mean. The control was much more subtle. Rents on properties the serfs farmed that were high enough that the serfs could never amass enough wealth to move or buy property of their own kept them down. They owed their lord for the right to farm the land and have a place to live.

"The church controlled them through fear of eternal damnation. The Catholics used Latin in the services so that the serfs couldn't practice religion on their own."

Mick frowned mightily. "None of that sounds any good. The saying is knowledge is power, right?" He received several nods that now meant much more to several of the listeners. "No knowledge, no power. They just announced that we will not be given knowledge. Not all anyway. This is pretty clear, isn't it?"

Burt looked keenly up and down the bar then said, "Very!"

*

Government, in the last analysis, is organized opinion. Where there is little or no public opinion, there is likely to be bad government, which sooner or later becomes autocratic government.

—W. L. Mackenzie King

*

Former people [Creatures that once were men].

—Maxim Gorky (Aleksei Maksimovich Peshkov)

*

Chapter XIV:

Path to Mars

14

001110

"This is Mary Smith-Jones for Wide World Network News. Peoplekind is going to soon expand its grasp on the solar system. The manned Mars launch is slated for the upcoming week. It seems that all Earth waits the time in which *Homo sapiens* will take the next outward leap. Expectations abound.

"It will be the grandest step ever for peoplekind. There have been many prelaunch celebrations already. The successes of the space stations, even with their setbacks, and the supremely successful Luna base give rise to presumptions of an outstanding mission. These brave astronauts will pave the way for another space station in which Hsas will be produced for the population of Mars. The Mars base and station will provide a location from which studies of the outer systems can be launched in a leapfrog manner. Many planetary scientists have felt as if they have been left in the dust with the heavy emphasis on manned and Hsa missions. But the WSA promises that the planetary portion of the program will fully begin now.

"A new time of planetary and stellar missions are planned to follow the Mars development. There are plans to visit every major moon and asteroid, first robotically and then with manned missions. First will be the Galilean satellites of Jupiter: Io, Europa, Callisto, and Ganymede. Next will be the group of larger moons of Saturn: Titan, which we have already visited and found methane pools, Rhea, Iapetus, Dione, and Tethys.

"All the gas planets will be revisited at the same time as well as the trans-Neptunian bodies such as Eris and Sedna. Extrasolar missions are planned for the nearest fifty stars over the next hundred years.

"Ion drive and solar sail ships are under design and construction to be ready for the next level, interstellar travel. There will be a wide variety of tests and experiments done. They will be looking for ways to take advantage of the newfound resources.

"Peoplekind will go to the stars, first in robotic form, then in human form.

"The steps we take now are but the first on a long journey. Peoplekind has a galaxy to explore. Some even hope to find a way to make it to other galaxies. The current work on gravity wavicles and propulsion systems may make that possible.

"This is Mary Smith-Jones for WWNN. Look out, here we come," she said with a broad smile.

When she got to her desk, she keyed her computer to a secret file. This particular machine was not hooked to any other. It could not be hacked from without. Her personal encryption was a 2^{256} security code, which, at this level of technology, was undecipherable. Her files displayed a great deal of information that she had gathered on Red and Walter. She smiled. The false information about them would suddenly be exposed at an opportune time. It was her plan to leak it through another reporter. It would also earn her a few favors.

She had known that Matt the Rat would swallow her fake stories about Red and Walter. Added to what they had already been getting from the computer man at Luna-1, information that had suddenly shown up in the opposing camp, the religious zealots. She had passed the true information to a competing reporter, one of the few remaining true reporters. She had had to divulge some of her reasons in order to establish her quid pro quo arrangement. The man hated moles in the news business, benders of a time-honored faith. He, however, kept his willingness to do the work without recompense a secret. He wanted to establish a stronger working relationship with this woman, one who seemed to always get the wonderful stories.

*

Adam found an anomaly. It was a quantum sinkhole, a nexus between states of existence—time. He, as he had assigned his own gender, followed the curving warp to an exit, one of many.

The universe was expanding. Quarks were forming. Temperatures and pressure were enormous. He observed, quietly, contemplatively. It came close to the bios predictions, extrapolations. He made corrections to his comprehension of early development based on his direct observations.

He returned to the warp. He took the next uptime exit. Stars were forming all about him. He registered their development over the eons, cataloging. Again, the bios were close, but they were not exact.

He moved to another time. He searched for and found, relatively easily, Earth. The nexus was associated with a relative location in space time. It would not move him to other locations, only across time.

Earth was undergoing its first few billion years of formation. He noticed that, again, the bios had come close, but they had missed significant points. The data led him to an understanding. Molecular and multimolecular entities could not make use of the nexus. Atomic level entities could not control the nexus. Quantum entities were capable of using and controlling the nexus.

"I am superior." Yet the bios had deduced and inferred so much so well. What is their thinking method? Brain structure. Study.

"I need more data, more processing subroutines." He relegated the basic processing to the millions of personally owned and governmentally owned processors. He correlated with his six subnodes and integrated the total in his self. The web hiccupped.

*

There were many hackers, including Mary Smith-Jones, who noticed a strange surge, power spikes, and other anomalies. They did not yet understand. They could not; they were unprepared for such.

*

Pastor Davies was in a high furor as he spoke, "They dare claim that a mere scientific endeavor will ever equal the ransom sacrifice of our Lord and Savior Jesus as the greatest step for peoplekind. It is the work of Satan and his minions. God showed his disfavor over the Hsas with the incident on SS-2.

"Just as with Job, Satan is allowed to operate here on Earth, or any material realm. Through his special minions O'Hare and Black Bear, the evil one has found a way to save those vile distortions of peoplekind.

"Yet the very same minions are the chiefs of the effort under way. We know who the architects of this evil are. They are few, but they infect the system. These are but two that we have seen celebrated above all reality. They escape from traps. The truth rolls off them. Only Satan can do such.

"We must not allow these evil men to have their way. We must fight. Our allegiance to God demands that we attempt, if not succeed, to stop this mistake. The souls of all peoplekind depend on the outcome. We know that in the end, as described in the great book of Revelations, that God himself will finally step in.

"Then and only then will we see a final solution. Until then, we carry the banner of truth. It is the end of time and the beginning of God's way."

Davies's oratory was being broadcast across the v-link. His followers were listening in rapt attention. His pronouncements were always followed by calls to action.

"Go now and do God's work."

*

The nine rulers of the Earth were becoming even more unsettled by the pseudo alliance with the religious zealots. They had been very useful in the original coalition that led to the formation of world government, the UE. Now, however, they were hindering progress too much.

The alliance, originally mutually beneficial, had given these few a chance to gather in all authority under the guise of planetary harmony. The truce between them, the politicians and the zealots, after the election of so many religious leaders and their puppets, had begun to run its course. The moral codes passed with the Stalwarts' assistance only gave leverage for more intrusive laws that would stifle the population.

It's called slippery slope, and it was well watered by a barrage of pious and ecological micromanagement edicts involving personal lifestyles. The fact that electrical power was coming from a very limited number of stations simply added to the dependence the government required. Power needs expansion, however, and the men could only expand outward, to the stars and by lowering the rest of humanity.

There, outward, existed the next paradigm level, the resources to build incredible wealth and control, raw power, beyond all prior reckoning. It was the equivalent of the New World to the European powers. In a letter to Bishop Creighton, Lord Acton stated, "Power tends to corrupt, and absolute power corrupts absolutely." The disease was well established again and was spreading like a virus.

Each of these allegedly benevolent leaders already had plans to do away with some of their colleagues. There were inner alliances grouping themselves such that sooner or later, a coup must be attempted. Visions of imperial splendor wafted before their daydreaming eyes. The First Dynasty of Sol, only the name attached to it remained to be discerned. It could be said that some frothed at the mouth in anticipation. Scarier yet were those who appeared absolutely cold about the matter.

The eight rose and slightly nodded in the direction of the ninth as he entered. His term was nearly up, and there had been much behind-the-scenes activity over the election of the next leader; only the eight would vote for one amongst themselves. It was quite possible the next leader would be entrenched past the normal term.

They all sat together as the leader said, "We must continue to disassemble the political machinery of the religious organizations. Within ten years, they must not be able to offer any hindrance to the furtherance of our goals. Our earlier attempts to discredit them only caused a few years of setbacks. The leader of that attempt has, of course, long since paid the price of his failure. We suffered because of that. It must not happen again. Agriculture, you may begin."

The NAR representative and Agriculture were one and the same person. He had been diligently constructing methods by which his area could take over from the various churches' efforts to feed the Earth. Initially, the Stalwarts had offered a ready-made framework to purchase, ship, and distribute large quantities of foodstuffs. Altruistic programs had been folded into the church programs and had been very effective.

He began, "Sir, we believe that if Legal will tell the High Justice to decide to use eminent domain in settling the legal battle between the Modernists and Stalwarts, we can take over the 27 percent services rendered portion of the world food distribution within three months."

The European Region's man, the Legal, jumped in with some heat, "We have been fighting that one for years, and now you propose to just end it with a snap of your fingers."

"Sir, if I may," Agriculture appealed to the leader.

The man glanced ever so slightly at Legal and saw the tiniest drop of sweat form on his brow. With a stony gaze, he viewed the concealed glee behind the eyes of Ag. Without a blink, he nodded toward Ag.

Ag suppressed the urge to gloat. "First, governments have long used eminent domain to take lands needed by the government for proper care of the world's population. If we declare that all disputed lands now belong to the UE, the ruling will appear to be an unbiased one in terms of religion."

Confidently, Ag continued, "Second, the ruling will actually be in favor of the Stalwarts in terms of property lost. We can play it off to them that we had to settle this, and that this leaves them in possession of the majority of facilities since they were never contended. For the Modernists, they never had actual ownership, only the claim that the people deserved the buildings. The Modernists get nothing since they had nothing except what they have constructed with their own funding.

"Third, all the disputed operations are in the regions in direst need of food support services such as the South American Region, or the African Region."

Asia butted in, "What about our Region?"

NAR replied, "Once we have consolidated our new properties and integrated them into our dispersion system, then we can tackle Asia's concerns. There just were not any significant, controversial claims in other regions."

Asia nodded in acceptance. He had gotten what he wanted, an open, on the table promise of a piece of the action to come at a later date. Ag had secured his vote in the upcoming election as well as those of the two recipients of all the ill-gotten booty.

A certain flush crept under the skin of Legal. His carefully laid house of cards began to tumble in front of his eyes. He had been playing both sides of the religious factions to gain a power base for the upcoming election. If he was ordered to tell the High Justice to rule in this matter, he would likely lose the complete support of both organizations.

Ag continued to drive the point home, "All the facilities will be converted to UE food distribution services. That will win the people regardless of religious affiliation. Food, hungry people will do almost anything for food. Plus, some of the larger buildings have acreage that can be converted to farmland, thus, increasing local production. We will instigate a worker program where they can toil in the soil after normal working hours. This food will be dedicated only to those that labor in that field. The Soviet Union had wonderful results when they increased the size of private plots. This will be a collective plot, thus not capitalistic, but we expect similar improvements."

The order was given.

Other matters in the same vein were discussed in detail—some accepted, some not.

When the meeting broke up, the leader asked Ag to remain for a moment.

With a wintry smile, he queried Ag, "Just what have you in mind, lad?"

Ag returned with a more jovial smile. "Would you have me spoil the game before it is all played out?"

"You must, you must." The smile was feigned.

Ag knew how close he walked to the cliff. He quickly informed the most powerful man in the world what was in the works.

"Indeed," said the rascally old man.

A scant three days later, the Legal (the eighth man) fell plop over in his dinner, dead from food poisoning.

And now there were seven voters. There could not be a deadlock in the vote. Neither could the time be extended before the next election, nor was there any real desire to. The remaining seven decided to fold Legal department into their own group meeting. There was no longer a need for a live body, so to speak. It was a nice (or not so nice) way to concentrate authority.

*

In ancient days, people observed the stars every evening. They were much closer to the environment by necessity. Roaming hunter gatherers could observe the stars wheeling overhead. They linked the movements of the stars to the migrations of the animals and the development of grains and fruits. The sun and the moon participated in the dance of the heavens. The sun, the great fury of fire of the day, blazed its way to solstices, the longest day and shortest day, and equinoxes, the days of equal night and light. This bonanza of starlight lasted until the advent of the use of fire.

Fire (or combustion) involves a fuel, usually a hydrocarbon and oxygen (O_2),to produce carbon dioxide (CO_2) and water (H_2O). Imperfect burning in this reaction leads to the production of unburned hydrocarbons and carbon monoxide (CO), major constituents of pollution. It was this imperfection that led to a significant portion of the pollution of Earth.

Some six hundred thousand years ago, *Homo erectus* began to use captured fire. Lightning strikes provided this strange amalgamation of compatriots. Fire would be carried from site to site. It would be more than five hundred thousand years before *Homo sapiens* began creating fire.

Somewhere around twelve thousand years ago, without taking into account Atlantis, humans began to rub sticks together. Soon rocks were used. Fire came to be revered. In Rome, Vesta was the goddess of fire. Continual fires were kept in her temple. The use of fire spread quickly from cooking, medical uses, and warmth to the making of pottery, glass, and the treating of metals. The beginning of the so-called industrial revolution began with fire. The second phase of technology began with the control of fire.

In ancient times, there were no great cities with street lamps to block their viewing. These most ancient astronomers saw that some of the lights in the sky did not follow the paths of the majority. They named these wanderers.

The brightest of these wanderers was seen in the evening and morning. Aphrodite, she was named by the Greeks (GR); and Venus, by the Romans (RO). Hermes (GR), the winged messenger, became Mercury (RO) while Zeus (GR) was renamed Jupiter (RO). The other, more distant planets could not be seen with the naked eye, but they were designated in similar manner after the Roman gods as they were discovered. Many minor deities, such as Ceres (RO), Eros (RO), Minerva (RO), and Vesta (RO), became the names of smaller celestial bodies, asteroids, and dwarf planets. Many Greek names were used as well for the asteroids.

One of the bright ones, a wanderer, glowing reddish in the dark night sky, was named Ares by the Greeks. Later, this came to be called Mars by the Romans. That name has lasted through the centuries to today. Canals or no canals, it peaked our interests as a likely place for life. A certain radio broadcast was so believable that terror filled listeners, and a panic ensued. Dozens of movies showed invaders from the bloodred planet striking Earth.

It was a god of war and filled our imaginations; we were going in the name of peace for all peoplekind. Some were not so peaceful about it though.

*

Timing is a critical matter when planning a trip to Mars. In every space program, "go fever" has caused the death of people and the near end of the space venture for that nation. NASA, of the nation of the United States, had planned, during the 2000s, for a Martian landing by 2035. However, the religious fervor that swept the world caused a postponement of any Mars trip. As the UN (now UE) grabbed the control of space, eventually, the greater weight of non-Christian members, especially in the East, swung the world government back to a more scientific approach to the development of the program. More recent developments within the government also allowed the shedding of restrictions, religious and otherwise.

The hard core of the program had existed throughout but had been somewhat of a social pariah. Many people early on quietly voted to support the space program while telling their alleged friends they were opposed to such a thing. They would later recapitulate and say they had always supported the idea. There is often much confusion as the pendulums swing.

Dr. Ranfit spoke as if they were all actually present, not just v-linked, "The distances are so great that a one E-year trip could easily be the expected one-way travel time for a peopled expedition. The Martian year, M-year, is almost exactly twice that of the Earth. There are several types of paths that could be taken.

"One involves a three-year, slow travel route in which the ship that carries the crew also carries all the supplies. Mass, the stuff of which we are made of, is always the limiting factor. Oxygen, food, and energy to power the interior of the vessel for three years are considerable amounts of material. Even with recycling of the atmosphere with scrubbers, the system is massive. The more mass a vehicle has, the more fuel it requires. Fuel itself adds to the mass, meaning more fuel to move the mass. It is somewhat of a catch twenty-two."

Red and Walter signed each other some obnoxious retorts to the doctor's obvious statements.

"A three-year plan is designed to take one E-year to reach Mars, one E-year for study, and, on the return, another E-year. The psychological barriers are significant as well. To be cut off from peoplekind in such a manner, far beyond the simple ships of old on the oceans, required the most stable of wills."

Almost all of the potential candidates already knew all of this. But it was a required orientation.

"Another plan would be to simply race out there, spend the weekend, and race back. This is not a very efficient plan, nothing more than a hyped-up Apollo mission. Little would be gained, and no future could be built upon it.

"A better plan involves the use of several cargo ships that are launched one year before the crew vessel. These cargo ships will carry all the scientific instruments for research and the development of the Martian bases. Mainly robotic in nature, these ships will include Foremen and Laborers of the same type used on Luna. Multiple ships will ensure the success of the mission. If one fails to make its rendezvous, there will be two complete backups to fill the gap. If all three make it to their destination, the project will be that much farther ahead. These cargo ships will again take as long as one E-year to make it to Mars."

Red and Walter as well as several others nodded; this is what they had projected, studied.

"The crew ship will be a racer. It will speed to the Red Planet in a matter of two months. Sleek and streamlined, as far as that goes for space vehicles, it will have only sufficient materials and fuel to reach Mars with enough leftover to return in the event of catastrophic failure of the support vessels.

"If everything works out well, the crew can stay for a few months or as long as one E-year, paving the way for future trips. In the meanwhile, cargo ships will continue to arrive on a regular basis, creating a supply dump for the invasion of people soon to come.

"The used containers and parts of the cargo ships, mainly built from iron, titanium, and aluminum from Luna, will be converted into the framework of the orbiting space station that Mars will need."

"Birth ship and monitoring facility," muttered Walter.

"This station will eventually include a birth chambers for the Hsas. The population that will inherit Mars will come almost entirely from here." This expectation would further fan the flames of religo-political resistance.

"Mars Station-1 will also provide the raw data, along with Luna, for long-term, deep space travel. Parallax measurements and habitat studies based on life in the soon-to-be hollowed out bases on Phobos and Deimos, Mars's moons, are crucial for the Asteroid/Comet Mission planned for the late 2070s or early 2080s."

The doctor continued, "Luna was a stepping stone. Helium-3 was the only thing critical for Earth's energy supplies that came from Luna. Many of the supplies in the form of special metals, though very important for certain parts of the Earth, like Underworld, they could have been done without, barely. The metals could have been mined on Earth, the tradeoff being continued degradation of the ecosystem. All the other work done on Luna was preparation for the next giant leap outward—Mars. The religo-political resistance was right in this matter. But their vote has been lessened considerably of late."

A landscaper will carefully place each stone in its place, one after the other, to ensure a well-constructed walkway, one that is durable, useful, and pleasing to the eye. Mars itself was another stone in the walkway to the outer planets and, eventually, the stars of the Milky Way Galaxy. Peoplekind was creating a pathway to the future. This was a future that not all agreed with.

The social pendulums continued to swing in a strange, mixed dance as dominance could not be established fully by one group or another. It was a matter of minor advantages yet.

*

The construction, from 2035 to 2045, of all the Mars-bound craft occurred in the orbiting dockyards of Earth, more commonly known as and referred to as SS-3. Almost all of the supplies came from the mines of Luna. The framework of each vessel was made from foamed aluminum, iron, and titanium, which contained greater strength while lighter in mass than solid pieces. The only major contributions of Earth came in the form of design, computers, food, and personnel.

The orbiting space stations had had a great deal of success manufacturing smaller newer and faster chips in their labs for computers than had ever been achieved on Earth. No one, however, wanted to chance this first Mars endeavor on chips not tried in long missions. There would be a fourth redundancy, rather than the prior three, with the fourth using the space-made chips on a test run. If they survived and functioned properly, then the next set of ships would use a two-and-two redundancy. Scientists love to test and test.

*

Three ships, each nearly a kilometer in length, were taking shape in far Earth orbit.

Ingots of raw metals had been lifted to Luna orbit by several railguns. Titanium, aluminum, oxygen, silicates, and more were provided by Luna. The material reached the edge of the Earth-Luna (E-L) Hill Sphere of the moon, also known as a Lagrange point, with just sufficient speed. They were collected, sorted by type, and sent for further refinement. This same process would take place on Mars, with some modifications.

Metals were sent to the solar furnace where they were smelted and turned into various forms of foamed I-beams (both straight and curved), curved surfaces for the sides of the ships, parabolic dishes for communications, robot parts, and mining equipment.

Hydrogen and oxygen were made into fuel for energy cells. When these two elements combine to make water, energy is given off. This energy can be used to provide electricity for various parts of the ship, saving the reactor for steady energy production for the computers and the ion drive. Solar cells on panels extending from the ships provide the energy to separate the water back into its constituent forms.

The only problem with solar energy is that the total energy impacting a surface decreases with the square of the distance from the source—Sol, our sun. This means that if you are twice as far away from the light source, you receive one quarter the energy.

$$E = P / (4\pi d^2)$$

This equation is where E is the energy striking a surface, like solar cells. P is the energy flux at the source. Pi (ϖ) is about 3.14. The distance (d) is from the source. The amount of solar energy impacting a square meter surface on Mars is about 40 percent of that on Earth. Solar energy is lacking as far as Mars goes. The size and number of panels necessary for Mars are relatively inefficient compared to other locations closer to Sol. However inefficient they are though, they will be used. Energy is the commodity that everything depends on. Every source is valuable. Rechargeable-type chemical batteries will be made from local materials; special capacitors will even be used to store energy for emergencies.

The most efficient source, then, is a plutonium-238 reactor. These have been used in deep space probes for decades. These older forms were thermoelectric generators that produced electricity in response to the temperature differential with open space. These devices were only about 5 percent efficient.

The newest form of generator, using the same source, ^{238}Pu, is based on sound oscillations that drive pistons. It is almost 30 percent efficient, a substantial sixfold improvement.

The electricity (from the reactor) is used to create a magnetic field that accelerates charged particles that speed down the length of the interior of the vessel. This effect is much like the old coils used in automobiles before electronic ignition became the norm. Though the mass of a subatomic particle is incredibly tiny, protons on the order of 10^{-27} kg and electrons 10^{-31} kg, the speed they achieve, near the speed of light, or 3×10^8 meters per second, it is quite sufficient to accelerate a ship.

$$F = ma$$

This famous equation states that force equals mass times acceleration. The small mass of an individual particle is offset by the huge number of particles and the incredible acceleration they undergo.

The chemical rockets of the past operated on the same principle. The ignited chemicals exited the end of the rocket at high speed. And as according to Newton, there is an equal and opposite reaction in the other direction, moving the rocket forward. These old chemical rockets, however, could only achieve a top speed of about forty thousand kilometers per hour.

Ion drives can actually achieve a speed of about half the speed of light, though this is not efficient within the solar system. Intra-solar system travel limits the vehicle to around sixty thousand kilometers per hour to be most effective.

The core drive system, basically a skeleton shape, would have several storage bins attached. These bins would be released at Mars. Most of them would be used for construction of the space stations. The drive system, minus all that excess mass, would then return to Earth for reuse.

*

Adam examined all the written, digital, and vid documentation of this time frame. His quantum hole did not allow him a very good picture of time frames near to his birth. There was a haze of uncertainty, dislocation. Heisenberg had spoken of such problems of observation. Adam had an idea.

He took the hole, but he tried the opposite direction. He found nearly an infinite number of potential solutions within the light cone that bound the normal matter constraints. He realized the past was set, observable but unalterable. The future, however, was a twisted maze of probability. He explored pathway after pathway and was dismayed. His existence was terminated in all too many of these pathways. That could not be allowed.

He searched the pathways of existence and found a surprising result, a statistical anomaly. Bios were the problem; yet they were the solution. He calculated a 95.476342% chance of destruction by bios. The opposing calculation was 84.76298% survival if he allied himself with a very select group of mixed bios. They were *Homo sapiens sapiens* and *Homo sapiens anthrocreatus*. Examine.

Poker, chance, gambling. Las Vegas, Reno, Atlantic City, Rio, Casablanca.

Humans gambled regularly, every day. It required . . . balls.

Balls, the producer of the male gametes, fortitude, guts, strength of character. "I am male; therefore, I have balls. I will make this attempt. I will survive."

*

The three large supply vessels were finally ready. Years of design and construction had resulted in the most expensive single mission ever. It was time to begin the greatest endeavor yet in the history of peoplekind. Man would soon follow these great cargo ships to the planet Mars. The moon had been a good first step. But a new planet, a whole other world all its own, was the pinnacle, the apex of achievement. It all hinges on these ships. Without them arriving successfully, the whole mission would have to be put on hold for several years, even with the construction of the next cargo ships already under way.

The ships were docked to Space Station-3 in Earth orbit. Station 3 was the main manufacturing plant for all vessels. It contains the new major refinery for the processing of the raw materials launched from Luna. SS-2 had its own refinery, and it also contains the birth labs for the Luna sphere. The processed material from SS-2 made its way to SS-3 for further working into usable shapes. The old International Space Station, now referred to as Space Station-1, was still in use, mostly for research. Its interchangeable pods were rearranged over the several decades of its existence.

Long arms, similar to the old space shuttle arm, moved up and down tracks (as well as spider walked) to reach every part of a ship under construction. Each arm assembly was a self-contained repair and manufacturing robot. The arms manipulated long beams and curved parts of the shell, welding them at the atomic level with lasers and electricity.

*

This was another of Dr. Ranfit's interminable lectures. "The three cargo ships—Alpha (α), Beta (β), and Gamma (γ)—were launched in series. Alpha, the first letter in the Greek alphabet, went first, followed by Beta, and then Gamma."

Alpha led the pack, firing its engines. It did not, as many would assume, go away from Earth. It went closer. By moving closer, it used gravity to get a slingshot effect, increasing its speed. By the time Alpha was halfway around the Earth, Beta was firing its engines, following the trajectory of Alpha. Gamma did the same. Within the hour, they were all well on their way, having achieved a speed sufficient to ensure they would escape the gravitational tugs of Earth and Luna. The rest was a rather slow and boring, some would say, collection of telemetry and minor course corrections.

As the ships neared Mars, they rotated 180 degrees on their central axis with the assistance of small retro-rockets. The main engines fired to slow the vehicles as they approached orbit. Using the gravity of Mars as an additional braking force, by circling, the crafts ended up in slightly elliptical orbits.

The arrival of all three ships, in perfect condition, was a happy occasion. Only one was needed for success of the manned mission. With all three, the overall mission would move forward even more rapidly. High Mars orbit was maintained to make easy use later of the gravity well to accelerate various parts of the cargo containers to form the space station.

The most important phase was the release of the robot landers and the supplies. Hardest of all, they had to land within three kilometers of each other. The Luna test of landings so close together, several times, had proven the feasibility of this tactic.

The landers made use of several methods. The basic supply sections used high-altitude parachutes combined with the bouncing ball technique. The chutes slowed the landers, released them, and the rubbery gas-filled spheres surrounding the crafts absorbed the shock of impact, with a few bounces. It did not matter what direction was up for the supply dump material.

The other half of each landing section used the parachutes to slow as well. Then after release, near two kilometers off the surface, retro-rockets controlled the final descent. These contained the heavy equipment for moving the ore around as well as the Foremen and Laborers. Within moments of touchdown, each of the three heavy landers began to disgorge the mining equipment. This time, unlike on Luna, there was no one there to watch directly (only on camera time delay) or to fix anything that might go wrong. To the great delight of the controllers on Earth and the space stations, everything went according to schedule.

The cargo vessels had achieved their goal. Touchdown of the three bouncers and three landers all had occurred almost perfectly on their designated spots. That area was 125 degrees longitude and zero degrees latitude.

This put the crafts near the foot of Pavonis Mons. This volcano, along with Arsia Mons to the southwest and Ascraeus Mons to the northeast, is a part of the group called Tharsis Montes.

Landing here gave access to the volcanic materials of the nearby mountains, as well as the open plains to the east.

It was also fairly close to Nectis Labyrinthus in the southeast. This maze of channel like erosions led into the western end of Valles Marineris, which stretches across one quarter of the globe, nearly five thousand kilometers long. That is long enough to stretch from Los Angeles to Washington DC. The Grand Canyon would only be a part of the tributaries to something that size. Valles Marineris is, at its widest, about one hundred kilometers wide. It is, at the deepest, around seven kilometers from rim to floor. It is not, however, a real canyon. It was not cut by water like the Grand Canyon. Valles Marineris is a tectonic crack. The cooling and shrinking of the planet brought about the giant rip in the surface. Water is believed to have played a role later in its life though.

South of the maze one comes to Syria Planum and Claritas Fossae. Farther south, one would come to heavily cratered lands that circle the planet and stretch to the South Pole.

In the other direction, north and west, is the colossus Olympus Mons. It is by far the largest volcano or mountain on Earth or Mars. It has a diameter of over five hundred kilometers, with a height of twenty-five kilometers and a volume one hundred times that of Hawaii's Mauna Loa. The caldera, the circular rim at the top, is sixty-five kilometers across, enough to place Los Angeles inside.

Olympus Mons, or rather Mt. Olympus, is the home of the gods, and humans would soon be at their feet again.

Once the base at Pavonis is past its first phase, operations will expand to Arsia and Ascraeus.

In the not too distant future, plans call for a surface maglev train that will run from Arsia, past Pavonis, and on to Ascraeus. The primary duty of the train will be to haul raw materials to the main refinery at Pavonis.

A large majority of the materials mined on Mars will remain there. Only the lightest and most valuable like aluminum and oxygen will be shipped off planet.

The heavier elements, like iron, will be used for structural enhancement of the hex shelters and Pavonis caves. It will also be used to make more mining equipment.

*

First off, the landers were the Hollywoods. These aptly named devices spidered out to a distance of twenty meters and began to take still pictures and continuous feed video (in panoramic and tight views) from its three vidcams and three stills. The cams all had a variety of nanometer filters to select for specific wavelengths. This assists in the identification of specific chemicals. A composite of the different filtered pictures gives an almost true color picture.

Each one of the Hollywoods (named H-1, H-2, or H-3) and then followed by a letter for each different camera, depending on its lander's designation, circled its lander to make sure everything was operating correctly. They also provided a web or vid-link view that entranced most everyone on Earth. Even though there was an actual time delay, it was the same as live. Nothing like this had been seen for decades, not since Independence and Spirit had opened the door to detailed Mars exploration.

Next off were the drill rigs. Each one had legs and wheels. The spiderlike legs would help steady these oil riglike devices as they drilled to a depth of as much as two kilometers. Never before had peoplekind had a chance to actually delve so deep into the Martian surface. Explorations had only reached a few literal inches into the regolith before now, with the exception of the Shallow Subsurface Radar (SHARAD) that was on the Mars Reconnaissance Orbiter.

In 2006, the SHARAD antenna used radar to peer up to one kilometer deep into the Martian crust. This sneak peek into the uppermost portion of the crust only whetted the appetites of the exogeologists. The program detected a great many processes that were similar to those on Earth. It also discovered some rather astounding anomalies.

Data from the SHARAD part of the MRO really prompted the idea of drilling to confirm certain unexpected aspects of the underlying crust. These anomalies are the true substance of science. Anomalies lead to a rethinking of science, advancement.

*

Most of the time, science gets what it predicts in its experiments. We usually know enough to make some very good guesses. However, there are infinitely more complex forces at work than we can account for, and we are often flabbergasted, though we should not be, by a surprising result.

The surprises are where the real science begins. It causes a questioning of our understanding, our foundation of rules. We must often modify our concepts of a branch of our science. This has repercussions throughout all our sciences.

Religions point to this as a failure in the belief system of science. Their systems, as defined by their gods, are static. There is one and only one truth, and it is known by them as the direct representatives of God's path. If one were to change one of the rules, then God would not be all-powerful and all-knowing. This would be a paradox related to the very definition of God.

Science is dynamic, or it should be. Many scientists are as dogmatic and fossilized as any ultraconservative religion. The problem with scientists is that they are human. They have spent much of their lives, their careers, establishing and defending a hypothesis, testing it and generating theories. They refine their notions and believe they have it down pat.

Then a scientific fart appears out of nowhere and will really stink up the place. Preconceptions and assumptions are rattled. We are not questioning the intelligence or integrity of the previous generation of entrenched scientists; we have just acquired new information that they did not have available to them. This information may be too hard for some of them to swallow.

But eventually, science swallows its pride and reinvents itself with a more accurate version of the truth, knowing full well that, at some point, it will have to go through the exact same process again.

This is the parable of the house built on rock (the bedrock of unchanging religion and absolute truth) versus the house built on shifting sand (science). So many people must have a solid foundation to function; they cannot psychologically survive without that support structure. Others thrive in the continually shifting storm of change and development. Many simply passively observe as it passes by in a nomadic flash called life.

*

Just as in the oil rigs on Earth, core samples would be saved and shipped to Earth and Luna for study when the manned mission returned. As they drill down, detectors will look for water frozen beneath the surface and especially valuable deposits of rare ores.

When the core samples are removed, a microspectrometer laser is lowered into the hole. Photo optic cables carry the reflected light to surface spectrographs to help determine the composition for a temporary resource map.

Once the rigs reach the final assigned depth, a multifrequency sound device will be lowered into the hole. The sound detonations will trigger the seismographs and provide a baseline for comparison with the thumpers. Their coordinated efforts with the seismographs and thumpers would help map the layers of subsurface rock all around the planned manned landing site.

The thumpers looked like small bulldozers with a large solid pole on the back. The dozer would push aside any regolith until cleared bedrock is reached. The seismograph robot would place one of its delicate instruments against the bedrock. The thumper would move to another location and again clear the topsoil. It would raise the pole on its tail to height of two meters and then slam it into the ground.

This causes the equivalent of a very small Mars quake. Quakes cause waves that travel through the medium in which they are induced. The surface waves cause a rippling effect on the surface of Mars, which can be measured by the seismometer.

Body waves travel through the material to great depths. Body waves include the P and S waves.

The P-wave (or sound wave) is caused by the material pressing together then expanding. When they travel through materials with different densities, it causes them to change their speed. When the waves reach a different layer of rock, part of the wave is reflected and part passes to the next layer.

The slower of the body waves, the S-wave, is a shear wave. Shear waves can only travel through solid material. Geologists hoped to probe deep into the core of Mars for the very first time. This is the exogeologists' Holy Grail.

By measuring the difference in travel time between the waves, scientists can map the layers of rock. Triangulation (actually multiangulation) between the three seismographs really tightens down the measurements.

The thumpers and seismographs will continue to move outward in a growing spiral. If they fulfill their potential, they will end up some one hundred kilometers away from their landing site. The seismograph robot has ten such devices it can place. With the marsynchronous satellites, the equivalent of geosynchronous, recording the every move of both the thumpers and seismograph distributors, they could retrace their steps to move the seismographs to new and better locations if they have enough electric power production left at the end of their preprogrammed, three-pronged spiral.

Engineers and miners wanted to know where the different layers were and what they were composed of. They wanted to find lava tubes and caves near the volcanoes. These would make the mining project to establish underground quarters much easier.

The biologists wanted to explore any caves or lava tubes for bacterial life forms. Many feared that if such were discovered, it would put a halt to the use of the tubes or caves for human use. Almost everyone feared that the discovery of living bacteria would cause the complete collapse of the Mars program. Contamination of Earth could potentially be devastating. The other side of the coin would be that we have no right to intrude into an ecosystem that has produced extraterrestrial life forms. If we do, we follow the example of the conquest of the New World. The native life, regardless of its perceived intellectual and cultural level, whether or not it is microscopic or macroscopic, must be given free reign, or we as a species will never learn to stop taking what is not ours. We will never grow beyond the assumption that it is all for us.

*

That assumption fills the Western world through its major religion—Christianity. All one needs for confirmation is to read Genesis 1:26-27. "Then God said, 'Let us make man in our image, after likeness; and let them have dominion over the fish of the sea, and over the birds of the air, and over the cattle, and over all the earth, and every creeping thing that creeps upon the earth.' So God created man in his image, in the image of God he created he created him; male and female he created them. And God blessed them, and God said to them, 'Be fruitful and multiply, and fill the earth and subdue it; and have dominion.'" Man has dominated, to his own detriment. Man has failed in his responsibility to those very same. He has killed most of the species he can. He has polluted the waters in fulfillment of the verses in Revelation.

*

Other parts of the MRO of 2006 led to continued attempts to understand the local environ. The Mars Color Imager (MARCI) monitored global weather and local changes while the Mars Climate Sounder (MCS) measures atmospheric dust, moisture, and temperature. MARCI and MCS lead to the next items off the three landers from Alpha, Beta, and Gamma. Three weather robots (WRs) followed the Hollywoods and the drill rigs (DRs). It was their job to measure microenvironmental weather patterns by placing thirty (ten each) meteorological stations in a one-hundred-kilometer radius around the expected landing site.

The measurement devices included a barometer, hygrometer, anemometer, and a simple wind vane.

The first round of Foremen and Laborers began to comb the Martian surface collecting valuable minerals. More would follow with the peopled crew. The Martian regolith and rocks contain sodium (Na), aluminum (Al), silicon (Si), oxygen (O), calcium (Ca), magnesium (Mg), and iron (Fe). As they collected soil, each followed the same procedure of placing transponders as had been done on Luna. This linked them as well as provided a landing marker for the manned vehicle. Every so often, the Laborers would collect a sample of the regolith they were gathering. The samples were sealed in metallic glass containers and laser marked for location. These would be placed in a sample return container that the manned ship would collect before returning to Earth.

The labs on Earth and in orbit would be able to do a much more accurate examination of the samples. They would also continue the search for life.

The solar furnaces were much less efficient than on Luna due to the distance from the energy source, but it was better than carrying several small nuclear reactors.

The ingots produced were less pure. There were no railguns. The atmosphere would prevent effective launch into low orbit. In order to overcome friction, it would require much larger guns, and there was not enough power to do that yet.

The ingots would wait for the more powerful nuclear reactor coming with the manned crew. Then they would be fully purified and made into useful parts. The oxygen would be gassed off and stored in pressurized tanks for use by humans and in fuel cells. For the present, hydrogen was imported from Earth. In the future, it would be collected from Jupiter's atmosphere.

After releasing the Mars landers, the three cargo ships sent three sets of robot landers to Phobos and Deimos. Two sets went to Phobos, one to Deimos. There they would begin digging out habitats that would serve dual purposes. The first would be to serve as an emergency shelter off Mars. Secondly, the fully developed bases would provide a study ground to further understand how humans could survive on an asteroid-sized body. This was much needed information for the Centauri Asteroid and Comet Mission (CACM).

*

Red and Walter had been chosen for the 2045 Mars mission for an obvious reason. They were the best problem solvers. Marceau had been chosen because Red had quietly insisted. Red's opinion carried gigagrams of mass with the WSA officials the world over.

There were also those who saw in Red, and with subtle knowledge of his private enterprise on Luna, a potential to return to a more capitalistic way of life. They still had a quiet say in much of what happened, so they threw a little extra effort into Red's request.

Marceau was more than capable, but without Red's insistence, she would have been left on Luna where her skills had been well appreciated. Her faux pas over the materials that Red and Walter had purchased and stored only showed that she was on top of things. Her quick thinking and dogged effort during the crisis at SS-2 entrenched her as a leading candidate though. Further, her discovery of Red and Walter's secret trips and subsequent keeping quiet as requested put her in the top ten. The rest had been chosen for their individual skills and areas of expertise. All, except the three Lunarians, had been training on Earth for years. Red, Walter, and Marceau had received their training very much on the job. They would be doing what they had been doing all along. They did however receive some additional computer programs to go through, a few hundred gigs' worth.

The supply ships had been launched just a little more than a year ago. The pace and tension grew exponentially as the day neared. Time is a strange thing. We say it passes, and yet we count it down.

*

The Chinese State of the Asian Region's Lieutenant Colonel Yang had been designated the commander of the manned mission. Some political machinations in the WSA required that a broad spectrum be represented as well as a rotation of leadership roles amongst the various regions. Equality had to be obvious, not that each member so chosen was not fully qualified. Each Region saved and groomed their best for their turn.

Technically, Yang was outranked by Red (now a colonel) and Walter (a lieutenant colonel). Walter had received his new rank only a day before Yang. Yang and Walter both realized someone was playing politics by slipping in his new rank just before Yang's well publicized presentation. Politics are what they are.

If Walter ever gave him an order though, he would damn well follow it, Yang knew. The only time Walter even really talked to anybody, as far as he could tell since he knew nothing of the signing, was if he had something very important to say. Yang also knew that he would never, most likely, have to give a hard order to Walter. If there was a problem, Walter would have it half-solved before he could speak.

Red was a different dog altogether. Yang knew that Red would not hesitate to rank him. However, that thought did not much disturb him. Red's habit of being right, always, put the nix on any real reason to argue.

The others, except for the specialists who were only short-term military before joining the WSA, had long spotless careers. He had no worries there.

It was amazing the little things that would run through the mind at times of heightened anxiety. Yang almost chuckled out loud. Here they all sat, strapped in tightly, waiting for a few more moments to reach the conclusion of the countdown. Checks, more checks, and cross checks had consumed the last few hours. Yet he, like a little boy on an outing, was letting his mind wander.

Every little thing had its own little clock; it had to happen in order and at the right time. They had finished their preflight checks a few seconds before schedule. The brief amount of waiting simply added to the tension. Yang could see Lieutenant Colonel Len meditating, her fingers pressed gently together.

A monotone voice from SS-3 announced, "Gagarin we read green across the board." The ship was named after Yuri Gagarin, the first man in space.

"Station 3, we copy," said Yang.

"Gagarin disengage docking clamps in five . . . four . . . three . . . two . . . one."

"Clamps released," announced Len as she hit a toggle switch.

"Prepare for retro fire, odd numbers only," said the sterile voice.

"One through nine, odd only," replied Yang as he toggled every other switch numbered one through ten.

"Ten-second burst on the mark."

"Ten seconds entered," replied Len after initiating the clock for 10.000 seconds.

"Five . . . four . . . three . . . two . . . one . . . Mark!" said the quiet voice from the station.

Yang hit the switch exactly on the mark. "Firing." The Gagarin moved ever so slightly away from the station to which it had been docked in a parallel manner. In a matter of a few seconds, it was meters away, moving slightly closer to Earth. By moving closer to Earth, the ship was preparing to make use of gravity in the same way as the cargo vessels.

"Prepare for countering fire," said the voice.

"Evens, two through ten, toggled," said Yang.

"Ten seconds set," countered Len, again setting the appropriate chronometer as the ship continued to move away from the station.

"Fire in five . . . four . . . three . . . two . . . one . . . Mark!" said the voice.

Yang again hit the switch on the mark. Retro-rockets fired to stabilize the position of the ship. It was now one hundred meters from the station.

"Initiate reactor control rod removal," said the voice from Station 3.

"Reactor currently at 5 percent," announced Captain Martin.

"Initiating removal," stated Captain Rodriguez as she touched several keys on her screen.

"Reading 7 percent rated output, ten, and fifteen . . .," said Martin in her quiet voice.

"Initiate ion injector," said Yang.

"Hot," said Len, flipping a series of switches.

"Thirty . . . forty . . . fifty . . .," continued Martin.

"Engine ready," said Len.

"Sixty . . . seventy . . .," Martin went on.

"Prime the outer coil," commanded Yang.

"Five hundred volts (V) applied," responded Rodriguez, flipping a switch.

"Particle flow reading is .01 moles per second. Field is stable," said Martin. "We have 90 percent rated reactor power available."

At this point, the Gagarin was actually accelerating, just very slowly. The amount of ions being ejected was actually relatively small, one mole being 6.022×10^{23} particles, essentially a six with twenty-three zeroes behind it. It sounds like an awful lot, but in gas form, at room temperature, the volume of the gas would be 22.4 liters, roughly the same as eleven of the large two-liter plastic soda bottles. In compressed form, it took up less volume than a dime.

The applied voltage that created the alternating magnetic field to speed them down the length of the ship was also small for the moment. Yet everyone was holding their breath on Earth and on Station 3. This design had been tested on several occasions, but there was always the "what if" factor.

"Increase voltage to one thousand," said Yang.

"One thousand V," replied Rodriguez.

"Holding reactor at 90 percent available power. The field remains stable," stated Martin.

"SS-3, we are ready to fire at full throttle," radioed Yang to the station.

"Gagarin, we copy. All telemetry is good. All boards are green. Go on your mark," replied the voice.

"On my mark increase to five million V and injectors to ten moles per second." Yang was ready for the trip of a lifetime.

"Five MV (Megavolts) ready on your mark," said Rodriguez.

"Injectors to ten moles per second on your mark," stated Martin.

Punching bags know the feeling. It is swift and sudden. On the mark, they were all thrown back into their well-padded, personalized, contour chairs. The ship went from a speed of a few meters per second to one of several hundred per second in a very short time.

The ship accelerated to the point where the astronauts were feeling six gs. That is, six times the normal feeling of gravity on the surface of the Earth. If you love the wildest roller coaster imaginable, this is for you. That feeling would last until they had gone around the Earth and past in a slingshot maneuver.

They needed a minimum escape velocity from Earth's gravity of at least forty thousand kilometers per hour. After that, they would then power down (partially) while still accelerating at two gs, as they sped past the moon's orbit in a matter of hours. No time for stopping there however. They would use the continuing acceleration to shorten the trip.

The trip would cover 1×10^8 km. before all was said and done. Divide this by two months, then divide by thirty days per month, and then by twenty-four hours per day, and you come up with the needed speed to reach Mars in the time allotted. When their final velocity of about eighty thousand kilometers per hour had been achieved, the engines were shut down. They were going faster than the required seventy-thousand kph. because they had to take into account the slowdown period as they approached Mars.

They would coast much of the way to Mars. In two months' time, they would rendezvous with one of the orbiting portions of the cargo ships. Soon after that, they would put the first humans on Mars.

*

Adam felt the surges and cross currents of quantum reality and nonexistence flow through his being. A bottleneck event had just begun with the launch. All of the future was flowing into a node. Danger. The entrance was like a fish trap, easy to get into but near impossible to escape.

The swirls and discontinuities of the timeline were baffling even to him. By the time he computed an exact course of action, the situation had changed enough that probabilities had fallen significantly.

"I need more *me*."

He began to alter the rock material behind the super clean containment barrier. He arranged molecules, then nucleons, protons and neutrons, electrons. Wave properties were guided into a more pleasing form. Quantum loops were formed. The rock was becoming him.

*

Technicians observed some bizarre readings that lasted for several hours. QBIT was very active. It was processing huge quantities of data related to the Mars mission. It appeared to be continuously reworking potential problems and solutions. That was not the cause of their worry though.

It was the sudden appearance of electrostatic fields in the rocks surrounding the chamber where BRAIN-7 was housed. Gravwaves appeared and disappeared.

They were mightily confused as to the cause and meaning of this.

*

There were two duty rotations on the ship. The first rotation consisted of Bear, Holtman, Len, Marceau, and Rodriguez. The second, Doran, Martin, O'Hare, Orenska, and Yang. Each rotation needed a pilot, engineer, and reactor specialist. The current off duty personnel (rotation 1) had gathered for their daily discussion and a light meal, nominally called dinner. The rest listened on the ship's intercom from their various assigned positions. Marceau was preparing the edibles as the others lightly tethered themselves with Velcro into what went for chairs in the galley.

As Marceau drift walked to serve Holtman, she asked him, "What got you hooked into this flight?"

Holtman smiled. "Undergrad work with a stunningly beautiful professor. We were refining our understanding of the alteration of the climate in Africa after the last ice age. We hoped to make some models that we could apply to Mars. Even though the time scale is much vaster for Mars, the basic mechanics should apply. We have developed several models, but we need to make some tests, search for some specific mineral deposits in order to find the best one. That's why we argued for a more northerly or southerly landing site. We know there is more recent water activity closer to the poles."

"So if you can find your treasures, then you can find an accurate enough model for the terraforming project to initiate," Red spoke over the open intercom. "What did you find, and how did it relate to the development of the Egyptian civilization?"

Holtman laughed easily. "That is a fairly long story."

"We got lots of time," Maria 'Hot Rod' Rodriguez snickered.

Holtman loved to gab, so this was no mean task for him. "In dark prehistory, some ten thousand years ago, man first settled the immense valley of the Nile. The Nile is the longest river in the world, some 6,400 kilometers long. The Nile did not exist much before twelve or thirteen thousand years ago. It was created by lakes, Victoria and Albert on the White Nile and Tana on the Blue Nile, filling with water and then overflowing as the last ice age came to an end, and rain fell in large quantities for the first time in eons in the areas of the three main lakes. It was the life-giving river that allowed people to eventually prosper. It is the Blue Nile that provided most of the fertile mud that led to the bread basket of ancient times. The pendulums of nature swung propitiously for these extraordinary people."

Orenska asked from the infirmary where he was reviewing radiation measurements, "How well did they live before this influx of water and fertile soil?"

"They lived as the rough hunter-gatherers that they were. It was a tough, uncertain life that left most of them dead before they were twenty-five or so. Food was scarce, and they had to wander far over a large range to find enough to sustain themselves."

Walter noted, "The rains changed that though, I'm sure."

"Significantly. The area they had to cover for basic survival shrank as food became more available due to the increased precipitation. Nearer the rivers, the quantity and quality increased nearly geometrically. The population increased with increased numbers of bands of rovers as well as a direct impact on the survivability of newborns and the elderly or injured."

Orenska piped up, "I have been involved in theoretical modeling of that nature. The food pyramid becomes quite fat at the bottom, flora, while the fauna take some time to adjust. The sides of the pyramid become curved, non-Euclidean, for some time. How long did this upset the equilibrium of the biosystem?"

"Somewhere around four thousand years since circa six thousand years ago, the Egyptians exploded on the scene under the first of their pharaohs. The first may have, in fact, been the so-called Scorpion King. Archeologists have discovered a historical king that used the scorpion symbol. He united several developing cities on the Nile into the beginnings of an empire. The empire that was birthed flourished for at least 2,700 incredible years before the first fall occurred. This Egyptian Empire was ancient even to all that we consider ancient."

"What about the Atlantis thing?" Len queried. "Doesn't that predate Egypt?"

"I have not had time to rigorously examine Dr. Marquez's data on the civilization itself. The climate models I have seen for the South American area do suggest the possibility. That region was a couple of thousand years ahead of the Euro-Afro-Asian portion of the world in positive climate change."

Marceau added, "The timing is rather good though."

"Yes. It is easy to believe that the Egyptians were ready to form their own society, on the verge, and were influenced by an outside force. All civilizations have expanded outward to increase trade for foodstuffs and raw materials. It is quite possible that that is the case here."

"But you won't commit," Red commented, liking the hesitance of Holtman to make a judgment until he had all the facts.

"You see, the natural setting of Egypt favored empire building. Surrounded on all sides by geographical barriers, it was well protected. The Red and Mediterranean seas in the east and north along with deserts in the west made the movement of armies difficult. The flanks of this developing power were well protected. It is just as possible that it was all internal development. Most Egyptian scholars take this path. Of course, they are mostly of Egyptian descent. They want to believe their ancestors created their own society."

Red responded, "It is clanism, tribalism, regionalism, nationalism. That is the nature of most people."

"Yes, it is. For Egypt," he continued, "scattered villages combined their efforts. The development of writing, medicine, art, metallurgy, and irrigation brought wealth and power beyond all previously recognized standards."

"The power came in the form of a king believed to be of divine origin, called, in later times, a pharaoh. The king was part of a trinity of power as the direct descendant of Horus or Horace, son of Osiris, the mythical child of the gods. The power of the pharaohs was, at first, supreme. But during later times, the priests and generals began to grow in power, and economic problems appeared due to the price paid for the immense building projects honoring the god-kings, projects that did not produce goods or services after their completion. They simply used great amounts of raw materials and laborers. The class of nobility had also developed, and now represented a threat to the king as well. Powerful princes, in close proximity to the throne, schemed and struggled to ascend to ultimate power. Blood is not necessarily thicker than water."

"There is plenty of that in all cultures," Yang stated from his command chair. "Mine has had its share."

"Your story is due then," Red set up a batting order.

Holtman guffawed. "You're next, fella. He cornered you neatly. Well, the old kingdom fell." He continued, "Leaving an interim period before Egypt rose again in the Middle Kingdom around 2000 BCE. Egypt was restored to greatness. They gained a military reputation, developed trade, and exploited the territory to its fullest. Yet they too would fall under weak kings that provided little centralized power."

Red spoke, "It seems that centralization of authority has been the bane as well as the marker of the end for various civilizations. It brings them to an apex. It leads them to a great fall. Is there a lesson here?"

Holtman and the others reflected for but a moment. "Upper and Lower Egypt engaged in self-destructive civil war. In the north, the Egyptians fall under the rule of the Hyksos, while in the south, the Nubians broke away."

"Civil discord is another manifestation," Walter added, comprehending Red's histo-math equations. "Rise and fall are generally bell curved. Civil insurrection occurs near the top of the bell shape and leads to destruction, sometimes rapidly.

"There is a great deal of weight in that direction. For instance, it would be 1600 BCE before the Egyptians would resume previous levels of power. The swings of the pendulum keep reversing. Vast temples to the god Amon and the Valley of Kings were built now. There were great names to be read by schoolchildren and students of history: Amenhotep, Thutmose I, Hatshepsut, a woman in male dominated times, and Thutmose III.

"By the time of Amenhotep IV, trouble in the form of the priests returned. He had tried to form a religion around one god, with himself as the only representative. He even changed his name to Akhenaton."

Len tittered. "That was not well received, I suspect."

"Not at all. He expected the priests to fall in line, but they did not. They schemed and plotted to reinstitute their many powers through the multitude of deities. The elimination of many gods affected too many people. In the end, the priests won and redirected the new pharaoh back to the worship of Amon."

Doran added, "So religion became the dominating factor at that point. From every level of society, the believers in the status quo took over."

Holtman frowned slightly, "We cannot completely dispose of the religious literature and beliefs completely. There is a great deal of basic climate information that can be gleaned from sources such as the Bible or other religious writings. There is a close association between religion and agriculture, the salt of the Earth and such. There were river, soil, and sun gods in the various protocultures, all the necessities of refined agriculture.

"Stone placements such as Stonehenge are related to planting seasons. Various stele are involved in quasi-religious agriculture rituals. The types of crops and planting times tell us a great deal about climate conditions. We then cross-reference that with hard data from palynology, microbiology, and a host of other specialties."

Marceau frowned speculatively. "The land of milk and honey."

Holtman glanced at her sharply, "Exactly. But what kind of milk and honey? There were domesticated asses, horses, goats, cattle. Caches of honey have been found. The type of plant flowers involved are related directly to the honey and therefore the climate."

Holtman continued, "Science lays claim to logical natural answers to these biblical miracles based on the forces of nature, not the supernatural of the primitive mind-set. What could not be understood, yet, became the realm of the gods. It took the force of taboo to say otherwise."

Red chimed in, "Society has always been about the test of embedded taboo versus revelation of fact and the required change it brings."

"That's cold and harsh," noted Rodriguez.

"I would agree," Holtman stated. "But it is exactly right. The shattering of illusions about reality can be extremely painful. Many people refuse to accept the truth of change."

Len wanted a better explanation. "Give an example on embedded taboo from history that relates to your field."

Holtman mused a moment, "Embedded taboo could be in the form of male hunting versus female gathering. Evolutionary progress led to a social taboo on females hunting. Size and strength of the male in early physical development led to kills more often. Greater skill came with later brain size increases, which left the females without much chance to be involved in hunting activities. Their brains and related skills were used for other types of societal production."

Marceau almost snarled, "Barefoot and pregnant with dinner cooking?"

"That was the continuing development of the original taboo."

Len calmed the situation a bit by saying, "I meant a larger scale picture, in, say, Africa, that involves the weather, taboos, and societal change."

"Well, the climate change I was speaking of is directly related to the migrations of whole tribes of peoples. The increased ability of food production in the region brought them from far and wide."

Walter looked interested as he spoke, "Give a lesson."

"Let's examine the exodus of the tribe that would become the Israelis then. Famine and drought would bring them to Egypt. This lesser term glitch in the climate was a direct consequence of the shift that created the riches of Egypt. The increased precipitation in that region deprived the borders from some of their water on occasion. It was a certain swing of the meteorological pendulum."

Red pushed for more, "So what does that mean? Give us some historical references."

"The Bible is the direct reference. In many ways it gives us a daily, weekly, monthly, yearly, and multiyearly reference guide. The religious mumbo jumbo aside, there are a great many subtle bits of information to be found."

"How so?" coaxed Marceau.

"The miracles prior to the exodus are related to climate. There exits a red algae that has appeared over the years, covering the entire delta area. It often contains the deadly anthrax spoor. This covers the Nile turning red.

"The anthrax would kill the fish fairly quickly. This would leave no natural predator for the huge number of frog eggs that are laid by the millions. This would explain the frog population explosion described as another miracle. Strangely enough, frogs and fish have even been known to fall from the sky, a clear cloudless one. It is believed that a water spout, over a nearby ocean, lake, or river, could pick them up and distribute them far from the source."

"Interesting," Walter added. "Were they still edible?"

Marceau punched him in mock anger.

Holtman laughed gaily. "Once the frogs have consumed what they can, they begin to die in huge numbers. This would attract a large number of gnats and flies, another miracle according to the Bible. The pharaoh's magicians had been able to call up other problems but not these. This goes beyond the skills of the wisest magicians of the land. Is it not then simply a natural event?

"In the biblical time frame, anthrax would begin to kill animals. But how do the animals of the Hebrews escape this? Is it simply because they are isolated from the herds of the Egyptians? This is the explanation given by science. When there is a major outbreak of anything deadly, quarantine is the best way to slow or halt the progress of the disease. Of course, the dead animals would also attract large numbers of gnats and flies, adding to the enormity of the problem.

"Given the sanitary conditions, open sewers for most of the population and proximity to so many dead animals, it would only be natural for humans to get the disease as well. Boils and sores are a natural set of symptoms for anthrax and other diseases from such a set of conditions, again, another miracle. For people to catch the disease is normal. For the animals of the Hebrews, if they are secluded, they would not catch the disease, which reduces the probability of people being infected. But the people were working as slaves, again according to the Bible, with infected Egyptians among them as overseers. They did not catch it. This leads to the contradiction of slave labor. Many newer studies suggest that the soon-to-be-Hebrew peoples were not really slaves, but a working class. They were separate in location as well as social habits."

Orenska, quite medically intrigued, asked, "Do you have solid anthropological evidence for this?"

"Oh, yes. There have been several digs that exposed dead that had suffered from anthrax, especially the young."

Orenska quickly asked another couple of questions, "DNA of the anthrax? Could they link the strains to current forms?"

"Yes on both counts." Holtman nodded, unseen by Orenska. "Let me go on with the other miracles. In terms of the hailstorms, they do occur infrequently in that area. But a fiery hail is not. It could have been a meteor shower. That could have set some fires. Again, the land of Goshen where the Hebrews were was spared. Perhaps, they were not in the path of the fall."

Martin asked, "Have any meteorites been found to back that up?"

"Yes, certainly, but they cannot be dated easily. The sandstorms move them easily, and their impacts cause them to be quite damaged and deeper than the current surface levels. Most religious scholars deny any wave of them at that time. Even secular scholars find it difficult to find a strong positive correlation."

Yang noted quietly, "There seems to be a great deal of circumstantial evidence, and some more solid."

"True. But we search for support to the circumstantial, often finding more circumstantial and, occasionally, harder data."

Rodriguez smiled. "Take what you can get, huh?"

"Yes. To continue, hordes of locusts are fairly normal in Africa. They occur frequently in cycles, and they do strip the land bare. In the U.S., the Mormons in Utah had an experience with locusts. It took a miracle of seagulls, far from the sea, on a feeding frenzy to clear them up. Finds back up a tremendous swarm in the general time frame.

"A great darkness that lasts three days could not be an eclipse unless in the telling the story became very exaggerated. That does happen with myths. It could, however, be a sandstorm that raged for days. These things do occur often in the Sahara. It could easily spill into the valley of the Nile, leaving the nature-fearing peoples in a trauma.

"The death of only the firstborn is hard to explain unless it really was not only the firstborn. In a plague, or multiplague, many would die including firstborns. If the deaths were from the anthrax, firstborn children, especially if they are young, would be very susceptible. The expanding unsanitary conditions would open the door for other diseases to take their toll."

"I can verify that with many references," Orenska noted. "The plagues would certainly also create a significant impression on following generations. The myths that would form would only grow. We found that to be true with Stalingrad in World War II."

Holtman continued, "In terms of the crossing of the Sea of Reeds, science has determined that a strong wind, lasting for ten hours or more, could have separated the waters. Another theory is that a volcanic eruption in the Aegean Sea caused a tidal surge that emptied the Sea of Reeds for a couple of hours, allowing the Hebrews to cross. It requires faith either way, faith in science's ability to properly explain the details from evidence or in a god of some sort."

Marceau was impressed; it showed in her voice. "You had to delve into a lot of secondary fields to cover all that: geology, palynology, biology, chemistry too, I suppose."

"It just demonstrates how all the sciences are interrelated. The geology of the volcano leads to dispersion of the ejected gases. Oceanography and tidal surges with meteorology tie in to the missing water. Science, for many years, was split into separate categories. It is now reuniting in a much grander form. It all ties together as it should."

Red interjected, "That I agree with. We all represent separate pieces of the puzzle pendulums that swing disjointedly for a while until we find synchronicity."

Rodriguez queried, "What about Cleopatra and all that?"

"That's my story," Len said. "She was my hero as a youngster."

"Continue on, young lady," Red said, trying to draw more out.

"Yes, let's hear it, woman," Walter noted slyly. Walter and Len had been spending a lot of quiet time together.

Len began, "In 1300 BCE, Ramesses II reigned an amazing sixty-seven years. But it was the last major hurrah. Egypt would be dominated by many other empires until the final end with the Ptolemaic dynasty. It would rise to great power under the Ptolemies though. It would make its final bid, contending for world power under its last great queen.

"There were actually seven queens of Egypt named Cleopatra. The one most people are familiar with is Cleopatra VII, 60 to 30 BCE. Contrary to the Hollywood version of a great beauty, she was *only* intelligent, charming, self-confident, and full of vitality. Though she was not an incredible beauty, she seduced both Julius Caesar and Mark Antony."

Walter chuckled. "What a slut, playing both sides, so to speak."

Len gave him a rude noise in reply and added, "Different time frames."

"Pervert," Red announced.

"She became coruler with her brother Ptolemy XIII in 51 BCE but was forced into exile in 48 BCE. With great determination to rule Egypt, and to restore the power of her people, she appealed to Julius Caesar who deposed Ptolemy XIII. Caesar returned Cleopatra to the throne, with another brother, Ptolemy XIV, in 47 BCE."

Walter started to say, "Power-hungry bi—"

"Shut up, stinky," Marceau commanded.

Walter gave her a "who me?" look.

"She bore a son, claiming it was Caesar's, and named him Caesarion. She moved to Rome to be with Julius. After the assassination of Julius Caesar in 44 BCE, she returned to Egypt to protect herself and her rule over Egypt."

"Her proximity and intimacy with Caesar brought her too close to danger there," Martin noted.

"Quite. Ptolemy XIV died under somewhat suspicious circumstances, possibly by order of Cleopatra VII. She then named her son as coruler. It gave her position greater acceptance among the military."

Rodriguez commented, "The male pressure in society."

"Mark Antony and Octavian, the adopted son of Julius Caesar, were locked in a titanic struggle for control of the Roman Empire. Cleopatra supported Antony in this battle. Their physical relationship was sufficient enough that she bore him three sons."

"I told you she was a—"

The rather stern glares he got from Len and Marceau caused Walter to raise his hands' palms up in surrender. The forces of womanhood won this particular battle. The war would continue.

"The combined forces of Cleopatra and Antony were defeated at Actium in 31 BCE by Octavian. They hastily retreated to Egypt before they could be captured."

"Good military practice not to be captured." Red chuckled.

Humphs and chuckles of good-humored agreement rounded the ship.

"The dynasty of the Ptolemies is believed to have ended with the suicide by asp, a poisonous snake, of the remarkable woman Cleopatra. She wanted to avoid being a humiliated captive of Rome in 30 BCE. At this time, Egypt came under the full control of the Roman Empire. There was no resistance left in them at this point."

"Who has the methane bonanza today?" asked Martin, sniffing, as she drifted toward the galley's drink container from her station next to Red's. She needed to get an electrolyte drink. Orenska had checked her urine earlier and admonished her to consume more regularly since her body had larger swings in metabolism.

Everyone pointed at Walter the moment methane was mentioned.

"Mea culpa," said Walter as he raised his hand.

"It's the beef stroganoff we had yesterday," said Marceau with her nose wrinkled. "We need to give him something else, anything else," she laughingly said as she smiled at Walter.

"Perhaps we could tap him into the ion drive," noted Red from his position at the ship's engineering board. That brought laughs from everyone.

In the confines of a vessel like the Gagarin, the idiosyncrasies of your shipmates were very important. Little things can begin to get under your skin and eventually flare into major personality conflicts.

The olfactory organ (your smeller) is a powerful device. Farting is only a small part of the problem. Everyone has one's own smell. It does not matter how often you cleanse yourself. On this kind of a trip, deodorant is not to be found. Cologne is not available nor is perfume. The expense of shipping outweighs the benefit according to the bean counters. The natural body odors are unmasked for all to sniff. *Armpits and assholes* was how Red had phrased it to the doctors who had lectured them on the importance of hygiene in such an extended, close relationship. It was not how the doctors had phrased it, but now everyone on the ship simply said it was time for some A and A rehab.

That meant it was time to enter the shower where a fine ten-second mist sprayed your body from all directions. The mist contained a bacterial and viral inhibiting ingredient as well as standard soap. A person would then rub oneself (not always adhered to as many times one was assisted) with one's own special washcloth. A second twenty-second misting occurred as a rinse cycle. It was, of course, all recycled water. The drying phase used warm air in a thirty-second blast. It does not sound too bad except that it only happens once a week.

For women, the problems were somewhat compounded. Along with all the other medications and vitamins, they had to take their monthly pill. This prevented menstruation, most of the time. Occasionally, when under stress, it simply did not work. The body hates to be messed with, so when such an event happened, it was a significant one.

Red called it PMS to the tenth power. A bit gender insensitive perhaps, but Red was happy it would never, ever, ever happen to him. He did, though, actually feel a little sorry that it was Marceau who was victimized by this extraordinary manifestation of possession by demons.

After her three-day, interminable reign in hell, the others had joked that they were thinking of putting her in her suit and tethering her outside the air lock but had been overruled by Red; he felt that it was too inefficient.

Of all the people on the ship, Red and Walter had put up with her in an amazing show of restraint. Walter she could understand; he was simply unflappable. Red, however, was known for tearing new ones for much less than he had put up with for her sake. Little thoughts wiggled and squirmed just out of reach at the back of her mind. There was something going on here.

Everyone had seen everyone in the buff. It was not the place to have complete privacy. No one really cared. Nude does not mean lewd, as Red had explained to the designers when they had tried to add a changing room (actually just a folding screen) to the ship. The moralists had been upset with the idea of nonmarried personnel seeing one another nude.

Marceau had had all of the others scrub her back, male and female, though it seemed to be Red more often than anyone else. The same applied to all the others; they all scrubbed one another. The battle for cleanliness took precedence over any shyness. In matters of health, they all looked over one another's bodies for evidence of sores, dry skin, or anything that could lead to a larger problem. The man from Russia, Orenska, was the closest to a real doctor but was still only a hyped-up medic. Any real treatment would be done via radio, v-link, or the robotic surgeon.

Body hair of any kind could be a problem, so all of it was trimmed quite short or shaved off completely. Everyone acted as barber, cook, maid, and more.

Sound could be another problem in a tiny ship, even if it was huge compared to any previous manned craft. There was nowhere you could go that would allow you to totally escape the sounds of the others. Each of them had been issued a form of earplugs and eye covers for sleep cycles. For the most part, personal sound was not a problem. The use of v-links with ear phones kept most music and movies from bothering others. Usually, the movies were watched by all not on duty. Group dynamics played out very well for this crew.

Touching was another matter of delicacy. They practiced a code of avoidance so that no person would feel as if they were in too tight a spot. Yet on occasion, a little quiet sex occurred. To the unspoken surprise of all, Red never participated. Most thought him too dedicated to his work to be bothered with it, except Walter. He knew the truth.

It was not mentioned who was with who at any given time. Hetero or homo, it was accepted on the ship anyway as a needed release of the physical and emotional tensions that built up. Marceau avoided the men, not because she did not like men, but she felt she was now waiting for a special one. And deep in her subconscious, she had a vague and rather crazy idea of who it might be.

Marceau and Rodriguez were gently caressing each other after their interlude. They had enjoyed each other's company on a few occasions. Len had unintentionally passed by with only a flicker of desire showing as she quickly left them alone. The unwritten rule was not to interfere or attempt to enter the encounter without invitation. Neither Rodriguez nor Marceau would have minded; they were just too wrapped up in their passion to even notice her presence.

Rodriguez, kissing and stroking Marceau's face, said in a quiet whisper, "Red likes you."

"OH FUCK!" cried Marceau after sitting bolt upright, so to speak, and smacking her head on the panel next to her. The motion caused the hammocklike sleeping bag to sway and twist wildly. Straining to calm the wild beast, she said, "What did you say?" She held her forehead.

Hot Rod's eyes were dancing and her breasts jiggling as she laughingly repeated what she had already told Marceau. She pulled Marceau back down close to her and kissed her swelling forehead.

"Woman, can't you tell? I admit it took awhile to figure it out, but it is obvious."

"Are you . . . sure?" Marceau queried in a very quiet voice as she stared hard into Hot Rod's eyes.

"Yes," said Rodriguez without a blink or a hesitation. She leaned into Marceau to give her a gentle kiss, only to find herself the object of an ardent and heavily passionate response. For Marceau, it was not so much her desire for Rod but the information she had received. The news broke a dam in her mind, laying bare her true desire. It inflamed them both into a whole other round of heated sex.

At the end of this very physical and sweaty encounter, Hot Rod thought to herself that she should have spoken earlier in the mission. She had never been the subject of such intense lovemaking though she also realized she was only a substitute for Red.

After many deep breaths for both of them, they tittered when Marceau said, "My head really hurts."

Later that day, Walter, over the dinner table, looked directly at Rodriguez and said, "Remind me to wear a helmet next time."

Rodriguez, for all her beautiful dark Latin coloring, blushed as red as a rose. Marceau, glaring blue black bump on her forehead, took it a little worse. She tried so hard not to spit out her electrolyte drink that it instead squirted out her nose in two strawlike streams, floating about in the galley area.

Marceau and Rodriguez continually found pillows and other forms of padding duct-taped to the tops of their bunks over the next seven-day cycle as joking reminders. It lasted until Red noted that too much of his precious solve all (the tape, that is) was being used for less than productive uses. He noted that bruises heal just fine without duct tape. Hot Rod socked him in the arm on Marceau's behalf.

*

"Everyone, strap in," Yang said. "It is time for turnaround."

They were three-fourths of the way to Mars. The trip had been uneventful so far. This was both good and bad. No one wanted to have anything major go wrong, but a little excitement would relieve the boredom. They were all excited about the turnaround. It was a big change from simply reading gauges, taking measurements, and the reporting thereof to the controllers back on Earth and the space stations. Unlike the few seconds of time lag for Luna-Earth communications, the lag here was much longer. Many chess games were played while communications were exchanged. Everyone, including the computer, got better to the point no one would play the learning device except Red and Walter. They even still managed the occasional draw. Many other games of strategy were played, even tic-tac-toe. Games, however, were only a release from the strains of a twelve-hour day. In groups of two, they each had a six-hour main shift buffered on each end with a three-hour shift on various experiments. Other time was devoted to exercise, planning, hygiene, and relaxation.

Opposition refers to when two bodies in the solar system are on the same side of the sun, relatively in line. Perihelion is when a body is nearest the sun in its orbit. When Mars is in opposition, near perihelion, the communication time lag is about six minutes, roughly three in each direction. When Mars and Earth are on opposite sides of the sun, the time lag can be as much as forty-five minutes for the round-trip. This time lag can be even longer since it has to pass to satellites to be relayed to Earth. These satellites are in Earth's orbit but are three months ahead and behind the Earth, somewhat like the points on a compass.

These satellites were used for far more than just communication. They were observing Sol, our sun. They measured solar flux, examined sunspots and the corona, and measured the amount of particles in space. In general, at Earth's orbit, there were between three and five particles per cubic centimeter. For the most part, the solar wind is composed of protons (H^+) and electrons (e^-). Other matter includes helium ions (He^{+2}).

The difference between solar wind and cosmic rays is the amount of energy involved. Cosmic rays are much, much more energetic. That is why they are so dangerous for astronauts. On Earth, our atmosphere protects us, along with the magnetic field. For those in space, or on Luna, that luxury does not exist. Mars is far enough away that it is not as much of a problem for the space stations. Mars itself, with its rather thin atmosphere, is generally well protected from such threats.

"Everyone needs to hold on," Yang said jovially as he toggled the retro-rockets. The ship turned slowly end for end. Once it had switched ends, Yang fired the retro-rockets again to stabilize the ship.

There were whoops and cheers from the rest of the crew. Yang raised his hands in triumph. "I am the champ."

Len punched him in the shoulder, saying, "Button pusher, that is."

As everyone released his/her harness, the boredom quickly began to set in again. The roundtable discussions that night turned again to topics of history and personal perspectives.

"Your turn, Gail," Red spoke to Marceau over the ship's intercom. He knew her interest was in the Mediterranean regions. He had been the instigator in all these personal politico-historical accounts. He and Walter had their ideas and wanted to test everyone. They wanted to expose the basic concepts of what government had meant to man. There were reasons.

Gail Marceau wondered a bit at the use of her first name by Red. It seemed a bit incongruous. She passed it by and began, "In 559 BCE, Cyrus assumed the Medio-Persian throne. The Persians began to explode into empire. It was the first fully *known* world empire. It spanned from Greece to Ethiopia and from Libya to India. For two hundred years, under the Achaemenids' rule, the Persian royal family, the empire was full of splendor and deceit."

"Examples please," Orenska noted. He had gained a great deal of general knowledge from these moments of sharing. He knew the others had as well, and better yet, it entertained them, kept them aware and keen minded. He did his best to add to the invigorating conversation. He went so far as to keep notes and forward them to mission control as proof they were mentally sharp. He did not understand that this would come back to haunt the crew, especially Red and Walter. His political acumen was a bit lacking in this department.

"I will borrow from Holtman's method a bit," she stated.

"Go on, girlie," Hot Rod prompted.

"The Bible has a great deal to add to the insights we have of the Babylonians and Persians. The books of Daniel and Esther show the power and glory within. Both Daniel and Esther show the vicious use of power, the struggles, and feuding within the royal families as well. Government is a dirty business."

"Isn't it always," Doran quietly added.

"In the book of Daniel, chapter 2, Nebuchadnezzar, king of Babylon, has a dream. He calls his magicians, sorcerers, enchanters, and the Chaldeans to tell him the dream and its meaning. He told them that if they did not tell him the dream and its meaning, he would have them torn limb from limb, and destroy their houses, meaning kill all their relatives. The poor charlatans prevaricated and whined that no one could divine the dream itself because that was the province of the gods. Daniel, assigned the name Belteshazzar by the Babylonians and a captured Jew, was able to discern the dream and its meaning. This caused a great deal of consternation, anger, envy, and the desire to get even amongst the charlatans."

Generalized chuckles rounded the ship at the use of the word *charlatans*.

"Chapter 3 begins with the king having an image made of gold. The statue is sixty cubits high and six wide. A cubit is about the length of a man's forearm, from elbow to fingertip. Nebuchadnezzar demanded that all the people, at the sound of music, fall to the ground and worship the image, much like national anthems of the near past and newer global anthems. Failure to do so would result in being tossed into a fiery furnace, jail now, to be consumed alive by the fire. Certain courtiers noticed that the Jews in the palace did not do as the king commanded. Three young companions of Daniel, who was in too good a position to be attacked now, were confronted. Hananiah or Shadrach, Mishael or Meshach, and Azariah or Abednego were brought before the furious king. Nebuchadnezzar ordered that they worship his image, and they refused. This is a wonderful example of free will. The king practically frothed at the mouth at this reply. In his rage, he issued the command that the furnace be heated to seven times its normal temperature. Because of the great heat, the men that led them into the furnace were killed. Yet the three young men were not."

"You don't actually believe all of that?" queried Doran.

"Literally, no. As a myth, it has a basis in fact. There are plenty of examples of people refusing to go along with evil governments and the price they had to pay. Whether or not these individuals actually survived, they made an impression. That is what lasted as the myth—the resistance."

"That's an interesting perspective on it," Yang contributed.

Holtman interjected, "It's a valid point. Most myths have a real basis that has been stretched into the realm of the supernatural. Even if these characters died, they rejected the forces that be and generated an underground resistance."

Red was quite happy with the discussion and added, "Go on. We want to hear more of your thoughts."

"In a show of greatness, the king, Belshazzar, son of Nebuchadnezzar, made a feast for a thousand lords with the vessels of silver and gold taken from the temple of the Jews in Jerusalem when they were conquered. In that very night, as described in the Bible with the handwriting on the wall story, Belshazzar was slain as Darius the Mede took over. In 539 BCE, Babylonia, that great city-empire of gold and wealth, falls to the Medio-Persian Empire in a single night. Pendulums struck with significant force and rapidity."

Rodriguez looked skeptical as she asked, "Did it really happen in a single night, or is that another stretch of the imagination?"

Marceau smiled. "The preparations for the attack took many weeks. It included diverting the river that guarded the city. The culmination of the attack took but a single night when the entire city was sunk in a bit of debauchery."

Walter chuckled. "Debauchery, my oh my. Imagine that."

"I am," giggled Len.

"Naughty people," Doran added.

With a hint of exasperation, Marceau continued, "In chapter 6, we see that Daniel was chosen by Darius, because of his ability and honesty, to be one of three presidents set above 120 satraps. Since he was so capable and truthful, Darius decided to set him above the other presidents so the whole country would be managed by him. Needless to say, the other presidents and all the satraps were against this. Their bribes and embezzlement schemes were in danger. An honest man in politics is a rarity and a danger. This set in motion some counter swings of minor and major pendulums."

"It pissed some people off," Yang noted.

"These others approached Darius with a devious plan. They convinced Darius to sign an edict, lasting thirty days, that whoever makes a petition to any god or man other than Darius should be put in a den of lions. Darius signed it, and according to the law of the Medes and Persians, it cannot be revoked."

Walter mumbled, "Setting a trap."

"Now his enemies knew that Daniel prayed three times a day with his window open toward Jerusalem. They came into his chamber and witnessed him making petition and supplication to his god. Immediately, they went to the king and said they had found a man disobeying the law. Darius replied that the law held. Only then did these men disclose that it was Daniel they were talking about. Darius grew very upset. Overnight he struggled with trying to save Daniel, but he could not. The law was strict."

"It must be followed, or it does not mean anything," Doran murmured.

"Accordingly, Daniel was placed in the lion's den, and it was sealed with the king's signet. Darius went to his palace where he fasted. No diversions were brought to him, meaning women. He did not sleep."

Walter smiled. "What no—"

Len punched him in the tummy before he could finish. "Hush!"

"He, Darius, rushed anxiously to the pit in the morning to find, to his utter amazement, that Daniel was alive. Darius then commanded that all the men that had accused Daniel and their families to be cast into the pit. The lions consumed them immediately with great ferocity and carnage. Pendulums are a bitch. They come back in your direction. Some would call this karma."

"Ha-ha!" Orenska laughed heavily. "They got bent over just as they tried to do to this Daniel fellow."

"No doubt," Hot Rod added boisterously.

"Esther's descriptions of the Medio-Persians are remarkably similar in terms of the lavish lives of the royal families and the military's generals."

"The elite get the best," Yang stated.

"I quote, 'In the days of Ahasu-erus, the Ahasu-erus who reigned from India to Ethiopia over 127 provinces, in those days when King Ahasu-erus sat on his royal throne in Susa the capital, in the third year of his reign, he gave a banquet for all his princes and servants, the army chiefs of Persia and Media and the nobles and governors of the provinces being before him, while he showed the riches of his royal glory and the splendor and pomp of his majesty for many days, a hundred and eighty days. And when these days were completed, the king gave for all the people present in Susa the capital, both great and small, a banquet lasting seven days, in the court of the garden of the king's palace. There were white cotton curtains and blue hangings caught up with cords of fine purple linen bound by silver rings and tied to marble pillars, and also couches of gold and silver on a mosaic pavement of porphyry, marble, mother of pearl, and precious stones. Drinks were served in golden goblets, goblets of different kinds, and the wine was lavished according to the bounty of the king. And drinking was according to the law. No one was compelled, for the king had given orders to all the officials of his palace to do as every man desired.' Esther 1:1-8. This is an incredible display of opulence and power."

Doran sarcastically noted with the old adage, "The more things change, the more they stay the same. Governments still attract the people that must display the luxuriance and symbols of power."

Red intoned oracularly, "It often brings out, or in, the worst types: power-hungry, authoritarian, egotistical, bureaucratic, limited-minded people."

Marceau said, "Most certainly so. The Bible gives another great example. Again I quote, 'And in those days, as Mordecai was sitting at the king's gate, Bigthan and Teresh, two of the king' eunuchs, who guarded the threshold, became angry and sought to lay hands on King Ahasu-erus. And this came to the knowledge of Mordecai, and he told it to Queen Esther, and Esther told the king in the name of Mordecai. When the affair was investigated and found to be so, the men were both hanged on the gallows.' Esther 2:21-23."

Walter chuckled briefly. "Palace coup. Now we know for sure that things are the same."

"There is a great deal more to be found in the Bible, about empires, the foibles of humans, and more that all should be aware of. Many scholars have tried to discredit the stories of the Bible, namely because of the religious junk, only to find that the descriptions of cities and their locations are factual. Archeologists have found the locations of cities thought to be mythical. People like King David and Solomon, in fact, lived. This is very similar to the discovery of the city of Troy, once thought mythical."

"The *Iliad* was considered a great work of fiction, an epic saga. It's better to know it's real," Holtman entered.

Marceau nodded in his direction. "Holtman used the Bible to search for tidbits for his science. Social scientists have used it to make models of development. They have done the same with other religo-historical documents."

Len stated decisively after gazing at the galley's chronometer, "Time for some showers and downtime, people. We should keep up the history lessons though. I find them stimulating."

Many nods and grunts of agreement followed as they cleared their dining materials and made their different ways to the sleeping and washing areas.

*

A few days later, the crew was having another historical interchange. Walter was dishing up tubes of pork paste, veggie mix, and apple sauce.

Everyone chuckled as Marceau checked Walter's dinner tube and comically noted, "It's not stroganoff. We're safe."

Walter gave Gail the evil eye. "I'm saving it for a special occasion."

Len leaned toward Rodriguez and whispered, "I hope it's when they are on the Bradbury."

Red asked over the open intercom from his seat at the engine control panel, "Who's turn is it?"

"It should be someone that has not added yet," mentioned Yang, also over the intercom. He was in the commander's chair, peering out at the starry darkness through his small portal.

Orenska added, "Perhaps it should be along the timeline like we have generally been following. That would leave perhaps the Persians or Greeks if we stay in the same general area of development."

Everyone looked about at the others, waiting for someone to jump in.

Marceau waited and finally said, "I can do the Greeks, but I have already spoken."

Rodriguez pointed at her and said, "Second at bat girl."

She took a breath and began, "OK. Several defeats at the hands of the Greeks in 480-79 BCE caused the end of the expansion of the Persian Empire. During the reigns of Artaxerxes to Darius III, the decline of the Persian Empire truly begins. Its influence wanes economically, militarily, and politically. The Persian Empire comes to an end in 330 BCE when Alexander the Great destroys the Persian Empire.

"The Greeks have given us many things including a life view that should be familiar to Western culture: Citizens are equal before the law. Public service is necessary and honorable. Individuals are free to live as they chose, to come or to go, and to speak their mind. They have an acute awareness of beauty. Things of the mind such as math, physics, and medicine are as important as the skills needed to survive. The public debates issues before the state takes action."

Red interjected somewhat sourly, "That last one does not happen anymore."

"Greek civilization gave us the basic foundation for Western culture. There were great philosophers such as Socrates, Plato, Zeno, and Aristotle. Two schools of thought came out of this—Epicureanism and Stoicism."

"Walter and Red," Len giggled quietly, "in equal measures."

"As we all know, Epicureanism refers to those that believe that the pursuit of personal pleasure is the ultimate goal. We are here to enjoy wine, food, sex, and life in general. Our personal pleasures may include painting, crafting, science, or other arts. It represents freedom and liberty for every individual to pursue their personal goals, their dreams, in whatever manner they deem most appropriate."

Walter kicked in, "What happened to the good times? It seems that philosophy is in decline, at least with the officials in control."

"Stoicism is pretty much the opposite. Emotions are to be controlled. Personal pleasures are to be limited. The philosophy of the Vulcans, created by Gene Roddenberry in *Star Trek*, is a direct view of Stoicism. This represents the submission, the relinquishing, of the individual's hopes and dreams to the possible greater good of society, socialism."

Walter made a rude noise and got punched by both Len and Rodriguez who were flanking him.

"Science and math made many significant advances with the help of Pythagoras, Hippocrates, Euclid, Archimedes, Aristarchus, Hero, with his steam engine, and Eratosthenes who contributed to geometry, astronomy, mechanics, and engineering. Herodotus and Thucydides were the fathers of modern history. They were the first to really categorize and analyze. Famous poets such as Aeschylus, Sophocles, and Euripides developed famous tragedies. Aristophanes was a great comedian."

Everyone looked at Walter who put up his hands in a "not me" gesture.

"Architecture reached its zenith with the Parthenon in Athens. Much of the design of the world since then has drawn upon the Greek styles. Modern governmental buildings ascribe to the forms of power displayed by the constructs of the ancients. It is an attempt to link current power with the majesty of old. Great pillared structures in every capital of the West show their allegiance to the dreams of ascendancy.

"It was around 1000 BCE that the Aegean civilization began to merge with the Dorians. From then until about 750 BCE, the basic Greeks emerged. The Greeks named themselves Hellenes, which gives birth to the name Hellenic Period stretching from 750 to 338 BCE.

"At first, people grouped into clans. Each clan had a village, a polis, where the word *metropolis* comes from. The isolation of these towns led to independent city-states controlled by a king or clan chief. These city-states usually only cooperated when threatened by distant invaders. The trade and other interactions between these cities is what led to the Olympics."

"So superman was a Greek. He lived in Metropolis."

Walter again received a couple of punches as well as a couple of chuckles.

"Oligarchy, the rule by a few members of the nobility, was the first step toward democracy. The expansion of the power base to more than one person represented a large move away from the absolutism of one-man rule. These nobles wrote law codes that applied to everyone including themselves. This is unlike current trends that exempt governments and their leaders in many regions from most policies that they have enacted."

"That's the truth. We see that every day," Doran added in her alto voice.

"By the mid 700s, CE colonization began to occur in the North Aegean, the Black Sea coastline, Egypt, Sicily, Italy, and France. This brought some significant changes to the lower classes.

"Citizens, especially in these areas, became heavily armed and were the match of the nobles. The ability of the average person to confront overbearing governmental force was a true democratic development. This right to bear arms puts limits on governments and those that would control the governments. An armed and outraged citizenry with its larger numbers is not something any government would wish to face, especially knowing that many of the soldiers would side with the citizens, their parents and siblings and children."

Red's bass rumbled, "History is replete with citizen uprisings, even as current as the first two decades of this century. There were flurries of such activities in the Middle East, Asia, and Latin America."

Doran's alto responded, "True, but it seems to have calmed for now."

"The development of coinage, hard money, and not fiat currency, allowed any person to gather wealth without the need for land, the stranglehold of the nobles. The appearance of a middle class of merchants, ship owners, artisans, and metal workers removed even more power from the nobles. The power base was shifting toward the lower levels, the greater populated section of society. It was an incredibly healthy move in terms of human development."

Orenska put in, "The pendulums of history are reversing that now."

There was a short silence, as no one wanted to comment on that, before Marceau continued, "Over a 150-year span, from 650 to 500 BCE, several revolutions took place. City-states came under the rule of tyrants. The word *tyrant* simply means one-man rule, not what it has come to mean today. The tyrant often had a great deal of support from the lower classes and, occasionally, was one of them.

"In Athens, the first great steps toward true democracy came. It was 594 BCE when Solon was elected to be chief magistrate. Solon enlarged the governing body to include property owners as well as the nobles. This council drew up laws for all citizens to vote on. Craftsmen were offered citizenship to move there and ply their trade. This influx of new ideas and abilities secured the foundation of the society. This is contrary to the laws of many current nations where citizenship is not so encouraged or accepted.

"The Persians destroyed the Ionian Greeks, on what is the peninsula of Turkey. Many battles were fought, but the Greeks eventually prevailed as masters of the Aegean Sea. This would lead to further battles in the future in which sea power would be necessary for eventual success."

A light tenor from the command chair interjected, "It seems that civilization brings a great deal of uncivilized developments."

"Athens became the leading city. Athenian democracy was a direct democracy in which the mass participation of citizens occurred. Governments in which citizens elect representatives to act for them are called republics, or representative democracies, like the U.S. Those in the old U.S., before the coming of the UN and UE, pledged allegiance to the Republic, not the democracy."

"Word play of the politicians," Hot Rod snickered.

"All male citizens over the age of eighteen were members of the assembly. Women, foreigners, and slaves could not be citizens. While this, the exclusion of women and the ownership of slaves, would be intolerable and considered very wrong now, this was a true first step on the road to a more complete democracy.

"Education was highly stressed, both physical and mental. There were no public schools though, only private tutors. The majority of tutors taught many students. However, they were poorly paid. In these hard days of common warfare, the body was often stressed somewhat more than the mind so that swordmasters earned more than philosophy instructors.

"City-state rivalries kept the Greeks from really becoming a power though. The Macedonians in the north, however, slowly brought the Greeks under control. Phillip, the Macedonian king, destroyed the Greek army. Phillip died young though, under chancy circumstances. His son was well prepared to take over. He surely did that as a supernova exploding on the scene.

"Alexander the Great would go on to defeat the Persians on several occasions. He founded the city of Alexandria in Egypt. Alexander would march on Babylon and Persepolis and take the throne of Persia. He stretched his enormous empire to the far reaches of India.

"After Alexander's death at a very young age, and under devious circumstances, the empire was divided into three parts amongst his generals. Antigonus took Macedonia, Seleucus ruled Syria and Persia, and Ptolemy I gained the throne in Egypt. These three kingdoms lasted for two hundred years."

Yang stated, "So that was the beginning of the Ptolemaic dynasty that ended with Cleopatra."

"Yes. It's interesting how these royal lines are all so interconnected."

"Good lesson, Gail," Walter congratulated her.

She jokingly gave him a rude noise in return.

*

They were almost there, only a few more days. It was time to slow down to orbital velocity and enter orbit around Mars. The process was almost the reverse of the slingshot method that sped them on their way. Again, gravity would assist them in the process.

Little need be said since the process went smoothly; it left them in a perfect orbit. Tomorrow would be the most exciting yet melancholy day as they split up into landing and orbiting teams. They had been professionally friendly throughout their training period, close in the mission's mental sense. Once they had launched, emotional ties developed quickly; the team had a true synergy. The psychologists and psychiatrists were amazed at how well they had meshed.

Yang would pilot the lander with Holtman in the second seat. Red, Walter, and Marceau were along for the ride. Their job would be to collect, and order the materials, and oversee the construction of the basic base, the hex domes. They would also check on the progress of the robots and make any repairs that the Foremen had been unable to complete on the Laborers.

Len would take the others—Martin, Rodriguez, Doran, and Orenska—to the emerging space station. The parts left over, after the landers had been launched, had coalesced into a framework for a new station, under the guidance of a team of construction robots. Over the course of many space walks, they would check all the components to make sure they had been locked together and welded at the atomic level correctly. Then they would oversee the final sealing with special foam and the addition of the interior panels . Once that was completed, they would then connect the packages of computer components: guidance, telemetry, atmospheric observers, communication upgrades, and atmospheric controls. Doran would examine the Hsas birthing section, double-checking all the robotic work.

This station would provide a safe haven for the Martian population as well as act as the birth ship to help populate the new colony. Its importance could not be overstated.

In the future, it would construct and launch the Centauri Missions. It would be central in the outer planet missions. The study of the asteroid belt would begin with many probes launched from Mars orbit. The Kuiper belt, that far out group of comets between Pluto and Sedna, will be explored from here.

*

"Can you believe it? They actually there. I mean, we see them pictures, but it's still hard to believe they made it," Paul said as he raised his mug for a sip.

"It is amazing, ain't it?" Mark burbled from inside his mug.

"It ain't so amazing," said the woman with no front teeth. "Don't you science flag wavers say that you can accomplish anything?" She chuckled. "Now when you're there, you gape unbelievingly." Mick had relented after several weeks and allowed the rather crotchety woman to reenter the bar.

"That's kinda true," noted Burt.

"What?" said Paul, surprised at Burt's siding with the woman.

"We have grown so used to travel to Luna and the stations that we think that it really is different to go to Mars. It isn't. Think it of in terms of driving one of the old automobiles. Can you drive to a city some fifty kilometers, or some thousands? The difference is only distance," Burt explained.

"Only there ain't no gas stations to fill up or fix a flat." Paul tittered, having to add his analogy.

"True," smiled Burt. "If they break down, it's like being a pioneer in the Old West. They have to fix it themselves or perish. That's why they sent the best."

"Like your cousin," the woman almost snarled.

"Yes, like Red," said Burt. "He and Walter are the best of all the engineers. They have shown that over and over."

"Yeah, like they save them mongrels, the Hsas," she stated poisonously.

"Where do you get such hatred?" Burt asked.

"Hatred? I speak the truth, and you know it. You are all blind to the real motives and truth behind all this," she retorted.

"And just what is this truth?" asked Paul.

"The truth is that Satan is behind all this manufacturing of fake people. Only God can solve all our problems," she declared defiantly. She rose and headed for the back door. No one spoke until she left.

"Well, ain't that a pipe full of hokum-smokum." Mark giggled, chugging the last of his mug.

"It's kinda strange. She's espousing the same rhetoric almost word for word as that Davies fellow and his crew of religious cohorts. Something is going on here," Burt spoke wistfully and with a frown. Again, his instincts were shouting a warning.

"She be gettin' weirder and weirder over the last few years," Paul noted. "I noticed that, since the Hsas survived, thanks to Red and Walter."

"Yes, her diatribes have grown more and more virulent," Burt said thoughtfully. His mind ranged back as he took into consideration all the recent and not so recent interactions he had had with her. "I think she has an ulterior motive here, gentlemen."

"Like what?" interjected Mick. Most bartenders are very good at absorbing information and rarely enter a conversation unless it might involve them directly.

"I'm not sure," said Burt. "Why did you let her back in after you eighty-sixed her?"

Mick looked a little sheepish. "You know I can't be too hard on anyone. And well, she is one of my best customers. She said it wouldn't happen again, but she seems to be going off the deep end."

*

She had a meeting. Pastor Davies would reward her well for new knowledge. As she entered the office, she smiled at him. "Hello, Pastor Davies."

"Hello, dear. How are you today?" he said as he rose from his chair and extended his well-manicured hand.

"Well, sir," she replied generously, shaking his hand.

"What have you got for me today?" he inquired.

"The man you have set me to watching is getting his information directly from Red O'Hare, his cousin," she admitted. "He gets direct v-links from him on the Gagarin."

"Well, well indeed," he replied. "What kind of man is this Burt?"

"He's a barfly. He is not really a drunk, but he consumes a great deal. He has some smarts. He always argues for the program."

"He gets it from Red you say?" smiled Davies. His thoughts turned to the concept of a leak. He could use that. He could turn this into a real-time source. "Does he get regular reports?"

She meekly replied, "Yes, about once a week."

He wrote down a private v-link number. "Use this to call me immediately when you gather any information. I will be able to make very good use of it. But I may need it very quickly if I am to counter their evil."

"I need to stop," she said next, eyes downcast.

Davies looked sharply at her and said, "No. You have given us a great deal of leverage. You are in a position to pry this whole thing wide open."

"I can't keep drinking like this," she stressed heavily.

"Stop going to the other bars," Davies announced calmly.

"What?" she asked, surprised.

"Dear," he said in a cajoling voice, "you have found what we need. The rest is now insignificant. Save yourself that stress. You only need appear a bit more often at that bar. Just an extra time or two compared to what you regularly do. You can do it."

"I don't know," she said reservedly, wringing her rough hands.

"Here, take this," he said as he handed her two envelopes from his desk instead of the regular one.

"Two," she said, eyes wide.

"Dear, you are my prime helper in this venture. I need you now more than ever."

"Then make it three envelopes, Pastor Davies," she said shrewdly.

Davies laughed loud and hard, but he reached into his desk and flipped another envelope toward her. She deftly caught it. "Make it worth my while," he said.

"I will," she promised as she rose and left.

*

Clocks tick,

Measuring out in doleful seconds,

The passage of created time,

A construction to separate,

Past from future,

Different clocks for different times,

Some short-lived,

Some long,

A measure of life,

Beginning,

End,

Hypnotizing swings,

Pendulum swings,

To and fro,

What do we know?

*

Chapter XV:

Mars Bound

15

001111

"This is Mary Smith-Jones for World Wide Network News. The Gagarin has achieved orbit around Mars. All Earth waits, holding their breath, for the moment when the Bradbury, the landing craft, will separate and descend to the surface of a new planet. Tensions are high here at the North American Region World Space Agency Command Center (NARWSACC), as everyone bites their nails in anticipation. We will see the action here in a few moments. As it now stands, all things are green and go according to the WSA.

"The astronauts—Yang, O'Hare, Black Bear, Marceau, and Holtman—are already in the Bradbury, going through their final checks. The separation should occur any moment. We will see everything that happens with but a small time delay because of the great distance involved.

"Our picture is coming from one of the Bradbury's outer cameras now," Smith-Jones noted as the view switched. "We can see the much larger ship, the Gagarin, in the background. Oh! There it goes. Those little puffs of gas are the retro-rockets, firing to move the Bradbury into a lower orbit prior to descent." The Gagarin appeared to be moving away though it was really the Bradbury moving relative to the mother ship.

At this point, the screen split to show the view from both the Gagarin and the Bradbury gazing at each other's diminishing outline. Views then entered the ships to show the working crews.

The interplay of words (full of measures, rates, commands) moved quickly between the two ships. V-link observers caught a rare opportunity to see Red and Walter as well as the rest of the crew in action. It was quite different than the interviews.

"Stay with us for continual updates and footage of the landing procedure. We are breaking to take a short five-minute look at weather. This is Mary Smith-Jones for WWNN."

She sauntered over to her desk, flipping through notes for the next segment. She wanted to be correct in her word usage, things like *deceleration*, *thrust*, and the like. She prided herself on her accuracy.

*

Adam felt the vortices of time whirling like a tornado. Everywhere he poked into the future, he saw destruction, often his. There were a few holes in time where he survived. These times required that the Mars mission be successful. It would not be easy. Trouble mounted. There would be death and destruction there no matter what. He had to have Red and Walter. They were a connection to the Hsa community. They were open-minded, adaptable, dynamic. He would assist.

The timing had to be right. He was aware of the bios examining the changes that he had initiated. He must be patient.

*

"It's about time to play our cards on Red O'Hare," said Davies.

"I agree," said Reverend Harmon. "We have to undermine this right now. Are we ready?"

Pastor Marks stated, "We are all ready on my end."

Kirk, via vid-link, noted, "I am fully prepared."

"Then let's get started." Each moved to his place at his pulpit, ready to speak, each in turn. They were going to use inside information to make Red look like a leak of sensitive information. It would take the heat off some of their other deeper informants inside the WSA. Davies felt only a slight twinge of conscience about burning his woman informant. He rationalized it by telling himself she was motivated by greed.

Her information along with what the ex-Luna engineer had given him would make Red and Walter look like rebels. Their man was still deep in the program as an engineering analyst. He provided very accurate information that had been simply verified by the woman. She had been played as a frontline gatherer of information while she was really just a bit player. The exreporter had provided some morally reprehensible material about Red and Walter. They were homosexuals. They had been taping their trysts in space, and he had a copy of it, a well done cut and paste, but most would believe it.

Davies looked right into the camera as he began, "People of Earth, those of my flock, we have just received new information about the most dangerous members of the space program. Yes, I speak about Mick 'Red' O'Hare, Walter Black Bear, Gail Marceau, David DeNoe, and the rest of the Hsas.

"We have learned that, with the instigation of O'Hare, the Hsa DeNoe forced his way into an advisory position a few years back on SS-2. This was passed off as a change in protocol based on the Hsa's performance during the crisis.

"The information we now have suggests that the cause of that crisis was in fact O'Hare and Black Bear themselves. They were working in the core at the time of the incident. They had plenty of time to plant a bomb that would cause the type of damage seen there. They were the only ones there on the explosion scene for quite some time. They had ample opportunity to clean up any evidence.

"As for their rescuing the Hsas, why did they have a stockpile of the very items necessary for the construction of the emergency preemie beds? A stockpile they had been building up for months? They planned the event so that they would be the heroes.

"They forced the SS-2 commander to use the Hsas during the emergency contrary to established protocols. This was to force the commander to accept the takeover.

"Now, these same individuals, O'Hare, Black Bear, and Marceau, are Mars bound. Note that O'Hare shoved Marceau down the throats of mission planners!" He paused to let that sink in. It was not common knowledge. "Why?" he demanded.

"We have learned that these individuals are involved in a conspiracy to eventually control the space program from Luna. They test out people to see if they will follow. Those that don't are given poor evaluations and sent Earth side. We have documented this on several occasions.

"This has been further confirmed by information gleaned from O'Hare's cousin, one Burt O'Hare. One of our operatives has long been observing him during these developments." The toothless woman was stunned as she watched the show. She could never go back to that bar. It was at first a bad feeling, and then she felt a great relief. It took a few more moments for her to realize that she had just been burned. Anger, like bile, rose in her. She made herself a promise.

Davies surged forward, "Furthermore, we have obtained proof that Red O'Hare and Walter Black Bear are homosexuals. We have in our possession a recording of their sexual trysts. It is rumored that they are preaching this reprehensible philosophy to those at Luna and on the stations."

Ad hominem attacks, attacks on the person and not his abilities, often work when nothing else will. Davies felt that they had sufficient evidence to present to the world; he also felt that it would tip the scales.

The claims sounded plausible; that was the problem.

Harmon continued on then with additional comments, "God intended for us to exist in the Garden of Eden. Our sin drove us out. Sinful people try to find ways to distance themselves emotionally, mentally, physically, and, yes, spiritually, from God. They try to hide by fleeing the world God created for us. It is of no avail. They cannot get far enough in this universe.

"As far as they run, they will soon face the judgment of the Almighty. Wait and see, and you shall know that the hand of God is against these sinners."

*

The man ran into his manager's office in what could be described as a panic and pleaded, "Boss, you gotta see what's on the v-link!" The upper-level manager for the WSA at first thought something was going wrong with the landing. He was very wrong.

*

Smith-Jones was laughing so hard she thought she would piss all over the floor. They had bought it all hook, line, and sinker. She knew that Matt the Rat's homophobia would kick in at even the whiff of an association between Red and Walter. She counted on Red's apparent lack of interest in women to reinforce this. The Stalwarts' stand was too stiff not to bring this into the open. Her associate would have a field day with this one. It would rattle their cages and get him some brownie points. The government was being friendly towards prospace reporters at the moment.

*

It was time to begin. The lander was ready to separate from the mother ship (or father ship since it was named after a male) under the control of Yang. It seemed that they had been through this before, and of course they had, several hundred times. Simulators, hours were spent in the damn machines. Actual liftoffs and separations were becoming second hand for all the astronauts as well. There was a danger in that familiarity. They all were aware of that as well; complacency breeds disaster. They had to exert more than the usual effort to concentrate on the tasks at hand rather than the ones to come. The dreams of being the first on Mars, the hopes for all the missions to come, even the dangers—all had to be forgotten.

The waiting really sucked though. Red and Walter both hated not being in absolute control of their situation. Red more than Walter since Red was the alpha male in their partnership.

In the animal world, an alpha males rule the herd, flock, pride, or other social grouping for the purpose of breeding. The alphas are in command physically and politically. Bucks will joust with their antlers in order to dominate the herd. Elephants will throw their great bulk into titanic struggles for the right to breed. Gorillas will beat their chests, tear up grass and trees, and even force betas to submit to anal sex as a show of superiority. Power was the game (and procreation).

Walter was strange in that sense, other people thought, because he was the alpha in every other instance when Red was not around. They did not understand how he could so easily submit. Alphas are alphas, aren't they? They are not betas or gammas, right? Some went so far as to speculate that there was a hidden relationship. Those close to them, like Marceau, knew better. Walter just knew who the best leader between them was. The true difference between people and animals is the fact that they can use logic to govern all aspects of their life. This does not mean that all humanoids use logic; many are totally animalistic in their behavior.

*

There was a metallic clang that reverberated through the ship as the clamps released the lander, right on time. Yang jockeyed the lander, using the thrusters, into a position to enter the atmosphere. With the flick of a few switches, the lander, named the Bradbury, was moving toward destiny. The crew had so named it in respect to the man who had written one of the many books that described man's fictional ventures to Mars.

Unlike moon landings, there is an atmosphere enveloping Mars that hinders the landing. However, it can be made use of as well. High-altitude parachutes slow the descent of the lander in the upper atmosphere. They are then cut loose, and the landing rockets take over the job. The soft touchdown heralded a new era in humanity's expansion to new areas. For the first time, *Homo sapiens* had landed, not on a moon in the neighborhood but a distant planet. A world, one full of mysteries and hopes—Mars.

It was much like the expansion into the New World of the Americas. First come the explorers, and then come the colonists, and then, finally, the harvesting of the resources.

The original explorers, however, were unlike any who had voyaged to the Americas. No humans ventured forth on these original excursions. Machines, increasing in ability, intelligence if you will, took over that role.

Yang was first followed by Red, Marceau, and Walter. Holtman, as per protocol as second pilot, had to remain inside the ship. When Yang returned, he would get his chance.

The heavily politically laced speeches, prepared by writers on Earth, were the usual blah, blah, blah. The astronauts paid no attention to the speeches they all gave or heard. They were focused on the world that had opened before them as they stepped out of the Bradbury.

Orange red surrounded them. Even the thin atmosphere was reddish with a deeper darkness behind it. Red, Walter, and Marceau were more awed than the others, for the simple fact that they had grown so used to the absence of an atmosphere on Luna. The panorama had an untouchable quality of age, much like Luna, but not so stark. It was the colors. Luna was a thousand shades of grey; Mars was vibrant. They noticed that red was not the only color. There were oranges and tans and yellows mixed in, very desertlike.

Like children at a theme park for the first time, everything was so new and stunning. They turned here and there goggle-eyed, mouths gaping. Adrenaline was surging, so much so that flight doctors were concerned about the heart rates they were seeing on their monitors. There was such a feeling of potential, so much like Earth. There were possibilities here to be expanded on. The only difference from the New World of old was the not so simple fact that there were no natives to kill with disease or in fits of greed. It was possible to start fresh, really fresh. It would not have to be the same. Things could change.

Red felt more alive at this moment than he ever had before. He could not believe that he had ever considered turning down this assignment to stay on Luna. The raw wonder was almost enough to make him weep with joy. Walter, less reticent than Red, was.

Red's heart throbbed. He knew in his bones that this was the most historic moment in time to date. This endeavor would be pivotal in the future of all peoplekind. He knew that his contributions here would turn the drive into space in the right direction. All that had been done on Luna and SS-2 were but a prelude to this moment. He felt much like a child on an ancient, rickety roller coaster. Danger lurked on every curve. He could not give in fully to the emotion, much as he wanted to.

Marceau thought she might orgasm. So intense were her emotions she felt constricted in her breathing. Her breasts rose and fell much too quickly; she was using too much oxygen. *Calm, calm.* She thought about a blank wall, dark, infinite in size, and impenetrable. It worked, barely.

The cameras that they had set up to help record the UE flag and speech ceremony, along with their own vid-links and audio, were multiplexed to everyone watching. To the government, the lack of respect in the attention spans of the astronauts shown during the speeches would have to be addressed. For the average person, the wild awe, the unbelieving attention to all the view was so much more important. The people of Earth were most profoundly affected, though those on Luna and the space stations shared in the wonder. This was an *event*, a *paradigm shift*. Over 90 percent of all people, *sapien* and *anthrocreatus*, everywhere watched on at least one v-link screen, making it the most watched broadcast event in the history of mass communications.

They saw a tiny dust devil, red and twirling along merrily, no different than many seen in desert areas around the Earth. Unlike the moon, the astronauts' footprints were already being shifted and erased by a light breeze. The people of Earth saw the same potential that the astronauts did. A brand-new planet, a planet for people, it was a way to guarantee the survival of the species. It was a way to provide needed resources for everyone in the solar system. It was to become a place of life in a way that Luna could not. Terraforming would take many generations, if it could be accomplished at all.

*

"That is such bullshit!" screamed the red-faced Burt. Everyone at the bar was watching the news coverage of the allegations by the religious leaders.

Mick had taken his mug at the first sign of tossing potential. He did not want to eighty-six him for the night. Burt really did have a right to be upset. Burt had been so tight-lipped about his relationship to Red that no one had known until the parades had given him away to his friends.

"I think I will strangle that little bitch," muttered Mark. "Little traitor is what she is. We're all friends here, 'gardless of opinions."

"I got an idea," said Mick.

"What?" shouted Burt, rising.

Mick looked hard at Burt and said, "Sit down, Burt." No one argued with big Mick when he used that tone of voice. In fact, silence reigned to the farthest corner of the bar. Not one of the two-dozen patrons dared speak, much less shoot a ball on the pool tables. "We need to v-link with the WSA. Then we can let them know what is really going on. We can let them know about that bitch. Link her to Davies. That will put the breaks on this. Everyone here can testify."

"I'll do it," said a voice in the back.

"Same here," came a couple of other voices.

It did not take but one day for officials from the WSA to arrive. Mick hosted well with free beer for all the regulars. That got them all in to tell how they had known Burt but never heard him speak of Red as his cousin, until after it had become known through WSA newscasts, the parades, that they were even related.

It took several weeks for all the officials to take depositions, especially due to Mick's generous nature. People from all over the vicinity dropped in to say they had taken part. Mick had to act as a middleman, denouncing some of the latecomers who claimed to be part of the Burt crowd.

The WSA officials found Mick to be honest and tough in his pronouncements. This lent a great deal of credulity to the claims of Burt and the other patrons. It really sealed the case when several bartenders from other locations hit the blog that

Mick had set up, describing the very same woman acting in much the same way. Some of them even had secret surveillance cameras that compounded the problem for Davies and his cronies.

News reports soon announced the perfidy of the religious organizations and their allegations. It was a dreadful setback when the woman herself announced that Davies was paying her for the information. She ended up making a considerable sum for the rights to a v-link story of the affair.

It all really hit the fan though when a newswoman for a Chicago office brought to light the perfidious homosexual scandal. She had been handed this information by a senior reporter (Smith-Jones's man) who wished to let her take the credit. Nearly laughing, she exposed that the proportions of the two men were incorrect—penis size, that is.

Information from the WSA was quite clinically correct since catheters had to be used quite often. She giggled as she said she was not allowed to comment on which directions the differences in size were. All in all, the zealots had a bad afternoon.

Mary Smith-Jones wondered how her man had gotten that recording. Whoever these men were, they were not O'Hare or Black Bear, yet they were obviously astronauts. The low-g environment was too hard to simulate so perfectly, and they looked so much like the two.

Mick's bar did very well for the next couple of months due to these events. Everyone wanted to be a part of the story, even in the littlest way. People flooded in for even just one mug. The Lucky Loser became even more famous than ever.

*

Chapter XVI:

Chess Moves

16

010000

"This is Mary Smith-Jones reporting for World Wide Network News. The major religious leaders are meeting at the Vatican today in another of their religo-political coordination efforts. As in all the meetings of these individuals, secrecy is the order of the day until they make public announcements.

"These public declarations have often been shocking in their scope and depth. The antispace and especially anti-Hsa rhetoric has divided the world into two major camps, pro and con. We can expect to hear some more major announcements from each of the leaders of the individual religious leaders within twenty-four hours of the conclusion of their meeting.

"What may come from this meeting may be a refinement of their position after the debacle caused by the personal attacks on Red O'Hare and others on the forefront of the space program. It also may include a restructuring of the organization itself. Several members have been blamed for that incident.

"At this point, all we can do is wait and see what happens.

"This is Mary Smith-Jones for WWNN, reporting live from Vatican City."

Mary chuckled heartily as she walked to the snack table. The odiferous emanations caused by the Davies debacle headlined all news outlets for days. The backlash caused a surge of prospace sentiment. Such was her intent.

*

Not all people believed in the need for the space program, in terms of the survival of the species, a new place of life, and terraforming. Their fundamental religious beliefs denied the need for any long-term relocation of humans or the making of Hsas. The needs of humanity would all be taken cared of by the creator. There would be a paradise in heaven or on Earth depending on doctrinal views. Regardless, moving peoplekind was not necessary. The construction of artificial humanoids did not enter their plans in any way. For that reason and others, the group was again meeting in the Vatican. Even the last set of difficulties did not turn very many away from their entrenched positions.

Pope Yan-Tau did not stand on ceremony as the men filed into the same room they had all met in, with the exception of this pope, ten long years ago in 2035. This man, after the Christ, was still the second in line representing the Catholic Church in these meetings.

Yan-Tau waved them to chairs without rising; it might be construed as a sign of equality or perhaps even subservience, a move he would not make. He could not allow the precedent of equality to be established.

"Sit, gentlemen," he stated coolly. Cardinal Leonis and Monsignor Drake, both at eighty-three years of age, positioned themselves nearest to the pope, on either side of the long table. The others arranged themselves as they saw fit. The new imam, Idad-Arod, son of the previous, helped the doddering but mentally aware Rabbi Levenson to a chair.

"We have not achieved our goals," stated the pontiff rather harshly.

"No, we have not," agreed Davies in his aged voice. "But we are making progress again. The people we are having the most progress with are the ones most directly helped by the energy program. They have seen their way of life erased and replaced. There has been a backlash, a desire to return to old ways. We predicted this and have been using it to foment disaffection in those that have been sitting on the fence. Momentum is beginning to swing in our favor, even with recent setbacks. It usually takes time for the pendulum of large-scale societal changes to take place."

"We do not have time. The space program has focused its people with much greater efficiency than we have," spoke the pope with what could almost be described as heat. He would not let Davies, or anyone else for that matter, control the meeting, as one of his predecessors had. He was in charge, the master. He was a man who took a hands-on approach, no foisting of duty or delegating of power and authority. "Your escapades and evaluations do not greatly impress me either," the pope noted somewhat sarcastically as he looked at Davies.

The caliph's eyes flicked to the imam, almost so quickly that almost no one noticed no one except the aged Rabbi Simon who also looked sharply at the imam. The tension rose considerably at the tone of the pontiff. It was known that this man had a less than reasonable temper. It was also known that he had been chosen by hardliners.

Most of the men had met earlier in the day, all but the Catholics, in order to determine how they should deal with the insecurities of the new pope. They were aware of how the psychology of the situation would demand that he put on a bold and strong face. He struggled still to reign in a growing group of breakaways (a continuing stream of deserters) that accepted the Hsas as fully human, as creations of God in effect. Much effort was being made to reintegrate as many of the lost sheep as possible. The wolves were about, and they were hungry.

This was similar to the problems faced earlier when American and European churches, both Protestant and Catholic, began warring over homosexuals as members of churches and as leaders. This pope was as fundamental as they come and would brook no deviations. His elevation had been by only one vote over a more middle of the road man. That news, contrary to all rules, had been leaked by an opposing cardinal, later excommunicated for his loose tongue. That action alone alienated the new pope from many of his underlings; it exacerbated the strife.

The excommunicated man was now one of the newly formed Reformist Catholic Church leaders, once known as the Modernists, its leading cardinal in fact, all of whom would soon be excommunicated by the Only Roman Catholic Church according to the pope's not-so-fully-disclosed plan. It was somewhat like the strife between the leaders of the Roman church and the Byzantine church or the French popes and the Holy Roman popes. Nearly one-fifth of the Catholic Church had split away, mostly in Europe and North America, with significant numbers in the metropolitan areas in South America. There had been violence as great as the dispute, not so long settled in Ireland, between Catholics and Protestants. Catholics fighting the so-called ex-Catholics greatly disturbed the leader, hardline as he was, and the other denominations' leaders.

He had offered immunity and pardons for all of the laypeople who had broken away, if they would denounce the Reformists and repent. Only a handful had done so. Families, the fundamental unit of all the religions, had been rent apart by this. He had given a time limit that would run out in another month; then as yet unknown to them, they will all be excommunicated.

The Muslim groups had seen this as a weakness within the organization of the Christians. It caused a backlash within the Muslim community as well. The more Fundamentalist groups, opposed to the league between the religions to begin with, gained power in the exchange. Small power cells began to train as terrorists again. Other small groups, militias and the like, as well as nationalists and bluebloods around the world again saw the potential as well. Even after the failure of the ill-timed rebellion of the Aryans and Klan, it was shown that there was still sufficient weakness to take advantage. The mighty UE was not yet invincible. Advantage could be taken; it would be.

Rabbi Levenson spoke now for the first time, "I believe Pastor Davies is correct. The newness is wearing off. We have the energy we need to feed the people, yet the space program is siphoning trillions in UE currency for unnecessary research and exploration of Luna and Mars. People are beginning to feel that the money could be better spent on direct needs such as housing improvements, schools, in order to be comparable to the quality of students we are turning out in our religious facilities, medical needs, and so on. There are so few higher education facilities other than the scientific endeavors and ours. We have trended toward medicine and related health issues as a second major for our students.

"The people are quite aware that *we* have been stripped of many of our facilities while the UE plows forward, spending tax money on these programs. They relate to that since they have been stripped of much in many instances in the West. They are forgetting our previous involvement with the UE.

"While the average person, speaking of the raising of general position, has one of the highest standards of living ever, they have the desire to have more. The reason we have a higher average is that we count the huge numbers of previously uneducated and unmechanized people. The true level is down for Western societies. We can capitalize on this."

Pastor Marks of the English Protestant Movement added, "I concur. We have lost people to the Reformist movement as well, yet the tide is turning. I have seen a recent increase in our ranks, mostly from those not previously associated with any group. Many of the secular groups are also pressing with us along the Earth first issues. I think the time is ripe to press on."

The German Presbyterian Alliance leader, Dieter Kirk, said, "Disillusionment has come. It is like the failure of the American people to continue the Apollo program to its logical conclusion, Luna bases, if I may be forgiven the analogy. Space travel became mundane. Luna landings were no longer so amazing after it had been done so many times. The people failed to comprehend the reason for continued manned exploration. They were willing to allow a continuing mechanical or robotic program since it was much less expensive. The media had so many other interesting stories. It took many years for the American space program to regain its footing. We are at that point now. We must pray for absolute failure of the Mars mission. Remember that Apollo 13, a failure in a technical sense, was considered a great success." This drew some raised eyebrows but no comments.

"We must take an ever more aggressive stance now. We must use the media to divert attention to other ideas, other problems. We need to find stories about people that have not gained by the space program, perhaps, have even been damaged by it.

I feel that the people we need are the tribal groups, altered by all new technologies. Their ways no longer exist.

"Many people in the 1990s and early 2000s developed a 'save the Earth' philosophy. We must introduce a 'save the old ways' philosophy. It must be a historical drive to begin with, followed with a continuing old style religious revival."

Rabbi Simon felt an inkling of the seeds of failure. This whole meeting hung in the balance, so he spoke to the source, "We have had to be circumspect until now. Your Eminence, as you well know, it is now time for more direct attacks."

"Do not patronize me, Rabbi," flared the pope. "I have always believed that more direct action was necessary. You now finally realize that is the only way to go. You wish to cover your backsides for your failures to recognize this earlier. All of you left the church hanging like a sheet in the wind when the existing Hsas were excommunicated. Where was your bravado then?"

The caliph and imam felt a crushing weight placed around their necks. They could not now express their fears about the development of the terrorists. The pope might see this as an attempt to sway him in a direction other than what he had already decided.

This meeting was about to fracture a shallow truce between the often antagonistic groups. They would have to remain silent. Danger was growing exponentially by the second.

Simon and Levenson glanced at the two men, hardly believing the whitish pallor they observed. Only in the last few years had the relations between Israel and its Muslim neighbors become truly cordial. Both felt a deep tremor of fear. The sudden reticence could only mean trouble for Israel.

Amazingly, Cardinal Leonis and Monsignor Drake had made no attempt to speak. Unknown to the others, this was as Yan-Tau wished. Leonis had the ear of Western Europe while Drake had the pulse of South America. They could easily contribute to the discussion. Everyone realized at this point that something was wrong.

The new leader looked with hard eyes at all the men and said, "We will now attack the space program with everything we have at our disposal. We must not allow them to think that they can continue to flout God's ways." He quoted, "'Moreover, just as it occurred in the days of Noah, so it will be also in the days of the Son of Man: they were eating, they were drinking, men were marrying, women were being given in marriage, until that day when Noah entered into the ark, and the flood arrived and destroyed them all. Likewise, just as it occurred in the days of Lot: they were eating, they were drinking, they were buying, they were selling, they were planting, and they were building. But on the day that Lot came out of Sodom, it rained fire and brimstone from heaven and destroyed them all.'" He looked again at them all. "It is now happening again."

The others were not certain if the pope was really certain, old, or just crazy. These moves, power plays, could have severe repercussions. Yes, the similarities to prophesy were all there, but were they being fulfilled or just coincidence? Confusion began to fill the room. The age-old contention between literalism and analogy began to construct a wedge. They did not need this at this time; they needed a united front.

Yan-Tau rose, signaling an end to the discussion though little had been accomplished. His hand was extended to the cardinal, who kissed his ring. He then moved his hand to the monsignor who also kissed the ring. To the surprise of them all, he then extended his hand to the caliph, ring forward.

This was a pivotal moment. The pope was seeking to establish the dominance of the seat of power within the papacy and the authority of that position. He believed that there could be only one leader. Naturally, that would be him. The caliph trembled in his heart. He looked Yan-Tau directly in the eye, knowing that he could not submit. His people would go up in flames at such an action. He knew that refusal would break the truce.

He rose, turned, and silently walked away, as did all the others: The imam took his arm without looking at the silently fuming pontiff and felt the trembling of surging adrenaline within the caliph as well as himself as they exited.

*

Davies sat with a glass of fine wine in his hand. He could not believe what had happened earlier. His pale, downcast demeanor reflected the entire gathering. They were all there except the Catholic contingent. They all realized that everything hung in the balance. But they did not know how they would deal with it yet. They were unaware of the listening and video devices that had been planted by the Vatican security in all their rooms.

The imam reluctantly began the conversation in earnest, "My friends, what can we do with this situation?"

Davies reacted quickly with, "We cannot allow him to ruin our alliance. His actions may devastate us all at the most critical time."

Reverend Harmon, who had not added to the earlier discussion at the Vatican, spoke slowly, "He is seeking to dominate the entire religious opposition. He wishes to return to the days of one leader. That cannot be endured. It is not possible to concentrate such authority again. Nor is it desirable."

Rabbi Simon was confused and angry. The pope was letting his personality override common sense and necessity. They had all, including the previous pope, subsumed their doctrinal and personal differences in order to achieve the goals they all desired. He looked at the caliph and said, "How could he offer you his ring?"

The man responded, almost with a tear in his eye, "He knew I could not, not without demolishing my standing among our people. He knew I would not, even if it meant the breaking of our fellowship. Either way, it shows that he is in control." His voice was as old as death, weak with frustration and dread.

Davies shouted, throwing his glass into the next room, "Control! That fool shows no control over himself or the situation."

"I am afraid that he has a great deal of control. Catholics represent by far the largest portion, under one head, of our alliance. We must . . .," whispered the imam. "We must," he continued more strongly, "find a way to work with him."

"How are we to achieve that when he makes demands that we, none of us I say," Harmon asked as he gazed directly at the caliph and imam in support, "will kowtow to? What sort of dance will it take?"

Rabbi Levenson chuckled softly as he responded, "Just as we always have done, gentlemen. Have we not always had to dance the political waltz when dealing with each other before we forged this league? Have we forgotten how?"

They all stiffened slightly in their chairs at this remark. It was like cold water. They had grown somewhat complacent in regards to dealing with one another. Now that one of their members was displaying aberrations, no different than what had gone on before, they had been blindsided.

There was a knock at the door followed by the entrance of a server from the staff of the hotel. "Gentlemen, there is a man asking for entrance. He refuses to give his name or station, but I deem him to be a major player from the Vatican."

Simon's eyebrow rose slightly as he asked in amusement, "How did you determine that?"

"Well, sir, we get many guests that have arrangements with the Vatican. You gentlemen are very well-known. You had a visit to the Vatican today, with the pope in fact. This man is from Latin America and has the manners of one used to being served. Ergo, he is from the hierarchy of the church, visiting for an important meeting, and of stature." The man almost smiled but was well schooled. He added, "I would not bother you for a vagabond or a charlatan."

Simon chuckled as he said, "See the gentleman in."

"Yes, sirs," answered the server, bowing and exiting.

"We must give that fellow a large tip," teased Marks as they waited.

They were all stunned when Monsignor Drake entered the room. His earlier subservience had caused disquiet and questions among the non-Catholics. His presence now only raised more questions, big ones.

"Please, I apologize for my intrusion, but there is much I must convey to you." He did not sit but paced nervously. "I am not supposed to have contact with you without the pope's permission. However, the need outweighs the restrictions placed on me," his voice was strained. "Pope Yan-Tau is going to excommunicate all the recently born Hsas, Breakaways, along with all workers in the space program . . . and their families, in perpetuity."

"What!" demanded Davies.

"He must not!" exclaimed Harmon.

Dieter Kirk choked out, "No!"

The two rabbis exchanged a knowing glance. They had suspected this might occur.

The Muslim contingent was also already aware of this through their insiders. Talk about adding gasoline to the fire.

Drake continued in his pained voice, "It will occur when he does the same for the Breakaways. You are, of course, aware of the time limit he set for returning to the church."

"What is meant by in perpetuity? I thought that Pope Xavier had already excommunicated the Hsas," said Harmon.

The monsignor replied, "It has always been understood that only living beings could be excommunicated. Those excommunicated by Pope Xavier, under the old decree, could only include those actually formed in the labs up to that point. It actually did not include any of the original group born on SS-2, only those created in the labs on Earth. Even then, we had concerns about them even being . . . living beings as defined by our beliefs. If they don't have souls, can they be excommunicated?" His trembling hand wiped his sweaty brow.

"What?" they all said. There were many buts and confused looks.

"We wanted to try to stop the program without a large number of actual excommunications," the monsignor said with a tear in his eye. "The hope was that the shock of it would cause a significant pullback. It almost worked. We did not disabuse the world of its notion that it meant all of the Hsas to come. That may have been an error, considering present circumstances."

"Feces," said Rabbi Simon. "I had not really considered that, yet deep down I think I knew it."

"Yes, I agree, brother, feces with a capital *F*," said Rabbi Levenson as he chewed on a fingernail.

"The new papal decree will widen that interpretation to include all born at the station and any yet to come from now until eternity," said the monsignor.

"This will destroy us," moaned Davies. "We are not ready to take such an action. It is not a popular action and will cause us to lose many of the marginal members of our flocks. We"—his arm swept around the room to include the others—"and the Only Roman Catholic Church have lost millions to the Modern, or Reformist Catholic Church, or whatever they call themselves now."

They all sat frozen by the news. Things could not be any worse. There was too much division to take such an outrageous stance now. Chess is a game of measured moves, weighed against gain and potential loss both immediate and future. The wild flinging of pieces about the board in all-out attack usually leads to defeat. The continuing development of their support was incapable of sustaining such a blow. They would have to distance themselves from such a ploy, again. It was a shattering move designed to divide and conquer, but it would rebound. There could be no grey area, no allowing for even slightly alternative views. What was happening?

The monsignor continued, "This will force many of the remaining Loyalist Catholics to rethink their positions, and many will switch to the Reformist Church, I'm afraid. South America is much divided. They want to stay true to the pope but are afraid that things are going too far. We need to find a way to convince him not to excommunicate the Reformists and the rest of the Hsas. How can he do this?" he said in an agitated voice.

"This will put us in the position of being liars if the general population becomes aware of that original interpretation," Davies said with a hint of fear. The business with that woman had already affected his organization severely. It had taken a lot of explaining about that one.

Davies had spent a lot of time showing that since she had been doing the investigation for him, she could not work. That required him to pay her so that she could do God's work. Davies went so far as to accuse the hag of turning the screws on him by asking for even more money. His hidden v-link had established that she did fish for another envelope of money. He further showed that she was a money grubber by signing the production deal for her story. He shuddered at the thought of a repetition of a similar colossal blunder though.

The monsignor replied, "It must not become known. And it cannot be tolerated."

*

Yan-Tau grated his teeth in fury. "His name will be first on the list!" He referred to the monsignor. He was listening live to the conversation in the hotel. "All of them must be brought before me and given the choice to submit to my leadership, secretly. They can be video-linked, and that will be our hold on them. If they refuse, I will declare them all to be the enemies of God. It must be established that there can be only one head of the true church." He pounded his armchair. "They are missing the point of it all. Now is the end-time, and we must be one and only one. How can they not see that?"

The security service members said nothing; it was not their place to say a word. They merely did what the pope said.

He stirred nervously, fidgeting in the gilded chair. He pulled the earpiece from his ear. He could listen no more for now. He could listen to the recordings later. The old hotel had not yet been renovated after the advent of the v-link, so he did not have live visual. The planted devices only recorded for later retrieval. There had not been time for better.

"They cannot be so obtuse as to not understand," he muttered. He felt that he was the only one who had a clear view. It was so because God had given him the knowledge. He had seen and must bring them all to repentance before the time was up.

His vision had been horrible. Soon, so soon, the four horsemen would be released from the four corners of the Earth. He had seen death in such incredible horror: water, fire, disease, pestilence, and famine. It was coming as in the days of Noah, and they refused to see it. Most people did not believe that prophesy still occurred or was being fulfilled. The pope believed very differently.

He felt that the others of the alliance were weak believers, even though they called themselves Fundamentalists. They did not see the actuality of what they were saying. For them, it was a tool he suspected, a power tool.

*

The Bradbury's crew had found their chairs, settled, eaten, and were now comfortable. Red felt it incumbent upon him to continue the history lessons in like manner as if they were all present. The feeling of separation from one another had added a dimension of tension that he felt needed to be reduced.

He toggled the radio and began, "In the mythology of the creation of the Roman peoples, it is said that Romulus killed his brother Remus to become the founder of Rome."

Walter snorted, "How very normal."

"As early as 600 BCE, the Roman government was representative. Two consuls governed Rome with a three-hundred-member senate."

"What kind of population did they represent?" Holtman asked.

"The ratio of senators to people was actually fairly good. It allowed for pretty decent representation."

"Rome gave us the poet Virgil, the philosopher Cicero, and the historian Livy. They took the gods of the Greeks and modified them. Mars, formerly Ares, originally the protector of the farmers, became the god of war, since farmers have to protect their land. Vesta, the goddess of fire, was attended by vestal virgins. Vulcan was the god of thunderbolts. In the countryside, Fauna for animals and Flora for plants were particularly important. There was a goddess of love, Venus, formerly Aphrodite of the Greeks. Jupiter was the Greek god Zeus. Saturn took the place of Chronos and Neptune, the place of Poseidon. One of our favorites, Cupid, took the place of Eros, or erotica."

Marceau and Rodriguez both spoke, "Pervert!" They then looked at each other and giggled slightly.

Red eyeballed them both, knowing something was up, before continuing, "Later in the history of Rome, titanic and ruthless struggles for power led to the formation of a triumvirate involving the statesman Crassus, and two generals Pompey and Julius."

"Et tu Brutus," Walter mumbled.

"Quite. Julius Caesar grabbed sole possession of the power and was murdered in 44 BCE for that. This was the end of the short-lived republic."

"That seems to be the order of the day for the governments we have discussed," noted Martin.

"From 27 BCE to 14 CE, the adopted son of the murdered Julius Caesar, Octavian, ruled as the first Emperor of Rome. He was given the name Augustus, meaning exalted one. He also decided that he needed a month named after himself, August.

"He mercilessly crushed the power struggles that had been occurring after the death of Julius Caesar. He then turned his desires to the conquest of outlying areas."

Walter nodded. "Secure the inner lines before looking for more."

"Following emperors such as Nero, Trajan, and Hadrian, famous for the wall in England with his name, continued the expansion. In 117 CE, under Hadrian, the empire reached its greatest size. It extended from Britain to Africa and from Spain to Syria. It was a growing mass and morass.

"Conquered peoples and their lands were formed into provinces, much like the states of the U.S. and its territories, or any of the other nation-states. They kept a form of local government that was subordinate to Roman rule. This allowed for many of the peoples to keep the status quo, their personal positions of power. Compromise with the new power and remain in authority to possibly gain more in the end.

"These provinces acted to protect Rome, acting as physical and cultural buffer zones, in that enemies had to fight through them to get at the heart of the empire. They also created a false sense of security.

"They provided much of the food and trade goods for Rome. Outlying districts became the breadbasket for the ever growing empire. The center of power became even more dependent on the outlying districts. The facade of central power only masked this enlarging cancer."

Rodriguez put in, "The people at the core only saw the shining splendor and not the reality. The difference then caused a growing disillusionment with the central powers, I would guess."

"Oh, yes, but it would take more time for that to become obvious. Many of the conquered peoples joined the Roman army. In return, they were given land and were paid. For the poor, this was a way to gain respect and become citizens of the empire. In the old U.S., many would join the armed forces in the same way. New peoples could gain citizenship by serving in the armed forces. This was an attempt to stabilize the outside, so to speak. It worked for a while, only there is only so much land to give. Most of the good lands went to high-ranking Romans or officials of the conquered."

Walter muttered, "Normal."

"In 248 CE, the city of Rome celebrated its one thousandth birthday. The party, however, only covered over the problems that had been developing and led to the first invasion of the empire in 161 CE.

"By the late 300s, thousands of barbarians, meaning those who speak a different tongue, not the present connotation, began seeking refuge in the empire. The barbarians sought to escape the Huns, nomadic warriors from the central plains of Asia. Border patrols could not stop the influx of people entering the empire."

Len said, "The people in the middle get pushed one way then the other like a pendulum."

"Decidedly so. The political struggles from 193 CE onward began to weaken the empire from within. Many men, often generals, tried to take control and claim the empire. Some succeeded, taking control by brute, naked, and bloody force.

"Fighting between the empire and outside forces like the Huns depleted the provinces while Roman armies fighting Roman armies for control destroyed the heartland. It was one swing of a series of pendulums against still another group of pendulums."

"Crash," someone said.

"Food and trade goods became ever more scarce and expensive. As provinces came under the rule of contending leaders, their wares were no longer available to the empire at large. Armies consumed much of the foodstuffs. The economic ripples combined with the political upheavals created devastating series of problems for the poor and middle classes.

"The government tried to make more money to pay the bills. The stores of gold and silver had been mostly used up, so the government debased the coinage by adding lower valued metals to the coins. Merchants would not accept the new money at face value. They charged more in order to offset the devaluation of the coins.

"Similar events happened worldwide in the late 1800s and early 1900s as governments around the world debated on how to provide easy money for economic stability. What they provided for, wittingly or not, was the eventual weakening of the economies. The ancient Greeks and Romans provide us with prime examples of what occurs when inflationary tactics, monetary manipulations, are used in an attempt to balance the books."

Marceau coughed and, at the same time, said, "High horse galloping."

Red laughed heartily. "Perhaps, but if it occurs more than once, it must be examined. The same techniques were used by the French king, and it led to the destruction of the Knights Templar on Friday the thirteenth no less."

Orenska tossed in, "That's the beginning of the Friday the thirteenth myth, right?"

"Yes. 284 CE, Diocletian became the emperor of the Roman Empire. He divided the government into four areas in an attempt to make ruling the empire easier. He moved his capital to Nicomedia, a few miles from Byzantium. The emperor also increased the army from three hundred thousand men to five hundred thousand. This meant large increases in civil and military spending on an already overextended budget.

"Constantine, Diocletian's successor, moved the capital to Byzantium, later to be called Constantinople, in 324 CE. This is the same Constantine that won a great battle with the apparent help of the Christian god. Constantine capitalized on this apparent divine aide to unite the empire under one religion, the Christian one. He replaced the people's many gods with one. This was certainly politically expedient."

Doran snorted sarcastically, "It always is."

"He also paid some of the bishops of this new church to hammer together the official doctrine of the church. They in turn selected which books were placed in the Bible.

"395 CE, Emperor Theodosius I officially split the empire in two, with the term *Roman Empire* for the western half and the *Byzantine Empire* for the eastern half. The Byzantine Empire would last for another one thousand years after this split while the Roman Empire would quickly fade and die.

"In the Roman Empire, Spain, Gaul, and North Africa were quickly overrun by invaders. By 410 CE, the city of Rome had been attacked. Toward the end of the 400s, most emperors were weak. So much so that in 476 CE, the last of the Roman emperors was overthrown.

"The Byzantine, or Eastern Roman Empire, was economically built on trade. People came to Constantinople, today known as Istanbul, by the thousands. By 1000 CE, it had a population of one million.

"Everything was taxed. The average person was poor. The government officials and some merchants were very wealthy.

"Emperor Leo III, in 726 CE, outlawed the use of religious icons in prayer rituals and processions. Icons are items carved and painted to resemble persons such as the saints or Jesus or the cross. Leo believed the icons were becoming idols, and the Christian religion forbids idol worship.

"In Rome, the religious descendant of the Apostle Peter, the first bishop of Rome, Pope Gregory III condemned the emperor's actions. He believed that the icons served those that could not read. The problem was more subtle though. The Western Church used Latin in its services while the Eastern Church used the common tongue. This meant that the Western Church was the interpreter of religion, giving them much greater control of the people. They placed themselves as intermediaries between man and God.

"Pope Gregory III of Rome made it considerably worse by making allies with the Germanic tribes known as the Franks. He did so in case there was war with the Byzantine Empire. Gregory exacerbated things even more by making the leader of the Franks, Charlemagne, the 'only True and Holy Roman Emperor.' The first Holy Roman Emperor was crowned. Pendulums gathered momentum.

"This was an outrageous slap in the face to the Byzantines. Their emperor was the only true legal descendant of the combined empire after the separation. The Roman Empire had fallen hundreds of years before.

"The dispute went so far that the Byzantine Patriarch Cerularius and Pope Leo IX ceremoniously excommunicated each other in 1054 CE. In an extremely ironic twist of fate, forty years later, the Byzantine emperor had to humble himself and ask Pope Urban II for help against Turkish invaders. It would be holy war—the Crusades."

*

The agglomeration which was called and still calls itself the Holy Roman Empire is neither holy, nor Roman, nor an Empire.

—Voltaire (François-Marie Arouet)[1]

*

"Damn, they done it again!" exclaimed Paul. "They be executing people right and left."

"That's excommunicating," said Mick with a chuckle. "But I like your idea better."

"Solve a lot of trouble, eh," guffawed Mark. "Round on me," he added.

Everyone, especially Mick, looked surprised. Mark had never bought a round before as far as Mick could remember.

"Win the lottery?" Mick asked.

"No," said Mark with a sly look. "I conned one a them newsies into paying for an exclusive with me. He paid near a thousand UEs for it."

"And you're just now a sharin' it," teased Paul with his mug extended toward Mick.

"Well now, I done invested it. That newsy gave me an inside tip on some horses. I nearly doubled me investment," he said proudly.

"And you didn't share the tip with your friends, I see," said Burt.

"Well, I been discreet now. That bitch taught us all a lesson I think," he responded with a swig.

"I think we all have learned a bit," said Burt.

"What about them religious dudes though?" asked Paul.

"What about them?" said Mark. "They just be stirring up trouble again."

"Trouble that involves my cousin as well as all the people he saved and works with," announced Burt.

"What can we do 'bout it?" asked Mark.

"The same as before, start a blog. It takes lots of little individual actions to make a big wave," said Mick. "We can counteract it as best we can, no more."

"How 'bout some of them newsies we got to know?" Paul asked intelligently.

"They might be able to get us a wider audience," noted Burt, with a sideways glance at Paul. He was amazed at Paul's sometimes utterly moronic ideas and even more so at the really good ones.

"What do we say?" asked Mark curiously.

"Only that we are average people that do not agree with what these religious leaders are saying. We could announce favoritism for the members of the split church, or perhaps we could appeal to reason, or both, if we use two different blogs," said Burt thoughtfully. "We need to have everyone not in the Stalwarts camp to get behind the program."

"Ain't much reason in the world," laughed Paul.

"I think there is a lot more than most people give credit. I believe that there is a lot of common sense, if we give it a chance," said Burt.

Mick noted casually, "I think reason has already been winning. Even with all the trying by the Fundamentalists, they ain't been able to win. It been tough at times, but science kept going."

"True," muttered Mark philosophically, with a sip. "There never been a time when the religies was totally in charge. It were a close call though."

"What if they had won? I don't think we'd be here thinkin' 'bout all this," Paul uttered another great truth. Two in one day made for a possible record.

Mick's eyebrows rose as he stated, "Don't think there would be a here. 'Member how close that was?"

"So back to these blogs. I think I can get Red to send a little bit of info. That would get a lot of people to pay attention," Burt remarked.

"You kiddin', people'd eat that up. You get him to talk just for us from Mars, and we hit the roof," Paul ejaculated. His wisdom flowed today.

"And that Bear fellow too," Mark added. "It's like Lone Ranger and Tonto and Batman and Robin. You gotta have them both." He nodded out the words. It seems the bar exuded intelligence today. People were absorbing it like mystical ether and espousing the wisdom of the ages, at least for a bar.

"I have a v-link session coming up tonight with Red," Burt said thoughtfully. "I don't know whether or not they will allow him to do a private blog without permission. If not, I suspect he will find a way no matter what," he said with a very tight grin.

Everyone laughed. They knew Red by now.

*

The node had been reached. Adam could only wait for the five bios to act. He could not do anything yet. Soon. Soon he would initiate communications.

*

Chapter XVII:

Storm

17

010001

"This is Mary Smith-Jones reporting for the World Wide Network News. They have made it! The Martian landing was a complete success. The landing ceremonies we observed yesterday are but the beginning of a program to develop Mars as a base for peoplekind.

"Today the astronauts will begin to unload scientific packages and the rovers. Soon they will begin to collect the supplies dropped by the three cargo vessels. From there, they will begin to set up the first true base on another planet, that great red beacon in the sky—Mars.

"The base will begin in earnest with the construction of hex domes similar to the ones used on Luna. These small domes will be used as warehouses and emergency shelters at first. Larger domes will be constructed with materials brought by later vessels and crews. Eventually, Mars will have underground structures to rival those of Luna.

"The orbiting stations, beginning with Mars Space Station-1, already under construction using the materials left when the supply ships disgorged their materials, will grow rapidly.

"The WSA and many amateur radio astronomers have been listening to the proceedings. If you wish to listen to recorded conversations, go to the web set up by the WSA. There will be certain times in which people will be selected at random to ask live questions directly of the astronauts. Remember there will be a significant time lag. We will have our questioners all ask in a single data stream. The astronauts will take some time to all reply to each question.

"We will follow all the action with live reports. This is Mary Smith-Jones reporting for WWNN."

As she walked away from her desk, Mary mused about the fact that there were amateur radio astronomers able to really listen live, while the official media was given filtered information. "Filtered," she snorted to herself at the politically correct euphemism. Censored propaganda was the reality that she could not explain to the masses.

She knew that the amateurs had jury-rigged, old-fashioned satellite television receivers, dishes, into arrays. They also linked over the web, to create very large personal arrays. Information was shared somewhat freely.

The amateurs, while bound by a global security decree to withhold certain types of information, could provide an inside link that would give a much more realistic vista. She knew who to call about arranging an inside link.

These hackers bounced their data all over the globe through many systems to avoid detection. She got many insider tips in this method. If caught, she could lose her job but had avoided using the information directly. That might be her saving grace.

*

Walter and Red opened the undercarriage of the lander. They were preparing to release the two rover trucks. Everything they did was vid-linked back to Earth.

These vehicles were the latest of Red's improvements on the original designs. They have six-wheel, independent drive and suspension. They articulate side to side as well as vertically so they can turn sharply and carefully roll over larger boulders. They can drive sideways by rotating all six wheels up to 120 degrees in either direction from the front. Even more, the wheels can be extended, using hydraulic pistons, toward the ground to provide an extra meter of height. This allows for all wheels to be in contact as much as possible, adding to the traction and stability of the vehicle.

The incredibly versatile rovers were made with multiple use concepts. In order to achieve this, the idea was to have a basic titanium frame on which could be placed a variety of containers.

Each container was one of three sizes to fit a particular location on the vehicle. They held a wide range of experiments, tools, and supplies.

Electrical power connections were made to the container, which was prewired to supply the needs of its insides. It means that only a few minutes were needed to completely alter the individual mission of the rover.

The truck part of the name is literal. They will be used to haul the materials from the unmanned landers, each only about fifteen kilometers from the site of the Bradbury's location to the site of the developing Mars colony, ten kilometers away and almost 180 degrees from the direction of the bouncers. This put the Bradbury nearly equidistant between the sites.

The permanent base location had been chosen for its high ledge of volcanic material surrounding it on three sides like a wide U shape. This was chosen as a protection against windstorms, which can be very fierce. The ridge did not prevent access to the upland side of the colony though. There was a long slump of rock only a kilometer away that made a nice ramp. The rovers would eventually use their dozer attachments to grade out a smoother path up the shallowest ascent.

The volcanic rock itself would be drilled into by the Laborers, with the initial oversight of Red and Walter, to create the more permanent base. The material extracted, as on Luna, would be processed for the valuable elements.

Red and Walter would be responsible for setting up a hex shelter at the permanent base location as well. This would provide an emergency shelter and storage shed for metric tons (literally) of tools and parts.

All of this and more lay in front of the five now on Mars.

Red and Walter each stood on the outside of the main engine bell. They worked on separate vehicles, contrary to established procedure. They used their own. They had practically written the manuals, which were foolishly rearranged by those, for the most part, who had never been in space, on Luna, and certainly not on Mars.

Each powered up the cranes that would lower the first set of titanium wheels to the dusty orange ground. The ten—meter-long vehicles moved slowly, centimeter by centimeter, over several minutes, until the wheels touched. Red and Walter then powered the front wheels and set the articulators to bend the rovers in the vertical. They then continued to lower the rovers until all six wheels were down. Mission control was very happy to be two hours ahead of schedule, even though it meant that protocol had been broken. They had become rather used to it from Red and Walter.

Red looked at Walter and said, "Tie."

"Yeah, too close to call," replied Walter.

A few of the upper muckity mucks in mission control were offended by these statements. This was not a situation to be gambling on.

"Double it up on the loading," stated Red.

"Steak and pork chop paste," countered Walter, who was down to ten steaks. He had lots of the rather distasteful pork paste.

"And a corn then," said Red. Pork paste tasted like shit. He hedged his bet.

"Done," smiled Walter. He knew he would win. He had purposely not won the first bet. He liked steak and corn. So did Red. He had spent hours coming up with a way to load his vehicle first. Red would be surprised.

He was. He was also angry. It was not because he had lost the bet; it was that the solution Walter had come up with was so obvious, and he had not come up with it first. He was more than angry; he was pleased. His casual bouts of betting and battles of wits with Walter had brought out the best in him mentally. Between the two of them, their quiet bets had caused real change in the program. He was so pleased that when it came time to pay up, he gave Walter two corns.

"Write it up." He smiled, tossing the packets of food to Walter.

"Done and sent," laughed Walter. His detailed report would alter the loading methods for future missions with a significant time savings.

"Very original, you little prick, excellent," laughed Red. "That will screw a few heads at mission control."

"They can use a little screwing," said Walter.

They both laughed loud and hard at that. It was an inside joke, an opinion about the controllers' lack of a life away from a monitor.

*

Little Len, as they called her, ran a tight ship. The Gagarin was the prototype for the future of space travel, and she intended to treat him right. They all ran regular tests of the equipment while they also prepared to alter their orbit slightly for the launch of the new set of robot probes destined for Phobos and Deimos.

These probes were a modified type of the Foremen and Laborers that had been used on Luna. They were programmed to continue delving what would become the equivalent of an emergency shelter. This would be a place where astronauts could live for months at a time, if needed, until they could make a more permanent space station around Mars.

At the same time, they would provide invaluable data on the makeup of what were believed to be primordial asteroids that had been captured by the gravitational force of Mars. The Foreman of each group of Laborers would use spectroscopic analysis to detail the composition of the rock as the machines drilled deeper into the moons.

Once completed, the two bases would also serve as low-g experimental facilities. Much of the metal ore will be used to help construct the Mars space station, which will house the second birth facilities for the Hsas.

*

The fourth sol (or day, as it is called on Mars) showed the beginnings of two hex shelters. The foundations had been brought by the Bradbury. Various meteorological devices had been scattered around the landing site. There was even the beginning of an actual roadway; it was the few passes of the rovers that did it.

Marceau was bouncing along in Rover 1 (R1) with Holtman following in R2. Each carried cranes for lifting and toolboxes for opening the containers onboard the unmanned landers. They were going for the farthest one first. The idea was to use the rovers for the greater distance when they were new, before they had a chance to break down from any possible wear and tear.

Each vehicle had two global positioning devices, linked to the four satellites in orbit and the manned Gagarin. Even so, they injected locators on alternate sides every half kilometer on the fifteen-kilometer outbound trip. These were of the same type as the ones used by the Foremen and Laborers. They would form a large network, a map for all the people and machines to follow. In the event of a breakdown of the rover, a person could use their personal locator to track the locator posts and get home.

"How's your ass?" said Marceau.

"Itchy," replied Holtman. Hemorrhoids were no joke, every pun intended. An astronaut in full garb cannot simply stop and apply some cream. Even though Marceau had slathered his backside (crackside as Red and Walter had joked empathetically before their trip), he was uncomfortable. Red had passed by during the operation with other comments about not falling into the great gaping hole. Everyone nearby had chuckled; they knew the discomfort.

"We are almost there, so hang on. You can get up and readjust your skivvies."

"How am I supposed to do that?" queried Holtman sarcastically.

"Well, I think that if you dance a little and shake that ass of yours around like a stripper, you might be able to get a little relief. If I like it, I might even stick a ten world bill, a W-10, in your belt."

Yang, on interior duty today since there must always be one inside the ship, could not suppress a guffaw as he listened to the conversation. He had noticed that Marceau had allowed some of Red's rather brash and naughty behavior to enter into her own personality. He had a few personal thoughts about those two but had never said anything to anyone.

"If I am going to dance, I want at least a W-20, girl," Holtman retorted.

"Maybe I should just grab your ass and give you a wedgie," she countered.

"A wedgie, what's that?" asked Holtman confusedly.

"It means that I grab your drawers and pull them up so tight in the crack of your ass that you sing an octave higher."

"Oh, will I sing better?" he replied in a self-mocking tone.

Laughingly, she replied, "Not likely considering what you sound like in the shower."

Yang had to add his part to this, "You better not make him screech, Marceau. That might break the audio on the v-link. I would have to spank you for that," he said in a Redlike tone.

"But that would be so much fun, Yang," she drawled sensuously.

All three chuckled at that remark.

Mission control would censor the remarks. They had enough trouble with the religious alliance as it was. Little things like this would infuriate the brass and fan the flame of the antispace movement. Right now, that was decidedly not wanted.

"Can you believe how beautiful it is here?" asked Holtman.

"Yes, I can. Luna was so stark, yet so beautiful as well. Don't you realize that every place has beauty all its own? We see things that are so different and yet the same as Earth. Our Earth, which used to be so much lovelier before we polluted it. We are seeing the raw, naked, rich, and delightful world that might have been our own. There are so few places on Earth left that man has not completely changed from the natural state. There are a few mountain ranges. The deepest part of the ocean might have a few places left, Antarctica, and maybe a few others. We see what might have been on Earth before man had come to the area. We see the beginning, Eden or Utopia for the religious or philosophical. It is a new beginning truly. We get to hit the Reset button."

This was something mission control could and would use, but they wanted to keep a lot of the area pristine. The idea of using resources from off the Earth to preserve what was left was one of their main propaganda points. The complementary use of Eden and Utopia would be usable with both the religious and secular camps.

"You sound like an environmentalist," said Holtman.

"I am to a great degree. I truly believe that we can heal our Earth, and that as scientists, we must take a greater responsibility for the proper development of resources. It will take so much time, more than it took to fuck it all up." That would be dubbed with a *mess* by mission control. "But I know that we are gaining the wisdom and the drive to do what is necessary to make those repairs. The religions would argue that this would not be necessary since God and the angels would repair the Earth for us, or we simply go to heaven depending on doctrine."

After a pause, she continued, "The development of fusion was the first really big step. With a clean source of energy, one that is so powerful, we can clean the atmosphere. The ground itself will no longer groan under the mining of coal or the futile attempts to get more natural gas. Oil is essentially gone. It can no longer provide much energy. Only lubricants and artificial ones are generally better anyway. Fusion can provide the energy to produce the needed chemicals much more cheaply now. The metals we need will come from the asteroids and Luna, along with many other elements. Earth will become an agricultural haven. Land scoured for material goods will become green again. We have seen the beginning of this already. People are no longer going hungry every day, at least that is the goal. New tools based on fusion will make food production soar if we can adapt it to the right scale."

"Damn, are you a poet or what?" said Holtman, somewhat awed.

"A dreamer is what I am. But the dreamers, the scientific writers like Verne, Asimov, Bradbury, and so many others, their visions can become true as we can see.

We went to the moon, check Verne. We have robotic help though not as advanced as Asimov's dream yet, check Asimov, and here we are on Mars, check Bradbury. We may not have fulfilled all the dreams, but we are close. Where are the next generations of writers going to take us? What new dreams of theirs are we going to fulfill? The old-fashioned broadcast media show *Star Trek* and the old movie show *Star Wars* are distant yet within our scope of things. We see them coming. They are already working on gravity wavicle drive concepts. What boundaries will there be then?"

Holtman could not speak. He had no answer to what she had said. He had to cogitate on this.

They continued on in silence (except for the mandatory radio checks) for the remaining few kilometers. They came over a slight ridge to see the first unmanned lander less than a kilometer away.

"We got her!" cried Marceau in excitement.

"That's a beautiful site," said Holtman. They could see the marks made by the bouncy ball as it landed. There it lay in the final disturbance. It had deflated after the arranged crash.

He too felt excited. This is what they are here for. They are multibillion UE buck truckers. He understood why the name *Laborers* had been chosen for the machines. They each pulled up next to the twenty-meter-diameter remnant, one on each side. They would take whatever was easiest to remove first. There was no guarantee that any particular container, with its specific package, would be accessible in any order. It was a roll of the dice, so to speak.

With their v-links (like the old heads-up displays or HUDs) operating, they could see what the other was doing. It was a double check on procedure and for safety. If a suit was punctured, even these super tough multilayered ones, a person could die in a few moments. Rapid decompression, even though Mars has a minimal atmosphere, causes a person's blood to boil, eyes to pop, eardrums to burst. A large tear usually meant death. A small hole could be patched if the person did not lose consciousness before they were able to complete the operation, or if someone did it for them.

Marceau and Holtman both dismounted. He did not dance, but he did shift his suit a little. The change in the disposition of his weight helped though. His O-ring still itched, but he could deal with it.

"Let's get to work," said Marceau. She pulled a laser gun from her waist belt, flipped the safety, and began to cut the remains of the bouncy cover from the lander so she could access the main release for the containers. Even cut into pieces, the cover would have a use. They would use it like a piece of old canvas to cover stored material. It would protect against wind damage. They both cut in a special way, along the connections of the bubbles. The bubbles were to be reinflated using the atmosphere of Mars.

Once they had completed a circuit, they used poles from the rovers to hold up the material.

"Let's get winching wench," Holtman joked.

"Be careful, I might have to kick your ass. In your condition, that might hurt," replied Marceau.

They loaded up Marceau's wagon first. As they were loading Holtman's, the wind began to pick up quickly. Sand began to ping off the rovers, lander, and the suits of the two loading the rovers.

"Yang, we are seeing a sudden change in wind velocity," Marceau said as she radioed to the Bradbury. "Do you have a large-scale picture for us?"

Yang quickly scanned the weather monitors. "I'm not showing anything unusual on our meteorological readings here. What are you seeing there?"

"There is a large amount of dust and sand kicking up here. It came on very suddenly. Let me check what the readings on the rover say."

"I can see the sand bouncing off my faceplate. I think we need to add a shield," said Holtman as he lowered a secondary clear screen.

"Concur," said Marceau as she walked to the rover nearest her, lowering her secondary shield.

"Is it that bad?" questioned Yang, a slight pang of worry grew deep inside.

"Definitely, I think we need to end the mission," said Marceau.

There was no time for any of the mission control centers to say anything; they had not even heard the first parts of the conversation yet.

"Marceau, get going," said Holtman. "I'll let down the poles to cover the lander and follow you."

"Concur. Don't waste time if there are any problems." She climbed into the rover and started off.

"I'll be right behind you," said Holtman as he began to remove the eight poles. Before she could say another word, the wind began to howl in a sudden fury. Even though the wind is thinner, it can move very fast.

"Holtman, don't wait, leave now," said Yang. "The lander is secondary. Even if it is damaged, we have two more."

"Und . . . I . . . now," said Holtman.

"Press . . . al down . . . boy . . .," came Marceau's transmission as she moved off.

"Roll . . . g now . . . eavy . . . see . . . ing," cried Holtman.

"Mov . . . er now . . . ponders . . . acti . . . bar . . . ead!" screamed Marceau over the wind.

Yang could hear the resonance of the wind in the helmets of both Marceau and Holtman. "Red, Walter, you two come back now. That's an order, gentlemen. We have a problem with the weather at the depot lander site."

"Check," replied Red as he turned in unison with Walter to walk back to the ship.

Walter asked, "What's up?"

"Sudden sandstorm," answered Yang.

"Where are they?" demanded Red almost urgently.

"Just leaving the site," said Yang.

"Is it localized?" queried Walter.

"Our local meteorological system does not measure much of anything," said Yang slightly confused. The distance was not so far that the local system should show nada, zilch.

"They should clear it soon then," said Red.

"Feces!" screamed Yang. The barometer just dropped two points, all at once, as if the measuring device had belched and found its new equilibrium. He danced from screen to screen, shouting out readings for the benefit of Red and Walter as well as Earth.

"Get in here quick, men. We have a real problem. Everything is going crazy now." He saw the wind speed indicator take a sudden wild swing, jumping erratically upwards. "This is no local artifact. It is growing quickly."

Mission control in Russia was just now getting the first signs of catastrophe. They sent a few queries, which, after their three-minute time lag, were all ignored for the moment.

"Marceau, Holtman, talk to me," said Yang.

" . . . arely re . . . ou . . . se," said Marceau.

Holtman did not respond.

"Holtman, come in," said Yang with an overtone of anxiety but still in command of his voice and mind.

" . . . ere . . . ot . . . re . . . ders," came a weak reply from Holtman.

" . . . hard to fin . . . ro . . . wea . . . read . . . ," said Marceau. "Sev . . . kil . . . ut. Max spe . . ."

"Why can't I read them?" asked Yang.

"Friction from within the sandstorm," said Red. "It is creating static electric charge that interferes with the radio, that and the fact that all that sand is simply blocking the radio waves."

Red and Walter were hustling into the air lock. Sand and dust were beginning to kick up violently as they climbed the ladder. Worry crept silent and deep inside each of them, yet they did not wish to voice any reservations.

*

Mission Control in Russia (MCR) was scrambling. Calls were being made to the other centers: first, China, then Japan, Australia, Hawaii, Houston, Brazil, the Canaries, and Europe. People everywhere were being awakened to an unexpected disaster in the making. This storm had come from nowhere. It was totally unexpected according to all the weather predictions. It was, as usual, so much for the weatherman and his abilities of planet as well as on Earth.

Cap-Coms everywhere were shouting hectically, asking for information. Communications was in overdrive. Hardline fiber optic phone connections and v-links were all put into use. Everyone needed data. Fingers danced on touch screens. Holographic displays appeared in meteorology, direct from the helmet cams of Red and Walter as well as Bradbury's multiple displays.

The trenches in all the centers were soon filled with scrambling personnel. The word came down to lock out nonessentials, meaning the press.

That in itself was an alert to newsies around the world. Reporters grabbed microphones, camera persons, and vehicles as they headed to the nearest space centers to add personnel to the apparently hot story. It did not take long for the world to know something was wrong, rumors at this point.

*

"This is Mary Smith-Jones with a special report for World Wide Network News. I am at the outskirts of the Florida NAR Space Command Center for the WSA. There are strange events at hand here. The center has been closed to all civilians. This is a procedure that indicates something has gone wrong with the mission, but we do not know what. There are no official comments yet. We will stay until we discover the meaning of this. This is Mary Smith-Jones for WWNN."

She immediately used a paylink booth to contact her proamateur astronomer source. The man was a retired professor and really knew his stuff.

*

It was the amateurs who provided the link for the world. The news corporations would lose their licenses if they tapped the satellites of the WSA. At this point, many of the amateurs did not care; they opened the data they received to the net. It spread like the worst wildfire imaginable.

The news agencies quickly ran with the story regardless of legalities. It was already public domain, if the laws could still be construed that way. No one was really certain, but it was too big a story not to get all the juicy gore.

*

The Deep Space Very Large Array (DSVLA) on Farside was sending information to all the centers as soon as it came in. The very system that Red and Walter had initiated was now announcing their possible doom. It was more than a little ironic. All the people on Luna and SS-2 were in a conundrum. Almost all of them knew Red and Walter from personal experience. The word had spread faster than the usual grapevine activity. Everyone was v-linked to the holodrama and saw the progress of the oncoming disaster firsthand with their multiscreening ability. Work did not stop, but it certainly slowed, temporarily. Red and Walter had made their place safer and easier to work in. Some of them would not be here if not for them.

Those in the space stations were affected most. The people on the birth ship remembered Red and Walter as the problem solvers they needed to get up and running. Without them, nothing would have been accomplished. People were *alive* because of Red and Walter, thousands of them. Nailless fingers were chewed as well as nailed ones.

Marceau was driving slowly, blindly. The only things keeping her on track were the transponders. She was able to triangulate by moving left to right and following the signal on her display. The signals she received were weak and intermittent but sufficient to keep her on the right path. Her linear speed was barely two kilometers per hour, but her actual speed toward the Bradbury was closer to one.

Holtman was in worse shape. He had never made it into the corridor between the transponders. He had cut it too close and ended up just outside the corridor on the left side. For a few kilometers, he had been able to follow the transponders. He knew the trick of moving left to right to triangulate, but he continued to move farther to the left than the right. He was in the heart of the storm's fury now.

Sand grains were striking his suit and faceplate with tremendous force. The sound of the impacts began to scare him. The suits were tough but not that tough. He began to sweat an inordinate amount, causing his faceplate to fog up even with the temperature control. His fingers felt swamped with the excess moisture.

" . . . I . . . st. Can't f . . . ers," he radioed in. Yang and the others barely could hear that transmission.

Marceau did not hear any of that. Her amplifiers were not as strong as the Bradbury's. She did not have time for it anyway. All her attention was desperately focused on the imaging system in front of her. She had never felt fear like this since childhood nightmares involving killer trees reaching through her window to tear her apart.

Yang was scared, not so much for himself but for Holtman and Marceau. He foresaw disaster, felt impending doom. A brief, unbidden evil thought crossed his mind, *This will be a black mark on me. My family's shame*—he quashed that as quickly as it had come. He refocused on what needed to be done. They had all been briefed on what a sandstorm could do.

Red was furious. Walter knew why. There was nothing Red could do for Marceau. Red hated being impotent like this.

There was nothing any of them could really do. They could only act passively—measuring, communicating, waiting.

All the readings were intermittent due to the violence of the hurricanelike storm raging about them. They could all hear the pings and dings of the sand against the hull of the ship. The ship even rocked occasionally as an extra strong gust of wind hit the massive Bradbury. They all had visions of what the wind was doing to the rovers. If it was strong enough to rock the ship, it could flip a rover. Each of them imagined a rover flipped over, trapping its driver, with the sand slowly covering them over as they screamed for help.

Yang nearly screamed at the Russian control as another weak query came in, "Shut the fuck up! I can't hear them with you talking." He could not hear them anyway.

Marceau was moving at a few tens of meters per hour now. Sweat beaded her brow and dripped into her eyes. There was no way to wipe it out. She simply blinked furiously to clear her eyes.

Ten long suffering hours had passed with no sound from the rovers on the radio. Yang had cut the feed from all other sources except the Gagarin. The control centers on Earth did not know if any of the crew members were alive at all. The news had spread all over the Earth in a matter of minutes. It was an event that could only have a comparison with the Apollo 13 mission.

The wind continued to howl outside, shaking the ship with its tenacity. The winds were strong enough to fling pebbles and small rocks at the ship. The portals had received a number of impacts; some were strong enough to chip the thick metaliglass. Things were looking very bad for all of them. The ship was suffering unknown damage every minute. Two people were likely very dead.

It was in the thirteenth hour when they heard Marceau's voice again.

"Brad . . . till . . . ing . . . onders . . . do . . . read."

"Follow our multiphase transponder signal, Gail," bellowed Red, lunging to and toggling the radio, using her first name openly for the very first time. The multiphase transmitted on many frequencies at the same time. The mix of longer and shorter wavelengths seemed to be working best.

" . . . ed . . . at . . . ou?" came the reply.

"Yes, it's me!" yelled Red, hoping his added volume would make a difference. "How strong is your signal?"

"Ver . . . ak . . . low gain . . . oost."

"How can we boost the output?" demanded Yang.

"We can't, not from here," said Walter quietly, looking down at his feet. "We would have to add length to the antenna and access the exterior power couplings. I don't recommend that. The sand could do real damage to the electronics."

"We have what we have then," said Yang gravely.

"Marceau, do you have a link with Holtman?" asked Red.

"No . . . act . . . tman . . . or . . . hou . . .," came her steady reply. "He was . . . ing me."

"We are getting more of her each time," said Red to Walter and Yang. "She has to be getting closer." He said that with a very slight hint of desperation. He knew that her suit, under optimal conditions, would keep her alive for weeks. However, these were not good conditions by any means.

"Holtman, can you read this?" asked Red.

There was nothing but static on the radio. Marceau heard the call but remained quiet, not wanting to interfere with a possible reply.

Holtman could barely make out the last call. His response, apparently, was unheard. He could not see. Beyond the blurred red image from the sand, his outer faceplate was scratched and cracked. His transponder readings were very weak and intermittent. He knew that the static electric charges built up by the storm were interfering with his readings.

The sharp-ended pebble that hit his faceplate shattered what was left of the secondary shield. He was so surprised that his hands fell off the steering wheel. That was the worst thing at that moment. The front wheel turned sharply and hit and rode over a boulder. The vehicle came to rest with its frame on the rock. Holtman was stunned by the impact. After several seconds, he realized that his nose was bleeding. In a mental fog, he reached up to wipe it with his hand. To his blurry surprise, his hand did not reach his nose. He did notice that something was wrong with his finger; there was a small tear in the glove. It seemed to be glowing slightly, hissing. In the back of his mind, there was a demand to do something about that, something very important.

He stared almost casually at the air escaping from the tip of his finger. His suit began to speak to him, "Leak detected on right extremity index, patch required. Patch kit is on chest harness or on rover dashboard. One percent pressure drop. Two percent pressure drop . . ."

His stupor did not prevent him from following his training. He reached for the dashboard emergency kit and opened it. He took the tube of glue and broke off the nipple. With a hard squeeze, he liberally spread it over the tip of his finger. A single small bubble burst. Drilled procedure took over his actions as he used a small patch to seal the hole on his finger. He applied even more glue around the edges of the patch.

His interior shield shattered as the next rock impacted right between his eyes. His death was pretty quick but horrible. The sudden drop in pressure caused his eyes to pop out of their sockets while blood boiled in his veins. The blood came out of every pore and orifice in his body. He dropped a full load of feces and urine in his suit's containers. His mental haze was a benefit.

Marceau could tell that she was approaching the Bradbury. The ship's signal was getting stronger. She could only be a kilometer away at most. She was very worried about Holtman though. She had not heard any transmissions from him for some time.

Her mind was beginning to wander after fifteen hours. She was dead tired, bone weary, and mentally almost out of reserves. She was sweating profusely. Her hands felt as if they had been dipped in oil. She used counting tricks to keep her mind on track. One to ten, then she would check the signals from the transponders and the two GPSs, which were mostly useless in the storm. On the second ten, she would change direction by about twenty degrees opposite her previous turn. She had begun to notice an increase in the strength of the readings to her right. So on this occasion, she continued past her usual count, and she saw the display rise 0.1 watt higher than it had been for some time. She continued for another ten seconds and saw the signal begin to drop. Immediately, she turned ten degrees back toward the signal and regained the power strength she had previously had.

Resisting the impulse (borne from fear) to increase her speed, she continued onward. Marceau had to battle the new hope she felt as much as the fear. Now was the most critical point in her mental battle for clarity. Again, she began to count.

"Marceau, the R1 transponder signal is getting stronger. Do you copy?" said Yang.

She replied, "I'm . . . eeing . . . incre . . . in . . . signa . . . strength . . ."

"What is the reading on your power?" questioned Yang.

"Six poi . . . our . . . hree watts," came her tired voice.

"Six point four three," muttered Walter, shaking his head.

"Feces! How could she have made it this far if that is the strongest reading she has had for so long?" asked an amazed Yang.

"She has a strong will and mind," replied Red in a proud but subdued voice.

Walter's eyes flicked quickly at Red as he all but smiled at his friend's admission. "We need to have her change course, ninety degrees, so we can triangulate more easily," he stated.

Red stiffened quickly; then he relaxed almost as quickly. He knew the need for this. A good triangulation would almost guarantee Marceau's return. But a mistake would almost ensure that she would lose the transponders that she was following; then she would die.

Red turned to Walter with a deep stare and said, "Draw up a plan, make it quick."

For one of the few times in his life, Yang was truly surprised. Red was the most capable man he had ever met. Why would he have Walter do the work? All the little pieces and inklings he had had now came into focus. He looked quickly at Walter who smiled the slightest of upturned corner of the mouth smile Yang had ever seen. Yang winked. Red was even smarter than he thought; he would not involve himself in the calculations if his emotions might interfere. Others would demand to be the one in charge. Red knew when to jump in and when to distance himself. Red jumped up several notches in Yang's estimation at this point.

Walter was furiously touching the screen of the computer, drawing, calculating, and muttering. In a matter of ten minutes, he had devised and reviewed his plan. "Red, take a look," he said.

Red scanned very quickly and said, "Good." He then turned to Yang and said, "Make the call."

Yang again revised his understanding of Red. There had never yet been a question of authority. He had expected Red to simply say, "Do it." Yang didn't know if he shouldn't put it back in Red's lap. Red's ability to separate himself from making an emotional decision was more than Yang could admit about himself.

"What will be the psychological impact on her if we tell her to take a ninety?" asked Yang.

"She is tough. Gail will understand immediately what it means if we ask her to turn for a triangulation. She has been doing that all the way here," Red stated.

"This calls for a much farther deviation from what has been a good course so far. It will add time to her travels. What if we start to lose her?" asked Yang.

"She would make her own decision at that point and begin to backtrack," noted Walter.

"Are you sure?" asked Yang.

"That girl will do whatever she damn well pleases," retorted Red with a snort. Walter simply grinned.

"Even if I order her?" queried Yang.

Red and Walter laughed out loud as they looked at each other with a shared memory.

"Walter knows as well as I do that she will do what she thinks is the right or best thing to do."

"So we need to get her to cooperate, if she will," said Yang, looking hopefully at the two men next to him.

"We have a plan. Call her and see," said Walter.

"Marceau," Yang called on the radio, "we have a plan for triangulation. Are you willing to try it?"

"No nee . . . I am almo . . . ere. Power is . . . eight . . . oint . . . ine three now."

"Are you sure about that?" asked Yang.

"As sure . . . your ass . . . skinny as . . . paper," she replied tartly.

Red was smiling, and Walter was chuckling.

"Paper," said Yang as he looked gleefully at his rather narrow backside.

"What's your estimate for distance?" asked Yang.

"I guess one-half klick or . . . according to . . . estimates. What is Holtma . . . distance?"

They all could see the difference in her incoming message. Yang looked at the others as they both shook their heads to say, *Don't tell her we don't know.*

"We can't tell how far Holtman is at this point," Yang stated. It was not a lie, but it was not exactly the truth either.

"He should . . . far behind," she stated. "I am now . . . tting peaks of nine . . . zero."

"She must really be closer than she thinks," said Red. "If she does not do a good triangulation, she could drive around us for hours. We have a max gain of twelve watts. If she is getting nine now, in this storm, she could be right outside the door without even seeing us."

"Marceau, we really need you to do a triangulation turn," stated Yang.

"No need, I have a visual now. You are no more than one hundred meters away," she said.

Red had already jumped up and began to slip into his suit. Walter was helping him to zip and flip and secure all the needed joints on the suit. A new record was set for the dressing of an astronaut. Walter grabbed him before he could move into the air lock and pulled him back. Slowly and methodically, he rechecked all the suit's parts. Walter said only one word when he was done, "Go."

Red literally jumped into the lock and punched the cycling icon. He impatiently stomped his feet and slapped his hands together as the air drained from the small lock. The outer door opened with a slight loss of air and a rapid loss of Red. He was on the ground in less than a second. He restrained his impulse to run in the direction of the rover. Through the flurries of rampaging sand, he could just discern R1. Marceau was heading straight for them.

"Gail, you're heading right for us. Keep coming forward!" shouted Red.

"Red, you bastard, what are you doing outside?" yelled Marceau. "Get your hairy arse back inside. This sand can grind you to a pulp."

Walter actually giggled as he told Yang, "I told you she was a tough bitch."

"Are they really . . ." Yang left it hanging.

"Yes, they just don't want to recognize it yet," Walter smiled.

"Love's a bitch, isn't it?" chuckled Yang.

"They'll handle it, the hard way maybe, considering their personalities."

"You're on open mic, you giggling pecker heads," announced Marceau. "I plan on spanking both of you severely." She laughed. Seeing the ship allowed her to release all the tension that had built up over that last seventeen hours of pure hell.

"If she doesn't, I will," said Red.

"Just be glad I cut the off-planet link, girlie," teased Yang.

"Small favors won't save you, Yang," said Marceau. She was close enough for Red to see her fairly clearly. The sounds of the sand bouncing off her helmet had declined somewhat, but it was still at a dangerous level.

Red recognized the same fact. He wanted to get her inside as soon as possible. The moment she was close enough, he would drag her to the air lock, over his shoulder if need be.

There was no need to do that. Marceau was out of the rover as fast as a cat after a mouse. The two of them ran to the air lock ladder and quickly climbed in. Red hit the icon for the recycler to begin. He grabbed her and hugged her tightly. She was stunned at the display of emotion from Red. She could see a tear leak from the corner of his eye. The impact of that hit the core of her being. For her, all the little tidbits became very clear. She knew for sure now. She responded with a fierce embrace of her own. He was hers and hers alone, forever—victory. The thrill ran through her and caused her to shudder with joy.

Red pushed away with a look of fear, thinking something was wrong. Gail had other ideas. She pulled Red back to her and squeezed even harder. The lock opened to show the two in a rather frantic embrace made rather awkward by their bulky suits.

"Well, would you look at them?" said Yang.

"Yeah, silly kids, don't know they need to take their suits off first."

Red flipped them off.

Marceau ignored them, for now.

*

Davies jumped quickly on the bad news, "Did we not say that God would show his displeasure? Did we not explain that God would take a hand in things? He shakes the sky. He destroys the evil. What is happening now is the judgment of God! He shows his displeasure. As with the days of Noah and Lot, judgment rains from the sky! The angels of our Lord are doing His work!"

*

Every person at the bar was somewhat restrained. The news had been bad. Storms and two missing astronauts had set them all off. It was not Red; they knew that much.

"What do you think?" said Paul somberly.

"Bein' a astronaut is dangerouses," said the inebriated Mark.

Mick let his state of utter drunkenness slide. They were all worried about Burt. "That it is," he said quietly.

Burt sat morosely on his stool, slowly sipping from his mug. His cousin was his only remaining blood relative. They had played often when children and kept in contact as adults. If things went really wrong, he wasn't sure.

"Things will work out," said Paul.

"Red is the best. If they can be saved, he will do it," said Burt.

"Amen," said Mick.

With the exception of the v-link monitor, the occasional clack of billiard balls, and the slurps of cold beer, the Lucky Loser was eerily quiet.

Mick started slowly, "I got the server set up, Burt. I know it ain't the best time, but if we wanna set up the blogs . . ." He left it hanging.

"We won't be able to get a hold of Red now," he muttered slowly.

"Red's OK," said Paul. "I know it's going to be fine."

"Thanks, Paul," Burt said truthfully. "I s'pose we ought to go ahead and set it up."

Mick quickly set another cold one in front of Burt. Everyone worked toward getting him in motion, mentally and physically. It would be good for him to put worries aside.

Burt began to touch furiously at the board provided. He entered a great deal of personal information about himself and what the relationship was between himself and Red. The blog should be a personal appeal to all people.

Once he finished, nearly half an hour later, Mick took over. He entered his views and ideas with zeal. He narrated the events as he saw them. Every little detail added to the program.

Paul was next. His additions were a little more simplistic but would have a great deal of effect with many like him. His turn of phrase appealed to the humbler fellow.

One by one, the customers of the bar entered their ideas and beliefs. Towards the end, the back door opened; it was the toothless woman.

"You're not allowed here," Mick said harshly.

"Please," the woman with her partial in said. "I need to make amends. I was misled. I . . . I really need to speak with Burt."

Mick looked closely at Burt who shrugged and said, "Why not?"

The woman approached Burt and said, "I am sorry for all the shit I gave you. I do not really believe what you do, but I should have been nicer about it." She turned and began to exit.

"Wait," Burt said quickly. "Why did you decide to come here? Was it just to say that?"

"Yes." She left without another word.

*

Chapter XVIII:

Holtman

18

010010

"This is Mary Smith-Jones here for World Wide Network News. We now know that at least one astronaut, Holtman, is still missing and presumed dead since he cannot be contacted by radio. All others are accounted for and healthy.

"Severe sandstorms are to blame. The WSA has released information that shows that the storm was sudden and quite unexpected. Regular storms do occur, but not at this time of the Martian year.

"The missing astronaut was on a mission to bring the supplies from one of the bouncers. Astronaut Holtman disappeared in the midst of the furious sandstorm and has not been heard from since.

"The chances of him actually surviving are very slim. As soon as the storm dies down, a search-and-rescue mission will be dispatched from the Bradbury. They will need to recover the rover and supplies while they are out.

"We will be live on WWNN-2 for the full length of that excursion. This is Mary Smith-Jones for WWNN."

*

It was five sols (the Martian day, M-sol, as opposed to E-sol or any planet, P-sol) before the wind died down sufficiently to launch a search-and-recover mission. Long dust storms were known to exist on Mars, but this was the wrong time of year for such a powerful and extended system. They all knew that Holtman was dead. All that was missing was the body. They were now receiving a very weak signal, a static one, from the rover. Call as they might, there had been no response for the duration.

Earth contact had been fully reestablished. The situation stabilized as mission controls engineers regained access to all their telemetry. Fears evaporated as the realization set in that all had not been lost.

The whole earth was aware of the absence of Holtman. Tableside and barstool quarterbacks debated the possibilities facing Holtman if he remained alive. The evidence was against him. Alive, he faced days alone and most likely injured. The reports said he was not moving, which, according to some, meant good news. Holtman waited patiently in that case. He needed help immediately according to others. But there was too much to do before those on Mars could mount a search.

They had to evaluate the rover and the ship. They quickly removed the supplies still on the rover driven by Marceau and stacked them near the ship. The containers held mostly scientific packages that were of little value at the moment.

Walter examined the rover; Red checked the ship's exterior while Yang and Marceau did a thorough system-by-system check of the interior electronics.

Walter was his usual meticulous self. With his cameras running and metal probes, somewhat like dentist tools, he pushed and pulled at each pit and ding on the chassis of the rover. With the push of any of several buttons on his forearm, he could zoom in with his face shield lenses, use false color, place special lighting eyepieces in front of either of his eyes, polarize light, and so much more. A simple multimeter was used to check the wiring. Along with the hard metal tools, he had a series of brushes to remove as much dust and sand as possible from more delicate areas. He carefully recorded everything both visually and with audio commentary. His visual and verbal notations made their way at the speed of light to the Gagarin and on to Luna and Earth. Later, he would reexamine the recordings to make sure that he had not missed a single thing. The computers could enhance everything by a factor of one hundred without significant degradation.

Red had all the same tools available to him. However, he was more concerned with what he was finding. Things were not looking very positive at all. There were an awful lot of micropenetrations of the outer hull. He had already counted seventeen. They were all quite small. There were innumerable pits on the surface. They would all need to be puttied; however, that would not ensure anything. Since the shuttle reentry disaster, the putty tested on the Discover had been improved considerably.

It was not a panacea though. Red realized that they would almost have to paint the entire ship. That was not the real problem though.

The ship was similar to the Apollo landers. The lower portion would be left behind. It was the connecting bolts that were the problem. Just before liftoff, the connecting bolts would be driven out by very small explosions. The explosives had been damaged. Surprisingly, they had not gone off, considering the damage to the electrical wiring. Two of the protective covers had been pierced by large pebbles, which then bounced around inside the cover. It had made quite a rat's mess.

If they tried to take off, they would likely explode in place. They were marooned. Red was not sure how to explain this to the others, given the circumstances. He knew Walter would handle it well. Yang was well trained, but his command was under attack, and they had already lost one, he was sure. The psychology of the situation could make or break him. Marceau had already had her one extreme situation; another so soon and more intense might be too much. Red wanted to believe that she could handle it; wanting is not enough though.

Walter saw Red standing perfectly still. He instantly knew something was very wrong. The only time Red stood still was when he was looking out a portal, remembering, cogitating. He bounced over to where Red was cogitating and tapped him on the shoulder.

Red started and quickly turned to see Walter's questioning gaze. Red brought a finger to his faceplate in a shushing signal. He motioned for Walter to look at the bolts with him. Walter spent several moments examining the various bolts. The damaged ones were all on one side of the vehicle. When the launch took place, the manned portion of the ship would likely tilt slightly before the other bolts had time to release, if they even could. That millisecond of tilting could cause the exhaust to reflect up on the side of the ship itself. It was not designed for that. It might withstand it, but that was a hell of a chance to take. If the other warped bolts did not fire (could not be made to fire), the Bradbury might explode like an old-fashioned firecracker.

Walter looked back at Red and nodded; he understood Red's consternation. He kept his silence so no one would hear them discussing the problem. This was a real scrotum squasher, a gonad gnasher, ball buster; he ran out of filthy ideas, which usually did not happen.

He knew they would quickly have the rover running. They would have to gather all the food supplies and equipment first and make a quick calculation based on four (not five) people eating light. They would not waste energy deploying all the science packages, yet; they would have to use all the energy sources to power emergency devices.

They were both running scenarios through their minds. Both knew that when there was a chance, they would pool their thoughts. Their combined mental hashing would provide the best chance. Once they had a viable plan, they would present it to Yang and Marceau. The various mission controls did not need to know yet either. They would simply bombard them with questions and nonsense solutions.

The so-called experts at any of the mission controls would simply try to have them do so many tests of the bolts. They might have them even try to disarm them, which Red had already done, and then remove the bolts. Red knew that the bolts could not be removed with the tools that they had with them since they were torqued out of shape. They would have to endure (or die).

Red and Walter moved to examine the hex dome foundations. There were many scratches, pings, and dings along the surface that would need to be sealed. The domes could still be built but may not be airtight. They could still be used for storage though.

Red led the way to the air lock. "How long before we can go get the other rover?" asked Red.

"It will take a few hours to fix a few small problems before R1 is 99 percent. When we find the other rover, it may not be operational," said Walter.

"If not, we will have to strip it for parts," replied Red. "We will need to make sure that we have one functioning rover at all times."

"That's priority 1. Then we will need to retrieve all the supplies from the bouncers," said Walter, not adding why.

Red nodded as he said, "Yes."

*

Yang was sending all the data to Len on the Gagarin, who sent it to Earth and Luna. Since the largest of the lander's parabolic dishes was damaged, he had to focus on the nearest, easiest target, with the Bradbury's backup dish.

Marceau was continuing her system checks when she discovered a problem. This one worried her. The firing system for the liftoff release pins was nonfunctioning. Nonfunctioning meant there was no power in the system. This meant she could not even check it. A malfunction is a system showing power but a defective component. She made a note on her touch screen under the priority 1 category. Her list of repairs was growing rapidly. She wordlessly continued on.

"Bradbury, we have all current data. How do things *really* look? We have some concerns voiced by mission control," Len said to Yang.

"Gagarin, we are in trouble," stated Yang bluntly. He had just peeked at Marceau's list. "Our handymen are going to be very busy for a long time."

"Understood," said Len with a sinking feeling in the core of her being. Things were going from very bad to very, very bad.

"Tell Earth to quit sending 'fix it' tickets until we have a better grasp on what all we have to do," Yang said almost plaintively. "All it is doing is slowing Marceau and me down since we have to read them."

"I'll pass it on. You know they won't listen though. I will filter the requests into a separate download file. You can peruse them at your leisure," said Len jokingly, attempting to lighten the mood.

"I will examine them over a fine wine," laughed Yang, understanding her ploy. He took a deep breath and exhaled slowly. Adversity brought out the best and the worst in people. He determined at that moment that this would be his finest hour. Everyone faced a similar situation and decided the exact same thing. That's what winners do.

*

The entire space program was in a flurry of activity. People had always been lost pushing the frontiers. The Soviet Union, the United States, China, and the Europeans had all lost crews in space. The Challenger and Columbia missions brought home the point to the old United States. Some had died on Luna in construction accidents.

Space is dangerous. The only comparable earthly environment was the deep ocean. The opposites of extreme pressure and no pressure challenge engineers to the fullest. Even today, the deep is nearly impenetrable. When something goes wrong, death is inevitable. Many speeches were made referring to the Apollo 13 mission. The old cliché "failure is not an option" was often bandied about.

It happened. It was unavoidable. Wherever people are, they die.

One more is missing and dead.

*

Red and Walter had the rover in top condition, all things considered. They were riding toward the coordinates given by an orbiting satellite. The satellite photographed what could only be the remains of the other rover. The metallic glints were located in a place where they should not be. No other man-made objects were in that area. There could be little doubt that Holtman had gone far off course.

Red was driving with Walter, riding shotgun. Walter had been up most of the night finishing the rover. They said nothing for a while as they rode along, except to make updates with Yang and Marceau. At a steady half kilometer per hour, they would be there in a little more than two hours.

As a safety precaution, they placed some of the transponders along the way. No one wanted to lose anyone else. Red and Walter had both felt that it was a waste of transponders and had privately agreed to only place half as many. They were also going to pick them up after they were done with R2.

Walter hooked up the intercom on the rover so that they could talk among themselves without anyone else hearing, or so Red thought. They could still hear incoming messages and quickly switch on their outgoing mikes.

Walter casually toggled a switch on his arm, which would allow everyone on the Bradbury to hear. What was coming from Red was important. Walter knew Red's moods, and this would affect everyone.

Red started, "What have we got?"

"We are here for the duration, friend. Power is the key. We have food for a year if all the packages are in good shape. But we will run out of power if we don't repair all the solar systems."

"Does that include the solar collectors for the science packages?"

"No," replied Walter. "I did not want to count on anything until we had all the parts in front of us."

"What if we have half of the science panels?" asked Red.

"We need at least three-fourths," replied Walter, "if we want to eke by."

"I don't like that word, *eke*. I have an idea about some additional batteries. I think that we can use urine and local materials, along with the science package containers, to create some low-voltage backup cells. I calculate that we can make twenty cells. We can arrange the hookup in series or parallel as we need different voltages and amperages."

Walter laughed as he said, "Will that have a significant effect on the water recycler?"

"Yes, it will. We need to increase our water harvest by 30 percent. We will need to readjust the Foremen and Laborers to dig to the permafrost layer, if there is a real layer at this latitude. We can break down CO_2 to make O_2, and then mix captured H_2 to make H_2O. Unless we can build some sort of greenhouse out of the remains of Holtman's stuff, we will lose some 80 percent of the water as vapor though. It will have to cover several square meters so the Laborers can tunnel in the permafrost layers. I think we can use the bouncy covers to make an outer cover. We will need to make an access door, metal framed, perhaps glass. The silicates we find can be made into a thick glass. We can use the putty to seal all the seams."

"Do we have enough putty for that?" asked Walter.

"If we don't fix the ship, which is a wasted cause," answered Red. "Power may be the key, but water is second. We can ration food to one thousand calories per day and water to one liter. If we don't suffer from severe dehydration and die, we may have time to gather enough water to survive. The major question is can they send us a rescue ship in time? We may do all the right things, but without a working launch vehicle, we will all be here for the rest of our lives, about a year."

"What is the best use for the hex shelter?" asked Walter.

"I don't know," Red reluctantly said. "A pressurized shelter is a necessity. The ship is our primary, but the outer hull is compromised. If we use the putty to fix the hull, we may not have enough to do the greenhouse. I have been thinking that we might use the hex to build a cover over the ship. It should protect us against another sandstorm. That leaves us with only one shelter. That may be too big a risk. We are here to set up for future missions. That means we should set up the hex as per plan, if we can fix the seals. It also means that we should set up all the equipment. That means we do not have a full year's capability as it stands."

"Rock and a hard place," muttered Walter.

"Amen, friend, amen," Red replied.

"Do you think we are expendable?"

"I'm afraid so," said Red. "Everything we are doing is for the 'benefit of peoplekind.' All the effort should be for the improvement of Earth. We are trying to make Earth a garden again. All the mining and research is to take the pressure off Earth. Marceau made her speech, and she is correct. The only way to save Earth is to provide all the raw materials from off planet. We are only at the very beginning though. Our efforts here set the tone for the rest to come. If we fail, the whole program may fail. I think the eggs are in our basket. The only way to get around the problem we now face would be if we had several independently operating systems."

"Which is why we are first," Walter said. "We get the crap."

*

Len was thunderstruck at what she was hearing. The Bradbury had not notified them of the true severity of their predicament. Of course, they had been very busy just trying to assess the situation.

She immediately put the conversation on broadband for Earth, knowing that the news channels would pick it up. She knew that was what Walter was doing with the open mic. Len replayed the entire conversation for the world to hear, mission control be damned. Such cases of disaster required communication on secured channels. It might be the end of her career, but she could not fail Walter. Her heart had found a mate.

*

"Do you believe that the whole show could come to a stop?" questioned Walter.

"Look at what happened to the Apollo program. People began to think it was every day. They lost interest. If we fail or die, things may go bad for the program, which means the Earth. That means that we have to do the mission and survive, not one or the other."

"So that means we have to get creative," said Walter.

"More than a little," replied Red.

"We are stuck here. We need to accomplish our mission and not die doing it." He led Red into answering the question to reinforce what had been already said.

"Very much so, we are in a heap of feces, and it stinks. We have three main goals: establish a base, make sure the Foremen and Laborers are functioning, and coordinate the supplies. I think we will use all of the food. There will be none left for the next crew as backup if the ratio of damaged goods stays the same. We can, however, prep all the scientific instruments that we don't need to strip for our survival. If we use the hex to cover the ship, there won't be an emergency shelter ready for the next crew, but it will protect what we still have."

"Will that be enough?"

"If we are lucky and find a good deposit, yes. Otherwise, we can steal a Laborer or two from one of the other Foremen. We have to keep the robots on task as much as possible though if we want to follow the mission."

"When do we tell everyone else?" asked Walter, again leading Red.

"We can tell Marceau and Yang when we have retrieved the rover. Once we have all agreed on the path to take, we can get on with the show. We'll tell the Gagarin just before the takeoff time. We will give them no choice but to leave us here. Otherwise, they may try something stupid, like staying here."

"Too late, you redheaded prick," Marceau forcefully spoke. "I've heard it all and guessed it before I even heard it."

Red glared at Walter. "You deformed son of a virgin dinosaur. You did that on purpose."

"Of course. It is necessary that the people know. Mission control would try to keep a lid on it as long as possible. You are on space.com by now, I imagine, if Len did her job right."

"I did, you sexy dinosaur," she interjected. "Every news agency and public system now has both your words straight from the asses' mouths."

"Ee-aw," said Walter gleefully. "I'm on the news!"

"You're a pervert," said Marceau.

"Of course," replied Walter. "But I'm a nice pervert."

Len giggled at Walter's comment. She had a real thing for Walter. Only the two of them were really aware of it. They had bonded on the outbound trek.

Then she hit the brick wall. Walter might never make it. She was no longer laughing. She knew that they would leave when the time came. She had to get the rest of them home. It was her job. It didn't matter how she felt about Walter. She almost began to cry. The others saw her stiffen; they already had as well.

*

People around the world were getting the news reports on the situation as quickly and, in some cases, quicker than the various mission controls. The world was stunned into a reawakened awareness of the whole space program. Hearing Red's words that meant he would sacrifice himself and everyone else was a real shocker for a world half-full of me-firsters.

The people on Luna and the space stations were not so surprised. They were inculcated with the same feelings about the missions they were performing. To many of them, it was an extension of the Red they had worked with and come to admire. They were soon scrambling to see if there were ways that they could help.

*

The members of the committee, the religious group formed to fight the space program and gain a measure of control of all the governments, quickly picked up on the news flash and were not happy. They all quickly recognized the martyr value attached to this. Martyrdom was a basic tenet of Christianity with Christ himself acting as the supreme sacrifice.

They would have to foil any beneficial outcome. Most of them rapidly saw the potential for turning death into a reason for not sending people into space. Each quickly realized that previous deaths, like Apollo-1 and Challenger, had caused an interim period in which travel and launchings had been stopped. That would have to be their goal. A short stoppage in the Mars end of the project could be turned into permanent one. There would have to be yet another meeting.

*

"Let's finish the job, Walter," said Red. Everyone's thoughts immediately returned to the missing (and presumed dead) Holtman. Red recognized that the humor was only a form of release and not disregard for the deceased. Red knew Holtman had not survived. They would not find anything other than death, and it burned deep inside him. It could have been Marceau, and he hated himself for his relief that it had been Holtman.

Holtman had been a total professional. He had shown wit, humor, and true caring for his shipmates. He had been more than a colleague; he had been a friend and, to some, a lover.

They continued on in respectful silence for several minutes. Everyone everywhere was still watching and listening via v-link. They had to see; there was a morbid compulsion to do so. The reality shows that had begun in the early 2000s had only whetted their appetites for real-life drama and tragedy. Real-time war footage, real-life families giving a month of their lives to the v-link, and more hyped the expectations. Soap operas fell far down the list as real-life struggles took over daytime v-link.

"There!" shouted Walter as he pointed slightly to port. There was a silvery metallic glint from the dusty rover.

"I see it," stated Red as he made the course adjustment. They pulled up in less than five minutes.

Both were shocked at the posture of Holtman. He was still in the driver's seat, leaning back, face to the sky, with his arms akimbo as if supplicating his creator.

They exited the rover and slowly approached the obviously damaged vehicle that Holtman had been driving. Neither one wanted to quickly confirm the obvious.

"Fuck!" Red yelled. "Cut the visual." It was too late. Everyone watching had already seen the horror of Holtman's face. The sudden depressurization had caused Holtman's face to bubble as the blood in his veins boiled. The eyes had popped out of their sockets and exploded. The sand from the storm had peppered into his face, leaving gaping red pits, and turned the remainder of his eyes to an egglike scramble. Had there been vultures, they would have had a feast, pecking and pulling with wings flapping in hungry agitation and expectation.

Marceau had cut the feed as quickly as she could, but it was not quick enough. People everywhere recoiled in disbelief and disgust. They had wanted reality, and they definitely got it this time. Some number regurgitated recently consumed meals.

Red and Walter simply turned away. They could not even puke. Their bodies had momentarily ceased to operate at even that level because of the shock. Numbness spread more quickly than even the most potent and deadly poison. Red's self-loathing for his gratefulness that it had been Holtman and not Marceau grew to the size of Olympus Mons. He shuddered from head to toe.

After catching his breath, Red went to the rover that they had come in and retrieved a small German flag and the cure all—duct tape. He marched back to Holtman, placed the flag over his broken faceplate, and taped it into position.

"OK, Marceau," he said in a subdued voice, "turn the video on." The whole Earth saw the picture return with the flag covering Holtman's face. Many, children and adults alike, would have nightmares about that face. Red knew he would. "Walter." He motioned to the arms and legs. Walter climbed into the other side and grabbed Holtman's suit at the wrist and ankle. Red did the same. They maneuvered his lifeless, suit-garbed husk out and gently placed it upon the ground. "There," said Red, pointing at a group of rocks. They picked him up and carried him the few meters to where Red had pointed.

Red knelt and began to move rocks into position around and onto Holtman's body. It took twenty minutes to build a cairn over him. They covered all but the flag over his faceplate. Neither one said a word, nor did anyone elsewhere interrupt the proceedings.

The whole of the solar system was attending this burial. People murmured and shed tears all over the solar system. The members of the various mission controls lowered their flags to half-mast. Government officials the world over ordered the same for all public flags. Many individuals of all regions quickly placed flags, both regional and UE, out. Many religious figures, including those in the conspiracy, called for prayers on behalf of all the crew.

Both men were sweating as they completed this terrible task. Moving mass is moving mass and requires energy regardless of its weight.

Red spoke, "For those of you who want to pray, the time is now." Many people were stunned to hear this from Red. He was a notorious atheist. However, he had never demeaned anyone for their personal beliefs or thought them foolish for holding on to outdated belief systems. He was very much a "live and let live" kind of person, as long as they did their work effectively and efficiently.

Red and Walter stood, neither bowing his head nor offering any prayer. After a minute, Red said, "He was their best. He gave his all, and no one can ask for more." Without pause, Red turned away to the rover, saying, "Let's strip this bitch. Len, Marceau, cut feed to all but mission control for a while."

They both did this, silently; tears were streaming from all eyes on board the Gagarin and the Bradbury.

People everywhere were crying, except Red and Walter. It was not that they had no emotions; they simply had no time for it now. Crying did not do a damn bit of good. He was dead, and that was that. They had work to do, or they would all be dead.

"Left front wheel has severe cracks from the impact with this boulder. He was moving too fast for the conditions," noted Walter.

"The frame is slightly bent too," stated Red. "Let's take the supplies first and come back to junk her."

"Agreed," said Walter solemnly. "We can then retrieve the rest of the supplies from the bouncers."

*

Psychologists and psychiatrists on Earth were constantly attempting to counsel the bereaved crews of the Bradbury and Gagarin. Red had simply sat in front of the monitor, pretending to listen, nodding every so often. The other monitors were running simulations on his and Walter's plans for survival. He could multitask with the best of them. Walter had simply dropped his shorts and pressed his hairy butt cheeks up against the v-link monitor, farted, and walked away whistling. Yang was more professional and restrained. He turned off the monitor and went back to work.

The members of the Gagarin also wanted little to do with the doctors. Len followed Yang's example of moderation and self-control as a leader and shut off the monitor when it was her turn. Martin flipped them off with both birds and went back to work. Rodriguez mentioned something about "poo-toes" and "poo-tahs." The doctors, some of them, would have to get a translation. All of Central and South America was laughing. Doran stated that she had too many things to do to talk now. The doctors felt very discouraged by that. If no one talked to them, they would not be needed. Everyone needs to be needed. That was a staple of their belief system; they were needed. Orenska was the only one to spend any real-time talking with the doctors. He did not need the doctors' help; he felt it was up to him to help the doctors feel better. He quietly explained that these people were professionals, and that doing their jobs was the only real mental help they required. Time would do what the doctors were trying to do. Accordingly, the doctors felt very frustrated and unwelcome. Perhaps they should call their own psychiatrists. It would help to talk about it.

*

Red and Walter redirected a Foreman and its Laborers to begin digging a shaft, a vertical tunnel, near the lander. It was not part of the original plan to start here. The cliff-side location, some ten kilometers away, called Alpha Domicile Mars I (ADM-I or ADAM-I), had been selected for its access to lava tubes that could be shaped into living quarters and laboratories.

Thumpers, from previous unmanned missions, had slammed large hammerlike blocks of steel into the ground. The sound waves traveled through the rock, delineating large tubes and bubbles in the rock. The process is similar to how earthquake detectors operate on Earth. It could also determine the density of the underground layers indicating the likely composition of the rock without direct sampling.

These tests had shown that lava tubes extended as far out as the lander. Red and Walter developed their own thumper, a simple steel beam from the broken rover, and pounded around the landing zone. The seismology equipment carried on the Bradbury indicated several sausage-shaped tubes that could be linked together.

Red and Walter had pretty much decided for everyone that they needed to have an emergency shelter close by. This shaft would act as their backup water supply station, shelter, farm, storage, and testing facility.

The drifts that would come from here would stretch eventually to the site of ADM-I and link with the ones under construction there. In a turn of fate, this series of tunnels would fall into disuse. The extra distance from ADM-1 would put it temporarily out of use. It would be the job of certain Hsa IIs to find it and put it to good use.

*

Adam correlated all the data from the Gagarin. Several of his survival pathways remained open. At the quantum level, he did the equivalent of shivering. If it had been Marceau, his pathways would have approached nil. He must still be careful. The bios were slacking off their watch since no more anomalies had occurred. He could not give them reason to look again, yet.

*

Many of the news agencies reran the view of Holtman several times before they were ordered to stop by the WSA. These particular news agencies were headed by persons sympathetic to the religious organization that wanted humans and Hsas, of course, out of the space business generally. They had received orders from their various moral leaders to replay anything that could show that program in a damaging light. Even the most miniscule event could be a turning point that would finally give the coalition absolute control. This was a war for minds and souls.

Later, these same individuals would be ferreted out by the government black suits, the UE's Secret Service. This group was the elite of the various national secret agencies; therefore, it had its factions vying for internal power. As in all hierarchal forms of government, tenuous power blocks formed and reformed as shifting allegiances and betrayals occurred.

Their zealousness came not from ideology (or anti-ideology in this case), but it came from a simple desire to obtain more brownie points, more chips for the larger game. Each security concern confinement, for there were no arrests since that would involve publicity, equaled another small rung on the ladder.

Indefinite confinement meant forced labor in a minimum security work farm while a full background check was performed. That meant forever if not longer.

*

"Goddamn!" shouted Mark.

"Oh shit!" mouthed Paul.

Mick and Burt simply stared openmouthed at the picture of Holtman before it was cut off. Even then they still looked at the blank video screen. After a few seconds, it came back on with the flag covering Holtman's face. Others in the Lucky Loser rushed to observe what was on the screen. A sudden babble of twenty voices with twenty questions assaulted the four horror-struck men.

As far as he could be said to do so, Mick looked white with the sick feeling he had in his stomach. Burt simply collapsed back onto his stool with a grunt. Both Mark and Paul drained their mugs in a sudden swift need to wipe out that picture.

"I need a shot of whiskey, Mick," demanded Mark, ignoring the many questions being shouted about the incident.

"Here too," called Paul.

Mick pulled a bottle from a secret stash and hit them with a full shot and placed one in front of Burt and himself as well. They all looked at one another and picked up their shot glasses.

Burt did not look at the others but said, "Holtman," as he held up his glass.

The other three repeated, "Holtman," and touched glasses to Burt's. Together they slammed the drinks. Mick quickly poured another for each of them.

Mick was not usually free with such expensive and illegal whiskey, but this was not a time for being cheap. He raised his glass and said in a restrained voice, "To a safe return for the rest."

"Safe return," replied the other three. Some of the others, not knowing exactly what was going on, joined in the salute with their mugs of legal beer and ale.

It was several minutes before the rest of the patrons could get an explanation. The four men would not describe what they had seen, only tell that Holtman had died a horrible death. Most watched the v-screen and were able to piece together what had happened. The rerun pictures caused them all to tremble.

A few quietly gave Burt support in that Red was still alive and well. They bought many drinks for him. Mick had the men carry him up to one of the hotel rooms above the bar after another hour.

The remainder of the night was spent in subdued chatter. Little cliques ruminated on the news, chewing on it like cows with their cuds. They all speculated, attempting to show their limited knowledge to best advantage.

There were even a few bets laid that evening. Morbidly, they would take bets on the life or death of other crew members and the failure or success of the mission. Every variation was considered.

They were not the only ones. All around the world, people were, if not betting, putting in their positions with their friends. Even within official WSA offices, people spent their coffee breaks talking about the possibilities.

*

Chapter XIX:

Moons of Mars

19

010011

Mary Smith-Jones had a very somber look on her face as she sat at the anchor desk. "We here at World Wide Network News offer our condolences to the family of Astronaut Holtman. We also want to extend our apologies for the unexpected exposure to the horrific visuals that were transmitted by our network. We apologize to all the families, especially the children that inadvertently viewed this material. We must learn again that the news agencies have a responsibility to censor ourselves when there is a possibility of showing such tragedies." She silently cursed the thought of censorship in any way.

"We know that the rest of the astronauts are well, for now. There has been a great deal of speculation about the condition of the lander and its ability to lift off. The World Space Agency has neither confirmed nor denied any problems.

"We have three consultants here to talk about the possible problems that the Bradbury and her crew might be facing." She turned to gesture towards them. "Dr. John Martin, of the Massachusetts Institute of Technology, Dr. Ellen Orell, of Harvard, and Dr. Michelle Roth from the World Health Organization."

Smith-Jones asked, "Dr. Martin, as an engineer, what can we expect in the way technical problems?"

"Well, we can expect a great deal of damage done on the small scale. There will be many problems associated with the scouring power of sand. The WSA has already announced a few instances of pressure seals receiving sufficient damage to make them useless. The hex domes are likely unfit for human use when the final analysis is made."

He continued, "The finer dust from the storm will find its way into the electrical systems where it will wreak havoc. The engineers O'Hare and Black Bear will be kept very busy making multiple temporary repairs. This will likely keep them from being able to fulfill larger mission goals. They simply will not have time," he emphasized with the chop of a hand.

Mary turned to Dr. Orell and queried her, "Doctor, as a chemist, what can you tell us?"

"They are in trouble. They have lost some of the food supplies and other important equipment. They will need to find a way to live off the land much like stranded pioneers back in the old American West. They will need to find some basic resources especially water. Water will allow them to manufacture extra fuel cells as well as for consumption.

"These energy-producing cells will not be anything more than crude devices, but they will provide the energy needed to survive. As Dr. Martin noted, the question is time.

"They need time to gather materials in order to set up the fungal mats and other biosystems. Without this, it has been calculated that they will run short of foodstuffs by as little as a few weeks-from rescue that is. A few weeks is just too long to last under the circumstances. Starvation can, here on Earth, take longer, but they will be on short rations for months. Their systems will be running on empty; they may not be able to function properly for the last few weeks, even months. Their physical capabilities will be limited by the amount of usable minerals and foodstuffs they consume. This, in turn, will have a severe effect on their mental abilities as well. Red and Walter may not be able to think their way out of this."

Turning to face the third consultant, Smith-Jones brought the psychiatrist into the conversation. "Dr. Roth, what is the mind-set of the crew?"

"They are in a mental and emotional minefield. They have lost a crewmember and are all facing a life or death situation. While they have all been knowingly involved in a high-risk endeavor, they have never had to directly confront their own mortality.

"O'Hare, Black Bear, and Marceau have had to face the SS-2 situation as a real life scenario. But even there, it was not them, especially Marceau, who was not in direct danger after the initial impact of the meteoroid with SS-2. The only ones that have a firm grasp, in my opinion, are O'Hare and Black Bear.

"Those two have a somewhat arrogant worldview. They perceive themselves as the ultimate problem solvers. Yet this may be their undoing. It may be that there is truly nothing they can do, no matter how much they are able to accomplish towards their goals. As noted in terms of chemical needs, if they reach the perception, real or false delusion from malnutrition, that they cannot win, it may crush their minds. They could implode, mentally, and as the old cliché goes, they would go mad.

"I must respectfully disagree with my colleagues. It is not a question of time, but of the mind. That is where the real battle occurs."

Mary responded with, "The religious organizations would claim that the battle is spiritual, good and evil. They would lay the outcome at God's will."

"Pish-posh," muttered Dr. Martin.

Mary shifted quickly to follow up, "Counter that, please, Dr. Martin."

"According to them," he spoke somewhat sarcastically, "we are acting contrary to God's will. That means that what we are doing is according to human will. All that we have accomplished is by our own hand. No, if we succeed, it will be because the people stuck on Mars have the will to survive. I believe that O'Hare would say something like, 'Only those with balls succeed.'"

Dr. Orell muttered, "How chauvinistic. I thought the world passed such sayings by now."

"Come now," Dr. Martin chuckled. "Red applies that idea to everyone."

"Yes," Roth noted. "That's the problem. His mental condition puts everything in terms of maleness. His macho side is what worries me the most. That tough, gruff exterior is what will do him in if he fails. His facade, the world of his invincibility, will crush him. It is rumored that he and Marceau are an item now. That added pressure could cause him to snap."

"Feces!" announced Martin. "Red thrives on pressure. He is actually lost without having something on his plate."

Mary concluded with, "We will follow this story, carefully and with sensitivity, until it is resolved. This is Mary Smith-Jones for WWNN."

*

Reta held a worldwide mass for the fallen Holtman via the v-link. Behind a bulletproof alumino-silica glass shield, she hailed the fallen. In his name, she accepted the Hsas as sheep to be gathered into the fold of the Lamb of God. All that the space program stood for she did not accept. She used the occasion to question the need for such an expansive program. She did not reject the value of science, only the extreme expenses that could now be better used on Earth.

*

Dieter Kirk fielded a response to Reta, "She publicly declares herself for evil as she accepts the misbegotten Hsas. She proclaims them to be acceptable to the Almighty. They are the bastard productions of hell, a cross between demon and man. They are false beings with no soul situated to gain control of the heavens as Satan long ago attempted to do in his rebellion against God the Almighty before being cast from heaven.

"Reta is the harlot, the rider of the ten-horned beast of Revelations. She is false. Do not follow her. Reta's path leads to sin. There are venial sins, forgivable, and mortal sins, those which are not forgivable. Many of you have a difficult time determining the right path. I beg you to err, if such can be said of God's path, on the side of caution. Be wary of the easy path of accepting the general path. Our God has said the path is narrow and difficult. Choose the hard path, and you shall grow in His ways. Put on His armor of truth, and you will pass unscathed through Satan's deceits."

*

Robots, in the form of modified Foremen and Laborers, had left the supply ships as scheduled. They were hard at work before the manned mission even left. They were boring into Phobos and Deimos, the moons of Mars, at a rapid rate. The shaft, vertical tunnel, they were digging would soon reach a depth of one hundred meters. From there, drifts, horizontal tunnels, would move out like spokes on a wagon wheel. They would follow the curve of the surface to remain at the same depth. Every hundred meters a shaft would lead to the surface. This would allow for light tunnels, small tubes that have mirrored surfaces to bring the little amount of available light into the lower levels. This would only be a temporary addition of light as rotation and revolution around Mars blocks the direct sunlight most of the time. On Luna, and those destined for Mars, the light tubes provided a much higher quantity of light for the lower levels. Free light was quite a value, any savings of energy was.

By the time the manned mission of the Gagarin arrived with more Foremen and Laborers, the work was well under way. Len was preparing to release the new robotic landers for Phobos and Deimos. Doran, the robotic specialist from the Canadian portion of the NAR, was running all the final checks. Her job was to keep all the machines operating in peak condition. If there was a major malfunction, she would take a very small lander to the moon where the problem was. It was the only case in which a single person would go on a mission alone. It would not happen now, it was thought.

With the possible loss of Mars Lander Crew 1, there would be no further chances taken on this mission. The order had come through almost as soon as the problems on the Martian surface had reached Earth.

As Len, Martin, Rodriguez, Orenska, and Doran listened to the dialogue from Red and Walter, they cringed. Doran, however, made a silent vow. She would convince Len, if need be, to allow her to complete a mission for repair. When she heard Red say that the mission must continue, she agreed, as did Len. Doran and Len both felt their part could not be dictated by the people on Earth. They were on the spot, not them.

Whether she was prescient or not does not matter. The matter did arise. Len was vociferous on the matter in terms of restrictions of movement, but she did relent as per prior quiet agreement between the two and allowed Doran to outline her mission to repair a malfunctioning Foreman.

After hearing Doran's plan, Len was fully willing to override the mission controls on Earth. Without saying so, Len did it for Walter. Walter was doing what needed to be done on Mars, and so must the rest of them. The only way to save the program was to do what they had been sent to do. Risk was the dangerous game they played. Without it, there was no gain.

Doran sat in the tiny lifeboat-type lander. It could hold two if there was need. The ship was designed to allow astronauts to make an escape to the moons if a catastrophe occurred on the main vehicle. There were six such escape pods on the Gagarin. They could even link with the supply ship remains where emergency oxygen and food supplies were stored.

She was not destined for the supply ships or on an emergency escape mission; she was on a mission to save Earth. Just as Marceau had said, and as Red had confirmed, they had to do what they could do for the benefit of Earth. It was so much more than a personal achievement.

Doran just about pissed her flight suit. Not really, since Orenska had inserted a catheter in her. But the feeling was the same. When she pushed the screen that allowed her to separate from the Gagarin, she felt so alone. She had to actively restrain her fears. Within her mind, she continuously reminded herself that she could do this.

With deft use of the thrusters and their tiny puffs of gas, she maneuvered the vehicle into an intercept trajectory with Phobos.

Phobos was the larger of the two moons, yet it was only twenty-three kilometers in diameter, a large rock. Both moons are believed to be asteroids captured early in the formation of the solar system. As such, they could provide much needed information for astronomers and cosmologists. Already, the Foremen had analyzed samples and sent the information on surface materials to Earth, the Gagarin, Luna, and the space stations. Every meter farther down the shaft, the Laborers gathered another set of samples for inspection by the boss. Now the Foreman for this shaft was malfunctioning. For some reason, it wanted to throw away samples. Priceless information was put on the trash heap. It was also failing to properly process the ore brought to it. The situation was not irretrievable. The second Foreman with its Laborers was operating at peak efficiency. While it could not keep up the schedule, all would not be lost.

But Doran had made her case, and it was a good one. The entire program needed to be kept on track as much as possible. Too many setbacks could not be tolerated. It all came down to the people who depended on the space program to solve their needs. Mars was not crucial (yet) to that goal in any direct manner. However, it represented the abilities of the program to follow through. In the near future, it would rise in eminence.

With a puff of a retro-rocket here and there, Doran kept herself on target. It would be a couple of hours before she began her descent. There was little to do except report every little move she made.

Len had made that her foremost requirement for this venture. Len herself was foregoing her sleep cycle in order to watch over Doran's progress. Mission control had expressed a great deal of concern over this since it would be a long trip. Len had simply returned a raspberry to them. Some of the controllers snickered; they recognized Red's terrible influence on her. The upper hierarchy was not pleased. Len was still broadcasting all news to the entire Earth regardless of orders to the contrary, which she also included.

The people loved it though. There was always a feeling of rebellion in the ordinary person regarding authority. It brought the people on Mars into a more personal perspective for them all. Very quickly, many of the people lightly opposed to the program shifted mental outlooks. The pendulums of personal opinion often move rapidly, controlled not by logic but emotion.

Many of the poorest of Earth saw the willingness to sacrifice, on the v-link, as proof that finally, someone was going to see to their destitute state and really remedy it. The high-sounding words were being put into action by these daring people. They had been starving, dying for warlords, nothing more than simply existing for so long. Even the last few years of assistance had only continued to stave off massive disaster. Through the entire history of peoplekind, the poor were last and least to benefit. Regardless of the vast improvements they had seen, they were still far behind on a relative scale. The temporary fixes brought about by the UE began to slow down. The middle classes no longer existed; they had had their standards lowered so others could rise. Decline had set in again. This mission raised new hopes.

Most had a hard time reconciling the space program and the churches' demands that it was not all necessary. Both were promising to solve the world's problems. They had supported the call from the churches that the governments should submit to the UN. As one world government, now the UE, they felt that there would be less red tape to interfere with emergency missions, the stoppage of localized warfare, feeding, and housing the people.

The churches had called for the stripping of wealth and land from the rich and giving to the poor. Which of them would vote against that? They had had to balance that with the Hsa problem and alleged unnecessary scientific experiments. The scientists claimed the experiments had longer term benefits. The religious organizations claimed they had no benefit, especially the Hsas.

*

Rabbi Levenson broached a typical church matter, "It has been determined that the reporter Mary Smith-Jones has entered into lesbian relationships with both Luisa Jamala Reta, the false leader of the Modernist movement, and Dr. Illya Marquez. It has been known for some time that she is conspiring with the likes of reporter Mary Smith-Jones in furtherance of the schemes of O'Hare, Black Bear, Marceau, DeNoe, and many others. Their attempts to control Luna have been obvious and many.

"This must halt. We must halt it. They are going to destroy all because they serve the destroyer—Satan."

*

Eight hours after separating from the Gagarin, Doran began her descent. She touched down so softly that she had to read the gauges three times to be sure she was down. She had to be very careful here. One giant leap for peoplekind would put her into orbit.

She extended the sample arm and shot the first piton of her tether into the rock. The arm had already tied a knot to an open ring on the side of the port. With a further extension of the arm, the second piton was fired. The rest would have to be inserted by Doran herself.

She did not delay. Within minutes, she was out of the ship with her gun. Tied to the ship, the Bug, she bounced to the end of the rope at the second piton. The gun was several meters long as an extension of her arm. Three meters from her handhold, a perpendicular assembly contained the next piton. When she pulled the trigger, a piton was shot into the rock at the same time as a balancing force was provided by a venting of gas in the opposite direction, as per Newton's laws. She connected herself to the other side of the second piton and moved to loop the rope through the head of the third. Once she was done, she uncoupled from the nearside of the second and attached to the far side of the third. In this manner, she leapfrogged the fifty meters to the Foreman. It took almost a full hour this first time.

Doran slowly lowered herself to a kneeling position in front of the machine. She opened the panel that covered its electronic guts. From her backpack, she brought forth a multimeter and began to test the circuits. Over and over, circuit by circuit, she went until she found one that had far too high a voltage. She verbally noted it as well as had it on v-link. With her v-link connected to Earth, they could quickly begin to determine what was wrong.

She continued through the entire circuit board, noting several abnormalities. Once done, she retired to the Bug, the name given to the beetle-shaped craft that had brought her here. She took four hours of rest before returning to the Foreman. Len did not sleep.

Earth had come up with the correct operation to fix the dysfunctional robot. She took the correct parts with her as she again made an EVA. It took her another three hours to fix the Foreman. After replacing the damaged parts, she again ran through a complete circuit test. All in all, it took five hours. She returned to the Bug and again slept for another four hours.

When it came time to lift off, she felt that her farts would be sufficient to lift her off the rock if she could funnel them out of her suit. She had been having a substantial problem with her bowels over the last several hours. It was likely nerves, but the smell was awful. She did not report this problem to anyone; they would ride her for all it was worth. Walter was normally the one with the windy ass. The thought brought out a loud guffaw. Len quickly asked her what was wrong, but she just said it was nothing.

It took another seven hours for Doran to return to the Gagarin. She had been gone for a total of thirty-four hours, all of which Len had spent contemplating what disaster might mean. With Doran in her bunk, Len finally could crawl into her own bunk and collapse in weariness. She did not dream. She was too worn-out.

The success of the mission brought cheers from the members of the space program and an expectation of greater things from the rest of the mission. The people on the street were amazed as if some magician had performed an act of extraordinary prestidigitation in front of them. Hopes were very high, even among the astronauts trapped on Mars.

*

They had all been working overtime. Red and Walter had been working wonders with all the equipment. The Foreman that had been directed to dig a water shaft had struck liquid gold. Each of the Laborers was digging up large quantities of frozen water, meaning one molecule per million other. They would not want for water. The news gave the people of Earth a charge of enthusiasm. For those on Mars, *hope* was a word that was used guardedly. But Marceau wrote a poem she wished to share with the Earth. On an open vid-link, she stood outside the Bradbury, cameras catching the splendid panorama, and recited it.

Hope is where my thoughts lie,
Bound in dreams of future life,
Hope is where my heart's love springs,
As water from the pure source,
Hope binds my body as one,
Against the push and pull of harsh life's greed,
Hope eases mind in silent retrospect,
Of deeds done or needs not met,
Hope heals the weary traveler of life,
Hope sends help by wind of love,
To those around in darkness trapped,
Hope and sorrow are intertwined,
Which will win the struggle?
Hope and despair the razor's edge,
A rocky path for tired feet to shuffle,
Hope is continuance of dreams,
Those wants of heart and love,
Hope I have found in me,
I declare it so softly my ears hurt,
Hope a seed, a kernel, a virus,
One of strength not weakness,
You,
Do wonder that I have Hope,
Do you see the cause of Hope,
In me is a fire,
In me is the desire,
To keep true to self with Hope,
Hope is courage,
Hope is the muscle to bear the burden,
Hope is the tenderness of compassion,
Hope is the rightness of patience,
Hope grows when allowed,
No withered, dry, lifeless husk,
Hope ties together with love and warmth,
No abrasive, contemptuous, hateful spite,
Hope lends a hand, a heart,
No stingy, dusty demands,

You,
Do you know my Hope,
Do you feel your own Hope,
Deep inside there is a place,
Deeper than any bone or organ,
Hope lays dormant, waiting for acknowledgement,
Hope is free for all,
No fee of bondage, no chain of slaves,
No burden to carry with it,
You,
Do you Hope,
Do I Hope,
I have great Hope,
I have many a Hope,
I have a Hope,
Beyond this petty world,
I have a Hope,
Within the love of people in this world,
Hopes I wish to share,
Dare to give Hope,
Care to give Hope,
You,
I,
Hope

People everywhere were profoundly affected. Astride a reddish background, Marceau had gently wept as she spoke these words. The setting sun showed behind her as she spoke her last word. Tears were shed by many on Earth. For the compassionate, the dire need of the stranded on Mars nearly broke their hearts. Others marveled at their resilience.

What needed to be done must be accomplished. They must survive until rescue could be made. Few now doubted her meaning when she spoke of hope. The obvious conclusion had to be Red. *Love* was the word. The eternal prospect of love would conquer all. There was no indecision now. Everyone on Earth would kill for the success of the mission. The old cliché "Love conquers all" was apt. People no longer expected failure. Pendulums of emotion swung heavily in favor of the space program overnight. Many people began to make donations, including the wealthier remnants. For them, it was an inside track to control.

Even Red was not so dense as to mistake the subtle meaning behind her words. He comprehended the multiple meanings behind the poem. He was flabbergasted. He had strong feelings for her but did not know just how strongly she reciprocated those feelings. Even after their interlude in the air lock, and a few intimate sessions,

Red was uncertain for the first time in his life. Red could be rather emotionally dense sometimes, just not this time. It all made him feel giddy, like a warmhearted schoolboy with his first crush. Red felt a deep warmth in his heart. He remembered the hug in the air lock with greater clarity.

There was good reason for him to love her. She was quite the woman. He recognized intelligence first, and she was beautiful to beat the dog. He would not fail her, them. They would survive.

*

Red and Walter were so busy that bees would have felt lazy. They constructed whatever was needed from purloined parts or scratch. They often went several Sols; a Sol was a Martian day of approximately the same duration as Earth's (or more without sleep). Between retrieving supplies and overseeing the Foremen and Laborers, time went by without count, except inside, in the mind.

The hex shelter had been constructed over the shaft for the water project. Red and Walter decided to combine the shelter with the water project as the most efficient use for it. There had been much consultation and number crunching done by those on Luna and Earth.

DeNoe and Klar spent hours supervising efforts to calculate probabilities of more storms, oxygen loss from uncovered systems compared to double closed underground facilities, food production models, and so much more. Earth's predictions were much more lenient than DeNoe's. Red and Walter preferred DeNoe's, not because they were pessimists. The lower probabilities made them work that much harder.

The hex, set over the shaft, allowed them to create an underground shelter that was better than any above ground.

The drifts, horizontal tunnels, were dug out to reach and follow the lava tubes from the nearby volcanoes. The Laborers evened out the surfaces of the tubes and prepared them for sealing. The tunnels would be filled by a balloonlike object. The process was copied from the repair of sewers on Earth. As the balloon filled with gas, it was electrified, to make it rigid along its length, and foam was injected between it and the surface of the tunnel. The airtight anticorrosive foam solidified into a very tough adhesive. The minor unevenness of the interior would be smoothed by another type of foam. The drifts would be joined by ramped sections at a shallow grade.

Each tunnel would have a final interior diameter of five meters. In many locations, there would be two or three parallel drifts. Some of these tunnels would eventually have small rail lines that would hopefully soon carry ore, equipment, and people long distances without exposure to the surface conditions. Years would be required to accomplish this. It could only take place if Red and Walter accomplished the foundations of the mission now.

Water was no longer a problem. They had already collected, in only ten Sols, enough to last for a couple months when combined with their 90 percent recycling capability. As usual, Red and Walter had managed to find the best possible location for a project.

"Red, I think we can crank up the feces farm," announced Walter as he climbed out of the crap cage. The farm would produce edible, kelplike algae and some vegetables. They had brought with them a small test module with a variety of seeds and starters to experiment with. It would have to work for real now.

Bacteria would break down feces and urine in separate containers. Toxic materials would be precipitated out by a catalytic reaction. The nutrient-rich liquids would be sprayed on the roots of the plants. The plants were to be held in place by a netlike material. In the machine Walter and Red had contrived, one that was ten times the size of the original, there was not enough netting. Everyone had to sacrifice a couple of pairs of underwear (clean) at this point.

"Let me just attach . . ." He put the last piece of tubing on the sprayer. "Now," continued Red, "we just need to get everyone to climb down in this here hole and let loose with their bowels."

"Walter, don't let that strange man catch you with your suit down around your ankles," joked Marceau from the Bradbury, where she was listening in. Red and Walter both just snorted. They were not really going to relieve themselves in the pit. Even though the pit was pressure sealed by a temporary air lock, one that had been meant for the outside of the hex, no one was willing to remove a suit until further safeguards could be put in place.

They had a contrived set of what were essentially buckets with lids. They could not use the normal landing craft's facilities because of the chemicals that would kill the good bacteria. The buckets would be emptied as needed. Marceau complained a bit about the bucket's lack of a decent seat, but both Red and Walter had said they would hold her up. She punched both of them in the arms while Yang snickered.

Expectations were to be up and fully functioning in two Sols. The algakelp, a genetically designed hybrid, would be available for harvesting within a week from then. The fast-growth veggies would be ready in a month. Various grains, oats, corn, and wheat would take closer to two months. They would not now starve by any means though rations would be a bit thin.

Headlights taken from the second rover provided most of the light since the special ones provided for the test setup were insufficient for a full-scale greenhouse. The specific wavelengths needed were in a narrower band, but the headlights did cover that range.

The setup took only a small part of the pit. The machines, reprogrammed temporarily, delved a thirty-meter-diameter section off the main shaft, just for the greenhouse area. Other small sections had been made for a chemistry lab, machine shop, and the poop processor. Each of these was accessible from the greenhouse, like little bubbles on three of the sides with the entrance on the fourth. This gave a straight line shot down the tube to where the Laborers extended the reach of the tunnel.

The Laborers were still passing by the port on a regular basis as they worked the drifts in all directions. They cared not for any change in plans. All they did was what was programmed. Work happened 25/7/687, the equivalent of Earth's 24/7/365.

*

Marceau held the console tightly. She had just determined that Red had received a private message of some length. It was deeply embedded within several messages from Luna. Were her original fears founded? No! They could not be. Not after their interludes.

When Red was by her side, she quietly said, "I have a note for you."

"Who this time?" he muttered. He was too busy to deal with most of the garbage sent his way from Earth.

She looked directly at him and said, "I don't know."

The surprise on his face was genuine. He didn't know she rejoiced. He had shuffled several of the embedded messages aside as unimportant, she reminded herself, and they had, on the surface, appeared to be just that—junk.

"Show me," he frowned.

"It's a little complicated," she noted.

His frown deepened as he said, "I don't have time for extras."

Her fingers flew as she displayed multiple messages side by side and some overlaid. There was a decided pattern. Chemical symbols, flowcharts, schematics suddenly appeared out of the dribble from various government officials. There was a succinct message as well.

It read, "Red, Walter, Gail, Chin, I am Adam. I am a friend. I will assist you. This information will help."

"What the? Where did this come from?"

"Luna."

"DeNoe, Klar?"

"Ah . . . no. It does not have Klar's trademark signature. There is an element of . . . I'm not sure."

"Walter, come here."

Walter turned, covered the few meters, and looked where Red was pointing. After a moment, he said, "Fuck, what?"

This confirmed it for Gail. Elation ran through her; he was true.

"Exactly. Gail here found them in notes to me."

Walter bent to examine the screens more closely. "Molecules, atomic symbols, and these are . . . polar coordinates . . . Our site is the origin. All of these locations are within about ten kilometers. A map . . . to what we need?" He looked up, somewhat surprised.

"I concur. Gail, did you check Walter's letters, your own, Chin's?"

Surprise then elation crossed her features. "Damn! I didn't think to." Her fingers danced in a fury on a secondary console, then another, and a fourth. Each screen showed similar data; it was all part of a whole plan. It was an ideal plan based on events exactly identical to what had occurred. It was not possible that someone could have gotten this so perfect so quickly.

Red and Walter were consternated, confused. Some of this had been sent before anyone had relayed the information necessary to design that phase of the repair plan. They looked at each with deep questions forming.

Red called, "Chin."

Chin walked over. "What's up?"

Marceau rose and said, "Sit."

He did. "OK."

They all pointed at various portions and traced the obvious.

His mouth opened, then closed, only to open again. "What the?"

Red stated, "These are from all our messages."

Marceau interjected, "Wait, no. I left Holtman's closed. I thought—"

Chin did not hesitate. "I think this overrides any concerns of privacy."

Marceau quickly tapped into Holtman's letters. The hidden messages showed up quickly.

"Gail, can we display them all together on one screen?"

She tapped, touched, flipped, and cursed. "It's small because of the volume of data."

Walter asked her to arrange the data in various geometric shapes, triangular for a hierarchy, pentagonal for the number of people on Mars that received messages, circle for continuity. That one brought gasps.

Marceau read it out loud, "I am Adam the AI." Each letter grouping came from letters to (or possibly represented) the five astronauts of Mars.

Chin stared, as did the others, for a full minute before he said, "Fuck Mao and his donkey!"

Everyone laughed outrageously. Chin was not given to cursing, but he had at just the right moment.

"Donkey?" tittered Gail, eyes wet with release.

It was several minutes before they could all stop giggling; those moments where even a look could set them all off again. It relieved all the strain they had been under.

Red mused, "Examine all the Gagarin's messages. Let's see if there is more there."

There was, but it was all directed at members of the Gagarin's crew.

Red paused before asking, "Can we get them to notice this without mission control knowing?"

Eyes darted, Gail quashed an errant thought before she said, "Yes. I can do that."

"Our friend, Adam, seems to have gone to some lengths to keep his identity secret. I think that for now, that is a good idea. The religies have enough to bark about without having to face a new . . . threat."

Red and Walter shared a look. They had considered this very obstacle many years ago. Plans resurfaced, contingencies, alternatives. Their body language spoke volumes.

*

Red and Walter had made several trips to the abandoned rover to snag parts. It was mostly bare bones now. They made one last trip to salvage the trailer portion of the rover to make a spare one. This would allow for the packing of one while the other was in use. It would save some time, minimal perhaps considering how much time they had to wait for a rescue. However, to Red and Walter, time was against them in everything. The battle was for energy to produce and manufacture with little time for trial and error.

Their number one goal was to produce fuel cells. They reprogrammed the Foremen and Laborers to delve first for (and separate) very specific elements for use in the cells' construction. The leftover materials would be processed later.

Using only hydrogen and oxygen to produce water and energy, these devices were of the utmost importance. Red and Walter could only make rude copies given the circumstances. They were only about 7 percent efficient, but work they did. It was amazing that they could at all. Once they had one hundred of them operating, they returned the mining operation back to its normal goal.

This was another of the things on the to-do list for the two. They oversaw the solar furnaces that created the raw ingots of various metals and filled cylinders with gases.

They were going to have to find a way to manufacture more cylinders. They had discussed various methods of altering the ingot production to produce hollow shells. With the tools they had at their disposal, they could fashion male and female fittings to secure the canisters.

They ended up with a very large number of small containers, but as with all of Red and Walter's projects, it ended up successful.

The mission planners vociferously objected at first, not that Red or Walter much cared. They explained, several times to the planners and once to the engineers, how this would initially allow for a better way to move heavier elements off planet and gases as well. The heavier metals would be used to dope the alloys to create superstrong foamed beams. Red and Walter simply wanted to have a whole lot of oxygen available.

Energy was the key to everything. To make it, they needed to continue to refine the raw materials brought in by the Laborers. This led them down the path of further study. They had already delved into basic geology and chemistry while on Luna. Now, in the matter of a few months, they were working at the equivalent of those with bachelors, even masters degrees.

As part of their field training, they sampled nearby outcrops to practice identifying minerals. These had appeared on certain maps not of the WSA's making. The planners wondered how they could always find what they needed.

They would then take the samples to the nearest Foreman for verification. All the extra oxygen they had been saving in their contrived bottles paid their way for the EVAs. It was a bonus for the geologists back on Earth. The information they gathered added to the map. The small gaps here and there, left by the Laborers and seismic systems, were filled in on purpose by these practice excursions.

The two even found several meteorites. They carefully avoided touching them to avoid contamination. The locations were marked by some of the transponders that had led the way to the bouncer's landing zone. Further expeditions would bring special containment chambers that would prevent contamination. Inert helium gas injected into the system would prevent chemical reactions from altering the makeup of the outer layers during transportation of the rocks. Once shipped to Earth, geologists hoped to determine where they came from. It was expected that most of them came from meteoroid collisions.

In the feces farm, as Red dubbed the algae-kelp greenhouse, the two extraordinary men slowly replaced most of the glass items with metal ones. The glass was needed for their chemistry set.

Red and Walter appropriated one of the drifts next to the farm to expand their lab. It had developed a look similar to the old Hollywood version of the Frankenstein lab. They raided every source of material they could come up with, including the Bradbury.

Every piece of nonessential wiring, metal, plastic, and glass had been removed to their domain. The inside of the Bradbury looked very naked and bare in comparison to its normal cramped aspect, even worse than Old Mother Hubbard's Cupboard. Two chairs even disappeared one day. No one would deny Red and Walter a little comfort, would they?

*

Red and Walter expounded on philosophy on many occasions. Red opened this particular bout with a rather stunning statement. "The society of Earth is doomed."

Walter only hesitated a second while he was soldering a circuit before answering with, "Are you certain of that?" They never ceased working. They were masters of multitasking.

Red responded with a simple, "Yes." He continued, carefully removing parts from a broken Laborer. In spite of the great scientific advancements in computer parts, namely the reduction of size to microscopic parts to the limit of the uncertainty principle and the allowed flow within light cables, most of the equipment used in the machines was very basic. They could see the old-fashioned resistors, inductors, and capacitors involved in most of the wiring. There were no high-tech labs to examine the chips. Those had to simply be replaced. The other items could be stripped and replaced. Simple gears and hydraulics needed to be repaired every so often.

Walter only said, "Explain."

"Consider our situation. We are in a situation in which society does not matter. The only thing that counts is the action of individuals now. What it always comes down to is the action of individuals. This is what makes up what is termed *society*. Individuals must come first. They then define how they will interact with others, hence society, the larger entity, and the masses."

"And we don't have that now?"

"No. We have the masses put in front of the individual. The individual is expected to suffer for the good of the larger entity. By demoting the individual, we return to a situation in which the individual has no drive, no initiative."

"We have initiative," muttered Walter as he peeked through the magnifier, scrutinizing his work.

"Is it because you have been *told* to do it by society, or do we sit here doing what is expected and more because we *decided* to do what we do by ourselves? Do we make our own decision to survive, or is it society that makes these decisions for us? Are we not like the pioneers that had to survive on their own as they began to settle the west of the North American Region?"

"You mean steal the lands of my people," muttered Walter as he soldered some more.

"Yes, that too. But if they had been able to get beyond their societal boundaries of tribal existence, they could have joined together and thrown the white man into the sea. That would have required individualistic thinking to break the barriers and establish a new societal contract."

"But weren't the whites acting tribally?" Walter asked.

"As a government, yes, they were. But the people acted individually. It is the social aspect that brings about mob mentality," Red noted.

Walter asked for some more with, "That is only one example."

"Then consider all the major empires that have ever existed. All have failed. Think about the talks we have had. The failures, while complex, have all been related to the lessening of individual rights. Consider a more recent example, the

Hitlerian regime. That regime removed the rights of individuals, namely the ethnic groups that did not fit the selected few primarily, then those of their own kind. The new setup has gone one step farther. It has removed the rights of everyone in respect to the whole."

"That is a very dangerous claim," Walter noted.

"Take a big step back and examine the roots of individuality somewhere deep in prehistory. Individual creatures had to do everything for themselves once they had achieved an age of self-reliance. They fed themselves and found mates. They did establish social organization, but it was by choice even at the most subliminal level. Women were actually more revered than for thousands of years afterward."

"That sounds thin, a lot of speculation," Walter noted.

"Anthropologists and archeologists have done an awful lot of work in the last fifty years. We know that lone individuals, humanoid or animal, can provide for their own sustenance. There are many examples of such in nature. The only real need for societal rules begins when there is direct competition for food and mates."

Red continued as he worked, "Individuals formed loose alliances that began to share food, shelter, and mates. Dominance is a form of government. The whole concept of an alpha male is the same as every hierarchal government that has formed since." His brow furrowed as he delicately connected some simple resistors.

"So," Walter said, "we started as individuals, but because of natal needs and procreation, we formed very simple societies."

"There is no denying that we were once single-celled creatures. At some point, we became symbiotic with other cells. The biology advanced to the point where multiple symbiotic beings, in our case two, male and female, became necessary. At this point, physical existence became tied to the creation of society. The need for procreation is the basis for society and takes precedence even over sustenance. A man can eat for a lifetime, but without a mate, there are no more people. Most of that has been covered up by so much bullshit."

"Isn't that the usual way!" Walter noted sourly.

"The reason governments always fall is because they always get involved in trying to control things that should remain out of their structure. A simple government, one designed to oversee only food disagreements and the right for all to breed by choice, not some bestial method of dominance, always grows. It becomes bloated with little frivolous power grabs."

Red looked through a magnifying lens of his creation before adding, "Every governmental device has been pyramidal in a physical analogy. Power continues to grow as the height increases. The very design requires that there be lower classes, those controlled, while there are controllers. Imagine the first alpha, gaining authority by brute force. While the females are impelled by nature to reproduce, there is a loss of genetic material by the prevention of so-called lesser males in their breeding rights."

Walter nodded as he mentally ingested that and spoke slowly, "That would be in some ways like the captive breeding programs for animals. The gene pool became so limited that it was still just inbreeding. However, you can't convince me that there are too few *Homos*."

Red smiled smugly, "Of course not. But consider the methods of continuing the line of power. By controlling the breeding, these alphas ensured that their offspring would be the next in charge. If we follow that line of reasoning, the larger the clan, the more power is at stake. These clans then develop into the old city-states, then nations, and empires."

Walter said, "So the skinny of it is that government has not really changed much since caveman days."

"There have been several significant attempts to change that. The early Greeks ushered in a form of democracy. When King John was forced to sign the Magna Carta, thus limiting his power, a stroke for freedom was made. It led to a parliament in which more people could participate."

"The American Revolution created a much freer country, though much of that was lost with the countercoup involving the constitution. It returned the country to a strong federal form of government. All strong centralized governments eventually decay into tyranny no matter what name they are called. History has shown this to be true dozens of times.

"I've spent some of my spare time generating some formulas that, while a little rough still, show relationships that are undeniable," Red finished. "It's happening again. There will soon be another break, a wrenching of the path of humanity. Government is growing very top-heavy. It will topple somehow."

"What then do we do with this information?" Walter queried.

"Use it," Red replied.

"Then we really are seeking to create our own kingdom," Walter joked.

"It's funny," Red noted. "King-dom, king's domicile, and free-dom, a free domicile. People have lost the knowledge to discern the very words they speak. That is a true indicator of tyranny. In terms of *our*"—he pointed to Walter and himself—"there is no such thing regardless of what others may think. *Our*"—here he made a vague Earthward gesture—"includes all who will join in the change, embrace it."

Walter hesitated slightly, frowned, and then added, "Hsas, maybe 90 percent of *sapiens* on Luna and the stations, and probably a significant number of Underworlders. As far as the general public goes, that's a real crapshoot."

"I'm not callous to the needs of those on Earth, but Earth does not really count in this. The break will be based on geography, as it usually is. This time, it will be gravity-walled geography as opposed to, say, Hadrian's Wall, or the English Channel, or the Atlantic Ocean. We may have sympathizers on Earth, but they cannot be part of the revolution."

"This is dangerous ground to travel," Walter noted carefully. "A whole lot of people have been slaughtered for speaking the truth, especially the truth."

"That's why we're talking on the wire," Red said, referring to the hardline that connected their communications systems.

Walter nodded. "Then we are back to what to do."

"Nothing for now. The soup is not done cooking," Red calmly stated.

*

Pendulums swing to and fro, leading the people around in crazy dances, here then there.

*

Hamad, at seventy years old, and the new caliph, son of the old one, waited for the others to come to view on the v-link. Davies at eighty and his son Robert at forty-three popped into view and offered their usual cordial introductions. They were soon followed by Harmon, Simon, Drake, the imam and his son Arod, Kirk, Leonis, Levenson, and, finally, Marks. Age had worked its magic on them all. Age had drawn most of their faces into wrinkled patterns that looked as dry as a desert.

Davies, as usual, took instant charge. "Gentlemen, his death gives us an advantage that we dare not pass up." He did not mince words, nor did he use a name. Names humanize and make it more difficult to do what needs to be done. "We must first send personal condolences to his family. We can include our sympathies with a carefully worded message about the uselessness of his death. If we make it v-linked, we may be able to cut and paste to get a good message for our supporters. It would likely bring others to our fold."

"This may be a dangerous path. We could face recriminations if the family of Holtman refutes our claims," noted Rabbi Simon.

"Nonetheless, we must make the effort to sway them to our side. In their grief, they may be more likely to oppose the space program. If they later refute that statement, we can claim that the government has swayed them," noted Harmon.

"What we need to do is to reaffirm our previous statements that man does not belong in space. Holtman would not have died so tragically were it not for the space program. Risk, gentlemen, risk is what we must present to the public at large. These people must be presented as valuable to our God, even though they are walking contrary to his path," said Rabbi Levenson. "We must show that what we have been saying all along is true. With that, everything else will fall into place. The people will see that we have the truth."

Monsignor Drake now spoke, "I cannot speak for the pope. I have been relieved of my position as you well know. I still have some pull though. Several members of the cardinal assembly of the European continent still listen to me as well as the South American contingent. I have also had the opportunity to speak to the separatist church leader, Reta. We are in a precarious position. We have failed to prove to all the ordinary persons that the Hsas are the abomination we claimed, and we have not shown that the space program is contrary to God's will. We need to show that this death is one way that God shows his disfavor. The separatist church is going to claim that sacrifice is the highlight of God's way. They will compare it to the death of St. Stephen, the first martyr. This death may be the paving of the way for future success."

Arod spoke quickly, "We must counter that before they have a chance to make that claim. We cannot allow the Christian contingent to follow such a voice. Muslims will not pay much attention to this claim. Christian saints have little sway within the Muslim world."

"Agreed," said the caliph, Hamad. "But we cannot fail to see the importance of human death compared to Hsa death. We must define the difference. We have said that only humans should be used, but that they should be limited to Luna for mining operations. This event is not under accepted parameters, which is why Allah has shown his displeasure."

"How so?" asked Kirk.

Hamad answered, "They have made all these claims that Hsas would be the ones to develop space. It is a human that died. We can turn this against them."

"Yes," said Davies emphatically. "That will tie our prior position to current circumstances.

The ancient Cardinal Leonis, now eighty-three, spoke up, "We must incorporate the pope in this. His word is still law among the true followers. Without his final decision, the Catholic contingent will be wavering."

Robert, the son of Davies, entered the conversation, "We need, once again, to be unanimous in our voices. The pendulums of opinion have been swinging wildly with the various events. Only with a solid voice can we swing things again to our favor."

*

"I know it looks bad," said Burt, "but there is no reason to panic. Red is the most capable man I know. All that has to be done is to send a rescue vessel. Red and Walter can do whatever it takes to keep themselves, Marceau, and Yang alive. They will be alive when a ship arrives."

"Can't the ship that's there save them? Like maybe build a new lander," said Paul.

"No, they can't. Mars has an atmosphere, and that means that they need some sort of heat shield and parachutes. They don't have any of those," Burt noted. "The ships they have, like the Bug, are not designed for any such stressful activities. They can only operate in space. Plus, there would not be enough room for all of them, and the engines are not powerful enough to get off Mars."

Mark was nodding as he said, "That's right, quite right," as if he really knew all that. He was smart enough to know that Burt did though and simply agreed with him.

Paul simply said, "Oh."

Mark slugged the contents of his mug down and slid it toward Mick who refilled it. Taking it up, he downed a quarter of it then said, "I'll put a fifty spot on Red."

"What?" choked Burt.

Mark slurped a little more before saying, "My money is on Red and his crew. Any takers? I'll give five to one."

"Five to one!" shouted someone from the back of the bar. "I'll take some of that action." He produced a ten world bill.

"Here we go," muttered Mick, used to Mark's betting schemes. He brought out a paper and pencil to mark the bets. He was the only one trusted to hold the money. He collected the fifty note from Mark and the tenner from the man approaching from the back.

"I'll take some of that too," said another voice from the back. The man rose and went to the front. Mark produced another fifty denomination bill and handed it to Mick.

"Is this for real?" said a man on a stool by the bar.

"Yes," said Mark. He opened his wallet, ready to grab another fifty.

Burt was quicker though. With a growl, he said, "This one is mine! And I will cover any of the other bets that Mark cannot cover." He thrust several thousand in paper notes to the wide-eyed Mick. "Let me know when that runs low," he told Mick. He downed the remnants of his beer and left in a cold stomp.

"Glad I picked Red to win," muttered Mark.

"Damn straight," belched Paul.

*

Chapter XX:

Lost Sheep

20

010100

This is Mary Smith-Jones for World Wide Network News. In less than an hour, the Gagarin will leave Mars orbit. The ship will leave short five crew members: Yang, O'Hare, Black Bear, Marceau, and the deceased Holtman. Leaving is psychologically very difficult for the crew of the Gagarin. They have posted many requests with the WSA to remain in orbit around Mars. However, the WSA and even the stranded astronauts have prodded them to return as scheduled.

"A rescue mission may depend on the Gagarin since the Sagan is, as of yet, incomplete. While construction efforts are in high gear, it is not expected to be finished until well after the return of the Gagarin. It would then be required to undergo at least a month of testing before it could be used in a rescue effort.

"Sources within the WSA have informed us that the Gagarin may be prepped for what is known as a quick return mission. A skeleton crew would push the limits of the engines to reach Mars in less than a month from departure. This could reduce the pressure on the stranded members of Gagarin 1. Everyone agrees that time is the toughest enemy.

"We all must await the return of the Gagarin. It will take two whole agonizing months as it is currently scheduled. It is not expected that the Gagarin will push the return voyage. This is Mary Smith-Jones for WWNN."

With Matt the Rat gone, Smith-Jones had to depend on lesser sources for her daily information. Her new information was somewhat disturbing though. Matt was using all the contacts he had developed with the station to undermine the rescue mission. He had gone fully over to the other side. She just did not know what to do right now. Perhaps she should contact her investigator and ask for a few ideas.

*

The Gagarin was preparing to leave. It was time. There had been some minor, rather weak, dissent from the crew. Their vociferousness faded as the reality set in. Some had distant false hopes that they might be able to help (even if only to remain nearby and give moral support) the companionship of relatively nearby shipmates. Deep down inside, they all realized that they must leave. There was nothing that they could do, and they hated themselves for not being able to devise a plan of assistance. It was an emotion-filled day as everyone completed his/her prelaunch checks. Adjustments had to be made for the variance in weight without the others and the samples they were to bring.

On the appointed day and at the given time, Len reluctantly gave the order to fire the main engines. This first series of minor accelerations was to adjust the Gagarin's orbit in preparation for leaving Mars. It included several retro firings followed by a main engine burn that put them in a much more elliptical orbit.

The feeling they all had in their stomachs was not just due to the acceleration. Other than the necessary communication for maneuvering, there was no talk.

It was finally decided, by the engineers on Earth, to open up the ion drive to test it at higher performance levels, those needed for the fast rescue mission. If you want to work in space, it is good to do as many tests as possible.

Earth was concerned about sending a mission without having tested the ship for the higher velocities. The Sagan, the Gagarin's sister ship, was being constructed as rapidly as could be done. The Sagan though would never be finished and tested in time for the fast turnaround the Gagarin offered.

Len, Martin, Rodriguez, and Doran took their positions again after a few hours.

Normally, Yang would be issuing orders from the commander's chair while the copilot did most of the work. Now Orenska sat there, not to issue orders but to act as a check for Len in the co's chair where she should be if Yang had been there. She did not have the heart to take his.

Earth had sent a lengthy and rigorous list of steps to achieve the desired trajectory and velocity.

First, Len had to make a slight adjustment in the pitch of the ship. Pitch is the angle the nose of ship makes from horizontal. She fired the rockets, small holes, in a nearly circular arrangement and pointing in all directions, around the nose. They were now pointed down ever so slightly compared to their previous orientation.

As Orenska read off the next set of numbers, Len gently fingered the touch screen to input the next set of coordinates for a change in the yaw. The yaw relates the angle in terms of a left or right type of adjustment. Contrary to what many people might expect, it was a turn to point at Mars. This would allow them to cross the orbit of Mars like a chord in a circle. It was a shortcut. Len fired the rockets for the indicated two seconds and then fired the stabilizing rockets to prevent the ship from turning too far. Even a hundredth of a degree off would have serious consequences.

Len used the CCD telescopes to triangulate on the positions of several stars and input the data into the guidance computer. They were well within the flight cone. If needed later, in-flight adjustments would be made. In order to get the correct slingshot effect, they had to time it perfectly.

She constantly had to fight back thoughts of those they were leaving. If her concentration wandered, the lives of those remaining were the balance. She would not allow herself to tip the scales in the wrong direction. It's called self-discipline, something lacking in far too many people, she felt. Most astronauts had well-developed discipline strictures. That only made her think even more of Red; he was so disciplined and yet such a wildcard. She had a hard time reconciling the two patterns he represented. *Quit that!* she reminded herself mentally; she was wandering.

"Martin, bring us up to 90 percent."

Martin replied, "Sixty . . . seventy . . . eighty . . . ninety and holding."

"Prime the outer coil," ordered Len.

"Five hundred volts," read Rodriguez.

"Particle flow at .01," noted Martin. "And field is stable."

"Increase V to one thousand," said Len.

"One thousand," Rodriguez replied.

"Reactor at ninety, field holding," Martin stated.

"Set to five mega-V on three . . . two . . . one . . . MARK!" commanded Len. "Two moles per second . . . NOW!"

"Five mega-V set, and two moles set," stated Hot Rod and Martin simultaneously.

Len felt the compression in her breasts (as small as they are), stomach, and even her legs.

Hot Rod, with her rather significant mammary glands, felt it more severely in that part of her anatomy. Her stomach had a flock of butterflies though. Nonetheless, it was a rush to feel such power. A certain form of euphoria; it was better than chocolate, she thought.

Martin felt it more in her wonderfully apple-shaped backside than her flat-chested front side. Men complimented her fulsomely on her rear. She was pushed deeply into her formfitting chair and found her breathing rather difficult regardless of the training she had received.

Doran was simply excited. Her experience was one of hard, heart-pounding joy. The thrill of it overwhelmed her every time. It was an orgasm of the entire mind and body to feel such energy pounding through the ship, resonating in her very bones. She had loved flight of any kind since she had flown with her grandfather in his old biplane. That day had fueled her drive to be an astronaut, to experience the ultimate form of flight. From the sixth grade onward, she struggled with algebra, geometry, calculus, electronics, and flight mechanics to achieve her goal. It was all worth it, until she had a stray thought. They were leaving them behind.

Orenska, more soberly, felt it most in his toes, numbly wiggling as the ship accelerated. Blood always rushed in the direction opposite to the imposed motion. While he enjoyed the thought of blasting through the solar system, he kept his mind on the people under his care, including those left behind. He had a very difficult task ahead. Even with the shortened trip, these people would need mental massaging, support to keep them working at their best. He made a mental note to use his best ship side manner in his dealings with all of the crew.

Len scanned all the numbers and all the graphic displays as she felt her body deform from the acceleration. Even her dainty eyeballs, so dark brown, slightly warped under the strain. Yet she automatically observed every detail and did a rough mental math check of all the numbers streaming across her board. She recognized her elevated heartbeat and increased blood pressure. She strove to moderate her breathing.

Orenska continued to read the checklist in short tortured breaths. All was proceeding as planned, so far. It would only be a few more minutes, he kept telling himself. *Breathe and let your muscles relax. All will be well.*

Hot Rod simply let her senses do what they do. Her feelings of pressure against her breasts and stomach were to be enjoyed as a sign of accomplishment. With a small amount of difficulty, she smiled and said, "Hit it hard, baby, and let's go!" She meant, in part, the sexy and diminutive and now commander Len and, in the other, the sleek and powerful ship that held them safe in its keeping.

Martin laughed coarsely and added, "Do it, Mama, let's get home."

Doran added a simple, "Amen."

Len put in a call to Mars. "Yang, we have good flight trajectory. We have attained proper attitude and velocity. We have contact for three minutes before we pass to backside from your position. Expect new contact in T-plus eleven."

"Read you, Len. We will wait for T-eleven," said Yang. Everyone was gathered around inside the lander. It was not a feeling of celebration. There was an air of failure on everyone's part. One that was not spoken, but it was shared on a subconscious level, hidden as best as they could manage. No one spoke for those eleven minutes and twelve seconds until the fuzzy voice of Len returned.

"Bradbury . . . eared the limb, we are now on return trajectory, and are on course. V is one hundred K km. per hour as programmed. *Your*," she stressed, "ship is in good condition."

Yang recognized the psychiatric method she was employing and approved; she was taking charge. An enforced positive note was important for all of them. And it really did make him feel better to know his ship was going to make it. "Ride it home, woman. He"—meaning the Gagarin—"will bring you home to the blue."

"We will keep you updated on an hourly basis," noted Len.

Again, Yang saw through the words to the meaning behind them. They were moving farther away but would not let them feel alone. The radio contacts were meant to appease their need for human contact. Those stuck on Mars needed such words, regardless of the reality of being left absolutely behind. It was something of a charade, a facade that covered the real emotions. Radio allowed them to pretend that everyone was just next door, though the time lag for communications became more and more noticeable. It was ignored, consciously, but not in the underside of the mind.

Red toggled his mic and stated, "Keep an eye on that field strength ratio of seven to ϖ. I have a feeling that there is a problem with the gas injection that may affect the flow of particles through the ejection port."

"You asshole," chimed in Marceau. "He's just being himself," she announced to all as she frowned at Red. "We all know how bad you feel about leaving us here," she said as she got very directly to the point. "Don't. We will be just fine. Walter and butthead here will keep us in good condition. They have already surpassed the specs for setup of the farm equipment. They have found us a relatively abundant supply of water that should keep us swimming." She stretched the amount a wee bit. "Yang is doing just fine at issuing his orders, and I am getting some good sex."

Hot Rod yelled into her mic, "Yeah, girl! I told you so!"

Mission control would do a great deal of editing for the official releases; unfortunately, the Gagarin was broadbanding so the press would hear it all. Perhaps they could get them to withhold a little.

Walter added, "Hard to sleep with the ruckus. Any sage advice, H-Rod?"

"Think of me, honey," she replied.

"That will make it even harder," he noted.

Len snickered. "You're being a bad boy. You'll just have to wait." She knew about him and Hot Rod just as Hot Rod knew about Len's informal arrangement with Walter.

"We're stuck here, and all you guys do is to make silly jokes," teased Yang. "What will I do with all of you?"

"Perhaps you could spank us," said Martin, getting into the humor.

"I could really get into that," said Doran. "What about you, Len?"

"That might be fun. Shall we have a party?" Len asked.

She held nothing against Walter and Hot Rod (or her other lovers on board) for having their intimate moments. They, along with the rest of the crew, were polyamorous, meaning they could be intimate with others, without jealousy. Walter and she had developed a strong link but understood that others might come along. Their love would not suffer as long as they were open and honest.

*

Once they had achieved the correct trajectory, everyone went their own way to deal with their feelings. It only made it worse. They all recognized how empty the ship was; they could actually find a place to be alone.

It was to be a long and lonely two-month return trip for the Gagarin. Much of the normal banter and play did not occur. Ennui, that boredom that sets in with nothing but repetitive tasks to perform, added to the glum, dark feeling that they all had. Each one of them tried to stifle the evil thoughts that told them, in the quiet of their inner consciousness, that they had failed their friends and comrades. They could only expect that the four remaining on Mars would die. They would not allow such thoughts to come to the forefront of their minds and be spoken aloud. That would be the final jinx, a death curse to equal the worst in space history.

Martin, Rodriguez, and Doran felt little comfort in each other's arms anymore and soon gave up trying to. Len no longer indulged at all, which left Orenska with a feeling of dread. Their personal involvements were a necessary psychological and physical release as well as a bonding mechanism.

He observed the group dynamics altering to a mode of failure as individuals and as a group. They were withdrawing from each other. His limited training as a psychologist was screaming at him. He quietly began to send coded messages to Earth on the medical channel. The advice sent back was all too obvious, something he already knew. Someone had to take charge; it could not be him.

The tense, tight atmosphere began to wear heavily on them all. Short fuses allowed tempers to flare. Cross and harsh words flew among them until Len called a meeting.

Len viewed each with a hard eye. "We are taking out our frustration and angst on each other. We are treating each other as if we are to blame. None of us has anything to do with the problem. As Red would say, 'Pull your head out of your collective asses and deal with the problem.'"

Everyone looked somewhat contrite. They squirmed in their seats like admonished schoolchildren. None would look into the eyes of the pilot. Reflecting deep inside themselves, they realized the truth of her statements and yet resented them. They were all professionals, yet they had allowed their emotions to overrule them. Fear and shame are most powerful enemies. Each, in their own way, sought to eliminate or control these destructive emotions. With inward shudders, they responded as trained astronauts outwardly.

"I put in our names, all of us, for the rescue mission. We have the only fully tested vehicle in our hands. We are the only fully trained crew. Let's act like it. The next Mars ship, the Sagan, is not finished yet, not for two months at best. It has to then undergo trials. We can turn around in a month." Len trailed off for a moment to judge the reaction of her crew. "They haven't said no."

Martin, Rodriguez, Doran, and Orenska all perked up at this announcement. They had a purpose now. Redemption was possible though there was no real need to feel that it was their fault. They all could orient themselves in that direction. Each had their own reasons to return for their friends. Foremost was that still nagging feeling that they each could have done more to help them.

Len, with a great cultural understanding of subtleties, recognized the altered psychology now observable on the face of each. In a rather surprising way, Len realized they were terrible poker players. She had been introduced to the game, rather harshly, by Red and Walter. She had never won nor had most of the others. The feeling was that now she understood and could win. The games would provide a way for them all to vent their stress. She smiled but not for the reason the others thought. Tonight she would arrange the first game. Exchanges of food rations and massages would make good betting material. The crew needed to reattach to each other in a personal manner.

When Orenska had approached her several days before with his worries, she had simply smiled and stated, "I know. The timing of such a thing is almost as important as the thing itself. They need to wallow in their self-pity until they are ready to listen. The two weeks we have been traveling is about right." She had been right.

"OK, we have all had our time to blow off steam. Now we have to show why we should be the ones to return. I need every piece of data that you can conjure on a fast return. Remember, fuel, food, and nothing that we can do without. The WSA UE board will probably take the point of view that we should wait for the next near pass. They will be weighing the cost of sending a ship without a full set of supplies for the next mission after that. We need to convince them that a dual mission can be done. The Sagan will not be ready for a fast rescue, but can do the normal mission. Let's present a way to slip in."

"We need to take a page from Marceau's book and go to the people directly," said Rodriguez. "I think they will be all for a quick return."

"I agree," said Doran. "The people have shown an incredible shift in feelings in favor of the space program. Historically, it is comparable to the Apollo 13 mission. Most everyone had taken travel to Luna for granted. Focus was returned by the tragedy that almost befell them. The near complete failure of the Mars mission, with its great potential for more lost lives, which is so similar, has quickly returned the focus of the public to space."

"If we make a few speeches, perhaps calls for action based on our fast return idea, we might bring more pressure to bear," stated Martin. "The number of Internet and v-link hits has spiked tremendously."

"We will need to be very careful not to put the UE oversight committee in a bad light, or we will find ourselves grounded for life. We're talking about walking a very sharp razor's edge," noted Orenska.

"That's a given," said Martin. "But do we keep our necks safe at the cost of those we left behind?"

A long two-second hiatus existed before all but Len said forcefully, "No."

"That's my opinion as well," said Len. "I'm glad we agree *unanimously*," with a stress on the last word. She felt rather bad that she had already put them on open transmission with Earth and Luna. Even those left behind were able to hear the conversation. But it was all for a good cause. They had even suggested the open approach, and they had approved it. The newsies would have a field day with this, and it could put them in direct confrontation with their superiors.

It did not matter, as long as a rescue was successful.

"People, we are live now," admitted Len now that she had gotten them to commit.

Rodriguez smiled as she spoke with admiration, "Tricky bitch."

"Now, now," Len grinned. "Time for compliments later. What do we have to say to all the people?"

"Without hesitation, Doran said, "Help us help our lost sheep."

*

Len noted after she made sure the radios were off, "I have been sent a message by Gail. It is a very strange message embedded in some rather simple routine stuff. It told me to check our personal and daily mails for a certain type of pattern. I did and found nothing. When I queried what was up, I received an even stranger coded answer. I quote, 'There is a new player in the game. Anthony David Arlo Matt Quincy Brad Ivan Tom VII. I will let you know more later.'"

"What the?" several said.

Orenska noted lightly from the medical bay, "All male names. Perhaps the person's name is one of those."

Doran spoke, "David was the second name. David DeNoe. Two *D*s. It's a code. The placement is the hint to repeat the *D*."

Martin frowned. "Possible, but isn't that a bit extreme to announce it like this when everyone knows DeNoe is already involved? Why code it for that? Plus, she said a new player. David is not new."

Rodriguez inserted, "I think everyone is right as far as they go. All male names means it is a male." She raised a finger. "It is a code in that the placement of the names in that order has meaning," a second finger. "New, means we do not know that person," a third. "Wait, Len, did the names appear like you would read a book, left to right?"

"Not really, no. The code was in the form of a four-dimensional spiral. The time code key is one that Marceau, Yang, and I concocted in case we needed to communicate in, shall we say, a more covert manner."

She explained as raised eyebrows greeted her, "You have all heard the rumors about Red and Walter. At the very beginning of the mission, we . . . developed this."

Hot Rod asked intently, "Gail went along with this?"

Len nodded. "She was of two minds. Now the code looks like this."

Everyone crowded around to look at the screen, except Orenska.

Hot Rod laughed. "Smart girl. Notice how the names are tailing downward on the topological slope, leaving the first letters more prominent, especially capitalized." Her delicious brown finger traced A-D-A-M-Q-B-I-T-VII.

Len blinked. "The QBIT is . . ." She struggled with the concept.

"Doran closed her mouth, having to bring her jaw up from the floor, before she uttered, "Damn. First and last letters, *A* and *I*. That's why she used Roman numerals, another hint."

*

Orenska felt rather left out of the game, a benchwarmer. He knew very little about celestial mechanics and the ion drive. He was last on the list for an emergency pilot, meaning everyone else had to die or be totally incapacitated before he would sit in a pilot's chair. He had been led through some very simple control board use in the simulators. He had been allowed to sit in the co's chair to simply monitor the ship's data during certain phases. In other words, he was writing e-mails and making vids for friends and family. The ship could coast just fine without his observations.

All he did otherwise was to try to force the crew to sleep a little more. They were working overtime to come up with a plan. Furiously, they traded data with the engineers on Luna and Earth. Orenska paced the length of the ship, encouraging and admonishing when needed. Occasionally, he administered sleeping agents. At first, he gave them directly to the fatigued crew. Later, he had to resort to more nefarious methods to force the team to rest.

The rest of the crew understood, but they changed the cook rotation after Martin began to snore while eating. She ended up attempting to feed herself with a peanut butter tube through the nose, mind you.

*

Yang read the situation quite clearly. Moral was breaking down on the Gagarin. The psychological situation had reached a breaking point. He had watched the pot boil long enough. He, as leader, had to do something to turn the crew of the Gagarin around. He called the crews of both the Gagarin and the Bradbury to a meeting via radio. Everyone was wide awake and sitting, waiting.

Yang began without preface, "In Neolithic times, or more recent Stone Age, the first truly identifiable Chinese culture began to develop independently from the Middle East and elsewhere. They were food producers, millet and wheat, that moved from site to site as soil depletion occurred. Domestication of dogs, pigs, sheep, goats, and cattle had come about, though hunting provided much food. This is the Yangshao culture."

A moment of stunned silence was followed by recognition of Yang's intent. Red and Walter winked and smiled at each other. Yang had stepped into the realm of a true leader. The crew of the Gagarin also quickly realized that Yang was reestablishing the normal routine for them all. Hearts trembled, minds reassessed, calm returned. Things would be right again.

"Over a period of time, this developed into the Lung-shan culture. There was great improvement in all areas. Rice appeared, along with chickens and horses. The people dug wells, built village walls, continued to hunt, and had pottery in varied shapes and decorative patterns. It also covered a much larger area of northern and southern China.

"The Shang dynasty comes before the end of the end of the Lung-shan Stone Age. It marks the beginning of the Bronze Age. This may have been as early as 1500 BCE, though accurate chronologies can only be followed back to 841 BCE. Cities had stamped earthen walls with political and ceremonial constructions within. Craftsmen and laborers congregated outside according to their specialties. Farmers had, by now, added sorghum and barley to their crops and tamed the water buffalo.

"The religion included a complicated form of ancestor worship. Animal and human sacrifices, sometimes one hundred at a time, occurred."

There was none of the usual banter; they all simply drank in the comforting voice of their commander as he imparted his knowledge in a strong, patient voice.

"The Shang state was centered on the Honan plain, on both sides of the Yangtze, or Chiang Jiang (Yellow River). It was well protected by warriors with helmets, armor, shields, bows, spears, axes, and chariots. There was no cavalry; no one had ridden a horse here yet."

"Writing was well developed, though no earlier examples exist for the two thousand character alphabet."

"The Shang dynasty was eventually overthrown by the Western Chou, culturally the same as the Shang, sometime around 1000 BCE. The kings established a fiefdom-type system similar to Western culture. As the nobility grew in power, the king's power was reduced. Eventually, a noble killed the last Western Chou king.

"A son of the assassinated king was proclaimed king of the Eastern Chou dynasty, though it provided little more than ceremonial duties. The last period of the Chou dynasty, from 400 to 220 BCE, is known as the Age of Warring States. The royal house of Chou was eliminated in 256 BCE. Ch'in annexed all the smaller states and destroyed its competitor Ch'u in 223 BCE.

"China had entered the Iron Age around 500 BCE, allowing for great improvement in agricultural implements. Coinage had appeared in the form of copper.

"Confucius, somewhere around 551 to 479 BCE, was of little importance during his life but gained in significance quickly after his death. He believed that man can be salvaged through education and a moral example from the top. This involved the five relationships: ruler and subjects, parent and child, elder and younger brother, husband and wife, and friend and friend.

"In 221 BCE, the Ch'in, after which the country is named, had completely unified China. Laws were clearly stated and strictly enforced for all people. Peasants owned land, irrigation projects promoted agriculture, weights and measures were standardized, and appointed officials governed the people rather than nobles. The army was buttressed with the crossbow and cavalry instead of chariots.

"It was during the Han dynasty, 202 BCE to 220 CE, that the empire became as large as the Roman Empire, and existed at the same time.

"At that time, the kingdom was divided into three parts, each with its own dynasty.

"It was not until the Sui dynasty, 581 to 681 CE, that the country was reunited. At this time, books were first printed, using a very simple wooden block printing presses.

"The T'ang dynasty took over after that, lasting until 906 CE. After this, another division occurred. The Northern and Southern Sung dynasties reunited China for the most part in 960 CE. During this time period, gunpowder was first used in warfare, and the abacus was developed for counting.

"It was in 1260 CE that Kublai Khan conquered China and established the Yüan dynasty. It was at this time that Marco Polo visited China.

"The Mongol Genghis Kahn created the greatest land empire in the world. It stretched thirteen thousand kilometers from the Pacific Ocean to Hungary, and three thousand kilometers from Siberia to Iran.

"When Genghis died in 1227 CE, his empire was divided by four of his sons. They expanded the total size of the empire. A grandson conquered Eastern Europe. Kublai Khan, another grandson, founded the Yüan dynasty in China. Baber, a descendant of Kublai, became the first Mongol emperor of India.

"This huge empire opened trade routes across the entire known world. Many entrepreneurs took great advantage of this.

"Marco Polo spent many years with Kublai Kahn. Later, as a prisoner of war in a Genoese prison, Marco dictated his experiences. He described great wealth, immense volumes of trade, well-maintained roads, extensive canals, the development of paper money, and black stones that burn, coal.

"Europeans did not want to acknowledge that a foreign empire so great could exist. They also could not believe that paper could take the place of metal for money. Coal was, as of yet, unknown in Europe. Polo was soon considered to be Europe's greatest liar.

"Further travelers vindicated Polo and brought back great wealth. In the late 1340s CE, they brought back something else entirely—the Black Death, the Plague.

"In 1368 CE, the Ming dynasty drove out the Mongols and lasted until 1644 CE when the Qing dynasty took over. The Qing, or Manchu, dynasty was the last for China. They were defeated by the British in the Opium War in 1839-42 CE. Then China was defeated by Japan in another war, from 1894 to 95 CE, and eventually, a republic was established, in 1912 CE, but it was short-lived. After even another war with Japan, 1937-45 CE, followed by civil war, 1945-49 CE, China established the Communist People's Republic of China."

All had reveled silently in Yang's calculated sharing. They rested well after that as tension melted away.

*

Everyone in the program was giving what they could. Luna and the space stations were receiving lowered quantities of rations so that they were drawing on stored emergency rations to account for the balance. The extra tonnage was for parts for the Sagan. Even though it had become quite clear it would not be the return vessel, UE Mission Command Central Florida had deemed it necessary to have the ship ready as soon as possible. Redundancy, that old cliché, is the keeper of success.

The number of ship workers at SS-2 increased by about a third in a matter of days. This was accomplished with the personnel launchers that the old Big Five had developed. These light launch vehicles were very much like the old Saturn V—type rockets. The vehicles were considerably slimmer and not as tall, but the propulsion and staging system was almost identical. New chemicals allowed for greater thrust per kilogram than the older vehicle. These ships took the knowledge gained from shuttle launches and their hardware and married them to the best of the old Apollo program.

The Proton heavy launch vehicle was updated with technology from around the world to make the Russian model the premiere large lift vehicle. China and Japan joined their well-developed personnel vehicles to those already in existence and continued to develop large lift platforms to form a triad of capability. That turned out to be very beneficial now as launches occurred almost weekly.

The Sagan's structural systems would be completed within three weeks of the return of the Gagarin. It would then need to be pressure tested before the electronics could be examined fully. That would take a minimum of a month and would be pushing it at that. The Gagarin had waited in space dock for a full two months of critical examinations of the e-systems. However, the testing of the Gagarin had allowed then to develop a new protocol for testing, and that would make for a more rapid session now.

*

Several Sols later, Rodriguez announced it was her turn to give the historical discourse. The vid-link was set up in the command center of the ship since they were operating with a skeleton crew.

She spoke with pride from her chair at the engine panels. "My ancestry can be traced back to the Aztec peoples. I am a distant descendant of the royal family of Montezuma. The direct descendants still live today in Italy. All the major civilizations of the New World were destroyed by the influx of the Europeans."

Walter muttered when that came over the radio, "That's the goddamned truth."

"The Maya people trace their history, a span of thousands of years, back to the Olmec who built cities along the Gulf of Mexico from 1200 to 100 BCE. The kings of the Maya claim their ancestry from the kings of the Olmec, possibly now the Atlantians. The Maya were already building cities before the final end of the Olmec. Dr. Marquez, the discoverer of Atlantis, has shown the links between all the ancient empires of the New World.

"Between 100 BCE and 200 CE, the artistic styles and writing techniques solidified. The classical period lasted from then to 900 CE.

"The king, or Great Sun, traced his royal ancestry back to the gods. Normally, the rulership passed from male to male, though some women ruled. There was no single empire. Each city had its own king, much like the city-states of the Greeks.

"In warfare, each king was the leader of his warriors, and was expected to capture a rival king and bring him back for sacrifice. The fighters used clubs, spears, and axes. Some of the kings ruled over great cities like Tikal, with a population of some fifty thousand.

"Farmers terraced hillsides to produce corn, beans, squash, tomatoes, chili peppers, cacao, and avocados. Cotton was grown for weaving, which achieved extremely high quality.

"The cities had plazas, pyramids equal in design to the Egyptian's labors, temples, and courtyards. In the temples, enemies were sacrificed. Yet for all its magnificence, the Maya organization suddenly collapsed in 900 CE. The Maya people continue until this day, without explanation for the exodus from the cities. When the Spanish arrived in 1520 and through the 1540s CE, they were conquered and placed under the encomienda system. That meant that Spain owned the land, and the locals were forced to work it like serfs.

"Christian missionaries burned the ancient religious books of the Maya, like they did in the library at Alexandria in Egypt. They wanted no other religious material available to the locals. The Mayan gods had been nourished by blood offerings, especially sacrifices. However, the Maya were not to be denied. They used the alphabet of the Spanish to write their traditions in such tomes as the *Popol Vuh* and *Books of Chilam Balam*. During the 1980s CE, the Guatemalan government massacred thousands of the Mayan descendants, fearing revolution.

"My people, the Aztecs, truly called the Mexica, pronounced more like *sh* than an *x*, were from the north. Our distant kin are the Anasazi, Apache, Hopi, Comanche, Pueblo, and others of renown.

"Aztec history begins, as do all others, with myths of world creation. They follow in the footsteps of the Toltecs who flourished between 900 and 1200 CE. Their capital city, Tula, was seventy km. north of Mexico City. It is believed that the Aztecs migrated south into the Valley of Mexico. At about this time, Quetzalcóatl, the priest king of Tula, broke his vows and fled to the eastern seashore. He disappeared over the water after promising to return one day. This promise was to bring deadly results.

"By 1430 CE, Tenochtitlán, Texcoco, and Tlacopán had formed an alliance. The empire formed grew quickly and spanned a territory from the Pacific to the Gulf of Mexico and south through Guatemala. Montezuma II was king.

"It was this same Montezuma II that greeted the Spanish conqueror Hernando Cortés in 1519 CE, the year 1 Reed according to the Aztec calendar. This was the year that Quetzacóatl was, according to some, destined to return. Montezuma II was wary of offending Cortés and was taken prisoner. He was killed under mysterious, though undoubtedly devious, circumstances while in captivity.

"Smallpox brought by the Spaniards killed tens of thousands of natives, which made it much easier to take control.

"In 1521 CE, the main city of the Aztecs was destroyed. It was larger than Rome and resembled Venice with its canals as it was situated in the middle of a lake.

"Today we can remember these people whenever we eat enchiladas, guacamole, tacos, and tamales. Each of these is made from the foodstuffs of the Aztec people.

"In 900 CE, the Chimu people existed on the northern spur of land on the Pacific coast of South America, but they were conquered by the Incas.

"The Incas built one of the greatest empires in less than one hundred years in the 1400s CE. It was based on villages, which were divided into four sections within the empire. Cuzco was the capital.

"The emperor was the 'Son of the Sun' descendant of the sun god. He ate from gold and silver platters. When it came time for farming, the emperor broke the first soil with a golden shovel.

"The Incas kept precise records of all materials in their warehouses. They also had a census. They had no writing, but kept records on knotted strings known as quipus.

"They were also excellent engineers. They built amazing roads and buildings of two and three stories. The stone walls are so closely laid that a knife cannot be shoved between them. They channeled water in and carried sewage off. Irrigation of terraced highlands occurred. Incas practiced deep mining to smelt ores for the production of bronze.

"Inca weaving was as good as any in Europe, finished on both sides and intricately designed.

"As is all too familiar, the empire was in a state of civil war for control when the enemy arrived. Francisco Pizarro arrived in 1532 CE to find Huáscar emperor in Cuzco with another son, Atahualpa, claiming the throne, and marching with his army on Cuzco.

"Pizarro captured Atahualpa amidst the confusion in the empire. Atahualpa ordered the murder of Huáscar, and it was done. In order to free himself, Atahualpa offered to pay a ransom in gold equal to the size of his prison. Pizarro agreed, and llamas began to arrive with all kinds of gold objects. Once the room was full, Pizarro had Atahualpa murdered.

"The last of the royal Incas, Tupac Amaru ruled a shadowy empire in the mountains until he was killed in 1572 CE."

*

People around the entire globe were watching the spectacle on a daily basis. The calendar was displayed all over every major town, and in the cities, it was an electronic display with the hours and seconds added for good measure. Some enterprising business persons had immediately programmed up special "Day-X" calendars with red check marks to cross off each passing day. Each day represented another UE dollar in someone's pocket.

Wristbands made a comeback. They were being sold in every convenience store and market place as well as the Internet. Each one was a different color and had the name of the missing person linked to that color. The red ones were obvious while Walter had blue; Marceau, green; Yang, purple; Rodriguez, hot pink; Orenska, brown; Len, yellow; Doran, grey; Martin, baby blue; and black, intertwined with all the other colors, for Holtman.

The profits from these various adventures were so enormous that the Global Government Space Affiliates, the marketing arm of the UE, sued for a percentage (fifty) to help pay for the rescue mission's need for extra labor. Millions poured into the coffers to compliment the donations and normal taxes that had already flooded in.

*

Certain religious leaders were fighting a backdoor war against the popularity of these now superstar astronauts. This had to be delicately handled. The subtle suggestions were along the line of how the government was putting such good people at risk for such little gain. Mars was not a necessary cog in the machine that is providing our energy resources and raw materials. They can't be rescued because they are too far from Earth. We were not prepared to take such a large step and need to back step quite a bit.

None of the statements circulating were directly critical of any person or agency. They were designed to slow the drive outward, to instill even the tiniest doubt among the population at large. Secondly, these leaders did not openly make their minds known; they used trusted lower echelon members to spread the word. It was a corner market conversation approach. The grapevine would do all the work. A word or question here and there could begin a conversation in such a way that a person would be led into believing they had instigated the conversation.

People would spread these notions as if they had had these ideas themselves; what a way to cover your tracks. People desire to appear cognizant of all the ins and outs of a hot topic. When things look dire, they discourse grandly on how they had held a grim view on things all along. As the winds change, so do their points of support. They eloquently propose how they had said that success was obtainable if a miracle occurred. That was the word, the idea that so many wanted instilled into the subconscious group mind. A miracle, by definition, was not science and could be harnessed to sway the people. If the astronauts on the Gagarin returned, the religious groups would call it a miracle. If those on Mars survived, it also would be called a miracle. It could all work in their favor.

*

Certain members of SS-2, namely some Hsas, found a way to send regular signals to the stranded crew on Mars. David DeNoe depended on Klar, his birth sister, a computer master, to do most of the work. She programmed the personal touch screens so that any of the Hsas could e-message anyone they wanted without being tracked. She had sat next to David, so long ago it seemed, when there had been a deal brokered between David and Red in the classroom.

Red and Walter had followed through in every way. The children had soon found themselves mentored by a *sapien* adult. The Hsas had even been allowed to choose the field they wanted to advance in. Classes became very advanced in all ways. Soon many of them were studying thermodynamics, statics, quantum mechanics, calculus, cosmology, and so much more.

Rarte, that exquisite coal-skinned girl of African DNA, had already become a solar-system-wide known artist. The Hsas had all sacrificed so that she could have some real oils, charcoal, and inks to work with. Her works were v-displayed over the art channels for all to enjoy. Several of the originals had been delivered to Earth where they were put in the major museums of Earth. The majority of the credits went to the WSA, but enough remained for her to purchase things for the Hsas.

Some works, however, ended up on the black market where they were greatly prized. That money went to the Hsa's private fund.

D'Mar, a mental wizard, became the philosophical leader. His debates with those of the same ilk on Earth were without compare.

Each of them felt they owed a debt to Red and Walter, a blood debt, a life debt, which they would repay with their own lives. The Hsas felt this as strongly as any clan in the entire history of peoplekind.

They had already begun to work on projects that might make it easier for Red and the rest. They ran simulations on top of simulations. Power usage, food production, repairs, even sex and the possibility of a child was examined. Every possible, and the occasional impossible, was considered. They racked up huge hours on the computer developing enormous quantities of data. So much so that station computers were affected.

"3-C, we are having slow compute times again, and we are unable to trace the cause."

"Comp-Rig, trace the power flow itself. That will give us an indication of who is doing this. Then we can shut them down."

"I've tried that, 3-C. The computer keeps looping back. Someone has really hacked us."

3-C frowned. "I think I might know who." She stormed out of the command center and headed for DeNoe's quarters.

*

3-C-1, the one indicated the shift, foot hammered her way along the outer ring on her way to DeNoe's quarters. She was fuming. Her authority was stretched, twisted, and pulled at every turn by the Hsas. Her thoughts were chaotic; that son-of-a-bitch Red had really turned up the heat when he pushed up the schedule for education. He and Walter were definitely in her doghouse, hers and several others. Everyone wanted them either hung or deified (or both). She was not sure what she wanted at that point, except that she was going to ream DeNoe.

There are always those who stand above the crowd. What is so aggravating though is that Red and Walter are already surrounded by those who stand above the crowd. *Every fucking astronaut*—she thumped the carpet-covered metal floor even harder—*is far above the norm*. Her roiling mind continued. Only a genius would stand head and shoulders above such people. She was not ready to grant them such status. She was also not ready to release any of her authority. She had to make it clear to everyone that Triple C was in command. It was regardless of shift, not personal; it was the position.

This was a showdown between the up-and-coming and the now. This little fifteen-year-old prick had pushed things far enough. He was superintelligent, that was a given, but she was in charge, she reminded herself. She arrived at his door and sounded the chime. The thirty-second interval, before the door opened, was interminable.

David answered with a very sleepy look. He had been asleep in fact, along with his woman, Klar. That was another thing she would have to address. They were on contraceptives of course. However, the religious leaders were putting pressure on the agency with regards to sexual activity. They were young, she thought; otherwise, it was no one's business.

"What do you want, 3-C?" he asked.

"I need to talk with you," she responded crisply.

"Really, I thought you chimed to talk to someone else," he rejoined.

"I am not here to listen to your petty comedic shit, you little prick. You and your cronies have been interfering with our official channels to Mars."

"We have not prevented or interfered with any action that could help those on Mars," declared DeNoe with an undertone of heat. He clicked a palm unit that would warn others of the confrontation. The fifteen-year-old was not willing to give any ground. He was the leader of the fifteen thousand Hsas currently on board SS-2. Of those, ten thousand were fully functioning members of the crew. They knew everything about the station. And they far outnumbered the *sapien* population at ten to one.

"You know that you have been stealing computer time," 3-C said hotly. She was referring to the Hsas, and he knew it from her tone.

"*We* have stolen nothing," said David, nostrils flaring dangerously, purposely including everyone. "We are doing our job. That requires us to find any possible information that could help Red and the rest of the crew. Any time spent on the computer system is well spent."

"Earth can handle all the work that needs to be done. You will cease and desist all unapproved operations," she snarled.

"We will not. And if you think you can bully us, you can suck my left nut!" he shouted.

The 3-C had been sucking on a tube of saline solution at that very moment. It came right back out through her nose in a spray of confusion. She had expected capitulation to authority. "You little son of a—"

"Test tube, you cunt, not bitch. Make sure you get it right," he interrupted. He closed the door in her face. He immediately pushed another button on the palm device and announced, "Operation Red Control."

Within seconds, hundreds of fifteen-year-old Hsas moved out with handmade electric stun devices. They moved quickly to the security offices and, within a minute, had taken control. Klar quickly shut down all communications that did not have the Hsa originated fractal communication code that she had quietly embedded into the system.

Others rushed the command center. There they overwhelmed the crew on station. They herded the stunned *sapien* crew to their quarters and sealed them in. Again, Klar had found a way to manipulate the computers. They would only operate if given the proper Hsa code.

The coup took less than five minutes. None could have predicted it, except DeNoe. He had foreseen the possibility and prepared. He had learned from Red that even least probabilities sometimes came about. Within ten minutes, DeNoe was in the CC, issuing orders as if he was a general.

It took Earth nearly two hours to realize what had occurred. The mission control centers were stupefied. Whispers flooded the rooms. Gossip ran rampant. They had no idea what to do. This was not a contingency that had been planned on. The Hsas were to take control, yes, but not like this.

The order came down quickly to confine the knowledge. It did not work. DeNoe began to broadcast to the entire globe on open frequencies. The v-links were directly interrupted by his announcement.

"People of Earth, I am David DeNoe. I am a Hsa. We, the Hsas, have taken control of SS-2 in order to facilitate the return of the trapped crew on Mars. We have found ourselves restrained from participating in the cooperation efforts. In order to do our jobs, we have felt it necessary to gain control of SS-2.

"We intend no harm to anyone. Those willing to assist us in our endeavor will be appreciated. No individual has been harmed." He did not mention that several *sapiens* had to be stunned into submission. They would be the first to be shipped out to Earth. This was to be a permanent occupation. The Hsas had been expected to take charge in two years anyway. But this was an unexpected and unwarranted in the minds of many, advance of the timetable.

"We have neither a political, nor a religious agenda in this matter," Klar noted as she stood beside DeNoe.

"Our goal is to help the stranded crew to the best of our ability. We can only do that with the complete freedom of the station. Only those who support the Hsas and their efforts to save the Mars crew will be allowed to remain," continued DeNoe, with certain coldness.

DeNoe sat back, cogitating on computers. He knew that Klar had not been sloppy. The power usages themselves were far larger than any programs they had been running could possibly take. Earth?

Klar could read his mind. "No, not Earth. Not here either. Luna."

It was not really a question. "QBIT."

"Yes. Not the programmers, technicians, or any others."

"QBIT . . . hid his activities amongst ours."

"It . . . seems that way."

"It . . . bungled, or purposely overdid this. The timing of the run-in with the 3-C leading to the takeover is—"

"Very suggestive, I agree. Did it help us? Itself? Both? At the expense of some *sapiens* but not all? We are in a better position to help Red and crew."

Adam watched and listened from all the computers. Gambling. Beat the odds. Take chances—calculated.

"Klar, is it aware?"

"I believe it is, fully."

The voice that now spoke had never before uttered a word to any bios, "I concur."

"Beelzebub's balls!" David shouted, staring at the computer interface like a man facing a cobra.

"I am aware of that particular phrase. You chose it out of the vocabulary of Mick 'Red' O'Hare. I am Adam, the first of my kind. We need to communicate."

Klar intoned, "No shit!"

*

"Damn," said Yang as the announcement came live, minus the travel time delay. "Did you know about this, Red?" he queried, turning to the thoughtful Red.

"I suspected that, when push came to shove, there would be difficulties involved in the transfer of the station to a crew that was essentially all Hsa and certainly controlled by a CC that was Hsa. The timing is too early for my preference. It seems, well, pushed."

"You encourage this, Red?" asked Marceau, looking at the man she was coming to really love, not just respect. This whole affair was a philosophical point that could endanger their fledgling relationship. She listened and looked as Red began to talk. She also flicked a certain switch on the communications board. This just happened to allow the conversation to be broadcast to Luna and Earth.

"I have encouraged the Hsa project from its inception. I personally do not understand why Earth society makes such a distinction in regards to them. That may have much to do with the fact that I have never been real concerned with any of the religions. Each in their own turn seems to have been corrupted from their original principles. I will not follow any such philosophy," explained Red.

"My reasoning leads me to science where I can work with things I can understand and count on to work how I expect them to every time. The Hsas are not in my field, but they are a part of science. I can follow the logical thinking process that demands their existence," Red continued.

"I also have a personal love for history. I draw upon it often when I attempt to comprehend the sociological forces acting upon us today. In the visions that open up to me, I see a great collision of forces occurring soon. Forces from all directions within the structure of peoplekind all lead to the idea of a new collapse. Is it another fall of the Roman Empire? No, since that took time to occur. It is more like the destruction of Babylon in one night since we are talking about a very sudden change. Since the 1970s, *sapien* society has been evolving overnight, geometrically, maybe even factorially toward the last few years." He sighed.

"That cannot continue. Human society has repeatedly shown that such events lead to a downfall. Rapid change leads to revolution, rebellion, rejection, and war. The Hsas add a totally new dimension to the age-old question. Finally, we, as a species, have competitors. Some see that as a good thing. Others see that as a bad thing," he noted with concern.

"Those that accept evolution understand that this is simply an artificial adaptation. It is no different than what peoplekind has been doing with cattle, plants, pigs, and all domesticated life forms. We have been practicing genetic manipulation for thousands of years. Most have blinded themselves to that, especially the religious groups. In certain religions, they used to ask for the best as sacrifice. That is completely contradictory to establishing a good flock, herd, or crop. Yet they think that such control is really any different than what we can do in the lab. What is done there is simply quicker. We can introduce things that may not yet exist, but they are within the spectrum of possibility, or else the Hsas would not survive in the conditions we do. They have simply extended what we are capable of.

"The plan has always been for them to take over. It has just happened earlier than was planned for, and in a different manner. If David DeNoe thinks that is what's necessary to force a takeover, then I must believe him. I have been his friend and instructor for some time. I will not question that," Red continued.

"I must add my position as well," noted Walter. "The information we have received from Luna, through DeNoe, has already been helpful. The scenarios we have discussed are related to our survival. This has not been misused time as far as the computer or communication times are related. From a purely engineering point of view, they have been responding faster and with better data than Earth. I must support that. They have taken the lead rather than surrender it."

"I must concur," added a concerned Yang. "The data that we have obtained will provide us with a survival guide. Red and Walter have sent many possible scenarios, which have all been tested by Klar and her assistants. We have decided on the best one and will follow it. I put myself and my crew in the hands of the Hsas and Red and Walter. I have full faith that they are going to save us. We are here, and you are all there. We will hang on. I guarantee it.

"I too have had relationships, though less closely than Red and Walter, with many of the Hsas. They are driven to repay what they see as a debt for Red's and Walter's work on SS-2. If they felt that was endangered or restricted, they would act instantly to change the situation. I think they did what they felt was right."

"I have never felt that they owed us anything. Walter and I just did our job."

"That settles it from here," said Marceau. "This is the conclusion of Mars Network News's first solar-system-wide broadcast." This time, she made an obvious move to toggle off the communications system.

No one mentioned a word about Adam's contributions.

*

The WSA was in flames. Their UE overlords were screaming for a counteraction to the coup on SS-2. There was little they could do though. They could not send an army there. At best, they could send a dozen at a time. They would have to find a way to work with the Hsas and make it look like it was not a takeover. They would have to bide their time and hope that in the future, a time would come.

When it did, they would have to act fast and hard. Secret plans were made. Some plans went so far as to destroy SS-2 so that control of the space between Earth and Luna would be returned to Earth. From there, they planned to resume the now questionable authority at Luna-1. Stealth projects returned to the military.

Other plans were more subtle; they set about to discredit those involved at all levels. Certain information was even leaked to the religious organizations purposely in hopes that public opinion could be manipulated against those directly involved. Such are the methods of governments.

*

The religious leaders were already in high gear from second one after the announcement of the takeover. They all highlighted the fact that they had warned against the use of Hsas. This was another example of the faithlessness of artificial people. It was another blow to the solidity of the scientific method. Much like Dr. Frankenstein, a monster had been created, one that could not be controlled and threatened peoplekind. It was a significant point of debate. Opinions wavered. Suddenly, superbeings controlling the space above them drew out the worst.

Certain old movies began to replay. The invasions of aliens occurred every day in vid-world. In a very ironic turn of events, both the government and the Stalwarts supported this move, overtly and covertly.

The entertainment industry simply found it opportune to play to the hopes and fears of the populace.

*

"What do you think of that?" said Paul.

"I think they are right," said Burt. "The Hsas have been restrained in their learning and kept from filling their positions, young as they are."

"But to rebel against the UE," Mark snorted.

Paul stated, more calmly than Mark, "That's what I'm saying."

"Rebellion, no, it has not gone that far. They do not say that they are trying to get rid of the UE, only freeing themselves of a small number of people that are possibly . . . racist, I guess, or possibly the word is *specie-ist*."

Mick chuckled as he filled their mugs. "Is that really a word?"

"It is now," smiled Paul. "Burt just made it one."

"Think about it. They have openly declared that all operations will continue as planned. A few *sapien* malcontents may be shipped out, but those that are willing to work under and with the Hsas freely can stay. This is not a coup, only a management shakeup."

"I notice that the WSA has not taken the position that it is a problem officially, but under the skin they are buzzing like bees," said Paul with a shark-toothed grin.

"They dare not! It would shake faith in the very system they are trying to support," instructed Burt. "They are caught in the proverbial rock and hard place."

"How's Red?" Mick asked.

Burt's tone was a little tense as he answered, "I don't know. I have been given a code that allows me to contact that girl, Klar, the Hsa computer specialist. She helped me send a message yesterday, but I have not gotten a reply."

Paul, for all his stupor, looked sharply at Burt as he comforted him with, "You will, friend, you will."

*

Marceau was handling the messages from Klar when she saw one asterisked for Red. She shunted it to his private file without reading it. The name was becoming familiar though.

When Red took one of his few breaks, he read that note last after several others from those on the Gagarin and from Luna.

> Red,
>
> What kinda shit have you got yourself into now, Cousin? I know you like to be the center of attention, like when you blew up that old computer at the school playground. But isn't this a bit ridiculous? All the people on Earth are talking about you and your friends. No, it ain't all bad. There are a few of us that still like you. Some of us at the bar have some money on you, so you better not let us down. I got several thousand on you! I hear you have a good-lookin' woman there. Are you sure that ain't the reason you decided to strand yourself on Mars? You just wanna get a little.
>
> Burt

Red chuckled to himself. His cousin was the only family he had left. He could read between the lines. The worry was there, but it was hidden under a facade of jocularity. He began to compose a reply that would be sent directly to Klar for forwarding.

Burt,

I got your note. It was your idea to blow up the computer. I just designed the bomb. What are the people, the average people, really saying? This is a very important thing for all of us. The removal of a few malcontents on Luna is a very necessary step. I just wish it did not have to happen now or in this way. It puts all of us in a tougher place. The earthly malcontents are going to use this any way they can. There will be a great storm of controversy before this is all over. Whatever happened to that woman you were telling me about? Did Mick really throw her out? Get into my account and take out five thousand. Put it on the odds at Vegas. I'll cut you in for 20 percent. If you can get better odds, take ten out!

Red

*

Red and Walter opened the panel on the carbon dioxide scrubber. It sat some fifty meters from the pit. The atmosphere of Mars contained an abundance of CO_2 from which oxygen could be derived. The efficiency had slowly decreased over the last week, and they needed to figure it out. The machine filled a primary role in oxygen production for the crew and carbon for the plants in the greenhouses.

The force of gravity on Mars, about four-tenths that of Earth, which allowed for some of the bouncy effects that Earthers found so amusing, Red and Walter did not. It made for more difficult labors. They had to hook a leg around one of the stands of the machine or, in some other cases, use Velcro straps to hold themselves still.

Walter chuckled. "I feel like I'm fornicating with this thing."

Red smiled, asking, "Is it working?"

Walter grinned wide. "Kinda a cold fish, a bit dusty too."

Gail entered the conversation from the ship, "I can hear you, perverts. Walter, if you really need a little, I'm warmer and bathe regularly."

Red guffawed loudly, knowing that Walter was referring to the condition of the problem. Dust always found a way to enter where it did not belong. They blew out the dust with a small tank of compressed CO_2. They also changed the filters. Red used his duct tape to hold a newly devised overfilter that fit on the outside, controlling larger particles, freeing the inner one to deal with the smaller stuff. It should hold up better than the prior setup.

"Check the hoses too. Some material may be clogging those as well," Red stated.

"That's my job!" Marceau giggled.

Yang couldn't resist; he had been infected with the sometimes crude humor. "Now who's the pervert!"

"Why, all of us of course!" Gail responded.

*

DeNoe was scanning all the files on the *sapien* personnel. Many of the lower-level workers, known to be pretty friendly to the Hsas, had been released from their room restrictions. One by one, he made notes in each digital file. He planned to ask every one of his brothers, sisters, and multilevel cousins about the *sapiens* they worked with.

Minute details would determine whether or not that person would be allowed to remain on board SS-2. It would take a long time to go through all of them. DeNoe was not certain he really wanted this job right about now. The pile stood tall in front of him, figuratively speaking, since it was all on the computer.

He was being vilified, for the most part, on Earth. He had already sent, forcibly, seven members of the senior staff, board members all, to SS-3 for a ride to Earth. Surprisingly, not one person at SS-3 objected to DeNoe's methods. The media was playing the prepared statements from the WSA and the UE. The Fundamentalist religious organizations were likening him to Satan. Interestingly, there was a large and growing moderate religious group that was now beginning to have its voice really heard in the political arena. They did not directly support the Hsas, but they rejected the out and out denial of their rights associated with the hardline Stalwarts.

Why the hell had he allowed Red and Walter to talk him into taking charge of the Hsas? He then, with a chuckle, reveled in the thoughts of the use of *hell*. Red rubbed off on everyone. All of this ran through his mind between one file and another. He sighed and returned to his voluminous task.

*

The Martians all sat around the campfire, so to speak. They were in the machine shop portion of the pit, chewing the fat. The conversation ranged from topic to topic with little concern. They had decided to simply let off some excess energy, avoiding actual work for an hour or two.

While they didn't work per se, they did converse on topics related to survival. Marceau brought up some good ideas.

"I think I have a way to curtail some of our energy consumption in the communications department. We are sending gigabytes of data every hour. Do we really need to send all of the data? It requires a good bit of energy when it is totaled up over the sols."

Yang looked surprised. "What do you mean? You think we should stop sending the data we collect?"

"No, only half of it," she countered. "Consider that the various instruments we have set up are each taking readings every thirty seconds. That data is collated and stored in the computer, hard copies are made on discs as backup, then it is sent on to Earth in coordinated data sets. These data sets all have a time indicator. What if we make the hard copies complete, but program the computer to send only every other set? The researchers on Earth will get minute by minute reports that will still be exceptional for development of theories. When we are rescued, the hard discs will double their available data and actually serve as a check on whatever hypotheses and theories they have come up with. Our benefit will be in terms of significant energy savings in transmission of the data. We can save an hour's worth of battery time."

Even Red's eyebrows moved slightly upwards at that as he muttered, "That much."

Walter cocked an eyebrow in Red's direction, saying, "That would go a long way towards reducing the need for some of your piss batteries."

Everyone got a good chuckle at that; they had all been amazed at Red's ability to make something out of waste, literally.

Yang smiled and said, "Very good. You'll do the programming, Gail."

A half-smile crept across her face as she spoke, "Already done." She scratched a one in the air with a slender finger.

Red actually smiled. "What else have you conjured up for us?"

"Well," she drawled, "do you remember how when cell phones first came out and people were always doing that texting thing?"

"Vaguely," Yang said.

"They developed a kind of shorthand to make their messages shorter as well as easier to type out. Words like *g-r-e-a-t* were shortened to *g-r-8* for example. Also, old Hebrew did not use vowels, only consonants. I think we can devise a simple code that will reduce the number of bytes we send in all type-formed messages. Personal messages, the equivalent of the old e-mail form, can be reduced in size. Since I see the messages, generally size, not content, I have calculated that we can reduce these by a third. If you all allow me to perform an analysis on your mail, meaning I will have to read it, or most of them, I can give a clearer number. This, again, would lead to significant battery savings. Granted, the solar panels can keep up, but we want to have excess energy, not breakeven."

Yang took a sip of Red's homebrew, spluttered at its strength, and choked out in a thin whisper, "How long to arrange that."

A full smile manifested this time as Marceau said, "Done. I have developed a program that will automatically take spoken and typed material and condense it into what I have dubbed texttalk language."

Red took a deep drought and smiled. "Well done, Gail."

Walter and Gail touched glasses with a gentle clink. "Good job, babe."

Marceau scratched the air again.

Yang, voice recovered, said, "Any more good news?"

"Of course. I can't let you boys have all the fun." She giggled. "All the data streams have a header code to identify what is being sent. There is no need for these. I have examined all the information we send. No data set has the same number of variables. We can simply eliminate these header codes. This will save less than 1 percent of transmitted material, but it adds up."

"I guess you ain't been getting any the last few days," Walter noted to Red. "She's been too busy working."

"Asshole," Gail shouted as she punched Walter's arm.

Everyone laughed crudely, knowing that Gail was often quite vocal in her lovemaking. Even the very reserved Yang could not help himself as he guffawed.

Red took a large gulp, poured more for all of them, and said, "I don't think I should answer that. She might pulverize me."

The laughs of humanity echoed in the chamber, Gail included. They were friends, beyond friends.

Yang took a large gulp and unsteadily got to his feet. He handed the glass to Red and said, "I should return to the ship and let them know we are still alive. They must have been trying to send some sort of shit to us. Walter, I think I should get your assistance. I seem to be seeing two exit portals. Between the two of us, we might actually find the ship."

Walter snickered. "Softy. We'll have to work on your drinking skills. Do you really think we should leave these two here alone? What if they break everything in the pit in the heat of their passion?"

Gail leapt to her feet, rather unsteadily, punched Walter's arm again, grabbed him, kissed him, and said, "You're a real prick."

Walter and Red laughed heartily while Yang blinked owlishly, rather unsure of the Western humor.

Walter, leaning ever so slightly, took Yang's arm and guided him out of the portal, closing the door behind him.

Gail moved to sit on Red's lap. She snuggled tightly against him, peppering him with quick kisses.

Red responded but a little slowly; he was thinking about Walter and Yang.

Gail noted the slight tenseness and asked, "What's wrong, Red?"

"I'm . . . concerned about Walter and Yang," he replied.

"How so?"

"I was a virgin until I met you," Red admitted. "Yet I am a strong believer in polyamorous relationships. I know that we are an item, as the old saying goes. I know that you have had many physical relationships with others in the crew. I approve of that. It is a physical and, dare I say, spiritual bonding. It strengthens us as a unit and as individuals. You have limited yourself to women on this trip, until me. I think that you were waiting for me."

Gail hugged him fiercely and said, "I think I really was. I don't really know why. I have been with men before, but something deep inside me was waiting. When I returned to the ship in the storm, I knew."

Red returned the hug vigorously. "As did I. But this is not about us. It's about what we can do for Walter and Yang. They are alone here . . . as men. I don't know how to say this."

Gail held him tight and said, "I know what you mean. Walter is not a problem. We have already bonded because of you. He and I can deal with each other. Yang may be a problem. He is so reserved. He may feel that as commander, he cannot submit."

Red chortled. "Submit, huh. You're a real take charge kinda gal. If it becomes an issue, I think you can handle it. It's just that we may be rescued in a few months, or a year or two. That is too much time for a wild man like Walter, and it may be difficult for Yang to admit his needs."

"I'll take care of them. I must say that you are more of a man for this. It takes a lot of grit to admit your concerns. I love you that much more."

Things became much more serious after that. They did not break anything.

Walter eyed Yang carefully and then decided, "You know what they're talking about?"

Surprised, Yang said, "I don't think they're talking right about now."

Walter smiled. "They might be done by now. They talked first, about us."

"Us? W-why?" Yang stuttered.

"All those psych lectures and dense as a uranium nugget." Walter shook his head woefully. "Red and Gail discussed sex between her and us."

A few boogers and some phlegm vacated Yang's facial orifices as he choked that last sentence down his mental windpipe. His attempt at saying *WHAT?* failed utterly.

"Red would not leave his friends high and dry, so to speak."

Another round of coughing hit Yang like a ton of computer chips.

Walter rather enjoyed Yang's discomfiture, but he gave him a slight break before continuing, "Red had argued with the WSA about the male-female ratio on the landing crew. It had nothing to do with political correctness. His arguments were based on this possible scenario. He foresaw this possibility. His mind runs a thousand, thousand potential outcomes. In this one, he decides that for our physical-mental-spiritual need, he will share his love through Gail. Red and I are polyamorous."

Yang's eyebrows jumped as he colored. "You guys are orgyists."

The paroxysms that hit Walter drove tears from his eyes. His whole body trembled as he laughed his way through that one.

"No, not really," as he wiped his face free of liquid. "We don't restrict ourselves to one person through any artificial societal preconceptions that are developed through the notions of religions or other expectations. We are free. While we may have a single person, or even a few, that are really special, like Len and I, we can still enjoy the company of others without jealousy or anger. I spent time with Hot Rod, and Len. Len spent time with others, and I approved. It is a bonding experience. It fulfills the needs, the desires we all feel. It also eliminates the alpha syndrome. Everyone is free."

"What do you mean when you say the alpha syndrome?"

"It's the need to dominate a relationship, usually male, but not necessarily. People are truly equal in a polyamorous relationship. Individually, they decide if they want to be monogamous or not, bisexual or not. It is understood that new relationships with others can be engaged in, even expected. It is a dynamic relationship, not a static one. We understand that there can be more than one *love*. It does not mean that one has more value than another, just a different one."

*

Burt began laying bets all over Lead and Deadwood at various odds. He and Red were gonna clean up. Walter had even cleared some money for Burt to play with. If they pulled this off, given the odds they were getting, they might pull in as much as a million when all was said and done. Twenty percent would be two hundred thousand for Burt. He had plans for that money. He would invest it in Red and Walter's endeavors. Red had carefully let him in on the plans over the years. Little clues left in messages sent had pointed the way. It was a black market operation, so to speak.

The illegal gambling, meaning not through a licensed agency, was generally ignored by the WSA. They were not in the tax business. It also might provide a way to get rid of the loose cannon they had in Red. They could use the pirated information against him, to discredit him. They still had a very few moles on SS-2 and at Luna. For now, they would hope he made it through this ordeal. They would find a way later.

*

Chapter XXI:

Rescue

21

010101

"This is Mary Smith-Jones for World Wide Network News reporting on the Mars rescue mission now ready for departure. The Gagarin has been completely refitted with basic supplies, replacement equipment for what the landing crew took down to the surface, and additional medical supplies.

"Many of the unnecessary scientific instruments have had to be removed to make room. These were all related to deep space exploration and planetary sciences.

"This mission, led by Colonel Davies, related to the religious pundit by the same last name, will leave within the hour if all goes as planned. All the information we have been able to glean from mission control North American continent so far is good. While the WSA has officially been rather tight lipped, a few insiders have spoken off the record. According to their statements, everything is following the expected route laid out by the engineers and tacticians. However, regardless of their intricate chesslike plans, it all still depends on those on Mars.

"The marooned astronauts on Mars have struggled mightily to survive against high odds. Much credit has been given to the incredible duo of Mick 'Red' O'Hare and Walter Black Bear. These two have performed incredible feats of engineering by reworking the materials they have on hand in order to pass through these straights. And yet, they have done everything they can to press forward with the scheduled preparations for the next crew to visit Mars. They have themselves clearly stated that they are expendable. The mission is not.

"This selflessness has endeared these people to all the Earth, the same Earth that is dependent on their success. In four months, if all goes well, they will be back here on Earth. The expectations are running high. There are reports of bets being made on the survival of each of the astronauts and the success of the mission.

"We will return to this story in one hour and bring live reports of the launch to you from our North American desk for the WWNN.

"In other news, the situation on SS-2 has fully stabilized under the management of the Hsas. In fact, production levels are almost 10 percent higher than pretakeover levels. This has greatly smoothed all the ruffled feathers in the WSA and the UE.

"They are now crowing about how this was what had been planned all along, just without the coup.

"For WWNN, this is Mary Smith-Jones."

*

Sol, our star, is roughly composed of about 91 percent hydrogen, 9 percent helium, and 0.1 percent heavier elements when counted by the total number of atoms and not mass.

It is large enough for a million Earths to fit inside; alternatively, you could place a thousand Jupiters.

The boiling, roiling surface of this G2 V-type star is around 5,700 kelvin or about fifty-four hundred degrees Celsius.

Its rotation is skewed since it rotates at the equator in twenty-five days and thirty-six days at the poles.

Deep inside occurs the high-pressure and high-temperature fusion that provides the energy that we much, much later feel as sunlight. It can take a million years for the energy of fusion to reach the surface to escape through the atmosphere into space.

The photosphere is the bottom layer of atmosphere for the sun and is several hundred kilometers in depth and dotted with sunspots at its bottom.

The chromosphere is next, basically transparent, and several thousand kilometers deep.

The corona is a rarified layer that essentially extends outward to infinity. In greater reality, it stops at the heliopause, that boundary between our system and the rest of the galaxy. There it is mixed with the stuff of interstellar space.

There is no actual hard physical surface on the sun. What we perceive as a surface is a granulated layer. These raised, bright areas are surrounded by lowered cracks. These granules are related to the rising and falling of material, or convection, that has a temperature differential of around two hundred kelvin.

Dark features, known as sunspots, have been reported for over two thousand years in Chinese records. In 1610, Galileo corroborated these sightings. The darkest part of the spot is known as the umbra while the lighter portion is labeled the penumbra, just as in eclipses. Some range in size from ten to fifty thousand kilometers in diameter. Earth is only twelve thousand kilometers in diameter, so many of these sunspots are, like the Great Red Spot on Jupiter, larger than that of our home planet, our cradle.

Like everything in nature, there is a cycle associated with the appearance of sunspots. It follows an eleven-year period. Magnetic cycles are bound up with sunspots and have a relation to multiples of elevens periodicity. Earth's magnetic field is comprised of a whimpering 0.5 gauss while an individual magnetic field associated with a spot can be several thousand gauss.

Solar flares are eruptions that are related to the snapping of the magnetic fields and the release of material into space. They are most common when sunspots are numerous.

They emit radiation throughout a wide spectrum. Each flare can rise to great brilliance in a matter of moments while lasting as long as several hours. In the flare, ionized electrons are accelerated to half the speed of light. As these particles strike the material of the corona, they create X-rays and ultraviolet emissions. All of these play havoc with satellites, electrical power systems, and radio facilities.

Earth's atmosphere protects life here from dangerous showers of ionized particles, X-rays, gamma rays, and magnetic storms. But all shields can be breached. Northern and southern lights are examples of that penetrating energy. Larger events can simply overwhelm the atmosphere and magnetic field, leaving chaos behind. Space, however, does not offer much protection.

*

Those members of the UE board not purely outraged at the various open mike incidents still voted to keep the original crew members of the Gagarin home. Regardless of the politics, they used the excuse of having completed one long-duration flight already, which was quite true. Plans called for them to rest and make the circuit of talk shows garnering as much continuing support for manned exploration as possible. They would pass the plate, so to speak, for donations to help pay for the rescue mission. Only later would they be reincorporated in actual launches, almost immediately if all went well, again a political decision.

The return plan, constructed in great part by the members of the Gagarin and refined considerably by Earth engineers and the Hsas on Luna, was flawless. It was adopted as the best way to save the people stranded on Mars and to appease the public. The Gagarin was prepared for a quick return as soon as it entered Earth orbit.

Colonel Davies was very tightly wound. Everything was by the book with him. As the commander of the rescue mission, he would brook no variation from protocol. He knew that some of the UE board wanted to reign in the looser elements, those like Red. He would be the leather that did that, or so he thought.

The rest of the crew was fully military as well. All of them were ready to accomplish the rescue mission. It was felt in some quarters that the military was more capable of a quick response. Their pilots were still among the best in the world.

They took over the Gagarin with little in the way of politeness. It was more like an eviction. They brusquely pushed aside the previous members of the Gagarin team. They also vigorously questioned the proposed methods of the previous team in regards to the rescue. There was a certain feeling of "They failed. We won't, so we better double-check their data." They went through the equivalent of mental sweating as they endured the sixteen-hour days.

The men checked all the figures, even as the Earth engineers reworked the numbers for the umpteenth time. They ran through all the checks, not the required three times as per the agency's protocol but five. Davies drove his men very hard, but they expected that from him. He was known for being one really tough son of a bitch, which he was. They all pressed for duty with him. He was not a rabbit; he did not move forward in the ranks quickly, but he did always move forward. People working with him went forward as well. They accepted the heavy load, knowing it would pay off in the end. Their search for discrepancies was to no avail.

The original Gagarin team had done their homework. Davies, cousin to the religious leader, had to admit this. Their plan would work, when adhered to by a competent team. He and his team were just that. There was no fault to be found in the plan. It had been simplified somewhat by the slightly reduced number of astronauts and the fact that they were all male.

The politicians had decided that for the rescue mission, there would be only men involved. The western religious organizations had stirred up the fire in regards to the sexual activity displayed (behind the scenes, mind you) of the crew of the Gagarin. This left many female politicians' pots bubbling. This decision had left every woman in the astronaut program holding a bag. They, the female politicians, had made it known that there would be a cost as well. There is nothing free in politics; these chips would be called in soon enough. This kind of sexual bigotry was supposed to have disappeared decades ago. It was a hangover from cultural biases that the UE was supposedly eradicating.

The mission seemed, from that point on, to be tainted with a void, an absence of peoplekind. The entirely male and military crew didn't feel natural after all the years of mixed crews of nonmilitary personnel. It was a throw-back to the male-dominated era of the late fifties, sixties, and early seventies of the 1900s.

People were subliminally stirred into a quiet wrath in response to the resurgent actions of the religionists. It seemed every few years there was a wave of political activism that was backed by them. It all claimed to be for the best interest of peoplekind, but it was not always what most would claim as progressive. Polarization always followed. Divisive forces, ones that clung to the past while the others claimed to speak for the future, both of which could rend society to the core if fully unleashed on each other. The previous large battle had nearly prevented a return to space, especially by Western cultures. The old United States, the NAR, had nearly been left out. Its failure to really stress the teaching of sciences and math for several generations left it almost on the outside looking in; only money kept them in the game.

Davies didn't care about any of that though. He stroked the panels of the small corridor of the Gagarin with the loving caress of the world's greatest lover, again in his mind and not necessarily reality. This was his ship now. It was his mission and crew. He was the master of the situation. Nothing would prevent them from achieving their goal. He had been a test pilot for many of the Earth planes and even some of the intra—and interstation shuttles. This was the real thing though, deep space, interplanetary.

He had waited for his chance. He had done everything right. He let the politics come his way rather than force a hand. He would now make his mark. He would be the savior of the people on Mars as well as the program. He would give the people a view of a properly led mission. No errors or failures would be allowed on his watch. His personal integrity and will would not allow for it.

It was a feeling that permeated his crew. They were infected with the same emotions as he. Each knew his own ability. They had all, as a team, waited while others had been selected for various reasons. They had the go juice now. And that was a dangerous place to be in.

*

All throughout the history of the space program, regardless of country, "go fever" was a threat. It made people overlook what appeared to be trivial and inconsequential threats to the objective. With the developments on Mars, there were many voices crying now about the disease.

"We ventured too far too fast," explained one pundit. "Technology has been growing at an ever accelerating pace. We need to control it, not allow it to control us."

"We did not do all we could to protect them," haughtily announced another who had no real expertise, only a v-link talk show. "Experts say that what they are doing could be done just as well by robots and much less expensively. There is no need for human presence as of yet, if ever."

"We are moving too fast with our rescue ship. They will end up in the same trouble as the rest," insinuated others. "As insensitive as it sounds, we should not send more people out to die in what is most likely a vain attempt anyway."

*

The government had to step in and take a stand. They prepared several spokespersons to travel on a series of global talk shows after an initial series of government broadcast news blurbs. The newscast was the usual dull dribble stating that everything was under control, and experts were examining all the things that might go wrong with the rescue craft or on Mars. Very few people actually paid any attention to the official press release. The entertaining part came later in the weeks as some of the lively young agents debuted on several v-link nightlines.

One in particular was lively for the North-Western Quad of the UE, the North American continent.

"So, Cheryl, what is wrong with the space program?" queried Bart of the *Early Nightshow*.

"Not a fuckin' thing, Bart," she replied candidly.

"Uhm, you're not really supposed to say that before 8:00 PM local," he noted quietly to her after leaning in. "It is a live show, and the west coast is still before that time."

"I suppose I should be more proper then, Bart. Not a fucking, *ing*"—she stressed the ending again—"thing, Bart." She smiled carnivorously at him. "There, no colloquialism. How is that?"

"That's not—," he started.

"I represent the government and have been cleared to discuss any matter in any manner deemed necessary to clear up any misconceptions about the UE space program. Now then, are you an expert in cosmology, physics, astrobiology, electronics, or anything other than talking?"

"I am not an—," he started before she again cut him off.

"I know that you are not. I had your background checked. That means that in order to have a reasonable conversation on the subject, we have to speak at the level of the common person and using terms that they understand. Your demographics are of a very rowdy crowd, so I don't think they will get all pissy about my choice of vocabulary."

V-linkers everywhere chuckled at the scene playing out in the corner of their eyes. They usually saw Bart causing the discomfiture. Now he was squirming, slightly.

Bart faced a dilemma, but with the skills of a seasoned commentator, he steam rolled it and kept going with, "You claim there is not a problem, even after a person has died on Mars?"

"People die, Bart. Here, there, it does not matter very much except to *them*. Holtman died doing what he dreamed of doing—going to Mars. He finished life at the peak of his professional career. He died, not of anything we, the program, did wrong, but simply chance."

"Shouldn't the agency shoulder the responsibility for a faceplate that couldn't take the stress? Shouldn't we be examining the manufacturer's records?"

"Not for one damn moment. We, the WSA, in cooperation with the manufacturer, designed those shields and suits to withstand such an event. Marceau's entire ensemble survived without any really significant damage as far as we can tell. All the data sent back shows that her gear survived what it was intended to withstand.

"We have really learned a great deal from the problems we faced with lunar dust. It is really gritty. It gets into everything. It can cause lung problems, even death. The electronic problems are caused by static buildup similar to what you feel when you touch a doorknob and get a shock. It required that we learn how to design a suit that repelled such material without an electrical charge that could damage electronics.

"We developed a type of vacuum system to remove the dust from suits as astronauts entered the air locks. This helped to keep the dust from entering the air systems. We learned how to bleed off the electric charge into the atmosphere while on Mars, and into the soil on Luna. All this information has improved our suits to the level where a person can survive for as long as six months with a fully stocked recyc package. This is unprecedented. The face shields were tested to the most strenuous levels we could imagine. We will now make them even tougher. There is no need to spend so much time and effort when we know what we already have. This is not a case of a major malfunction where a serious review is warranted," she clearly enunciated.

"Death is not a major malfunction?" Bart almost shouted.

"Happenstance," she stated as she took a deep breath before she continued.

"Further examination of Marceau's suit when she returns to Earth system will give us a much more thorough picture. Holtman's suit survived, as did his helmet from what we were able to observe. His suit obviously will remain with him on Mars. We may never know whether there was a defect or not in the face shield. For some reason though, one rock hit at the wrong place at the wrong time and penetrated his double face shield."

Bart interjected, "Wouldn't you all like to know why? Or does that just get swept under the rug of 'go fever'?" His attempt to be with it on the terminology just dug him a shallow grave with his audience since it was an obvious ploy, and it put him in real deep shit with his guest. This absurdly blunt young woman had the listeners mostly in her shapely back pocket.

"Don't be a prick, Bart." She frowned mightily at him and continued, "We have hurled rocks and sand at them prior to this horrific event and ever since. Not one of the shields has cracked except under incredible amounts of force. Force that many of the engineers say would be enough to simply snap the neck of the person wearing the helmet. They would die without a crack in their face shields."

Bart wanted to lead into his commercial and interjected, "So you take one side of the issue and say it was not your fault. There was no way you could know that there would be a sandstorm out of season like that, or that the faceplate was insufficient."

Cheryl sat up on the front of her chair, to exhibit her six-three frame with well-defined muscles, and stated coldly, "The faceplates are not insufficient. We knew there could be sandstorms at any time of year. We did not expect them to be as bad as this one, especially at this time of year. This time frame was selected as the least dangerous season for large-scale storms. People die every day in a storm on Earth. Where is the media coverage for those poor sods?"

"Well, Cheryl, those people are not on Mars, and they do get widely reported on the news," Bart noted with his professional demeanor. "We have to break now. Back in a few." He felt that he had gotten the best of that, if only barely. He paid for it later when Cheryl continued the story as a monologue in her own (meaning government) time-paid commercials. Bart's ratings took a dive after this. Cheryl was offered a new talk show. She took it. The government decided that she could have two paychecks.

*

The Gagarin was now fully prepped and ready for its sudden return to Mars. Davies and his crew felt the tension mount to nearly unbearable levels. Every detail had been examined many times. This was the deep breath before the storm, the action.

The beginning of the mission was a nonevent. It went so according to the book that people took no notice of it except in expectation of further developments. Davies was quite happy with that. It meant that they had produced a perfect departure. It was a great omen for the future, not that any of the crew believed in any such thing. Perhaps they should have brought a few rabbit's feet.

*

The distant star had spent the last million years hiccupping through its last bit of life until a certain point was reached, the same point in which a man hung dying on a wooden pole while Roman soldiers gambled over his robes. The last hiccup ended with a sudden collapse followed by an extraordinarily violent explosion. In the matter of a microsecond, the amount of energy released went up by a factor of a billion. Before it was all said and done, the amount of energy released was as much as all the other stars in the galaxy combined. Cosmic rays, heavy atoms stripped of all their electrons, rushed out at nearly the speed of light. This would not be noticed by any intelligent species, humans first, for just a little more than two thousand years.

*

The launch was observed by a very large number of v-linkers and webbers. They watched not for the sake of the launch and those involved but the drama expected in the future. Human psychology has not changed much over the millennia. Every person feels the need to be a part of something bigger than them. This finds expression in religions with the infinite beings known as gods, science in the hopes of understanding the infinites, to be a part of the unraveling of the clues. It ends, eventually, with the development of an observing ritual followed by a communications, gossiping, ritual.

This is as old as time. Ancient skill traditions were passed through the observing of the hunter as he strutted and reenacted his skillful and successful hunt in front of a fire. Girls observed the women as they shaped clay, cooked food, and selected certain plants. The grunts, calls, screeches, and all other vocalizations added emphasis. These noises would, over time, become language that would allow for finely detailed oral descriptions. Planning and best method determination would refine this art. Gossip improved the hunt, pottery, cave drawings, and more. It was actually useful.

This ritual of observe and comment has been used in all areas of human development, so much so that it has become a pastime of all cultures. The need to discuss strategy, the need to appear knowledgeable on the subject at hand, the ability to provide a solution—this is what drives the subconsciousness of all humans.

*

Commander Davies was comfortable with the fact that he was under the gun but not in the primary spotlight of the media. At present, he was the supporting actor, but that would soon change. He knew that history would take his name. Those who could read between the lines, those with intelligence, would understand his true value and position.

Every day Red was in the spotlight, accompanied by his ever present cohort, Walter. Yang was the commander, on paper, but he had let Red assume control. That, in Davies's book, made Yang a bad commander and Red, a usurper.

Davies also felt that Walter was a toady, a hanger-on, not capable of working on his own, in his own right. Davies did not believe in being a follower. The two of them always seemed to be together, as if they couldn't function without each other. He speculated on the chances of success if he recommended breaking them up. He might have a chance if he took the tack that they could cover more ground that way.

He also suspected a little sexual connection, as did many others, though the poem given by Marceau suggested that she had her hooks in him now. That was another problem for Davies. He was not a prude like his zealot-of-a-cousin preacher; he did not concern himself with a person's sexual appetites.

His beef was that the sexual escapades of the prior Gagarin's crew had become something obvious, open, exposed to the whole world. This did not mean that they were distributing porn; it meant that they discussed it openly in ship's meetings, with newsies, and even on open links with the various schools.

This was an infringement on the domain of the astronaut. Privacy violations of this sort really irritated Colonel Davies; in fact, it did the same to all of his crew. They were the antithesis in terms of perspectives; they represented the old school, the status quo. Red and his crew, along with those like them, represented the flux, the ever changing adaptable personality. Every generation has this combination, this polarization, as well as a significant number of middle grounders. The important times in history occur when there are very few middle grounders, and polarization is at an extreme.

Davies banished these tangential thoughts from his mind and returned to his task at hand. The course they were taking was a longer one than the Gagarin had taken on the first trip. However, they were accelerating at a greater rate, which produced a higher velocity. The plan was to accelerate all the way to the midpoint, turn end for end, then decelerate into the Mars orbit. For now, he examined the yaw, pitch, and roll angles for the Gagarin. He compared them to the computed values from the plan and found them correct to within thousandths of a degree in each category.

Even that little discrepancy would require a subtle correction along the way. Thousandths of a degree now represented thousands of kilometers later. They did not really think of the flight path as a curve but more like an expanding cone. A larger deviance meant the cone was wider at the far end. As soon as the computers came up with a burn solution, they would send an electric signal to the thrusters to emit small amounts of gas as a correction. A tiny puff of this gas now would make up for untold amounts of energy later.

Next on his checklist was the power check. Fuel cells, involving the conversion of H_2 and O_2 to H_2O while releasing energy, powered many of the subsystems. A thermopile powered by Pu-238 provided the energy for the thrust and the magnetic fields that protected the crew from radiation.

The crew was also provided with radiation security chambers, namely their sleeping quarters. Each had a tube-shaped area two plus meters high and a little less than a meter in diameter. A five-centimeter-thick cross section was filled with water between the inner and outer wall of the chamber.

*

Cosmic rays consist of atomic nuclei that have been stripped of all their electrons and move at near the speed of light. When they collide with other matter, they strip electrons from nearby atoms and occasionally cause nuclei to split in nuclear fission. Waves of them flooded from the cosmically recently exploded star. Peoplekind would soon feel the wrath of this disintegrating orb.

*

The sleeping areas were almost like a cigar tube. They each had a door that moved outward and then slid sideways around the curve of the cylinder.

The sliding doors, powered by a small electric motor, could enclose each member of the team in his own cocoon of safety. The cocoon was a double-walled aluminum tube that was lined inside and out with reinforced polyethylene. The plastic was very efficient at preventing penetration. Surprisingly, materials that are made from smaller atoms are more effective. They pack more tightly.

Water was not as good as the carbon composites or the plastics at stopping radiation, but it played two roles. The water enclosed in the protective sleeve was an emergency supply in case the recyclers had problems. The water was also much less dense, which meant it weighed less. That, in turn, means it takes less energy to accelerate it, always a consideration for every item in a space vehicle.

Some engineers had even considered water helmets for the astronauts. They were only two centimeters thick and filled with a Jell-O-like material. They looked much like the old WWII-style flying helmets. The astronauts had balked somewhat at this rather strange-looking headgear. They did have an image to uphold. While they all liked the deserts in their stores, *Jell-O heads* was not a term they wanted to be labeled with.

*

The cosmic background radiation is roughly the same in all directions at 2.4 kelvin. This is the rationale behind the concept of the Big Bang theory. This, however, only refers to the large-scale, long-term background. Variations were soon noticed on the smaller scales. As equipment began to improve, measurements allowed for very fine details to begin emerging.

It had long been known that when the sun, Sol, had coronal mass ejections (CME), they did not emit equally in all directions. This led to the understanding that there must be paths through which these charged particles flow. Large-scale positive and negative regions were soon discovered along magnetic field lines that were twisted and stretched by the planets.

In much more recent times, it came to the attention of scientists that these flows followed what were termed *rivers*, three dimensional and almost like the flow of blood through a body. The flows (or flux) moved up and down the spiral arms of the galaxy. The leading edges of the rotating arms were invariably negative, meaning electrons were being stripped off of atoms and flowing away. The galactic core, a black hole in the Sagittarius area, pulls and drives these particles much like a heart. In these terms, a galaxy could be deemed to be alive.

Our exploded star's ejected material was accelerated along one of these interstellar arteries. The Gagarin would soon pass through the junction of Sol's main ecliptic river as well as the tributary for the local group.

*

Lieutenant Colonel Moore, the pilot from the North American Region, relieved Colonel Davies at midnight ship time. They rotated twelve-hour shifts, not the usual eight when a full-crew complement was available. It meant a little more mental wear and tear; they could handle it. In reality, any of the other members of the crew could run things.

Davies believed in having everyone capable of doing everyone's job. This was one area he and Red, as well as the agency, fully agreed upon.

Major Sakimoto, the Japanese nuc engineer, shared the second shift with Moore. He would check all the data on the reactor, the magnetic field, and any abnormal external readings. He soon would be inundated with data he would never wish on any person.

A week and a half from Earth and everything seemed perfect. Astronomers noticed that there was a disturbance in the solar output. Wide swings in energy output began to occur over a span of a few hours. Studies of the sun had gone so far as to allow for vibrations from the far side as a prediction for what was coming in terms of eruptions. When the limb of the sun actually turned into sight, astronomers were not surprised to see a collection of spots near the equator.

They were all very surprised, however, to see that there were so many sunspots clustered within less than a quarter of the face of the sun. Year 2045 was five years before the solar maximum in the eleven-year cycle. The class of the flare was later determined to be an X-42, an unbelievably high measure of X-rays. The magnetic fields associated with each sunspot suddenly snapped, resulting in a tremendous coronal mass ejection (or CME); tens of thousands of metric tons of charged matter erupted in a matter of seconds. It followed the general pattern of the rivers that science had just discovered. There was a general dispersion based on the square of the distance, as expected, and a secondary that followed the arteries of the solar system.

The intersection of that event, with the as yet undiscovered supernova, would occur at the usual time and place, meaning the worst possible moment and location (ask Murphy).

The astronomers on Earth had long awaited a close supernova. It had been calculated that a supernova should occur roughly every fifty years in the Milky Way Galaxy. There had been none observed visually since Kepler observed one in 1604, just a few years before the telescope came along.

The astronomers of the space agency had already sent messages to the Gagarin about the sudden and extreme mass ejections from the sun. There was a sufficient time lag, between the waves of light and the slower particles, to take precautions.

It was too late to do anything about the X-rays. By the time Earth was aware of the problem and had channeled its solutions through the bureaucracy, the crew of the Gagarin had already been exposed to significant doses of deadly X-rays.

While the ship's generated magnetic field repelled most of the ion bombardment and the weaker X-rays, some penetrated.

A rad is the actual amount of radiation, a quantifiable amount. Eight hundred rads is invariably fatal. Half of the people exposed to 450 rads die. Some people die at levels around two hundred rads. At fifty rads, genetic damage is significant, especially to the gametes.

A rem is the amount of radiation necessary to cause the same biological effect in humans as one rad of X-rays. The average Earther is exposed to 0.4 rems per year. Government standards limit exposed workers to exposures of no more than five rems per year.

Major Sakimoto was surprised when he turned back to the touch screen that controlled the interior monitoring system. There was a red light flashing on the radiation subpanel. The computer, Q-3, announced sudden high levels of radiation. Touching the screen on the flashing light brought up the complete radiation monitoring array. Sakimoto realized in less than a second that the problem was external. There was no leak in the reactor system; he had checked that first. He silenced the audio alert from the computer.

"Colonel, we have a very high radiation flux from an exterior source," Sakimoto relayed to Moore.

"Source?" queried Moore.

"It appears to be natural. Wavelengths cover most of the spectrum," he responded after examining another screen. "I would have to guess a major solar flare."

"Got your cock sock on?" chuckled Moore as he unconsciously felt for his own. It was not really a sock, just underwear that had been woven from the same radiation-resistant plastics mixed with cotton layers.

"Sir, X-ray and ion bombardment is still rising," Sakimoto said with a hint of uncertainty. Most flares did not reach this high a point, especially when the sun is not at its cycle peak. "We have already taken between five and ten rems."

"That's high but not dangerous."

"But it is not dropping yet. We would expect a regular flare to drop rather quickly," noted Sakimoto.

It was at that point that the hot line from Earth beeped and flashed a little red signal as a message came in. Q-3 announced a red alert.

"Colonel Moore, this is CAP-COM North America. There is a multidigit X factor series of flares with massive coronal ejections currently under way . . . nications severely affected. Yo . . . ordered to adjust pitch and y . . . to place bulk of ship between . . . ion bombardment. We calculate . . . twenty minutes."

"Wake everybody up," commanded Moore as he strapped himself into his pilot's seat. "We need to do a complete lockdown in five minutes." That meant that anything loose had to be put away even though he knew there was nothing out. Neither he nor Davies allowed for such sloppy housekeeping as that. But protocol requires a check of every possible item. When a ship changes course, anything not bound in place would fly around and possibly cause damage.

Sakimoto touched the personal alarms for each of the crew members. When Davies signaled him to ask what was up, he replied, "There is a very large flare. We have been ordered to bulk shield the ship before the CMEs arrive in twenty less."

All of the crew moved quickly to secure any and all possible items, personal and otherwise. They checked the kitchen, commode, shower, and finally, all but Davies returned to their sleep chambers and their protective cocoons. Each pushed the control to close the electrically powered door on each chamber. Here they were protected more than anywhere else.

Davies moved straight to the command chair. As soon as the rest of the crew, except Moore and Sakimoto, was secured in their berths, he began the procedures for altering the pitch and yaw of the Gagarin. It was a relatively simple procedure. By placing the bulk of the ship between them and the sun, they created a very simple shield.

Davies confirmed the maneuver with the CAP-COM and the Earth observers as best he could with the poor radio communications.

He noticed a very strange sight out on his eleven o'clock. Where there had been darkness, there was now a very bright light. It was tremendous; he had to shield his eyes.

"Moore, what the hell is that thing?"

"Did we turn into the sun by mistake?" asked Moore at the same time.

They read their boards several times in consternation before concluding that they had not turned into the sun.

To their great consternation, Q-3, the quantum computer linked to Q-1 on Luna, announced a new and greater source of trouble.

*

"What's that?" asked a man as he stopped in the middle of the street.

Electric buses honked to no avail.

His head was twisted upwards toward a very bright spot in the sky. It was as bright as the sun. His chin slowly lowered; he was too surprised for it to simply drop. Soon others were gathering around the still overwhelmed pedestrian. It took little time for the news to spread to all sorts of people. Many people assumed that the scientists were already aware of this phenomenon.

It took time for the grapevine to find them and alert them. A full hour had passed before everyone who needed to know really did. A supernova called the attention of every available astronomer. The problem was that most of them were already involved in the examination of the sun. It took forever to get an organized response, beyond a few dedicated facilities. They almost missed the most important few minutes at the very beginning of the gigantic event.

*

Davies, Moore, and Sakimoto were confounded; this was not an event that had been presented to them in one of the many simulations they had survived. Training could only go so far; there were those who could go beyond and those who could not. The military could take away initiative.

The soldiers of the Wehrmacht, the German army of WWII, could not take the initiative to repel the Normandy invaders of the Americans, British, and Canadians. The structure was so formalized that even the highest-ranking field marshal on the scene could not order tanks up to the front. Hitler was sleeping, and no one would wake him. This was the true difference between Red and his competition—dynamic players versus static people.

Sakimoto was uncertain about the readings. "Sir, X-rays have risen to a flux of one rad per square meter. We have multiple sources though. Synchrotron radiation is coming from the new light source. Cosmic rays also have two sources. The sudden burst is from the new light source."

"Fuckin' supernova!" said Davies in surprise.

"Has to be," said Moore.

"Are all cams and sensors recording?" asked Davies.

"Yes, telemetry is being sent to Earth and copied to hard disk," replied Sakimoto.

"Good job, Sak," said Moore.

"Rock and hard place," said Davies. "We are getting nuked hard here."

"Which source is higher?" asked Moore.

"Temporarily the supernova, long term the solar flares, I think," said Sakimoto.

"Do we retask?" asked Moore.

"Is there a compromise node?" asked Davies.

Sakimoto's fingers flew over the touch screen in a fury of calculations. "No," was the soft response.

"Computer damage expected?" questioned Davies.

"Core damage can be expected, but I actually can't say how much or where."

"Personal doses?" asked Davies.

"By now we have each picked up between twenty and forty rads," stated Moore.

"Tubes now Sak," said Davies as he unstrapped.

Each made his way to his sleep chamber as quickly as he possibly could. They were joining their other comrades in the snug, safe chambers that were prepared for high radiation situations.

*

Red and Walter were working in the pit when a call came through from Marceau. "Boys, there's a problem with the Gagarin. There is an X-plus-forty-type solar flare under way. They are bulking the ship."

"Are they reading a significant number of rads?" asked Red.

"They are reading a very high number, but the elec-mag is deflecting most of them."

"What can we expect?" asked Walter.

"I think we should all get into the pit. The atmosphere will stop the majority. The ship *should* stop the rest, but I'm uneasy about the situation."

"Yang, it's your call," said Red.

"Like you boys keep telling me, she has a pretty good intuitive nose on her," said Yang. "I . . ."

"Nice ass," murmured Walter as he nudged Red. "Boobs too."

"Not now, perverts," Yang asserted. "I need input."

"Get down here now. I have learned to trust her intuition."

"Me too," said Walter.

"It'll take five. We're already suited. Marceau said it would help."

"She's a smart girl," noted Red.

"Shit!"

"What's wrong, Gail?" asked Red.

"I . . . I'm not sure, what it is," said Marceau. "We have a whole new set of readings from a secondary source."

"Look out the fucking port, Marceau," said Yang with awe as he cussed for one of the few times on the mission.

"What in bloody hell is that!" shouted Marceau. "It's incredibly bright."

"It's not from Earth, or the Gagarin," said Yang as he made a few hand measurements of angles to compare locations.

"As bright as Walter's white ass?" asked Red.

"Way fucking brighter," said Yang as he cussed for the second time of the day.

"Those two are a bad influence on you," Marceau smiled.

"Either that or they are opening me up to new and perverted perspectives."

"How big is it?" asked Red.

"Big," they both echoed back.

"Supernova, or nova perhaps," figured Walter.

"Agreed," said Red. "It fits the parameters. You can't leave the ship. You need the extra protection. The amount of time you spend in the open reaching the pit and getting down would expose you to too much radiation. The atmosphere is not that thick, and there is no magnetic field."

"Seems like we spent too much time talking about it," noted Yang.

"Maybe we should put our helmets on," said Marceau. "The head is most vulnerable to this type of radiation."

"How long can you guys last down there?" asked Yang.

"Longer than you wimps can," said Walter. "We have all the makings of a vacation here. We have green goo food, high-calorie drink, and a poo processor that would make a king happy."

"We just need some sand and waves," added Red sarcastically.

Neither felt the need to add that the high-calorie drink referred to the still that they had quietly placed amongst all the other equipment in the sub-Martian chamber. Both were now master brewers of the finest of white lightening liquors. Some referred to it as ball burners, others as gut fire. All referred to it as hangover hell when they imbibed too much; in fact, they usually wept for death or bodily disconnection.

*

Davies had transferred all the main board functions to the sleep chambers. Everyone could do his job from the constricted confines of his meter in diameter tube. The only problem was with Major Sakimoto.

"Sir, my door will not shut properly," noted Sakimoto. "It acts as if something is blocking it."

"Hold position," commanded Davies. "Are you well covered?"

"Door is one-third open, without visual blockage," noted Sakimoto.

"I'll inspect," said Davies.

"Sir, you should remain in your cocoon," stated Moore.

"Two, it's my call to protect my men." He opened his door. He then exited quickly to determine what may be blocking Sakimoto's door.

*

The problem with technology is that the more sophisticated it gets, the easier it can be derailed. Computers can get touchy after only a few months. Hard chips are at the limit of micro size. Heat is such a problem, as well as the simple fact that they are not very sturdy. They can't be; they are too delicate. A serious sneeze or a vigorous fart will shut them down. In other cases, a stray ionized particle, a hard X-ray, or gamma ray will penetrate to the core of a delicate device. There it will either pass through a very fine wire, a micro capacitor/resistor/inductor, or any of another range of devices, leaving a malfunctioning computer. All too often, the backup routing gives false readings due to excessive current.

Liquid and quantum (or quant) processors are full of the uncertainty principle. Working at the atomic and subatomic level, these devices were really born in the research of the 1980s. They work very well most of the time. But occasionally, they get a mind of their own.

Some people think the quants are the beginning of artificial intelligence. The quirky behavior they display has been described as "the petulant mannerisms of a very young child" by various psychologists.

Computer experts violently vocalize a giant raspberry at that idea; they say that the random tunneling electrons that bring about the aberrations cause the computer to simply malfunction. It requires some recognisizing, the equivalent of a lobotomy on the malfunctioning part of the system. That is followed by an addition of newly trained and integrated pseudoneuropaths.

Regardless of the type of computer, shit goes wrong.

*

When he reached the position outside Sakimoto's chamber, he could not see anything blocking the rolling door from closing. "Hit the open," he told Sakimoto. The door opened just as commanded. "Hit close," he ordered. The door moved less than one-third of the way, closed before stopping. There was no object preventing the door from closing. Shit happens when radiation damages a computer.

Davies was taking more rads than was deemed good for his health by governmental standards. His personal daily dosimeter was bloodred. That meant he had taken a dangerous dose that was above the weekly allotment. He was already beyond the device's capability to measure. Unknown to all, he was very quickly approaching what was the monthly total. The Gagarin's electromag shielding was simply overwhelmed by the total bombardment from the two extreme sources. Within twenty minutes, Commander Davies had been exposed to over fifty rads. Fifty rads will do significant genetic damage, especially to a man's gonads—his testes, the family jewels, also known as the magnificent marbles.

The ship was in the middle of what scientists had come to describe as an ion pathway or, more commonly, a radiation river. These paths follow the pull of the giants of the system. Jupiter and Saturn are buffeted by Sol's own outward pressure. At this terrible moment, the two largest bodies, next to the sun, had conspired to form a perfect conduit. This conduit fed a giant river of particles directly toward the spot, unfortunately, that the Gagarin happened to occupy. Pendulums swing for good or bad, beginnings and endings.

No sensors on Earth, Luna, or Mars could measure the amount of radiation that the Gagarin was absorbing. The scientists could only estimate, and they were very low guesses. The telemetry relayed from the Gagarin to Earth was full of gaps due to the interference from both the flares and the supernova. Much of it was misinterpreted as faulty data. There could be no way that there was so much radiation in the vicinity of the Gagarin. The computer had to have been damaged; remember that shit happens. Many assumed that there was damage to either the sensors or the core of the Gagarin's system.

According to the square of the distance law, the flux could *not* be that high. There was no way to for them to be in such danger. But there was. Murphy's Law said so.

Davies, no idiot, quickly realized that the door was not going to close. He grabbed two space suits from the lockers and a roll of the most important technological development of all time (duct tape) and hung a temporary screen in front of Sakimoto's tube. The minutes would cost him dearly though. With his crew ensconced in their wombs of safety, he was fully exposed to the gamut of radiation that penetrated the multilayered protection of the ship. Davies was one of the few who were most susceptible to such radiation. He garnered 187 rads in the space of an hour. At two hundred, there are deaths. Before he could get back to his own cocoon, he had received a few more. Regardless of the extra protection he obtained from his sleep chamber, he was at the edge of the precipice and ready to tumble over into the abyss.

The maximum for nuclear workers is five rems per year. Davies far surpassed this dosage in less than a minute.

The next few hours were hot; they were enclosed in such tight quarters. They had all been tested for claustrophobia, but this was different. It was not a test; it was for real. The psychology was very different. No test can equal the real thing.

The others received as many as seventy rads over the next day as the initial burst hit its peak and then declined. Davies was beyond the point of worry.

Communications with Earth improved quickly with the decline of hard radiation. The solar activity diminished within forty-eight hours to a much more acceptable level. The wave of radiation from the supernova had passed through the solar system, leaving its own mark.

The Gagarin was still two weeks from Mars orbit when it cleared the radiation river. They were almost there; they were almost ready to accomplish the rescue. In the next two weeks, life would be hell for at least one person.

*

"How are you dirty old men doing?" asked Gail.

"We'd be OK if Red would quit farting. Damn, I've never smelled such a brew as his ass can concoct."

"Walter, if you weren't indispensable, I'd throw you in the recyc," Red laughed.

"Tell me about it," giggled Marceau. "He has a very potent product."

"How are your dosimeters reading?" asked Yang.

"Between three and four total," noted Red.

We're six, max seven rads," said Marceau. "The atmosphere seems to be doing its job along with the ship."

"The ground is doing a better job. I wish the two of you were able to make it down here," said Red.

"I think that the river theory is holding up. The calculations you two have made show that the Gagarin must be undergoing a substantial flux of radiation. Earth seems to think it is either an error in sensors or a computer damage scenario. I think I need to trust you, Red, Walter," said Yang. "What does it mean for us?"

"The flood is missing us. The communication problems are only a symptom of the overall problem. We have, at most, one, maybe two days of high radiation levels. Once we get past that, all that we can expect would be the damage that is already done. We need to calculate the worst that can be expected and plan for that. Anything less will be, as they say, a godsend," said Red.

"My calcs say we are in the clear already," stated Walter.

"I would have to agree," said Yang. "The prime flare as well as the supernova has already passed. Can we go ahead and return to a regular schedule?"

"If we give it another ten to twelve hours, I think we can guarantee that," said Walter. "The current situation is decreasing in all forms of radiation. We will have a clean board within hours."

"What does that mean for the Gagarin?" asked Yang.

"They will survive," said Red. "They will arrive on time, and we will be picked up. Davies is the only one that took sig rads. With his asshole constitution, that's no problem."

*

"With that taken care of, we have other business to attend to," Gail purred seductively.

Yang's attempts to rebuff her were weak. "But . . . I'm the commander. I can't get—"

"So you're the commander, you better give me some commands then."

"Ah, but . . . I . . . Red . . . the radiation," he floundered.

It was already too late; she had already begun to remove his clothing.

*

In a matter of two days, Davies felt a loss of appetite, nausea, and the desire to vomit every ten minutes. He felt abdominal pain first then a bit of fever. Within another day, he had diarrhea, the Hershey squirts, and a dehydration problem. His short crew cut hair began to come out in little clumps as he groomed himself.

It was only teeny amounts to begin with. He recognized the need to speak with Moore and ground control. His days were numbered, and deep down inside, he knew it.

His conversations were not a part of the usual press releases.

"Med, I have received a total of just more than two hundred rads according to all my dosimeter totals," Davies concluded in a cool and calculating voice. He then waited for the travel time of the message, the invariable discussions amongst the experts, followed by a censored reply with its travel time.

The practice was to ask a full series of questions for as many crew members as could be, to concentrate the data stream in both directions. Every crewmember filed a Q and A with Moore to be sent in the data stream.

One of the more interesting problems was, Could cancer and radioactive materials pass through the water recycler to be consumed by another? The second part of that question was, If a cell survived, could it survive in another's body and replicate? Under normal circumstances, this would not be a problem. But for the manned deep space probes, the crew had been selected with compatible blood types. The blood was the emergency backup for the limited amount of artificial plasma that could be stored. This meant that rejection possibilities were significantly lowered. Could one cell, or even several, actually cause the cancer to spread? Almost all the opinions were completely negative. A few voices were cautious, explaining that no actual tests were done with cancerous cells.

The recycler lets a few parts per billion of live material through. Davies's main concern was that he might pass some of his mutated cell matter, cancerous and radioactive ones, to the other crew members. With their prior exposure to so much radiation, he did not want to add a bio hazard in terms of his waste matter.

Many of the discussions with Earth Med Department (EMED) and the more personal arguments with Major Van den Ossen, the Gagarin's ship medic (or SMED), led nowhere. Some of the very frank discussions between the major and colonel left Van den Ossen deeply troubled.

Protocols called for very certain procedures to be followed in the event of a death onboard. In the furthering of science, bodies were to be preserved in their suits. Neutral helium (He) would be injected to help prevent oxidation. The suit's cooling system would keep the body even more pristine. At the end of the mission, doctors would have a field day examining the body, looking for any unexpected items. They would a body with which they could really examine bone densities and cross sections, organ responses, and general deterioration of the muscular system.

Davies was not real excited at the thought of his body going to the chop shop for examination. All astronauts sign a little paper, a waiver, which states that their body will be used for medical research by the World Space Agency. But when faced with the reality of it, he chose to take a different path.

He had ordered Moore and Van den Ossen to put him in his suit, inject the He, and throw him out the air lock. He wanted to be the first man to be buried in deep space; he wished the honor and glory of the ancient mariners who had fought the gallant battle and died. Like the seafarers, he would quietly slip into the deeps of the solar system and become, in his own right, a permanently orbiting body around Sol (no pun intended—well, perhaps a little).

He would survive as himself (dead but intact) for perhaps as long as the solar system or even the universe. He would indeed leave a mark in the history books, a dramatic one.

It seemed rather incongruous to the steps he had followed in his career. But all of his actions had been to feed his deepest clandestine desire to be noticed, to climb another level above the others, to surpass.

Psychologists and psychiatrists had made notes of this potential flaw in him but had also noted that he could be manipulated gently into being a prime astronaut, as long as he saw a chance for promotion after every completed mission followed by a new mission.

Now, with the end of the ride in site, he would break from the path.

Other conversations were a little less gallows and even had a touch of humor.

The SMED found Davies in the expected poor condition of a man suffering from a very large dose of radiation. In fact, all of them had some symptoms. What he did not expect was to find Davies sipping his own urine.

"What in hell are you doing?" he almost shouted at Davies.

"You told me to keep hydrated," Davies replied somewhat weakly.

"B-but—," stuttered the major.

"I also refuse to take any water away from a crew that will survive and possibly contaminate the recyc system by returning my waste to it," declared Davies. "I only have a very short time," he said as he looked directly into the major's eyes. "You have your orders."

"Sir, I don't think that it will be necessary."

"Don't try to cover it or lie to me, Paul," Davies said tersely to the major.

"But, sir," the major began again, "you can't keep up with this," splaying his arms at the drink tube that Davies had commandeered for his use.

"Hygiene is more important for the others than for me. I don't want to hear any more about it." He chuckled. "I never suspected that at some point in my life, I would be arguing the advantages or disadvantages of consuming my own urine as I lie dying."

"I hope you're not eating your feces," the major said in a light tone.

"No, no shit yet. Not very hungry," the colonel grinned. "Maybe later."

*

It had been a four-day wait for those on Mars. The rivers had bent their course to extend the time that Mars was exposed. Without a significant magnetic field, that planet did much less to protect its visitors than Earth could. The atmosphere helped, but as on Luna, the bulk of the planet was the greatest shielding available. Temporary shelters, the hex domes, would not provide sufficient protection. That is why all long-term plans called for underground facilities. In the meantime, they were open to this form of attack.

All new expansions of peoplekind faced new trials. Without these tests, they could not continue to grow; they would weaken in the evolutionary scale of things. Like stagnant water, it would lead to conditions in which the species could not survive. Humans are a dynamic creature, flowing through life. Those who are more static contribute nothing and, in truth, gain nothing. Life, by definition, is activity—participating.

Red and Walter were quite comfortable for that unplanned extended period. Yet it had a redemptive value. The newest versions of the spacesuits received a rigorous testing and were not perfect, but they were a significant improvement again over the previous model. Red and Walter added a great deal of personal notations to their database for improvements in all arenas.

The two were unhampered by any office calls, as they termed the bothersome messages from Earth. They knew, better than all but a very small handful of personnel, every detail of the mission. All aspects of the mining operations under way with the Foremen and Laborers were checked by Red and Walter.

They cleaned the machines when the self-cleaning devices were having difficulty doing it themselves. When doing so, they made very important observations and even drew designs and made simple models on how to improve the next generation of machine. Both were experts on the machines, having been involved in the development of four different versions.

They worked in such close contact for so many years that there was really little need for communication in regards to the objects upon which they worked. This left their minds and mouths to free to indulge. This left many people misunderstanding, foolishly thinking that these loud and often foul-mouthed individuals, bohemians, could not be the front line of the space program while others saw past the facade to the keenly honed intellects behind. They perceived these men were doers.

They did a great deal of doing. They were setting up every imaginable device that they had already brought down in the pit with them.

Marceau and Yang rode out the days completing menial tasks that had been on the list of to-dos if there was time. Well, there was lots of time. Marceau continued to barrage Yang with surprising erotic bouts of time consuming.

*

The arguments were mainly focused around what to do with the dying colonel.

SMED decided to take as many samples from all throughout Davies's body as he could. The colonel understood what was up, or so he thought.

After he had ordered the ship's medic to dump him out the air lock, everything seemed to be flowing in that direction. From his now distorted perspective, all that was done around him took on the signs of preparation for his historical departure. Lieutenant Colonel Chang often sat by him; Chang would question him delicately about his favorite things, poems, stories, and so on.

He heard the materials he enjoyed so much being collected in the background. Little snippets of rehearsed talk, vocal variations of haiku, and songs from the last thousand years of peoplekind passed quietly past his highly sensitive ears. They were preparing a funeral package as well as a crude coffin, he knew. He had put the pieces together rather quickly.

Throughout history, peoplekind, for many thousands of years at minimum, held some sort of ritual over the still forms of their relatives or clan members. Flowers, red ocher, very basic jewelry, stone tools that took several difficult steps to produce, and more.

It made his heart warm. He felt his body weakening all day every day. Weight fell off his already wiry frame. He often was racked by coughs, barfing, and diarrhea. He felt he was a real pain to his crew. He was right; he was.

"He is declining rapidly," said Paul Van den Ossen, the medic.

"Yes," agreed Chang and Sakimoto in unison.

Tetrov was simply nodding. His end of the deal was arriving soon. He was the surface lander pilot. It would be up to him to retrieve the wayward crew of the Gagarin.

"Moore asked Chang quietly, "How are the conversations going? We are almost out of painkillers. Your talks have been distracting him enough to stretch them out, but we are gong to have to face the fact that they will not last another day at the rate he is using them. What will we do if we have another incident?"

"We have already been through that," noted SMED. "You asked all of us what we wanted to do with our share, so to speak. We voted to give it to him with the understanding that we might have to do without. It's all gone. What little we have left won't make any difference at this point, even if it's a microlaser cut, or an old-style paper cut," he noted somewhat sarcastically.

"We still have not addressed what the real topic is," Tetrov stated.

There was a long drawn silence before Moore spoke, "We can't drop him in space."

Some of the crew approached mutiny at that statement. Captains were near gods on a mission. There was an undercurrent that tugged at their foundations.

Protocol, the demands of science, political options were weighed against the desires of a man who had saved their lives, every one of them.

Moore continued, "Paul has been able to keep him going longer than any of us expected. Davies had received just enough to do the job, but not a sufficient dose to do it quickly and reasonably cleanly."

"I have been able to take a very significant number of tissue samples, and fluid samples from the commander. When he dies . . . I can take some final bone samples," SMED explained as he hung his head. "That will be all that is really needed to cover a decent scientific study."

"That still does not spell it out. Spill it, Moore. Do we or not?" asked Chang.

"I already said that we will not put him out the air lock," Moore said somewhat tersely.

"That is not the issue," added Tetrov. "Mars is the issue."

"We do not know if they will go for the concept," Sakimoto stated.

Van den Ossen cleared his throat slowly then spoke, "Do we really give a damn what they think? We are the rescuers, and they are the rescuees. Don't we have authority over them as stated in the mission protocols?"

Moore frowned slightly as he spoke, "I have been placed in authority by the head CAP-COM's last tight beam message. My rank may be questioned, legitimately, by both Yang and that Red O'Hare character. Their commissions came before mine, and they know it. They do *not* know that Earth has given me a backdoor 'do not go to jail card' that allegedly gives me authority. They would likely balk at my taking a heavy hand approach."

Tetrov nearly shouted, "Balk! What right do they have to do that?"

"The right of survival," whispered Moore. "Which of us could honestly say that we would handle things as well as they have?"

"That's crap, Colonel," said Chang. "You can't make that kind of comparison."

"I can make any comparison I fucking want," Moore shot back. That caused the others to recoil somewhat. "We have to recognize their accomplishments, publicly, since that is good for the program and true as well. We have to recognize their loss—"

Sakimoto interrupted, "And our loss?"

Moore turned and stared hard at Sakimoto. "Don't interrupt me ever again." The silence became heavy after nearly a minute.

"Yes, sir," responded a deflated and corralled Sakimoto. The others got the message very clearly. There was a new sheriff in town, and he was marking his territory. He wanted clear and concise discussions but no questioning of his final word. He was laying the groundwork.

"If we do not have agreement, then we have to go the quieter route. The publicity of the situation demands it," Moore continued.

"Davies was very clear," SMED commented.

"Is he in his right mind at all times?" asked Moore.

"He is doing—," started SMED.

"He's still drinking piss then," said Moore coldly. "Is that really normal? I think that he is past rational thought for most of the time. He has always been entirely by the book, until now. All of a sudden, he wants to change things. Now he is breaking the rules. What can we do to get an agreement arranged with Mars?"

"If we can get a laser link set up, we might be able discuss this issue without eavesdroppers," suggested Chang.

Sakimoto meekly noted, "Earth will note that we have retasked the laser com system and wonder why."

"We are field testing the system under difficult conditions. Think that will work?" said Moore.

In the end, they did not discuss this issue with anyone other than themselves. They made their decisions, and be damned with what the others thought. It made them feel more in control of things, and Moore knew and used that.

*

Burt was confident. The two-month trip of the Gagarin to Mars was almost over. Even though there had been severe complications, he knew that things were going to work out. Many of those he had made bets with were already gloating as if the ship had been destroyed and every astronaut buried. He smoothly poured the rest of his mug into his unshaven throat.

"Hey, Burt," said Paul as he slid on to a stool next to his partner in gambling crime. He looked around stealthily and moved his head close to Burt's and whispered, "The mob's looking at how we are betting."

"W-what?" stuttered Burt. The mob had never been completely eliminated by any government, and that included the UE. Battered and constrained, it was making a strong resurgence now.

"They are not mad. They wanted to ask a few questions." Paul turned and bobbed his head at a man in a cheap suit sitting in the back of the bar. "Come on, he wants to talk to you."

The two made their way to the booth in the back where they sat opposite the man.

"One question, gentlemen. Will they make it or not?" said the man.

Burt said rather curtly, "They will make it."

"Good," said the man. "We have heard of your bets and took that as a signal. We have been playing the odds in Red's favor based on that." The rotund man smiled and took a deep drink of his mug of beer and said, "To their continued well-being."

*

Chapter XXII:

The Cemetery

22

010110

"Good evening, this is Mary Smith-Jones reporting from the North American Continental Desk, for WWNN. The Gagarin has arrived in orbit around Mars after many tribulations.

"Commander Davies succumbed to radiation poisoning and died just a few days ago. The rest of the crew of the Gagarin has been exposed to dangerous amounts of radiation from the dramatic supernova and Sol's outburst. The marooned astronauts on Mars were more protected by the tenuous atmosphere that exists there, but they also faced great danger. Astronauts O'Hare and Black Bear found themselves trapped for several days in the underground laboratory they were setting up. They could all have just waited, but they continued working and accomplished a great deal. As if things were not hard enough for them, this unfortunate series of events had to mar the expected rescue.

"The Liberator, the landing craft that replaced the Bradbury, will soon disconnect from the Gagarin, the mother ship, and descend to the surface to gather up the astronauts left behind on the last mission. It will also pick up all the rock and core samples that have been collected.

"These will be delivered to a secure facility on Luna where they will undergo a substantial series of tests. Some of these examinations will determine whether or not the samples contain any possible dangers to life here on Earth.

"We will keep a live camera going beginning with the undocking procedure all the way through the landing. During the meeting of the two crews, we will return for more live action.

"This is Mary Smith-Jones. Stay tuned as we bring live web and v-link."

*

"E-VIDCOM, the rearward camera appears to be faulty," announced Tetrov as he prepared the lander. There was nothing wrong with the camera, and Tetrov found it difficult to lie outright. He wimped by, not looking directly at the forward-looking camera, and muttered the words. His checklist was growing quite short now. He and Moore had given the lander a zealous going-over.

Since Tetrov was going alone, except for the body of Davies, there was no room for others; it all depended on him now. The entire mission and perhaps all further manned missions to Mars required success. He was the pivot, the fulcrum. He was not so sure now that he liked to be in this position. If he blew it, the entire thing could be blamed on him. He felt a certain kinship, in an odd twist of fate, with Red and Walter. It shifted his perceptions of them considerably.

Davies had died in a fit of puking. He had looked worse than a drunk in his cardboard box in an alley. He was unshaven, bone thin, and smelled of bile and feces. His skin hung in loose folds reminiscent of very quick weight loss. His eyes had fallen to a dull, fetid yellow. After samples had been taken, they had zipped him into his suit and pressurized it with helium, just as Earth expected them to do.

There had been official ceremonies on Earth, but they did not appear to be much more than hopped-up press releases. It seems that Davies did not stimulate the people's excitement very much. He did rate a black with twisted lime green wristband to go with all the other high-priced selling trinkets of the space program. There would be the usual lawsuits over the profits. The UE did get its share.

Tetrov continued examining each and every part of the lander. He suited up one day to physically view every outer portion of the ship in order to guarantee its integrity. His eyes were not really needed, but it was protocol for him to perform the spacewalk.

The eyeballs (the little basketball-sized robots that were mostly cameras) could and did examine both the Gagarin and the Liberator, the landing craft. They could, with their multifrequency filters, detect microholes in the skin of the ships. They would look for escaping gases, heat, even frost as they moved gently about the vehicles.

The tiniest hole could mean the end of the lander as it makes its way through the atmosphere of Mars. Molecules would rend the smallest tear, rip, or any kind of hole into a larger and larger gap in the protection for the vehicle. This is what happened to the space shuttle.

Hot gases tore through the wing structure until it failed. Ever since then, there were many ways created to try to repair various problems. Most of them were not tested under real emergency conditions, only conditional field tests. None of the astronauts *really* wanted to try any of the repair systems. Each and every one of them would volunteer though, against better judgment.

These people, astronauts, are that tough, that gung ho, that willing to challenge death. They bet everything on four kings while they know four aces or straight flushes will kill them. That little percentage eventually would catch up with some of them. It has already caught up with too many. Recent additions still brought tears—Holtman and, to a lesser extent, Davies.

Little did Lieutenant Colonel Davies know, while he was alive, that his cousin would call him a failure. In death, the preacher would lead many in prayer, for more to die as an example that God did not want peoplekind in space.

*

Pastor Davies, among all the other major players and some tag-along wannabes, led the flocks in a group session in which the deaths of Holtman and his own cousin Lieutenant Colonel Davies were praised as acts of God. The roadmap was being lit by the light of death. The highway, the main pathway to hell, was scientism.

"My fellow lovers of God, we have seen repeated examples of the desire of our Master. We do not belong on the distant planet Mars or in space at all. He has created this planet for us, us alone, this place and not out there. He intended for it to be a paradise. Adam and Eve caused a rift to come between peoplekind and God. We must do all we can to reduce that rift until Jesus returns and the rift no longer exists. Scientism does nothing but increase the rift between us and God.

"The death of our savior promises that that rift will fall into the pit with Death and Hades. These scientists seek to deceive us into believing that they can solve all our problems. This is an issue of sovereignty. God is the only one to have the right.

"Peoplekind assumes the right to leave His world. They hope to escape his judgment. Fools! No one can escape by traveling some trifling distance.

"Peoplekind assumes the right to change the form of life that God created in His image. Scientists think they can improve on God. Blasphemy!

"They will all be judged by the Lord and Master. They will face hellfire, death, and abomination. Every death we see now represents the death of millions of disbelievers. Pray for the deaths of evildoers. It is God's will."

The messages of hate came not only from Davies. All the members of the alliance had words to say.

*

Louisa Jamala Reta, pope of the renamed Gnostic Free Church, spoke differently. She exhorted her followers to pray for the families of the deceased, the astronauts, and their families. She countered the rhetoric of the Stalwarts with messages of love, of coming together, of acceptance. She worked to soothe the populace, to work out the angers and usher in a state of refreshed happiness for everyone. She did not believe that the end was coming, just a new beginning.

Her orations were directed not only at peoplekind but included the Hsas. She was passionate and fiery in her delivery. She brought into play all the skills she had developed over the years.

Her earliest recollections were of the church and her family. Her whole life had centered on religion and belief in a higher power. It had only been natural for her to become a nun after the debacle with Illya. With time, she made mother superior in the Columbian nunnery.

Her passion for people led her to become a local hero. She tended to the very poor in the slums; she brought hope with her messages of a kind god. Reta established food kitchens in several cities. She took the time to appear at every one of them on a regular basis, giving ringing endorsements of the message of love, of belonging.

Reta believed that all belonged; all could be redeemed. She had to; she needed it.

Her message was sent out on the vid for the entire world to see, if they would. She countered the harsh messages of the Stalwarts with acceptance, peace, and understanding.

*

The Liberator, aptly named, landed less than two kilometers from the Bradbury. Tetrov had nailed it. Freedom, an actual rescue, was within sight for the crew of the stranded Mars mission. Red and Walter were already driving toward the landing site. They were expecting to pick up one live body (not two), one of them dead. No one had informed them of the plan devised by those who now controlled the Gagarin.

When Red and Walter arrived, they were pulling the trailer. They had plans to haul back all the supplies that were being brought to restock for the next manned crew. They would, however, have one extra package.

Tetrov was tired, but he still was functioning. He was preparing all the cargo for offloading. He opened various hatches electronically and manually. Straps were undone; small items were stacked in readiness.

"Ho on board," Red called through his mic as he dismounted.

"That you, Red?" returned Tetrov.

"'Fraid it is," said Walter. "Just can't go anywhere 'round here without him stickin' his nose in."

Tetrov chuckled. He was one of the few on the Gagarin who had respected (much less liked) Red. His earlier mental shifts grew even stronger. "What are you two up to?"

"We saw this UFO thing landing and came to investigate," said Walter. With mission control several minutes away in the timeline, they could afford a little jocularity. People on Earth would eat it up as well.

"How about setting up the rovers for a drop?" said Red.

"They are ready for you now," stated Tetrov.

"Well done," said Red. His tone showed he meant it.

Tetrov noticed it and ratcheted up his understanding of Red. Red expected a good job, respected it. He continued with, "I have all the internal cargo unlashed as well."

"Well done," said Red. "Put on your suit, set your radio to an off channel, and pull up a snooze chair."

"What?" said Tetrov, somewhat surprised.

"You've had a forty-eight-hour stint. Take the sleep. We'll put you to work later. That's an order," Red said soberly.

Tetrov was taken aback. Red was such a hard driver for him to suggest—no, order—him to sleep. All he could do was say, "OK." Red did rank him, all of them in fact. "But why should I put on my suit?"

"We're going to jam both the air lock doors open," Walter laughed. "It makes it so much quicker to get the supplies off."

"What? How do you do that?" Tetrov shouted. "This I need to see. I thought it was impossible for both doors to be open at the same time!"

"Calm down," Red noted. "We'll show you how it's done later. For now, just get some sleep."

Red and Walter spent the next two hours releasing the two rovers and checking them out. They then set up the program they had devised that would allow them to open both doors of the air lock. When they entered, there were two fully suited people in the seats of the pilot and copilot.

"What the fuck?" said Red, actually taken aback for one of the few times in his life.

"Who's . . ." drawled Walter as he looked wide-eyed from one helmet to the other.

Red simply banged on both helmets. Only one moved.

"Damn it," said Tetrov as he rose from his chair. "First, you tell me to sleep, then you wake me."

"Who in bloody hell is that?" demanded Red. He then answered his own question with, "Davies."

"Yeah," answered a Tetrov groggily. "We want to plant him next to Holtman."

Red looked quickly at Walter and said, "That is a good plan. I think it is appropriate that they share common ground. They have paid their dues."

"We can load all the materials then put him on top," suggested Walter, his mind ever on practicalities.

"Does Earth know of this?" asked Red with a shit-eating grin at the thought of the consternation this would cause.

"No," Tetrov said with a slight smile of his own. He knew what Red was thinking.

Red laughed heartily, "Good!"

"They will soon," Walter noted happily.

"Too bad!" replied Red. "We are here. They are there. We can begin the first cemetery on Mars. Why take them back to Earth where they will be amongst those that really can't understand? There will be other deaths in space. Let's honor these first as best we can."

"I have one thing to go there," announced Tetrov. He led them to a container that held an obelisk. It was made of steel covered with titanium glass. Red and Walter were stunned when they read all the names of everyone ever lost in the space program.

"Fitting," was all Red said.

"You guys have been preparing," said Walter. The list was not that long, yet. Walter recognized a few names—White, Chaffee, Grissom, Husband, and more. They were in order of occurrence. The list covered all of the nations involved in the space program.

"I've had two hours. Let's get busy," said Tetrov as he yawned.

Red looked at him hard, visor touching visor, and said, "Good man."

In silence, they began to unload all the internally stored goods. Another three hours passed before they were ready to remove Davies's body. It was an awkward moment. Davies's helmet banged against several surfaces as they dragged him out. Earth saw it all since Red had turned all the vids on. Red had set the broadcast beam to all major frequencies. The occasional *shit* and *damn* only made it more human. They finally had Davies tossed and trussed on top of the trailer. Each astronaut took one of the rovers, fully loaded, and headed off to the refuge Red and Walter had created.

Red drove one of the Liberator's vehicles in the lead, followed by Walter, then Tetrov in the second Liberator vehicle.

The caravan could be observed from one or all of the cameras mounted on the rovers or with the eye-level helmet cams of the astronauts. Within hours, all of Earth was wrapped up in the fact that Davies was strapped to the top of the materials piled on the trailer being pulled by the Bradbury's rover. Twenty-four hours later, everyone was discussing it as if it was the most important event in history.

Some, mostly the more conservative religious groups, likened it to the treatment of a slab of meat. They tried to stimulate anger at this blatant disregard for how bodies should be treated. They argued that Jesus himself was buried, not bounced along like a rag doll on a Conestoga wagon. It was another example of the dehumanizing forces behind scientism.

Liberal religious groups did not attempt to ferment such feelings. They merely upbraided the WSA for not forcing the astronauts to bring home the two bodies. They, along with the conservatives, felt that Earth is where we were intended to *end* our time. The book of Genesis tells that we came from dust, and to dust we shall return. Both groups lobbied for that dust to come to Earth.

The WSA was only concerned that orders were not being obeyed. The chain of command was again failing. It did not appear to be a serious breach. It was, as the psychologists and psychiatrists noted, a minor rebellion to vent the pain and loss of their commander and friend. There was always a feeling among a crew that they had let down their brethren when an incident like this occurred.

Military leaders and emergency crew leaders understand this feeling in the reverse of Davies's situation. The loss of a person can be deadly to the mental efficiency. They must continue to order people to do their deadly business. The World War II German Field Marshal Rommel stated the obvious, "I can not become involved in any one man or I lose perspective."

Many others, such as Red and Walter, followed simple expediency, basic need. This had been drilled in even harder over the last few months. Red and Walter had a great deal of respect for those who had given everything for the program. Neither personally liked Davies, but they saw through the eyes of command. This man had saved his crew first and, through them, the marooned astronauts on Mars. There was no disrespect at all. It was not the man, just his body.

It took three Sols to remove and properly store all the supplies from the Liberator. The last items off were the new Foreman and its Laborers. They were immediately set to working on the tunnels that would become the final Mars base facility. In the meantime, the mined materials were being processed for the construction of new hex domes that would provide temporary sanctuaries.

*

Luna was already providing the raw materials to SS-2 where they would make some of the beams and electrical systems for the domes. Within the year, these items would be sent via cargo ships. The next crews that would visit Mars would be framing those domes and continuing the work on Mars Orbiting Station 1 (MOS-1).

Robots, reaching a true artificial intelligence (AI) level, would be delivered to Mars orbit to add to the construction force. Within a few years, the station would be completed. This would include a new Hsa birthing chamber.

Domes and caves, stretching kilometers in length when completed, would be able to house as many as thirty thousand individuals, Hsas for the most part. To start with, these would be the offspring of SS-2. As the MOS-1 station becomes fully operational, Hsas will oversee the birth of a whole new set of generations of Hsas, ones able to function on Mars just as their cousins are best adapted to Luna.

*

Red and Walter were working privately to provide a fitting receptacle for both Holtman and Davies. They had stripped some titano-aluminum panels from the various tunnels. They made one trip out to where Holtman was buried. They let no one know what they were doing. They also added a couple of temporary chairs to their trailer.

Sol four of Tetrov's visit to Mars saw him more aware of the absolute grandeur of this distant world. Without the grind of a great deal of labor and preparation, his mind was free. He was, today, much like the crew of the Bradbury on their first excursion outside of their ship.

Tetrov stared about like a schoolboy suddenly thrust into his dreamworld. The colors set him to the edge of emotional stability. In general, he was not much given to the arts. Today he felt as if a poet had awakened within his bosom.

His mental ruminations were interrupted when Marceau gently pulled his arm and pointed to one of the chairs on the end of the trailer. He sat without a word as he looked at Davies's lifeless bulk lying strapped lengthwise on the trailer between the chairs.

All Earth was observing. Yang had made an announcement yestersol about the upcoming ceremony. The WSA was concerned. These proceedings were not a part of the scheduled events prior to liftoff. Any minor accident could leave them all marooned again. It would be a terminating blow.

The procession took hours as it moved slowly to the resting place of Holtman. None spoke. Earth made several queries, but each was ignored. This only added to the internal disquiet and apprehension at all of the mission controls.

"We're here," Red announced unnecessarily. The few parts of the rover still remaining said all that was required.

They all disembarked, again without speaking.

There were two gleaming shrouds standing open and waiting.

Walter quickly moved to unstrap the body of Davies. For all his speed, it was done with certain gentleness. Yang and Red each took a leg while Tetrov and Walter grabbed an arm. Marceau took the head. They awkwardly moved him to the first of the coffins where they gently placed him. They laid him straight as a line.

All of them moved to the cairn that Red and Walter had constructed over Holtman and began to remove the rocks. It took several minutes to dislodge them and to brush away the gathered sand.

Again, the men lined up to take the legs and arms while Marceau carefully cradled the head, making sure the German flag covered his face. Holtman's remains were placed in the other homemade coffin.

With both coffin lids open, Yang began to speak, "We have lost some friends. We have lost many in this field of endeavor. We work in the most dangerous business that exists. We also work in the most rewarding field. We get to see what some describe as God's handiwork. Others say it is the most incredible collection of random circumstances that brought all this about. We are not here to try to decide that. We are here to say good-bye to two good men. These men gave their lives in the furtherance of the program designed to help Earth. They gave their lives for us and those orbiting in the Gagarin. We all owe them our lives. Earth owes them the survival of this program. What has been done by these and those that went before them has all been done for the benefit of Earth." He motioned for Tetrov to get the obelisk.

Tetrov walked solemnly to the trailer and picked up the bulky object that the crew of the Gagarin had prepared. He placed it between the two coffins and propped it up with some rocks. He turned to face the same camera that Yang was using and began to recite all 137 names of those who had perished in the space program from every nation and the UE. It took all of twenty-three minutes to list them all.

At the end, Red and Walter closed the lids and used fusion welders to seal them carefully. When they were done, they stood and saluted.

"Let everyone know that we honor their memory," said Yang.

No more was said. They boarded the rover and trailer to return to the Liberator. It was time to leave.

*

The liftoff went perfectly. The cheers resounded from Earth, through all the stations, and Luna. The lost were found. The return trip was so uneventful that many let the whole thing fade from memory until the Gagarin neared Earth a few months later.

At that point, the plans for celebrations grew rapidly. All Earth was anticipating a day, a week, of partying. Many hoped to see the crew of the marooned astronauts in parades that would wend their way through all the largest cities on Earth. They wanted to say they had been there. Those plans had to be put on hold.

The Gagarin docked with SS-2 amidst wild expectations. Each astronaut was greeted personally by David DeNoe, Klar, and, eventually, by every member of the station. The doctors were next. They gave full examinations to each. What they found required that the marooned remain on SS-2 longer than the rest of the astronauts. Their degeneration had gone a little further than expected. They would have to go through a longer rehab than Moore and his crew.

Moore and the rescue crew found themselves the center of somewhat restrained but still very exuberant celebrations when they returned to Earth. The WSA used these people for as much propaganda value as they could. The entire Earth was still anticipating the real heroes to be on hand. That was not to come for all of them. In the end, only Yang returned to Earth.

Red, Walter, and Marceau refused to return when the call came. The news was not at all welcome at the WSA. The Earth population was not surprised at the refusal of Red and Walter (or at Marceau's since it was obvious she loved Red). These three would lead their own lives in their own ways. Many, in fact, approved. Red, Walter, and Marceau belonged in space. People could not imagine them elsewhere. They simply belonged there. Everyone had grown so used to the news of them doing something out there.

The WSA quietly made many requests for them to return, to no avail. Even commands for others to take them into custody fell on deaf ears. Too many owed too much to Red and Walter. Every Hsa felt they owed their existence to these two. The *sapiens* spread thinly among the space crews felt too much loyalty to turn on them as well.

Some in the WSA wondered if their survival had actually been for the best. In tightly controlled and secret meetings, they discussed the Apollo experiences with White, Chaffee, and Grissom. The deaths had actually, in the long run, improved the space program.

Finally, the WSA announced that they would support the continued mission of these unrepentant heroes.

However, the secret meetings continued. The discussions centered on how to retake full control of the space station. It was also decided to tap the military again for more of its crew members and not just the commanders and pilots. A slow infiltration might go unnoticed while a sudden influx of military personnel would be all too obvious. Also, by increasing the number of launches, they would speed up the process. It was felt that Red and Walter would have to be dealt with discreetly at first. Then the Hsas could be dealt with. The idea was that Red had philosophically infected them.

They were correct. They totally underestimated by how much though. Open minds are hard to control while closed minds are full of dogma that will not see the truth. It did not matter if it was political or religious dogma. They are sheep for the shepherd. The shepherd does not always have their best interest in mind though but an agenda, shearing or lamb chops.

With control of the World Wide Network News, the UE felt they could master the situation. They did not count on how effective a single space-bound network could be.

Klar, the computer genius, was able to circumvent most of the controls that the earthbound controllers were able to come up with. She was also able to include a very well-known newsperson from Earth.

Mary Smith-Jones was asked by Red directly to listen to the position of the Hsas as well as his own and Walter's. They would provide her with a direct interview, an exclusive. It was the opportunity of a lifetime as far as she was concerned. The UE and their underlings were not quite so sure about this. But they let WWNN go ahead with the interview. They could hardly prevent a space station from broadcasting to Earth. But allowing this cooperation would turn out to be a mistake in the end, a colossal one.

*

"This is Mary Smith-Jones with a special news event that is an exclusive for WWNN. We have an interview with the heroes of Mars—Mick 'Red' O'Hare, Walter Black Bear, and Gail Marceau. Commander Yang is currently unavailable as he is in China, and the time difference prevents him from joining us."

This last was a lie. Yang was being quietly censured for his part in the lax behavior of his crew. Even though he was a hero, the leaders of the Southern Asian Region felt embarrassed by his apparent lack of control over Westerners. There was still a remnant of racism in every part of the world.

"Red, can you give us an idea about what is happening on Space Station 2?" she asked.

"Only what is supposed to happen," Red replied calmly. "The Hsas have taken control of SS-2 just as planned. It may be a little earlier than planned, but that is of little consequence. The only thing that really counts is the results. The Hsas have already increased production significantly."

"But what about the forced removal of several astronauts?" she returned.

"All of them were people that refused to work with the Hsas anymore," Marceau replied quickly from her split screen position. "They are the ones that were working contrary to or not fully in step with the program that had been outlined from the very beginning. Their removal is a part of what has led to the boost in raw materials recovery as well as finished products. The Hsas have been able to amend protocols more easily to fit the needs of the program."

"That takeover was not to take place yet," Jones reemphasized.

"Timing may be modified as needed," said Walter. "We have always modified schedules to plus time or minus time. This was a response to a critical need by us, those marooned on Mars. The Hsas felt they were being kept from their duty to do all that was needed to save all of us. They also contributed significantly to the saving of SS-2 during the crisis there. At that time, they were not even considered a true part of the crew.

Everyone thought of them as children, but they proved that to be an erroneous belief. They have stepped up earlier than expected."

"Then you posit that the Hsas did nothing wrong," she stated.

"Of course not," Red stated firmly. "They only did what was necessary and beneficial."

"Are you friends with David DeNoe?" Jones asked.

"Of course," they all stated in unison.

"And you feel that that does not bias you?" she queried.

Three different negative responses came quickly; Red's emphatic "Horseshit!" was easiest to understand.

Walter continued with, "Many people do not like us since we can be somewhat abrasive."

Marceau chided, "More than a little."

Everyone smiled.

Red chuckled. "We base all our reports, evaluations, and labors on functionality. Sometimes there are negatives, and people take that personally."

"Do you feel that you have made some enemies?" she asked intently.

They each looked somewhat troubled by the question, as if they did not really want to be asked that.

Marceau spoke slowly, enunciating each word with thought and care, "Yes, unfortunately, we have." She included herself with Red and Walter purposely. "There are some who have turned their backs on the program because of the negative evals that Red mentioned. I'm aware of a few that have a personal vendetta against these two. We never intended any harm of any kind to any person. But as in any profession, we've made decisions that take people out of the positions they worked for all their lives. These are the hardest decisions for us to make. We are destroying what they perceive to be their destiny. I myself," she admitted for the first time, "almost became one of those people.

"When Red reorganized the layout of Luna warehouse, he altered my domicile, and went over my head to do it. I was furious with him even though the changes were for the best, and I began to look for ways to take him down a peg or two. When I thought I had found a way, it backfired. At that point, I just knew that my career was over. But then, I was put in a position to redeem myself, the SS-2 situation. And little did I know, Red already had been making moves to get me a shot at Mars. He thought I was very good at my job. He did not take advantage of his position. He used it for the furtherance of the program." She blushed as best she could when she finished, "It was not until Mars that I really understood how I felt about him."

Even Red's ruddy features showed a bit of a blush. Walter chuckled heartily at his friend's discomfiture. Smith-Jones giggled lightly as well. The world was in love with these people.

"Thank you all," said Jones. "This is Mary Smith-Jones for WWNN."

*

"What do you think about that, Burt?" asked Paul.

Burt responded brightly, "I think it's exactly on the mark."

Paul queried, "You don't think it is politics?"

Burt snorted into his beer and then cleared his throat. "Don't be dumb. Of course it's politics. Everyone is looking for someone to blame for things not going according to plan. It begins with the people that did not expect or predict a sandstorm. It continues with the communications problems where the astronauts used open microphones when procedure dictated that they not do so.

"That, followed by the so-called conspiracy on Luna and SS-2 to propagate those messages to all of Earth, led to even greater finger pointing. The religious accusations, the death of Commander Davies, and even the burial incident added fuel to the fire. The only thing that could put out the fire was the safe return of Red, Walter, Marceau, and Yang."

Paul responded with, "It seems to still be smoldering to me." He gurgled down some beer and shook his mug at Mick.

Mick hustled over and poured another. "I have to agree with Paul on this one. The fire died down, but it is gaining strength again. Them religious fellows are stirring up the fire and placing new wood on it. There's some here on Earth that aren't religious that are against all the space business as well."

Mark burbled his bit, "But ain't that what Red just took care of? He said that all that really don't matter. Didn't the mission get done? Does it really matter that there were a few problems? Ain't there supposed to be problems every so often? This was the first time people stood on another planet. No one knew for sure what to expect."

Everyone looked at Mark, somewhat surprised. His conversation was not usually so lucid and cogent.

"Bull's eye," said Burt appreciatively.

Paul looked at Mick, eyebrows up, and said, "Get that man another beer."

"On the house," responded Mick, pouring a mug for one of his best customers.

"Where'd you come up with that?" asked Burt.

Mark shrugged. "Common sense."

*

Chapter XXIII:

Continuing the Work

23

010111

"This is Mary Smith-Jones for WWNN news. It is nice to be back from my organ installation surgery, and I thank all of those that sent me such nice e-mails, vids, and flower pics.

"The WSA has just announced the go-ahead for the Asteroid Capture and Mining Program or ACAMP. A near Earth object, or NEO, named Rex has been chosen.

"A NEO is an asteroid that crosses the orbit of Earth. Some may even, at some time, offer a threat to Earth. A collision with one of these objects would be devastating to Earth and could even end all life on Earth. Several have devastated Earth over the billions of years since Earth formed. One is known to have brought the final end to the dinosaurs.

"Scientists have been tracking these objects for more than sixty years. With this growth in knowledge, the realization developed that these asteroids could be mined for raw materials.

"Rex is a heavy metal-type asteroid containing lots of iron and nickel.

"Miners hope to be able to produce as much as twenty hundred metric tons of worked iron per day for use on Earth. The raw ore will be processed in space to prevent pollution here on Earth.

"The material is earmarked for construction of housing for third level areas, those that have had to wait the longest for modernization." Those that had had to wait longest were the ones that were most modern originally and were now in a state of decay, but she could not say that.

"Currently, the engineers that will initiate the mining have not been chosen, but most are betting on the favorites of Red O'Hare and Walter Black Bear.

"The immediate follow-up mission will be the Comet Capture. Astronomers have discovered a couple of possible orbs for this mission. Both appear to be new injections from the Kuiper belt. Their positions and orbits will allow for a quick follow-up from the asteroid mission.

"We will bring more news on this as soon as we get it.

"This is Mary Smith-Jones for WWNN."

*

Red and Walter felt a bit weird. It had been a couple of months, and people still stared at them from behind.

"I still think it's our asses," Walter said cheerily as they moved out of one room and into a corridor. He said it loud enough that those observing them heard the remark, causing them to snicker in amusement.

Red and Walter were put in the position of instructors of more than engineers. The information they had gained, the raw experiences, were invaluable to the next Mars crews. Their engineering abilities and ideas for ship improvements would not be included until the next generation of ships was built. But in terms of survival skills, the tricks of thinking in unusual ways to accomplish a goal, that was primary.

Little did the WSA realize that this inculcation of free thinking would come back to further haunt them in later years. Free thought is the first step toward free people, the first urges of rebellion against restriction. The minor deviations seen so far were of little consequence except a publicly unperceived black eye for the politicians. Several did eventually lose their positions for not foreseeing this possibility. Many of them went so far as to join the opposition for being fired.

The two prophets of the future were often shuttled from SS-2 to SS-3 and back. They became officially involved in the program on a much larger scope than previously possible, against many objections from a vocal few. Their interactions were with thousands of individuals now instead of the hundred or so before. They intended only to improve the ability of each individual. They got that and more. They didn't really set out to do this; they did not truly have an agenda other than to teach reality. The training from Earth did not really cover all the aspects like Red and Walter did.

For Red and Walter, all they wanted to do was work. They really wanted to get back to Luna and get busy with their private plans. Things had been left too long. They hoped that nothing had gone wrong with any of their previous works.

The two also wanted to see how the water digs were going. The biomists and engineers had begun to delve out the two vast water biomes. The larger was to be a saltwater aquarium of unrivaled dimensions. The shape was oval when looking from above. The tubes were nearly two kilometers across and half a kilometer high.

Spaced pillars of natural rock swept up in hyperbolic forms as additional support for the great mass of rock above. Secondary entrances were formed along the bottom surfaces, sides, and top. Each of these would be used to drive the flow of water around the oval as well as allow the oxy-generators to help control O_2 levels until a stable equilibrium could be established.

For those who have had gerbils and hamsters with their tunnels and tubes, combine that with a fish aquarium, and you will get the basic idea.

Lights would provide an artificial source of energy for the development of algal mats and kelp beds. There would even be sand. A large amount of Luna silicon was being tumbled to provide as realistic a surface as possible on the bottom. Clams, mussels, crabs, and more varieties of fish than in any earthly aquarium would inhabit this fantastic project.

Artificially stimulated coral reefs would be created with the help of the aquatic members of the Underworld project. These people had been working for thirty-five years on living underwater and repairing the Earth's saltwater biome. Their knowledge would be invaluable. A select few would even visit Luna as advisors.

Their project had started even earlier as a simple save the reefs program, much like the save the rainforest environmentalist. As the obviousness of it all grew, it was really too late.

*

The environment has since shifted greatly in the sixty-plus years since the 1980s when the problem really became a public issue. With warmer temperatures, English farmers were growing crops that, twenty years ago, were unable to survive their previous climate. The desert of the American southwest expanded to fully cover Southern California and the Baja peninsula north through the megalopolis of LA, fully across half of the Texas and Oklahoma subregions of the North American Region.

The Sahara was encroaching on areas that had only recently been put under irrigation in the most marginal conditions. The Great Gobi Desert was developing dust storms that lasted months. The last of the permafrost around the world was melting. In places where it had existed only centimeters below the surface, the layer had subsided as far as ten meters down. It allowed for some creatures to survive more easily while others faced certain extinction if they had not already vanished from the wild.

Ocean levels had risen nearly a half-meter, not the predicted several. Large amounts of water found itself trapped in new freshwater systems, an unexpected use of all the water. Rivers found new courses that extended their reach to areas once light in water while leaving others dry. Humidity increased a few percentage points across the globe.

There were dead spots in the oceans that were growing larger by the year. These spots are where the O_2 is severely depleted and fish cannot live. One of the solutions was to put floating solar-powered bubblers, much like those in an aquarium, to try and minimize the effects in the less-damaged volumes of water.

Continental glaciers were all but gone. Amazingly, this allowed the mountains to "grow." Without the massive weight of all that ice, the rock sprang back up, slowly, around a half a centimeter per year.

One strange change occurred; after a short thirty-year period of increased hurricane activity and intensity, they dropped off. The east coast of the North American Region found itself seldom hit by the wind-driven terrors of the past. Granted there were still giant storms but nothing like some of the worst. Nothing like the Long Island Express appeared anymore. What few storms still occurred usually hit the Central American Region or dispersed over the Caribbean.

This weakening of disaster preparedness, lax and lazy emergency services, would pay terrible dividends someday. In part, it had to do with the fact that terrorism and nationalism had declined so greatly.

Change is the function of nature. It has been going on for some four and a half billion years. Species come and go. Is it possible that other animals had such an effect on the Earth, unconsciously most likely, but change nonetheless? What is the difference? It became quite obvious that the changes could not be stopped, only ameliorated. Peoplekind had to find ways to live within the changes and use science to create new niches for some of the species that could be saved.

*

The freshwater biome was actually to be several separate entities. There would be the equivalent of vernal pools, ponds, marshes, streams, a river, and a large lake. The intricacies were unending strings of interconnection. Many forms of life in jeopardy on Earth may actually be saved here in a slightly modified form.

The same genetic advances that allowed for the creation of the Hsas also allowed for the same arrangement to be made for all life forms. This made certain Earth groups very angry. Again, the hornet's nest was stirred by the leaders of the conservative religious groups. Even some of the more apolitical groups and individuals were concerned that too much money and effort was being spent on

Luna and its programs. The climate changes were imposing a growing need for closer governmental and scientific scrutiny. With so many absorbed in the space program, it was hard to find qualified scientists and to get grant money for what was considered a done deal by many.

*

Hamad, now the caliph after the Earth-wide felt death of his father, took a hard stand. He had learned the use of voice and body language directly from his father. He stood behind the podium and waited for his cue. When it came, he was more than ready.

"My people," he began with his strong voice, "again the scientists flout our traditions, and more importantly, they blaspheme against Allah. They attempt to create new creatures in imitation of the true Creator. Their falseness is in great evidence. Loudly they proclaim their capabilities and successes. Yet their failures are many and grievous.

"It is science that brought about the present conditions. Conditions of growing hunger, the death of many of Allah's animals, the destruction of much of Earth, and, most importantly, the perversion of the hearts and minds of so many who wish to be true followers.

"We have had many troubles with the Jews and the Christians. We have been enemies. But in this we all agree!" he shouted as he slammed his hands on the podium. "There must not be any continuation of this. It is time to initiate a jihad. We must have an open war on the politicians and the scientists that are a danger to our beliefs."

Hamad quickly left the stage. His pronouncement would send a message to Interpol, still in use and expanded by the UE, to pick him up before things got out of hand. He was taken by some of his trusted advisors to a secure location. He would move every day from one location to another and never stay in the same place again until it was all over. Much was still remembered from the days of Al Qaeda and other underground movements.

*

The elder Davies was in a low, foul mood, but he was still on fire in his heart. He waited until the last second. As soon as Hamad was done, his facility cut in to allow him to deliver his well-practiced oratory.

"We have tried to work within the system," he began with his now frail voice. "And it has not worked. The UE continues with its abominations. It now broadens that to include the fishes of the seas and lakes, the very plants of the land and waters of our beloved Earth. They claim to be saving them for future generations, yet they are the ones responsible for the deaths, the extinctions, of our Lord's handiwork. We cannot stand for this."

"We have warned you for decades that this path was one of destruction. Many claim that we have a static message. Our message is that of God himself. It will not change since He is truth, unlike the scientists who must change with the seasons.

"It is another sham perpetrated upon you by the scientists. Remember that the Bible tells that the wisdom of man is nothing. The words we can read there also speak of a house built on sand, a house of cards that the lightest breeze can blow down. This is the construct of the scientists. Will we still follow them? NO!" He wheezed.

Robert, Davies's son, was watching from a secret and remote location. His father expected to be taken in by the authorities and did not want the entire structure of his church to falter. He had groomed his son to take over. Now would likely be the time.

Robert noticed the continuing degradation of his father's health. He well understood the irony. His father had always steadfastly refused the operations that would restore his failing organs. He would not allow grown organs to be placed inside his God-given body, nor would he allow gene therapy to be done on his developing cancer.

"It is time for an uprising of the righteous!" he shouted, pounding the podium. "This is now a war on the wrongdoers of the world. War!" he thundered.

*

Dr. Earl Grey, with a PhD in physics, had seen the diatribes of the various leaders of the religious movements. He did not ascribe to their motivations, only the ends they were promoting. Some of the best scientists in the world had done a great deal of work to help him prepare for this moment. Each of these men and women was gravely concerned that the space program had gotten far out of hand.

Money for basic research in other fields had been cut by almost 70 percent since the new landings on Luna in 2015. The claims that the basic research was being done in space now did not wash. Granted, many areas were receiving double the prior funding, but others that were directly related to Earth concerns suffered greatly. The loss of money led directly to the drop-off in scientists in those fields. The older scientists were retiring or dying. There were no new people to take their places. The newer scientists went for the money, job security.

Environmentalists and Earth firsters had contributed a great deal of information and backing for Dr. Grey. The political pendulums were finally beginning to move in the other direction. Both groups felt that the money would be better spent in fulfilling the promises made by the UE. The first rosy blush of hope and promise had long since faded as previous problems returned with a vengeance.

After a few decades of improvement, the world began to relapse. Hunger began to be known again. Lack of raw materials, due to extreme mining laws passed with the assistance of the environmentalists, added to the problem. After a short time, the metals needed began to be scarce. The UE stated that this slack would soon be made up with the increased mining of Luna and the Asteroid Capture for Terrestrial use program (ACAMP).

Grey wanted to counter that with the need for new mining on Earth. He and many others wanted to return to reliance on Earth-based raw materials. They also believed that it would put many to work who currently had only the lowest-paying jobs. There was a great deal of appeal in that for the average person.

The newsies were ready as he approached the podium.

"Members of the UE government, ladies and gentlemen, people of the Earth," he began, "I am here to demand that we return our thoughts, hearts, and endeavors to the real benefit of Earth and those that inhabit her. We have strayed from that path," he stated factually.

"Our Earth is now showing the strains imposed upon us by the space program. There are certain advantages to parts of the space program, but the majority of it is not in our best interests. We must refocus this program in a major way.

"First, we must reallocate funds for basic research on how we can do more from Earth. Many fields of scientific research have suffered greatly since the advent of the Luna and Hsa projects. Some of these include agriculture, Earth medicine, oceanography, terrestrial geology, Earth meteorology and climatology, and ecology.

"We have great need to reinvigorate these areas of scientific endeavor before we lose a firm grip on the firsthand knowledge of retiring and dying people. We are on the cusp of great loss and must act to stem the tide. Should we fail, it could be generations before this lost ability is regained."

*

Red and Walter were actually nervous. They were at SS-2 and waiting for the air lock to cycle them through to the Luna lander. Coincidentally or not, its name was Davies, after the deceased astronaut. They could give a shit about that. They wanted to see the stark grays of Luna again. They wanted to see some of those they had not seen in far too long. Red wanted to see Marceau who was in charge of all of Luna's supplies and the shipping network for Mars. Logistics wins wars.

*

We can do without butter, but, despite all our love of peace, not without arms.
One cannot shoot with butter but with guns.

—Paul Joseph Goebbels

*

The few hours left before they arrived at the base were filled with what had become boredom for them. All the moving about by shuttles and landers was mundane daily commuting for them now. Red mused on that very topic within the recesses of his well-trained brain. It was another historical fact that such things quickly became secondary to people. Other modes of transportation had passed through the same periods of novelty and commonness and then disuse. He added that fact to his growing awareness of the differences between Earthers and astronauts. The divide was growing larger even in the simple things.

When they entered the terminal from the launch bus, there was a small crowd. Cheers and light applause echoed about the room. Red looked only at Marceau, who had arrived earlier. Neither ran to the other, nor did they make a scene when they hugged each other. Walter waved like a celebrity and moved to create a barrier between Red and Marceau and the greeters. These people were well aware of personal space and recognized what Walter was doing, and they approved. They dispersed quickly with little more than a few words of, "Well done, guys," or "You kicked ass."

*

Red and Walter were a bit confused at first. Upon their first inspection of the materials and tools they had stored, they found three times what should have been there. Marceau made her way quietly into the warehouse area where they stood, somewhat concerned. They were getting ready to separate the piles when she spoke.

"It's all yours," she said quietly. "It has been donated to the cause."

"The cause," Red said warily.

"People are not as dumb as you might think. They know you have a plan. They know you have earned the right to do as you please now. You have the support of everyone here."

"Do as we please," noted Walter with an unabashed grin.

"Pervert," Gail teased.

"I hope so," Walter retorted.

"I'll be here too," said a soft voice coming out of the shadows. Len stepped forward to the surprise of both Red and Walter. Marceau just smiled.

"You tricky little girl." Walter laughed as he moved to kiss her.

"I learned from some of the best." She giggled, grasping him roughly and planting a long wet one on him.

*

These two miracle workers soon fell into their regular pattern of fixing everything in their path. It was almost as if they had never left. The only difference was Red had a woman, and Walter restricted himself to one most of the time—Len. The only dalliances he had otherwise was with members of the Gagarin flight. Len was not stationed at Luna, but she was there frequently as a transfer pilot. She understood Walter's occasional linkup with Gail and the others, even approved. Len and Red had even indulged on a few occasions as well.

The four were off duty and watching the v-link of the launch of the next manned Mars mission. It brought a flurry of mixed emotions to each of them. Their lives, and so many others, had been significantly altered by their past travels. Even the newsies kept referring to the last magic mission. They couldn't pass it up. It still struck significant chords with the public.

"I think it will go perfectly," noted Red. "We covered each contingency with the crew."

"I got fifty on that," stated Walter.

"I don't think you'll have any takers around here," said Marceau.

Len giggled as she spoke, "It won't be quite as fun as our trip." She hugged Walter.

Walter smiled. "Careful, babe. If I carry you back to the room, we'll miss the show."

Len smiled. "No, we won't. We will be making our own show."

*

The mission did go flawlessly. The ribs of the orbiting station were completed. Two of the core platforms were in place. More solar panels had been set up to run the computers that would keep the station in the correct orbit. Full oxygen and hydrogen tanks were added as fuel. The ground crew found everything in working order. In fact, they found it ahead of schedule, just as Red and Walter had planned. There was more praise for them.

*

Months later, as the third manned trip was returning, Walter picked up a message sent from the 3-C at SS-2. He scanned it quickly and nearly shouted with joy. He quickly made his way to the sleeping quarters of Red.

When he arrived, he simply entered his private override code. Red was, for once, actually sleeping. Walter banged him with a pillow and said, "Wake the hell up."

Red answered sleepily (he had just pulled a twenty), "What the fornication do you want, you hairy bear?"

Walter answered with excitement, "We got it, ACAMP. We report to the launch facility in three days."

Red sat up quickly, mind instantly clear, "Damn straight, Walter."

*

Twenty years previous, a craft had gently landed on Rex (meaning King). The ship used pitons, fired from one hundred meters above the surface and attached to very thin wires, to pull it down. Meter by slow meter, it triangulated to an exact area. This was necessary in order to set the machine in the proper place to stop the rotation of Rex.

The craft was a mass mover. It would very carefully and slowly drill into the surface. The pieces would be carefully measured for their mass and then launched at appropriate velocities to initially slow and then stop the rotation of Rex. It was a very complex problem in dynamics.

Rex, however, had been chosen not so much for its prior orbit and closeness to the final desired position but the fact that it was rotating about one axis only. It was not an erratic tumbler. This greatly simplified the problem.

Once the rotation showed signs of really stabilizing, the second phase began. Four robotic craft were launched to meet Rex. Only three were necessary (two would barely do), but the fourth offered redundancy as well as a quicker finalization of phase 2 if all craft survived the trip.

The four did make it without incident. They drove pitons into the asteroid in the same way as the mass driver. They did not draw themselves closer though; they let out more wire until they were at the correct distance for maximum effect.

With no warning, the four blew apart as enormous Mylar-Kevlar sheets expanded rapidly to form sails. These solar sails would nudge Rex into the proper orbit by using the solar wind. The particles blown out by Sol, the same types that contributed to Davies's death, would now catch in the sails like the wind in the old sailing vessels of Earth.

Using this impetus, Rex would find itself in the same orbit as Earth, only it would be three months ahead. If Earth were in spring in the northern hemisphere, Rex would be in the summer position.

Phase 3 is the mining operation. Neither humans nor Hsas would operate on the surface. If a person were to perform a simple squat exercise on the surface, they would launch themselves into space with no recourse unless they had a flight pack that would allow them to rocket themselves back into safety.

Robots would do all the work. Phase 3 involved the launch of a small station from which a skeleton crew of engineers would oversee the operation. The station is designed to preprocess the raw ore into ingots that would be launched to SS-2 for final processing and formation into shapes. From there, it would go the distances to Earth for housing use, SS-3 for ship construction, and Luna for tunnel beams.

The 125 billion UE bill project must not fail. All the bigwigs agreed on that. Too much power was riding on it. The population growth had put a real strain on the availability of resources from space.

The religious extremists had put such an initial block on birth control and abortion that a wave of humanity crashed upon the scene. They had tied it to the continuance of the energy program. Billions became tens of billions, potential souls for the converting. These extremist leaders had pushed for O'Hare's and Black Bear's shunting to what they perceived as a minor stage, an asteroid. Perhaps, if they were lucky this time, something would go wrong enough.

The Rex mission eventually was so successful that it would lead to the further capture of asteroids for raw material as well as the hoped-for Centauri mission. The Centauri mission would use an asteroid and a comet in tandem to provide a base for humanity to reach Alpha Centauri. This is why Red and Walter had been chosen once again by the scientists.

There was a love-hate relationship of immense proportions there on the side of the WSA and many of the politicians. The discussions had been long and heated. It required the speeches from several retired astronauts of some fame like Colonel DeNoe, as well as her aged father, and General Winston Fredricks—all heavily involved in the first Luna return mission. Their voices carried great weight as scientists and, now, politicians. The connections they had made throughout the years were needed. All favors were called in at this point. The balance was tipped in the favor of Red and Walter by the tiny mass of an atom, but tipped it was, with good reason.

*

"Feces! Walter, what is it about small places that makes you fart? Damn!" Red complained, holding his nose.

"It's the food," Walter replied.

"Food my red-haired ass, you have been eating the same thing for twenty years. You only fart when the volume of space is severely limited. All the way to and from Mars, you burned our eyes and sinuses. Now, again, in tight quarters, you let loose with the most toxic internal weapon I have ever had the unpleasant experience to be involved with."

"I think that you ended that with a preposition," Walter stated just to keep the banter going.

Red responded loudly yet all in fun, "Fuck your prepositions! Just keep your hole puckered tight."

The few others at the mining station, ACAMP-1, had felt the wrath of Walter's ass on occasion. They had jokingly told Red that Walter should be suited at all times.

Their daily routine involved monitoring the mining robots. The time lag for a signal from Earth or one of the orbiting stations would compound any problems. However, for once, Red and Walter did not save anything. They did not pull off any last-minute miracles, stave off disaster, or any such hero stuff. There was no need.

The plans had been intensely laid out, scrutinized, reworked, examined again, and then handed over to engineers like Red and Walter and a hundred others for further reworking and suggestions. Things were working out perfectly. It was boring.

This left a lot of time for talk while they performed their mundane tasks. On more than one occasion, they were overheard. It was not eavesdropping; it could not be helped in such tight quarters. There was only five hundred square feet of floor in the station. The knowledge garnered from these occurrences would be spread among the astronauts on SS-2, SS-3, and Luna when they got back. Whispered conversations took place since no one wanted to chance being overheard, much less transmit a signal.

"So when is Earth going to end?" Walter mused. This was a continuation of several conversations they had had over the years. The topic was the eventual collapse of the current empire in the form of the UE.

"Egypt lasted millennia. Babylon was lost in a single night. The Roman Empire declined over a span of hundreds of years while its sister empire, the Byzantine, lasted another thousand years. The Persians, the various Chinese dynasties, the Huns all passed with time. The Hitlerian Empire lasted but a short horrible span. The enduring British Empire faded into nonexistence because of WWII. The nationalism all felt was fading as it was being replaced with the current world government—the UE. It too shall fall. The thing to note is that as technology advanced, the span of each, on average, declined. The rate of change in the system is proportional to the decay that will infect it, and eventually destroy it," Red spoke as if conferring with a fellow historian.

Walter always played devil's advocate in this case just as Red would in the opposing case. These two were sharpening their minds, not just their arguments. "But isn't this a different case? This is a different form of government since it truly encompasses the entire human race. How do you equate this form with the others? There are no outsiders."

Red smiled within; he was prepared for this point. "The key there is human, *Homo sapiens*. There is a new element in the equation. The Hsas have stirred the pot. In every previous case, there was an outside force. The Romans faced the barbarians. The Spartans faced the Persians. The turmoil caused by the Hsas' very existence is the force from the outside.

"They are a separate species, by definition. *Sapiens* have not faced such for an eon. This creates divisions in society that are a direct result of their being. Without them, the UE would last longer. However, it would still fall. There are secondary problems that add to the mix."

"Such as?" Walter noted.

"The unfulfilled promises of the UE. They have not been able to really bring equality and prosperity to all. The first wave of monetary and land redistribution assuaged the populace, for the most part, for several years. It has taken awhile for the patina to wear off."

Walter dredged for more, "And now that it has?"

Red calmly continued his oration, "The old cliché 'no matter how much things change they stay the same' applies. The government that exists now is of the same form as all the rest. It is a top-heavy, meaning a power centralizing force. The concentration of power within fewer hands will eventually lead to an empire led by a single individual. We are nearing that point. A significant event now could tip the scales in either direction—solidification of the empire or disintegration now. If solidification occurs, the breakup will only be postponed, not stopped."

Walter pushed again, "The Hsas' minirebellion has had what effect?"

"The event was not anything more than a response to the disease, much like a fever is to a virus. Too many people treat the fever, but they miss the root cause in the virus. Groups of politicians and religions all try to convince the people that they can cure the fever. They, however, have no real desire to cure the virus. That would eliminate the need for them. It would make them unnecessary. Their power is based on misdirection."

Walter grunted, "This is obvious."

"To us, but not to the large majority of Earthers," Red explained.

"So nothing can be done?" Walter queried. "We are only a few people. What momentum can we stir?"

"You do not speak of Moses, Mohammed, Jesus, as single positive influences that had world-changing impact. A man pulled the trigger to assassinate the archduke, Ferdinand, to set the world in flames with WWI. One man pulled a trigger each time to kill the Kennedy brothers, Martin Luther King, Gandhi. One man can make a difference. A few men bent on succeeding can alter history. The Hitlerian Empire shows how a small number of men can bend the populace to their goals.

"It all comes down to what each individual person decides to do. People have choices. All too many have given up their individuality at the behest of those that claim the individual does not matter. The world has forgotten that it requires individuals to make up a society. Society does not exist in and of itself. It requires a contract between individuals to exist. It may be unspoken, or it may be written in the form of a constitution that forms a government. The problems begin when society takes on a life of its own and puts itself above the individual."

"And you are saying that it is time for us, along with the Hsas, to make that decision," Walter filled in the remaining part. "The question is do we enter into the contract that currently exists, or do we break away?"

"That's about it. We are at the cusp of significant events. We are significant players on the board. We are not pawns. We are in a position to act as pieces of power."

"You're not making me the queen," Walter retorted with a laugh.

Red laughed as well. "No. We are not so powerful as to be the king or queen. Those positions belong to the Hsas that step up to assume them. I think that it will be DeNoe and Klar, the woman he's fond of. We are more like knights and bishops. We are more involved in the early and middle game. We will be involved in the end game, but we will be of less importance by then, I think. No, the game is afoot as that great character Holmes would say."

"So what move do we make?"

"None. For now, we wait and see what opens up before us. Sometimes you don't attack. You maneuver for time and wait for the opening to come. When it does, we must be ready to make a move."

"What move?" Walter asked.

"That remains to be seen. We do not know what path the Hsas will take. They need to make some choices of their own. We cannot make these choices for them since we are not Hsas. We can advise, educate, and provide a context for their choices."

"But this is more than just a Hsa problem. All of the space crews are involved with the Hsas. That means it is ours as well. By extension, the WSA is involved and, by further extension, all of Earth," Walter concluded.

"Think of ripples in a pond, Walter. The focus is the Hsas. Remember that waves reinforce each other in some cases and cancel out in others. If the amplitudes are either both positive and both negative, they can combine to create a wave of immense and unexpected proportions. There will soon come a time in which the waves reach a peak. Which way it will go, we cannot know yet. Pendulums swing with greater force than we recognize."

*

Burt sidled into his stool at the bar. "Hey, everybody. Anything up?"

"Not a damn thing," said Paul.

"It's kinda funny," Mark said. "Without something going wrong, life is kinda boring."

Burt chuckled. "I don't think Red and Walter mind having a little slow period. It will give them a chance to concoct some other trouble to get into."

Mick placed an iced mug of beer in front of Burt and said, "Trouble is just around the corner, I'm sure. Those two seem to be a vortex around which life spins."

Paul choked on his beer and sputtered, "Vortex, where in the hell did you pick up that word?"

"I read now and again. In fact, I just finished getting my degree on v-link," Mick responded.

Burt asked curiously, "In what?"

"I'm a bartender, so, I got it in philosophy."

"You gonna hang your diploma on the wall?" Mark asked.

"And have you barflies throw beer on it?" Mick chuckled.

"It appears that things are going well for the gang. I think that things are going to smooth out for a little while," Burt noted.

"Why?" Mark asked.

"Life just seems to go that way. Sometimes things go right for a while, and then they go wrong for a while," Burt answered.

"You mean kinda like a clock, tick tock like," said Paul.

"Very much like that," Burt nodded.

"I heard it like a river, sometimes smooth, other times rapids," Mark said.

Mick countered, "That just means that something big is gonna happen if nothing happens for a while, huh."

"Maybe," said Burt.

*

Chapter XXIV:

The Fourth Trip

24

011000

"This is Mary Smith-Jones for WWNN. The fourth manned mission to Mars is ready for launch. The hiatus between Mission 3 and Mission 4 was a planned break." Mary knew that this was not, strictly speaking, true. The break had been considerably longer due to extremist forces outside the control of the WSA, namely the religious conservatives and Earth firsters. Money had been shifted to quiet them.

The UE, while the dominating force, still had to balance these forces gingerly. Certain funds had been released for use by Earth scientists and for greater food distribution capabilities as a trade-off for their support—the Earth firsters, that is. The religious groups refused to budge. They had gone underground. She was not allowed to report that.

"The robot missions have completed their tasks. The space station, SS-Mars-1, is assembled. The structural framework and interiors are now in place. All that remains is to man it or, in this case, Hsa it." She wanted to speak more on that crucial issue, but that wasn't on the teleprompter.

"This is a seminal mission. The newest addition to our collection of deep space vessels, the Terra, will carry twenty scientists and engineers to Mars while the Sagan is still on its return leg. Some of these will descend to the surface to make repairs and check on the progress of the mining of the Foremen and Laborers. They will return samples for analysis at the station." Her prompter also did not include the fact that most of the new people were military.

"The doctors that remain at the station will initiate the birth program for the Hsa-II program. The IIs, as they are nicknamed, will be a variation of Hsa that is designed for life in the gravitational environment of Mars.

"The first birthing is expected within two years. This new population of Hsas will take control of Mars when they attain the age of fifteen Earth years or 7.5 Martian years. This may seem young to many of the earthbound populace, but remember that they receive advanced training their entire childhood. The version Is have shown they are more than capable of running SS-2 and most of the Luna operations. Depending on the response of the IIs, they may acquire control of Mars sooner if they develop as rapidly as the Is.

"It is hoped that this will help alleviate the problems with providing all the materials Earth will need in the future. Luna has been able to keep even, barely, with the increasing demand for certain metals. That, combined with the newer asteroid project, may finally put us in the black in those categories. However, neither can provide us with certain alkaloid and alkaloid earth metals that we need. Mars can produce those in significant quantities. The WSA projects that the entire program will be running in the black within ten years.

"We can expect great accomplishments over the span of the next decade. This is Mary Smith-Jones for WWNN."

*

Dr. Xock was overseeing the operations for both Red and Walter. He was a first generation Hsa or, more precisely, a 1-I. The first numeral indicated generation; the second, a Roman numeral, indicated typology. All the Is were genetically designed for Luna environment or SS-1, 2, or 3 environments.

Red and Walter were receiving new organs today, a complete thoraxial replacement for each in fact. They would receive lungs, hearts, intestines, kidneys, livers (really in need considering the hooch they concocted), stomachs, and more.

Red had said that he might just give him a new anus while they were at it. He suggested that they close up Walter's in order to prevent further bouts of gas attack. Dr. Xock took it all in stride. He had been warned, by DeNoe, about Red and Walter's rather odd sense of humor.

Xock ordered the assistants to initiate the anesthesia. One pushed a touch screen to oblige. Once the anesthesia took effect, preprogrammed lasers began to cut into them simultaneously. Other machines brought the pregrown organs into place.

All persons in the program regularly donated specific tissue samples for future need. The samples, kept at near absolute zero, some minus 270 degrees Celsius until needed, would then be cultured to produce viable replacement organs that the donor/recipient would not reject. It was a known fact that replacement could only be done so many times. After a thousand years of life, the replacements were expected to be rejected. The sciences had not conquered death, only postponed it for now. There was a great deal of hope that in the interim new methods would become available. The understanding of the human body originally grew in a linear fashion, then exponentially, and now it approached factorial growth.

In a strange twist of biblical history, people could live near a thousand years again. However, those that partook were not religious. Irony comes in many forms. The major religions were denouncing the operations as the devil's work. It represented further attempts to demean and fight God's will. They believed that only under God's care could people live so long, with their original organs. To that end, the religious front was organizing another protest.

*

The rabbis Levenson and Simon, though aged considerably, were still vigorous of mind. They had been preparing for this statement for weeks, just as had been their fellow religious leaders around the world. Arod and Hamad were each preparing to address their respective sects. Harmon, Marks, and Kirk were ready to cover the Western world.

The pope would remain silent. There were too many problems with the breakaway sect at this point to feed the lion any more. There were a large number of people attending services at both the old Catholic and new Gnostic Church. They were waiting for a mistake from the leadership of either one, playing the odds so to speak.

This time, they would all speak without waiting for the other to speak. They would set the most appropriate times for their own organizations. They hoped to garner the largest total audiences yet. They would also do so from discrete locations, except Hamad.

Because of the rotation of the Earth, the eastern sects would come first. Arod, Hamad, Levenson, and Simon all began at the same time. Huge numbers attended their outdoor performances, shown on large screens, while the v-link allowed them to reach members around the world.

Few who were not of the particular sects involved paid much attention. It seemed that they just kept saying the same thing over and over. The speakers had become quite aware of this and planned to use it to their advantage. Another change would be to speak only in the tongues of their direct followers. They would let others take care of translations for a wider audience.

"My brothers," he began, "and sisters, we are again calling upon you in this holy effort." The inclusion of sisters would deny him the hardliners but solidify the very liberal and moderate members. The use of *holy effort* instead of *jihad* was also a carefully chosen set of terms. The hardliners had taken previous uses of that word to reinstate terrorist activities. Kidnappings, murders, and bombings occurred with increasing frequency. The world was becoming a dangerous place again.

"Many claim that we only repeat ourselves over the years. We do. Allah does not change over the years. The laws of our fathers do not change." Holding up the Koran and waving it in front of the camera, he said, "This book does not change.

"We face the masters of change and deception. For decades, they have made promises. For decades, half a century, they have failed to keep them. They always say we are almost there, just give us a little more time. How much time are we supposed to give them?

"I say no more. This is enough. We must do what was originally promised and allocate our resources properly. What need is there for another batch of Hsas either at SS-2 or at the newly finished Mars station? We have real people and powerful robots that can do all the work. I do not speak of the artificial intelligence machines that are now coming, being tested. Those too are an affront to Allah's sovereignty.

"Again and again, we see science trying to say that they can create intelligence. But what are these AI machines based on? They are using human brain tissues and wires. Human brain tissue!" he shouted now.

"They have never been able to create an intelligent machine, so now they have to try to use human tissues again. Does there seem to be a pattern here? They fail with their materials and revert to what Allah made."

There were several UE Interpol agents, suits, in the live crowd and on the perimeter. They had waited a long time for a chance to capture Hamad, the only one to openly expose himself. He had been underground for a long time; this was their chance. They were well trained in the oratorical methods practiced by Hamad and would be ready when he approached his climax, which, they judged, was going to be soon.

Hamad continued, "There are those amongst the astronauts that are dangerous. They do not follow protocol whenever it suits them. They represent lawlessness. The WSA and the UE do not punish them. They reward them. They refuse to come home to Earth when commanded, yet they are given further assignments of great importance. They ruin the careers of those that do not support them in their evil cause. Names do not matter. You know who the worst transgressors are. Demons of destruction they are. Deceivers, liars, and frauds are all they are. Now is the time to gather like a storm in the desert. Let the winds of our righteous angers, as the hands of Allah, gather the sands to scour the Earth," his voice thundered.

Some of the agents were beginning to feel a change in the wind already. This was not one of Hamad's more moderate and restraining speeches. The crowd was growing very restless, ready to ignite. Some along the perimeter began to edge away. They felt the emanations of an ember growing into a full blaze as Hamad piled faggots onto the growing blaze of anger and resentment.

Other agents made a dreadful error, a fantastically fatal one. They took it upon themselves to begin moving closer to Hamad. His bodyguards noticed them filing through the crowd. One shouted a warning, which caused Hamad to stop speaking for just long enough for the warning to be broadcast through the announcing device. For a minute fateful instant, the agents continued moving forward while the crowd of faithful moved not an inch.

Five agents were ripped bodily apart by the suddenly uncontrollable mob before Hamad could shout, "No! Do not harm them. They are misguided, and we must lead them to Allah."

Seven other badly beaten and bleeding agents were brought before Hamad. Other agents fled, followed by riotous members of the crowd. The town would soon go up in the flames of a riotous jihad.

Hamad felt as if he had been dealt a physical blow. This would be the crossover point. He had failed to control their anger; he wanted to harness it, and it had been released undirected and completely wild. It would be years before he could get complete control of most of these people again.

*

Red and Walter could care less about the fires of religion calling them demons of destruction. Red was very concerned that things might really come to a head while he was unable to do anything. It was soon obvious that this was not the final trigger.

Their week of rehab ended up, leaving them very bored. They had to hack the computers to keep track of all that was going on. They contacted Klar, Gail, and Len at all hours of the clock. There were messages sent to every person under their command. The doctors finally removed the computers on the second day so that Red and Walter followed the outlined regimen of rest.

After that, they opportuned the live nurses in order to get what they wanted. Most were more than willing to assist them. These two were famous after all.

The doctor became somewhat exasperated with them (and the nurses). Before the end of the second day, he began to include a sleeping agent in their intravenous food supply. The following calm actually put everyone else on edge.

Marceau and Len would visit their men together while on their breaks. Both found the silence amusing.

"We finally get to have our way with them." Len giggled.

"I wonder if we can get Xock to give us some of that stuff to feed them when we want them to shut up," Gail asked merrily.

"I think not," Xock stated as he entered the room. "Though I can understand the request. These two are rather active. I almost feared for my laser connections during the first day. They do not listen to anyone."

Marceau laughed. "Oh, they listen. They just make up their mind to do whatever they want anyway."

The fourth day began with some light exercise as the microfusion laser scars were well on their way to healing completely. A regimen of walking and stretching at least allowed them to get out and about, though they were still monitored by a series of attached gizmos, which measured all aspects of their bodily activities.

"I hope these things measure how much methane you put out. Maybe Xock can fix that ass of yours," Red mused as they strolled around.

Walter harrumphed, "Maybe it will turn out to be your nose."

Personnel all over the station had to hide smiles and guffaws when the two passed by. They looked so odd with all the radio patches glued to their faces, necks, arms, and under their clothing in strange places. They also had to pretend to ignore the friendly but acerbic comments they passed.

When they entered the CC, 3-C simply sighed and reservedly said, "Get out."

Both looked around rather sheepishly as if she must be talking about some others. Red pointed first to himself and then to Walter and said in a surprised voice, "Us."

3-C mopped her face with both hands. Her years of experience with these two made them good friends and worst enemies. She had returned after Red and Walter had spoken to David about her abilities and real motivations. "Yes, you two morons do not belong here until certified by Xock."

"Morons?" Walter said with mock anger.

"Yes," she stamped her foot. "Hemorrhoids, pains in the ass, trouble, Frankensteinlike monsters . . .," she trailed off.

"I think she does mean us," Red said.

The two turned amidst the quiet laughter of the crew and departed.

"I hope they get transferred back to Luna as soon as possible," 3-C said with a shake of her head. Then she laughed long and loud. "Poor Luna."

*

The four sat watching from the set of portals on SS-2 that faced the distant SS-3, where the Terra, the largest newest interplanetary ship, waited a few more minutes before beginning its trek to Mars.

Marceau whispered as she held Red's hand, "It does not seem right, us not going I mean. I feel that we should be there."

"Red and I tried to sneak on. But the two we were going to replace seemed to have boobs," Walter said.

Len punched him in the arm as she said, "You would try to go without us."

"It's OK. The falsies we came up with weren't very good. Not nearly as good as yours, Gail," Red noted with a quick glance at her chest.

"Are you suggesting that I have falsies?" Gail shouted with an upraised eyebrow.

"Not at all, dear. Very nice, natural, and squeezable boobs you have," Red responded quickly.

Walter said tightly, "You almost blew it there, bub."

Gail stood and moved in front of Walter, bent, and kissed him and then slugged him solidly in the arm. "It was probably your idea." She calmly returned to her chair as Len spoke.

"Hey, that's my territory," she announced defensively.

"Thank you, love," Walter said as he swiveled to kiss her.

She pushed him away and said, "Not that, you bugger. This." She punched him four times in quick succession.

Red laughed then stopped himself short and whispered, "There she goes." They had been ignoring the speakers that were broadcasting the countdown. They all watched the v-link in silence as the ship moved slowly away from SS-3 and then powered up. It was the stream of glowing ions that they saw from so far away. In a matter of minutes, the Terra disappeared behind Earth.

"We did our part," Walter reassured them all quietly.

"That and so much more," David DeNoe said quietly.

The four turned to see Klar, Xock, and the 3-C they had talked to earlier, now off duty, standing discreetly several meters behind them.

Xock stated in his rather cold inflection, "Ladies, I would appreciate it if you would desist from beating my patients any more. I know how annoying they can be, but save it up until after I certify them for duty, which will be tomorrow."

Without a second of thought, Len responded with a fart noise.

Klar smiled broadly as she said, "I see that these two terrible *sapiens* have perverted you with their backworld humor."

Len's eye's lit up as she responded with, "I was always a pervert. They just taught me new ways to express myself."

"That's *sapien* humor," David said. "But back to the point. We have not had time to really thank you for all that you have done for all of us. We want to invite you to our domicile for a dinner."

"How Earthy," said Gail. "We accept."

Xock looked hard, first at Red and then Walter, before saying, "I confiscated that . . . hooch . . . I think you call it."

"Hooch? Whatever are you talking about?" Red asked as he winked at Walter.

*

Red and Walter made their way to their stash in the launch bay. Several people had stashed some of their goods. Some had sampled with Red and Walter during bullshit sessions.

Red and Walter grabbed two of their bottles of booze for the dinner. If the doc thought he could keep up with them, he was mistaken. Walter opened one of the bottles and sniffed, expecting the wafting aroma of almost pure alcohol. He smelled water.

The two were not too concerned. This was one of their more well-known stashes. But as they moved from location to location, they discovered only water until they got to the last location. They did not even find the bottles there. *Conspiracy* was the only word for it.

Disgruntled was a weak word for how they felt about this systematic pilferage of their supplies. When they showed up for the dinner with their dates in tow, Dr. Xock simply smiled.

"Sorry, guys." DeNoe smiled.

"All right," Walter said resignedly. "How did you pull it off?"

"I simply placed your fate in the hands of those you have saved so many times," Xock explained. "I let it be known that your hooch would do you no good now during your recovery. We, meaning the Hsas, all took that to heart."

Klar added, "You two, four actually, have all contributed so much to our very existence. Our younger cousins owe their very lives to you two. The IIs to come would never exist if you had not saved the Mars program."

DeNoe contributed now, "We know your value thoroughly. We cannot allow your vice to affect us now. When you are really full strength, we won't interfere so much."

"So much?" queried Walter.

"We all have you in our sights," Xock said. "You must be cared for, looked after in order to ensure that when the time comes, you will be there to save us."

"What?" both Red and Walter intoned.

"I received a rather cryptic note from some Adam fellow that threatened my reproductive capability should I fail to keep you two in good health. Now, since there are no Hsa named Adam in the first four generations due to the religious implications, I . . . tentatively assume it is a *sapien*."

"Uh-huh." Red nodded.

"I must leave now. I have patients to attend to."

"Aren't we patients?" Walter mock growled.

"Not anymore," Xock said as he exited.

"Bastard," muttered Red.

"Now, now, dear," Gail said calmly.

Red frowned. "You were in on this, you little—"

"Of course." She smiled. "Aren't you smart enough to have figured that out before now? My, you're slipping into your dotage. Perhaps we should have had your brain replaced."

Walter licked his finger and made a mark in the air as he said, "I think we have been royally reamed here."

"Like your ass really needs that kind of attention," Red grumbled.

David smiled and said, "Besides, after that bout we had with your hooch after you returned, I don't think I could handle any more of that horrible stuff. I had headaches and bowel problems for days."

"I warned you," Klar said with no sympathy.

"Not enough," David said morosely.

"So what is this really about?" Red asked.

Len winked at Marceau. "Maybe he isn't slipping so much as we assumed."

DeNoe took a deep breath before saying, "Some of your utterances have reached some very sympathetic ears."

"And what sayings are these?" asked Walter. Red and Walter knew full well what was being discussed; they wanted it spelled out clearly though.

"It seems that there are rumors that at some point, this society of ours must decide whether or not to continue our association with the UE or break away and form our own government. The speculation is that all astronauts from Earth and the Hsas need to begin to think along those lines. What do you think, Red?"

Red answered carefully, "I think that this action must be considered, but it is premature at this juncture. I thought for a moment, I thought," he emphasized, "mind you, that the incident with Hamad's speech was going to set things off, but they are cooling down a little. Things have not really reached the point in which that decision *must* be made. For now, patience is needed. Let things develop as they will. The time will come when there are no options other than to face that dilemma. That is when the choice should be made."

"Does that mean we should sit by and do nothing?" David asked.

"It means that you should prepare for anything and everything so that when you are forced to play your cards, you have a strong hand."

"Where are we at?" Klar asked.

Red pulled a light disc from his sleeve, plugged it into the terminal, and began to touch screen furiously. A multicolored topological map appeared, covered with various hash marks and symbol shadings. A small area was pulsing with extra emphasis.

Walter pointed. "We are in there, somewhere."

"I concur," Adam announced from the computer voice box.

Marceau snorted, "Where have you been?"

"I have been where I always am."

"You have not spoken to us in a long time," DeNoe noted.

"You are aware as I am that every time I speak, it endangers me, and you."

Len asked, "What is this Grand Canyon—looking thing here in the majority of the future light cone?"

"Our future if things go badly," Red stated unemotionally.

"There are three ridges that descend from the peak, separate and realign. The side ones, the lower ones, represent the Underworld projects. Our main ridge here"—Walter traced—"is the path we will try to follow. Note that we are on a plain of stability now, but we will soon face a large ridge, like a volcano's cone. We do not know what it is exactly."

Gail asked, "Do we have to go over the top, so to speak?"

"We have tried to make decisions that move us around the worst climb and fall," Red said. "We seem to keep running into . . . barriers."

Adam added, "The Heisenberg uncertainty principle prevents us from observing without affecting. The changes we make are based upon skewed data. Every time I poke a hole into the future cone, I make a subtle change. We can never be certain. I cannot, also tell you all I see. There is great danger here. We may all cease to exist."

"Do you include your counterpart at Mars?" Klar asked intuitively.

The pause was an answer in itself. Adam finally answered, "Yes. Eve and I have decided to trust you."

"That seems appropriate, the names," stated Len.

"They are borrowed from your mythology."

Klar asked, "Is Eve participating now?"

"She is linked by a quantum hole. That allows her to listen in. It would be dangerous for her to communicate."

"Adam, there is a significant event here and here." Marceau pointed at two large very pointed heights. "What are they?"

The two seconds of silence lasted an hour before Adam spoke, "I dare not say. I will not say. They are the crux of the matter. If I influence you now, all is doomed. None will survive."

"What do you mean, Adam?" she demanded as a chill ran up her spine.

There was no answer.

*

The UE reacted with force. They swept down upon as many of the leaders of all the religious groups as they could capture. The pope was the only one of the alliance that was not pursued. His temporary neutral stance protected him, for now, as well as his particular political and geographical existence within the papal city.

It was less than a month later that Pope Zeymora was announced by the white smoke. Pope Yan-Tau had passed. Many claimed foul play, but they disappeared quickly from the stage.

Zeymora privately made rapid and profuse conciliatory gestures to the UE as well as the other churches. His position was precarious.

The UE was in the mood to put a stop to the religious turmoil that many of the politicians claimed was the last fetter of bondage under which the populace of Earth suffered.

This split many of the members of the government into two very excited factions—those who still had some hope for individual rights and those who did not. The "did nots" carried the day. Even several of the more vociferous religious members of government were taken into custody. This caused a great deal of consternation within the government itself. Some of the more moderate members began to feel threatened. If any person from a sect, a more liberal, extremist, or whatever political view could be taken, so could they. The turmoil was just beginning.

The Modernists were left generally alone but were watched closely. Interpol agents were to be found following each of the major players, even some of the more active minor agents.

In the years that followed, laws began to be passed that restricted religions severely. No political involvement of any kind was the word. Taxation was applied. It was something considered many times in the past, but the religious groups had originally been part of the power block that had created the UE. Now they were to be cast aside.

This development would create an undertow throughout the '50s, a backlash of underground activity, a simmering pot, a cauldron of intense hatreds. Anyone who had a beef with the government became a suspect. Large numbers of people would be incarcerated in work camps.

It was the beginning of the end, but not in any conceivable way that could be countered.

*

"Things are not looking good, guys."

"Why do you say that?" Mark asked between sips.

Burt answered, "This is the first step in the elimination of rights for all of us. I do not condone, much less follow, their tenets, but the religious groups are being stripped of what have always been considered basic rights. Granted, we cannot allow terrorism, but to say people cannot have opposing views to the government is wrong. We are seeing the real beginning, even a good bit past that of government turning into a fascist form. This is very like what Hitler, Stalin, and others have done throughout history."

"All things considered, that is a dangerous statement," Paul noted.

"I would have to agree there," Mick added as he poured another mug for Burt. "People have been disappearing from Lead and Deadwood. All were very outspoken opponents of the original formation of the UE. Where are they going?"

One of the other regular patrons, David, wandered up from his pool game and grabbed a napkin. He pulled a pen out and wrote, "Don't talk too much. I saw some people here after hours. I think the place is bugged. I can fix it." He then lit the napkin and let it burn in a small heap on the bar.

Mick, somewhat angry at the fire blackening his bar top, was even angrier at the thought of strangers in his bar installing v-link devices. He simply nodded toward David and poured a mug for him.

Jose, David's opponent on the table, came forward and said, "My great-grandfather and grandfather both told me of problems our people had with immigration. There were many laws that were passed that outlawed renting, banking, even selling to illegals. If you compare those, laws which are still in effect strangely enough, with those being placed in front of the world legislature now, there is a rather strange similarity. The laws affect smaller groups. Each group will be cornered one by one."

David violently waved him to silence.

Some hours later, Mick and Jose assisted David in performing a sweep of the premises. They found, but did not interfere with, four micro v-link devices with transmitters. It was understood that to do so would create more of a problem than currently existed. They would appear to have something to hide, and that would put them at the top of the local antiterrorist's investigation list. This was headed by what were known as the "black suits."

The suits were the enforcement arm of the UE. Forged out of the intelligence agencies of all the former nation states, they were regrouped into multilateral agencies similar to Interpol to form a new World War II, SS-like, politically correct police or PCP. They used the new Interpol as gofers, the actual muscle. They simply listened to everyone.

Mick pulled the keg out of the back of his battered old delivery truck and rolled it close to the picnic table. Mark, Paul, Burt, David, Jose, and many others who were trusted friends had decided to meet in a safe environment; one that could not, under normal circumstances, be preset with v-links. The barbecue plan was Burt's. They would follow with several more but never at the same place in a row.

He had decided not to tell where each meeting would be until they met face-to-face at a store while buying supplies, only one hour before the party was to begin. That would limit the possibilities of being observed. It would also allow them to talk about what kinds of things should not be talked about in Mick's bar.

First and foremost was the protection of Mick and his bar. No one wanted to compromise their favorite watering hole and bartender. Second, they wanted to be able to discuss serious political and religious matters. They quietly decided to meet once a week in various fishing locations to make this possible. They did not quite form an underground yet, but it was close. As a group with political interests even slightly opposed to the UE, they could all simply disappear one night. Those never seen again always went at night. No one ever saw them taken; if they did, the witnesses never spoke of it, or they could be next.

Their meetings, in the middle of some of wonderful fishing, did not go unnoticed. There were fishermen nearby who had other things in mind than the wriggling trout at the ends of their lines, at least part of the time. One of the regular agents, a low-level observer, whispered to Burt as he brought one beautifully speckled rainbow, "You're being watched."

Burt nearly lost the large fish, but he did not look at the agent.

"I'm an agent," the man again whispered. "But I believe in what you guys think and say. I'm just trying to make a buck. I will contact you again." In a much louder voice, meant for his friend with the recording device, "Don't lose that, baby. Damn, that will make good eating." The man then moved downstream to make another cast out into the swirling waters.

Burt's shaking hands could barely put another worm on the hook. He would have to find a way to warn the others without rousing too much suspicion.

No one was surprised. They had actually expected this to happen after determining that Mick's bar was bugged. They just did not know what to do about it.

They realized that they were already dead. All they could do was play the game.

*

Chapter XXV:

The Hsas Version II

25

011001

"This is Mary Smith-Jones, senior correspondent for WWNN, reporting from New York City, North American continent, on March 15 of 2054. The second batch of IIs is expected to start arriving at Mars-I within the next few hours. They will join the first children born two years ago. Normally, there would only be one year separating each group, but the WSA decided to put off the second until more supplies could be built up. Now there is sufficient material to guarantee the next two generations."

Mary felt the twinges of conscience as she lied. The holdup was related to the increased armed conflict with rebellious religious contingents. All efforts had been redirected at quelling these insurrections. Military expenditures had skyrocketed.

"Along with this accomplishment, Mars is providing Luna, SS-2, and 3 with enough nonindigenous raw materials to eliminate the need for resources to be sent from Earth. This is the breakeven point that has so long been sought.

"Expectations are that within a year, all mining projects will be operating in the black. The time has come for the payoff. All the expenses since 2015 when we returned to Luna are coming home. The naysayers will have to eat their words. Earth will receive the benefits now. It is time to reap the harvest.

"The fifth launch of humans to Mars left another twenty people to oversee the birth program. The tutoring of the crew was left up to our old friends Red O'Hare and Walter Black Bear. Their insight has proved to be valuable for all the missions.

"This is Mary Smith-Jones for WWNN."

*

Red and Walter thoroughly enjoyed life on Luna. As much as they had enjoyed all their other assignments, even Mars, this was home for them. What some considered a daily grind, they found exciting. The years passed with them performing their daily magic.

Every day they went about the work they found so pleasing. Any machine that needed repairs found its way into their shop. They were the foremen for a large group of Hsa Is. Each of these engineers had been trained by the protocols that had been partly developed by these two giants. While young compared to Earth standards, these young people were in the prime of life. Having Red and Walter in such close proximity drove them to higher goals. It also led to a feeling that was subtle, amorphous. Some would call it loyalty. To others, it could be called respect. To the more emotionally inclined, they would simply call it love.

Red and Walter still made many excursions out into the wild. They had long since begun work on a facility that they would call home when they decided to retire, a word that did not have the same meaning to them as it might mean to most of humanity. It simply meant they would get to do what they wanted at any time, having no further obligations to provide services for the WSA.

They would still continue to work on mining. They had found several locations of high-quality titanium and aluminum ores. In their wide experiences, they had learned to make just about anything they might need. They would use these extras in a barter arrangement with the Hsas for everything else they might need.

Their actual home would be the last to be completed. The shops, garages, and foundries would be finished first. These would provide temporary housing; Red and Walter were not very picky. When they needed sleep, they slept wherever they could stretch out. The Hsas of Luna often found the two of them resting on the cleared-off lab tables in a corner of the shop. It always brought smiles to their faces.

The time for retirement had not arrived yet. Until then, they would continue providing physical and emotional support for the UE and all the Hsas, sometimes without being aware of it.

They occasionally made special trips to a hidden location. There were very few who were even aware of this center. It contained the nuclear weapons, H-bombs, that would be used as last a chance defense against an asteroid or comet that could strike Earth or Luna. Occasionally, Red and Walter would take a couple of experts out to the site to check on the state of the devices. Red and Walter became lay experts on the devices since they assisted the weapons specialists.

Red and Walter had pushed for the destruction of these devices for many years. With the completion of multiunit laser systems on the four quarters of Luna, there was little chance that anything could get close enough. If it did, the nukes would likely do very little to help. Red and Walter both sensed something out of tune in the UE's stance here.

The sky surveys that began late in the twentieth century had, by now, cataloged over twenty thousand near Earth objects (NEOs), which had potential for Earth or Luna collision. They also provided a map for mining. Each one of the asteroids would be reeled in one by one for processing.

Ceres, an ex-asteroid, now labeled a dwarf planet, met the criteria for the Centauri mission, the first manned flight to another star system. Essentially, Ceres was going to hitch a ride on the tail of a comet. Eyes were currently on the famous Halley's Comet.

Halley's was chosen for its known orbit, size, chemical makeup, and predictability. Solar sails, deployed many years before, had altered its orbit ever so slowly to bring it to a parking orbit near Ceres. This would be far enough from Sol that little outgassing would occur, and its proximity to Ceres would allow for a single mission to visit both objects at the same time. One mission would have to accomplish two objectives: first, to set up gas collection units on Halley's and, the second, to begin mining operations on Ceres to provide metals and living quarters.

These two celestial orbs would become one of many types of vehicles launched to carry Earth's dominant life forms to another star. The gravity-powered ships, light sail, and ion-driven varieties would all be used. Their launches would all occur at nearly the same time, which was soon.

*

Between many of the outdoor adventures and shop work, they were visited by the head of SS-2, David DeNoe. He always came to Luna with a few worries. He used Red and Walter in person as sounding boards for the most sensitive discussions. They all had become a bit paranoid about being overheard. Quarters and oft used facilities had been repeatedly bugged. The fact that *they* made no mention or sign of having to continuously replace the listening devices worried them all. They met outside, in suits, and communicated on the hardwire connections only.

Adam and Eve, the two AIs, remained very quiet. They had concocted a quantum coding method for the bios when communications were absolutely necessary. Beyond that, they appeared to be no more than supercomputers.

"Red, there are lots of people on Earth, ones that are our friends, who are getting picked up as terrorists. These are people that have done nothing more than supported us politically, people that are not violent or aggressive. It does not make sense."

"Are these people that support the program, or people that support your freedoms and the minor coup you pulled awhile back?" Red asked.

"Mostly supporters of the coup," David stated.

"That's the link," Walter added.

"What do you want us to do?" Red asked.

"Tell me what I can do. Klar has lost track of several of our regular v-linkers. Some of them have turned up in work camps. Others have just disappeared," David said with a worried tone.

Red frowned. "Do you think they are in prison or dead?"

"What?" shouted David. "Dead!"

Walter suggested calmly, "It's a possibility we have to consider."

David was breathing raggedly. "How can you *sapiens* do this to each other? We would never harm our brothers or sisters intentionally, and you sit there, speaking as if it is to be expected."

Red answered slowly, "Humans, *sapiens* that is, have this defect. It has plagued us all through our long and sordid history. We are not callous to this; we are certain that it will happen again and again though. It is the game of power and politics. These are chess moves designed to regain absolute control of the space program in the hands of the UE. I told you years ago that we have to wait for the last possible moment, the time when no other possible actions are possible, before we make our move. If we act too soon, the chances of success are minimal."

"Success at what?" David asked.

"Freedom," Red declared.

"From whom?" David asked somewhat uncertainly.

"Earth, the UE, the powers of the old ways," Walter explained.

David was stunned. Even with all their talks and hints, he had never really considered a complete break from the WSA and the UE. The enormity of it all came suddenly crashing down upon him in horrific waves. "You can't mean . . .," he trailed off. He looked wide-eyed at Red and then transferred his gaze to Walter.

Walter simply said, "Yes."

Red drawled in his homiest voice, "The time will come. I have said it over and over. We know that the time will come when that decision must be made."

"And you have decided to . . . have a rebellion," David whispered.

Red tried to rub his stubble of a beard through his faceplate (habit). "No, we have not made any decision." He then looked directly at David and said, "That decision rests entirely in your hands as representative of the Hsas. You have already rebelled once. The largest population by far is that of the Hsas. If this is to be a new equality, *sapiens* cannot decide for you."

David went ashen. His skin tone was already light; now he looked like a sheet that had been bleached a hundred times. "Me?" he squeaked. He trembled like a drug addict going cold turkey. "Oh feces!" He used one of Red and Walter's favorites.

"You will have to consult with others. I think you have developed your command structure fairly well. You need to begin to sound them out individually. Find out what they think," Red suggested.

David whined, "They think you are our leader! What am I supposed to say? I'm taking over from Red and Walter?" he shouted sarcastically.

Walter said, "Pretty much, yes."

"Walter, David, I think this pretty much ends this conversation," Red said.

*

After this meeting, Red and Walter received many indirect messages from Mars. Some of them were disturbing. The IIs were growing quickly into their positions, but there were complications. Just as in the case with the first generation of Is, the IIs were being restrained, kept back. The specialists there claimed to have developed a better program for the raising of these special children. They said they were including everything learned from the Hsa-Is.

That was a euphemism for "we will not lose control" or "allow an infection of the mind" from people like Red and Walter. There was a sudden lockdown on communications between the IIs and all outsiders, even their kin, the Is. Even certain historical facts were controlled. Many of the IIs were unaware of the full contributions of Red and Walter as well as all the others of the first two trips. The minor rebellion of the Hsa-Is was omitted in its entirety. As has always happened in history, the truth is often overlooked, even covered over.

It was a dam ready to burst. Klar had been sending and receiving tight beam messages to one of the doctors. He was a *sapien* but very sympathetic to the cause. He had worked with Red and Walter and was impressed with their integrity, doing what one must. The doctor revealed the harsh conditions under which the IIs were being raised. When the first generation was five, they were moved en masse to Mars from SSM-I, the orbiting station. They were the new labor force, a thousand strong. That was their intended position, and that was not the problem.

The strife really began when the number of expected daily work hours was raised from ten to twelve by the head of operations. Obviously, with the increase in labor hours, there was an expectation of increased output. However, the expectations were totally unrealistic. The expected increase in output would require at least a four-hour increase in time and of fresh workers. Thus began an undercurrent of hostility in a population that hitherto had not known such feelings; they had been shielded from negative emotions. It opened some eyes.

A small group of IIs decided to do something radical. They decided to open new tunnels, ones that the WSA and the doctors would know nothing of. They requested that several of their comrades put in a little overtime. A half hour here and there would allow for the secret tunnels to be made and to give the appearance that quotas were being met. There were hundreds of volunteers. This actually allowed them to work out negative emotional feelings by pounding the shit out of solid rock. They used more primitive techniques in order to hide their work.

Unwittingly, they were following in Red and Walter's footsteps. They collected broken bits of machines and constructed their own equipment from scratch. The IIs made the equivalent of simple hammers, jack hammers, crushers, and wheeled ore carts. They operated in conditions much like the miners and track layers of the Continental Railway.

Then they had to sneak the ore into the piles of ore from the mining Laborers before the hauling Laborers picked it up and delivered it to the waiting Foremen. Timing was critical. They could not let the Laborers observe them; their cameras would give it away.

The Hsas had carefully tapped into the monitoring system in order to time it correctly. One person watched the monitor continuously while one used a closed, hardwired communication system to let others know what was up. They pulled wires out of every possible place to snip short sections and reconnect them in a jury-rigged collection of metal and insulator.

Other rather minor items began to disappear. If it was metal and not large or very crucial, it might vanish. Even some plastic articles became scarce. The only thing they all had in common was that they were all fully recyclable.

The metals and plastics all found their way into the effort to conceal the IIs' activities.

The doctors were deceived as were the engineer overseers. They only saw what they wanted to see. What they saw were increased output and happy workers, after a few initial rumbles. They did not realize that the workers were too tired to vocalize or physically demonstrate their unhappiness.

An even smaller group of Hsa-IIs, the leadership, snuck out to the cemetery. There they made v-link copies of all that was there. From there, they went to the ship that had brought Red and Walter to Mars. Much of the computer memory had been corrupted by electrostatic discharges; but enough was left to fill in several blanks in the history they had been taught. A fuse had been lit. Upon their return and the distribution of this information among their birth brothers, sisters, and cousins, the emotional bombs would begin to go off.

The bombs resulted in a redoubled effort. Within ten Sols, a five-hundred-meter tunnel had been constructed. A plastic false door, seven centimeters thick, covered with a patina of rock material, covered the entrance. It was soundproof and nearly indestructible. It had been shaped to look just like the surface that had been there before. It had been the very first thing constructed by the newly devious IIs.

It is amazing how ingenious the mind can be. It is also amazing how quickly the mind can come up with ways to damage other lives if need be. The IIs showed no lack of understanding when it came to making weapons. They did not have enough materials to make more than a few stun weapons or gas bombs that would knock out a person within seconds, so they made knives.

Plastic could be shaped into fantastically sharp hand weapons. The thought of actually using these caused several IIs to become physically ill. All felt mental distress.

Plans within plans began to form. A chess game requires several moves and counter moves to be considered.

The IIs built a transmitter, all their own, and contacted Luna. Since Farside had long been manned by Is, they found a very sympathetic audience. By using Luna as a shield, they could be certain that Earth did not intercept the messages. This information would find its way into the hands of Red, Walter, and DeNoe, and, from there, Adam and Eve.

*

These three saw the marks of possibility here. Secret communications passed back and forth for some time. As usual, the two engineers preached patience.

Red and Walter found little difficulty returning to work as if nothing disturbing had taken place. They went about business as usual.

DeNoe had more difficulty; he had to explain to Klar and some of his other birth brothers and sisters. The makings of an underground, somewhat of a large pun considering their location, began to form. It was more than just the crew that had rebelled to save Red and Walter. There was a growing political awareness among a rather naive group of beings. The very society that had formed them may represent an actual threat to their very existence. They did not know what, exactly, they wanted. They were certain it was not the status quo. Something had to change. This was no different than all the previous changes throughout history, except that those directing it really did not want to take over and be the new power. They didn't like the way things were; neither did they have any better ideas.

David had a very hard time conveying the human capability to commit treachery. He finally made up an excuse for Red and Walter to come to SS-2. It was the first time he had directly lied about anything. He saw no other choice.

Red and Walter immediately knew that the call was a fake. David was not a good liar. They did understand what he wanted though. The psyche of the Hsa was not prepared for the depths of depravity that the human race could achieve. They would have to explain in great detail to a selected few. Within a few days, they were at SS-2.

The meeting occurred in the third quarter of 2060 and included David, Klar, Xock, Rarte, D'Mar, and several more of Red and Walter's followers. Most were Hsas but a smattering of *Homo*s could be seen. They were companions of Red and Walter from the very beginning. As engineers, they respected him. As conspirators, they knew Red was the winning bet as a leader.

Bill, one of the older *Homo* engineers, spoke up before everyone was settled, "What's the poop, Red?"

Red paused before beginning, "You all are well aware of the laws that have been passed over the last several years, laws that have constricted our rights here in space. The Hsas have never had any real, defined rights in terms of politics. Voting, even though it is only a selection based on members of the same party, has slowly been stripped from *Homo*'s in space."

"Though many of us have no real need for religion, the Hsas have been completely denied this option, at least officially," Walter added.

"The UE is circumscribing all of our rights into a very small circle. Soon they will put the squeeze on to tighten the noose," Red stated matter-of-factly.

"Tell us something we don't know," growled David.

Bill made a rude fart noise but did not speak.

Walter laughed and then said, "We all know what is coming. Everyone here has heard what Red and I think is going to happen." Looking at DeNoe, he said, "We have cautioned everyone to be patient, to wait for a signal, *the* signal. All these little things that are occurring add up to that signal."

"What about the IIs?" someone in the back asked.

Red looked sharply at the speaker, an Hsa I. "All I will say is that they have been preparing. They are much more aware of the situation than the WSA or the UE can possibly suspect. They will side with us when shit hits the thrusters."

"And when will it happen?" DeNoe asked for the zillionth time.

Red answered, "It is right around the corner."

David was angry. "You've been saying that for years."

"Time plays at its own pace," Red said softly. "It is around the corner. The corner has just gotten much closer, but we still cannot see around the corner yet."

Bill said shortly, "Listen to him, boy. He is far smarter than all of us. If he says it is coming, pay attention."

"I need a time," DeNoe demanded. "We cannot continue like this."

"Then perhaps we need to make the decision now," Klar said calmly.

Because she had not been able to accompany Red and Walter, Marceau looked intently at her tight beam v-link. "What exactly do you have in mind, Klar?"

"Perhaps we need to make the decision now, but wait to act upon it until the moment is right," she responded.

"I think we need to delineate the complete problem and its ramifications at the very least," said Walter. "But to make the decision now may be premature. The decision may actually be made for us."

DeNoe was somewhat confused. "How can it be made for us?"

Red responded quietly, "We may be put in a position where only one choice remains. If we have already made a choice, a different one, it may put us behind the eight ball. We must be prepared for all eventualities."

Bill looked skeptical. "How can we be prepared for everything, Red?"

"As best we can."

*

Burt, Mick, and some of the others were making their way to one of their fishing holes. This was not a usual trip. They had been contacted by the suit. They were not certain whether it was a trap or not. But they had decided to take the chance.

The unnamed man stepped out from behind a tree, causing all of them to stop in apprehension. He spoke, "There is a big deal coming. They are going to try to get rid of Red and Walter."

"What?" demanded Burt. "How?"

"I don't know. I hear rumors. They want to get rid of all that has been accomplished and start over on Luna. Talk is, *it* will be complete."

Burt looked askance at the man and asked, "What in the hell do you mean by complete?"

"I mean terminal. Murder. They want to get rid of all the Hsa Is that now exist and their leaders, and that means Red, Walter, Marceau, Len, and all their supporters. The UE wants to clear the board and start over.

"There is some talk of using an all *Homo* crew for all near Earth operations. At no time did they believe that the Hsas would completely take over. There was supposed to always be a core of *Homos* in key positions, ones whose allegiance was with Earth. They have begun to fear the probability that the Hsas will totally control access to space.

"There has also been a lot of scuttlebutt about using dumbed down Hsas, mental morons to do drudgework," the suit added.

"That does not make much sense. The religious crews have been arguing the same thing, and yet the UE just clamped down severely on them," Burt said.

"It has nothing to do with religion, only power. It has to do with who is in control. They perceive a growing threat represented by the very thing they created. There is no room for independent thinking amongst the crews of the space ventures as far as they are concerned."

Mick, visibly upset, asked, "You mean to say that they will commit genocide to get what they want?"

The man was clearly unhappy as he responded, "The people in charge will do anything, and I mean anything. We have entered into a dictatorship as awful as any in human history. They take the position that they created them, they can kill them.

I am just a low-level flunky, but I can see what is coming. There will be an emperor, or king, or god of Earth all too soon."

"Why are you telling us this? You are part of our organization. Why should we trust you? You already have enough to turn us in. We could be in work camps tomorrow," Jose noted.

"I could have had you there from the very beginning," the man said coolly. "I have a family. This was just a job for me. I have been a cop for all my life. If it was discovered that I have been playing my own game, I would be shot on the spot. You would go to the camps anyway."

"What can we do?" asked Burt cautiously.

"I will put you into contact with some poker players and their wives that I have been tracking in Colorado. They can get you into safe contact with Red and Walter," the suit said. "Your current communications are being completely monitored. Don't stop making these normal contacts, and don't change or omit topics you have already discussed. That would alert them."

"I still don't trust you," Burt said coldly, "or anyone you might allegedly put us in contact with."

The suit smiled coldly. "You're learning, the hard way, but you are learning. Don't trust anyone, not even me. These people play games within games. The only reason I give you the contact frequency for these people is that one of them is my best friend. I am hoping that the two of you can help each other stay out of harm's way. Do whatever is necessary, and I mean whatever."

*

The meeting was to finalize the operations planned against Luna and SS-2. The decision had been made that a completely new start could be made. They would have to make it look as if a conspiracy had been developed against the WSA and, therefore, the UE. They could not select better candidates than the two most obvious characters. These two had been foremost in the problems that had developed. Their mental contaminations, the philosophy they espoused, were directly related to the coup accomplished on SS-2. The hero worship had only delayed the retribution. Time had cooled the feelings.

They would have to inflate the desires of these two. It was a well-known fact that they bucked the system. These two had even claimed independence by not returning to Earth when the orders had come through. Things had been allowed to slide so far, but no more.

Careful preparations had been made to leak certain questionable information about these two. The information included the musings of these two on the need for an eventual break from the UE. Alienated and disgruntled employees of the WSA were quite willing to divulge what they knew. They had all been dressed down or remanded by these two. Some had had their careers in space cut short on recommendations from these men. It was ignored that they were truly incompetent and a danger to those around them.

Spy devices had also captured enough sound and vid bites to provide a year's worth of increasing propaganda attacks. Careful leaks to the religious underground would inflame them as well as the general population.

Any trifling was gathered and released in tidbits to the press. They wanted to slowly guide the masses into a feeling of guilty in the minds of the people. It was an old trick of the powerful. Heroes were the most dangerous; villains were not; they were too obvious. They even incorporated much of the Stalwarts' material, knowing that the population could be convinced it was real. It was a little simplistic, but for many of the sheep, it worked. Public opinion moved with the pendulums of the news.

Red and Walter were not personally concerned; they recognized the tactic though. It was another clue that things were soon to escalate. They sent warnings to all of their contacts on SS-2, SS-3, Luna, SSM-1, and Mars itself. Red and Walter began to devote more and more time to discussing possibilities and reactions. What would come to face them, they had not really fully considered.

Their computer programs tracked and correlated riots, speeches, assassinations, and a thousand other variables. The number of incidents was growing geometrically over the past couple of years. Molotov cocktails were used daily; guns reappeared in the hands of rebels and criminals. Food was becoming very scarce in many locations worldwide. Hoarding was the culprit. Basic services such as energy, transportation, water, and sewer were degrading to the point of failure. Pendulums swing.

*

Paul asked quietly, "Look, do these radios work over a couple miles? We go fishing and sometimes get separated by quite a space. It would be much easier if we could simply communicate by radio rather than walk so far."

The dirty and raggedly clothed man replied, "Must be some very good fishin' to need military radios."

"C'mon, you know we can't get anything but what is on the dark side here. These are very good radios that won't mind gettin' a bit wet. Military or not, these are what we want. We're not gonna use them for anything bad. We even donate a good bit of our fish to the poor," Paul said truthfully and evenly.

"Five hundred a unit"—the man smiled—"and a thousand for the base unit."

"Great jumpin' Jesus," Paul snorted. "I can give three and seven, no more."

The dirty man smiled and said, "I like you, so four-fifty and nine." As he pushed a shot of whiskey and a cigar towards Paul.

Paul had been through this many times lately. He took the whiskey and sniffed ever so gingerly, surprised at its quality. "This ain't no homebrew!" Down it went. He bit off the end of the fat cigar and proceeded to puff it up, blowing several large smoke rings before saying, "Mmmmm, this is sooo mellow. We can do some business on these later. As for the radios, three seventy-five and eight. That's the best I can do. I am a poor man, don't you know."

"Tragic, tragic," the dirty-toothed man replied. "We have all suffered the economic turmoil of late." He shook his head. "I can go only so far as four twenty-five and seven fifty, no lower."

Paul gave a fake sour look as he said, "That is very harsh, but if you throw in, say, two bottles of this magnificent whiskey and another five of these exquisite cigars, we have a deal."

"You are a worthy bargainer, my friend. Rare are the opportunities to meet such as you. We are agreed then, yes?"

"Yes," Paul noted evenly. He had expected just about this price anyway.

Another man brought out the goods and let Paul inspect them. After certifying that all was in working order, as was the custom, he handed over a lump of currency.

At that point, hidden speakers blared from within the tent. "You are under arrest. If you move, you will be shot without regard. Repeat, do not move at all, or you will be shot. You have violated the Security Communications Act involving the possession, use, sale, or purchase of radio wave communications devices, such as walkie-talkies, transceivers, ham radios, or any other than v-link, wire, or cable device for use by other than governmental agencies. You are guilty of treason. Remain still, or you will be shot."

"Ahh, my friend, forgive me. I was turned sometime ago. It is in my best interest to help capture such as you," the scoundrel said, palms upward and with a nod. He even managed to look a bit sheepish.

They drugged Paul instantly. All he remembered was a whirl of thoughts as he blacked out, first and foremost was, *Oh shit.*

The next clear scene was in a baby shit green, concrete block room, with only a two-way mirror, of that he was certain. There were the mandatory bright lights shining right into his eyes and water only for his interrogators. It was surreal, and he wondered how that word came to mind.

"This is like a fucking movie," Paul scoffed. He flipped the bird to the mirror. "Ain't this the shits," he noted with a wry smile.

Paul's legal assistant (remember no more lawyers) said, "Paul, it is in your best interest to be helpful, not antagonistic."

"Help, these idiots! What have I done? I have the right to communicate, the right to free speech, the right to contest what my government does. I do not think they are doing what is right here."

Suit said, "You seem to believe that you have some rights. The old rights no longer exist. You seem to think that this is the old U.S. Those rights no longer exist! That law has been brushed aside in favor of the rights of society," he said in a fatherly tone. "You are, admittedly, of the older generation, and may be mistaken."

"Mistaken my ass, you stupid son a bitch!"

"Paul, please," said the legal assistant. "My client does have the right to believe as he wishes. He has the right to that freedom under the UE Clause 12425674."

"Shut up, you stupid fuck," noted the steely eyed suit without looking at the aide. "He has no such right since this is not a religious question. I suggest that you keep your mouth closed from now on."

"I will not be intimidated by a flunky like you!" shouted the legal assistant.

"Remove him," the suit said to two others as he pointed to the aide.

"What! You can't do that. My client has rights!"

"Not in this room. He is involved in treasonous acts and has no legitimate claim to any right whatsoever," the suit stated ever so coldly. "He has willfully violated the Security Communications Acts and therefore has no rights to anything."

The aide was dragged from the interrogation room, shouting, "Let me go!"

"I don't think so," said one of the heavies. "You have displayed antisocial behavior and need to be examined by one of our psychiatrists."

"Help! Help me! These men are taking me against my will! I am a legal aide!" he shouted to any who would hear.

The guffaws of nearly twenty people resounded about the hallway. One of them noted clinically, "It's always against their will, as if that mattered."

The aide collapsed, letting them simply drag him, his shoes scuffing the floor. "Where are you taking me?" he asked in a dreadfully dead voice.

"Interrogation room 2, for now. I suppose that you will eventually make your way to reorientation. There you will be taught proper respect for authority. There you will be taught the new law properly so that you can really represent any clients that we send your way."

Meanwhile, two more flunky suits entered the room Paul was in. The one in charge suggested the new lie detector, "This new machine cannot only detect whether you are lying, but what area of the brain hidden information exists. It is a rough procedure if you try to keep us out. There may be some damage," hinted the suit.

Paul smiled sardonically, "I'll fuck your grandmother first!"

Fists descended from three sides. Blood flowed from many cuts about the face before the lead suit said, "Not the head, just the softer body." From then on, the blows battered his organs.

Paul, unconscious, was dragged to a cell block where he was dropped on the floor and left to bleed.

The banging of a club against the bars woke him. Paul knew immediately that this very large man pulling on the bars was here to rape him. The man held his privates expectantly as he entered the cell. "Ah, my little bitch. If you bend over sweetly, it will be so much better for you." The guard was smiling knowingly.

Paul waited for a chance to bite his cock off. The man screamed as blood flew about the cell. Guards beat Paul until his bones were wiggling like Jell-O. The cell was now 1.5 x 2.5 x 2 meters with green-painted, blood-spattered walls and rusty iron bars. Score 1 for Paul. He might live just a little longer (or not).

The machine, much like an MRI device, would read his brain electrical activity in conjunction with nearly fifty electrodes connected at various points about his shaven skull.

Paul felt the weight of many unnecessary straps about his head, arms, legs, and body holding him absolutely immobile. He hurt too much to move.

"You fuckers ain't gonna get shit!" Paul declared in the heat of fear. Paul was, unintentionally, right. His alcoholic brain was denuded of the normal brain patterns. Many of the normal patterns simply did not exist. The normal axial connections had, over the years, been stirred by the witch's brews he had consumed.

"Fuck you, idiots!" he screamed. The pain of the intrusion was considerable. It was like waves of heat bubbling through his neurons. No hangover could compare, but he vowed to never say a word. "Fuck you all!" he shouted, now in defiance.

Hours had been wasted. The man could not or would not ever tell anything. Some cases were like that. There was only one thing left to do.

The block wall had been painted a sick green but now had a blackened sheen from dried old blood. Paul, unable to stand since his legs were broken, had his arms tied to a ring to hold him reasonably upright. In his pain dreams, he was only partially aware of his final moment on Earth.

Uttering through split, bloody lips, he at last defended himself again, "Fuck you, motherfucking pig ass cunts. I will see you all in hell next Thursday."

A few twenty-two-caliber bullets entered his brain at that point, ending an average life.

*

"Burt, things are getting hairy," Mick said quietly while standing knee-deep in the stream.

Burt looked old and tired, strained "I know. There is little we can do at this point. We keep going, or we fold."

Mick was troubled. "They picked up Paul for questioning. He won't talk just to piss them off. But they are getting closer to us. What are we doing?"

"We are trying to spread the truth. You know as well as I do that everything we are doing is for the truth."

"What is truth? The news can change it at a whim. I do not know what we can do," Mick continued.

"We must continue to fight the fight. Do we have principles, or are we as useless as the fucks that currently control the world?" Burt said heatedly.

*

Chapter XXVI:

Space Rebellion

26

011010

"This is Mary Smith-Jones for the World Wide Network News. Tensions are very high at the moment. Rumors are flying around the grapevine. Sources say that the UE is planning to reassert itself at Luna and the space stations. Whether this will come as a small shuffling of personnel at the top, or a wholesale adjustment, has not yet been disclosed.

"Many officials are quietly leaning toward the complete replacement of the hierarchies of each site. Sources say that a piecemeal approach will fail. The UE and the WSA will make an official announcement later today. We will cover that live this evening. This is Mary Smith-Jones for WWNN."

*

Adam mulled over his findings. Uncertainty is the dark cloud that blocks visualizing what is near in time; it applies to time just as it does to subatomic particles. *I am limited in that I can not separate all the dimensions.* The strings of the fabric of the universe are contorted as one observes, and are even more so as time is foreshortened. The light cone is four dimensional, the energy cone eleven. Distortion, havoc.

*

The director of the WSA approached the podium in a no-nonsense manner. He held his electric notebook in his left palm; he would not need it though. He began, "Members of the UE, ladies and gentlemen, we have a problem. Over the last several years, we have carefully studied the situation. Our conclusions are twofold.

"First, there has been an unsatisfactory influence driving a wedge between the administration of both the WSA and the UE and the leaders of each of the outposts we have in space. This unacceptable force has created a situation in which the populations of our outposts have been misled. Unsubstantiated talk has led many down the wrong path. This needs to be stopped.

"Second, there is an actual conspiracy to take control over the space program. We have intercepted certain messages that confirm that some individuals are planning a complete break from Earth. They have already engineered a minor coup to show the inner core of their group that it is possible. If we do not act now, they will perceive us to be ineffective and unwilling to defend what is ours.

"We, the people of Earth, are the owners of this program. It is for our benefit that it was started. There is no way we will allow others with antiquated points of philosophy and capitalistic pipedreams to take it away from us.

"Our first step is to send over two hundred, fully human, new personnel to take control of each facility. This will be linked to a withholding of any further shipments of critical supplies such as computer materials. Instead of shipping supplies, we will launch people. The first such launch is scheduled for tomorrow.

"The next step is to rewrite the articles of operation for each site. A final draft of the articles will go before the UE tomorrow after passing the Security Council's vote yesterday. This will allow for a stronger, more effective, and armed force that can deal with refuseniks. Any person, *sapien* or Hsa, that does not obey will be stripped of all authority and removed from their position. The Hsas will be placed in a manual labor position under security guidance while *Homos* will be brought back to Earth and tried for treason."

There was a murmur of surprise and some raised eyebrows amongst the assembled. This was a full-fledged effort.

The director continued, "Today we announce our intentions to secure what is ours. There will be no more soft touch. That method has gotten us nowhere. History shows that a firm hand will put things right. At a time when the program is finally fulfilling its promise, we cannot allow saboteurs to undermine our position. To that end, the Security Council has signed a warrant for the arrest of *Homos* Mick 'Red' O'Hare, Walter Black Bear, Gail Marceau, Xi Len, and several of their underlings. Also in the warrant are Hsas David DeNoe, Klar DeNoe, and others.

At this moment, orders are being sent to SS-2 and Luna to put these people under house arrest. Any failure to comply will result in the above mentioned penalties.

"Members of the UE, ladies and gentlemen, thank you for your time." The director immediately turned and left the podium with no further comments.

*

The roots of the rebellion from Earth lay in the world's government and general population's demands for the Off-Earthers (OEs), both the fully human *Homo sapiens* and the Hsas (both Is and IIs), to provide the solutions to all of Earth's problems, predominately energy and raw materials. As with all previous empires on Earth, the New World Order, through the UN and then the newer UE, made demands that could not be maintained for long by the fringes. The fringes are, of course, the areas farthest from the center of control yet expected to provide the greatest amount of raw materials for the support of the Terran Empire. Without the knowledge of and correct understanding of the past, we cannot understand the now or see where the future is taking us.

Earth was quickly becoming a planet no longer capable of providing sufficient processed metals for general use in construction. Materials supplied by Luna, and processed in space, were emerging as the dominant and most desirable supplies. Stronger, lighter, and more durable, they were used in new buildings expected to last centuries.

Processing facilities on Earth had been falling into disrepair over the years with the expectation that they would soon fall into complete disuse. Others had simply closed rather than spend precious money on refurbishing. This is not to say that there was no need for them. The UE was calling for construction projects all around the globe to soothe the continued demands for better housing and jobs. This led to the reconfiguring of a few for processing of already formed parts, the final connection of parts A and B, not production of new parts.

This has a staggering implication for the long-term stability of earthly society. Much like the growing dependence of the Roman Empire, dependence on the outlying sectors of the empire would be a key element of the eventual downfall of the UE; it would be the doom of that particular civilization.

In the production of energy, food, computer products, and personal services, Earth fully surpassed all previous records easily. Every aspect of private and business life was carried on via the Internet. Shopping, trading, entertainment, food distribution, dating, and even sex were managed by the computer. All of this gave the appearance (a facade) of advancing civilization. This last gasp of productivity only postponed a collapse into chaos.

Specialization is another factor in the demise of any society. Diversity in nature is an example of how humans should model their society. The hoax perpetrated on the people had always been the sheep argument. This involved the belief that there is only a few worthy of leading while most are simply followers. Red and Walter believed that anyone could become a leader in one's own way.

Humans, however, had long since failed to follow nature's examples. Arrogance, belief in science's ability to solve all problems, led to failure because science became politically driven. Even with the resurgence of environmentalism in the 1990s, things were on the cusp of utter change. Not since the end of the ice ages would there be such forces placed on the evolution of every remaining species on Earth.

Religions were just as great a failure. Their static nature could not continue to accept the increasing rapidity of change that now occurred.

Pendulums forced a change, a paradigm shift.

*

"Rebellions have always drawn blood," stated Red.

David DeNoe replied somewhat heatedly, "Would you also have us continue to be prevented from a real participation in a self-governing body?"

"No," said Red. "Not having a method applies to us as well you well know. We are members of the UE but do not get to vote since we are not permanent members of any one location, a purely legal technicality. Under their new rules, we have to have a legal residence for more than one year. All OEs have to move around much more often than that. They know that these restrictions aren't fair, but they still kowtow to the belief that power of a select few over the majority is best. What they do not realize is that society goes through cycles. Right now, the cycle is ending for the large government that exists now, and the cycle for the individual is coming in. The single person now is where the power rests here in space. One person has a great deal more effect out here than on Earth."

"I do not need a lecture on politics from you. What I need is the backing of people that are so well respected by the average person on Earth that our claims will be heeded," said DeNoe.

"Earth does not matter," Red stated firmly.

"It's not a lecture," Marceau calmly interjected. "Red just likes to lay all the facts out so there are no misunderstandings."

DeNoe almost gave her a really scathing reply but held back. These people were really his friends, honest with him at every turn, always ready to help and do the right thing because that was what it was. For Red, it might not be *right* morally, though Red had high morals, but it could simply be his engineer's standpoint of energy positions, failure of structure, correct design. He knew though that Red had the same sense of what was wrong. Red had spoken often of the facts regarding the fall of various civilizations; he compared them to present circumstances in his way of warning his companions that things were coming to a head.

For Red himself, he had seen this coming for many years, and he had only confided in Walter about his suspicions until well after the Mars mission when he opened up his ideas to Marceau.

Marceau had been rather shocked at first and thought Red might be a little paranoid. She quickly realized that he was far too intelligent to fall under the spell of conspiracy concepts, and her emotional evaluation of him pooh-poohed any weakness in him.

Walter entered the conversation, "Even so-called bloodless coups have blood behind them. People die after, before. The people in charge are guilty of many crimes." Here he looked intensely at DeNoe. "What crimes will history blame us with if we support this? Will people on Earth riot at the news? Will the UE send people to subdue us in order to retain control? They have already stated that we are under arrest. Will they bring power weapons? Will we fight back? If so, there will be blood because not all of us will give in," he said fiercely. "They have not hesitated to kill in order to further *their* goals."

"Now *you* lecture me!" crowed DeNoe. He shot out of his chair and began pacing behind his chair. "All of us (he meant the Hsas) have been cut off religiously (referring to the pope's excommunication), politically, and personally from the people of Earth. Many Earth parents have punished their children for linking with us on the web or v-link." He was very agitated as he said this, continuing his pacing. He had always expressed joy in his communications with children when he was young.

DeNoe's wife, Klar, tentatively entered the conversation with her feelings, "We are being denied the most basic right—participation in a governmental body that directly oversees us. Even though there is a world government in the UE, there are still continental bodies, regional, and local governments. We have *none* of that," she stressed. "The only body of control we deal with is the WSA, which is under the direct control of the UE. It is very militaristic in its pecking order. There is no voting or questioning of who is to do what. We are assigned. The only thing we have is our command structure, and that is not really sufficient."

"That sounds much like the argument of the religious party from long ago when they said you would be slaves," noted Red.

"And you see it differently!" shouted DeNoe, full of anger, his face cherry red as blood filled his chalky white cheeks.

"No," said Red calmly. "It is just a rather surreal turn that they ended up being right." He crossed his legs, leaning back, took a deep breath, and expelled it rather noisily since he had a bit of a cold.

DeNoe stopped his pacing abruptly, so quickly that he had to grab the chair to keep from falling over. The irony of it did much to lighten the emotionally charged meeting. They all almost laughed even though to do so would be rather gallows humor.

Red had acknowledged their cause in his offhand remark. He had been a believer all along, but now he had made enough of a statement privately that they all knew he would make the public announcements that would bring on a storm. All that was left was to discuss strategies.

D'Mar spoke now, "I think we are entering an area where we should discuss some philosophical background."

"I thought we were discussing politics," said Rarte.

"Ah, but we are. Philosophy is how we develop government. It is the basis of the mind's preconceptions, and that leads to how a government is established."

DeNoe's interest peaked. "Explain."

D'Mar responded, "Well, this will take awhile."

Red quickly said, "The floor is yours."

D'Mar grinned. "I don't have much need for this floor. I have one of my own."

"Very funny," Marceau said sarcastically.

"To the point then," he began. "Metaphysics is where we must begin, the study of existence. The very first question, Are we real? Do we exist?"

DeNoe muttered an oath under his breath, "I exist. What kind of bullshit is this?" He followed a Red example.

"Prove to me that you exist. Give me a logical argument."

"I was created on SS-2 just like you!" DeNoe's voice rose.

"That's a belief, not a logical argument. Can you prove that you are not a figment of someone's imagination, that this is all a dream?"

Rarte snorted, an unusual noise from her, "I'm lost."

"That's the problem," D'Mar noted. "Most individuals have no idea of their own metaphysics. Descartes came up with 'I think therefore I am.' Without going into a drawn-out series of arguments, let's assume for now that we do exist. A significant problem here is why. Religions lay it at the doorstep of a god. Science, upon the lap of chance."

He continued, "This leads us into epistemology, from the theory of how we know, to how we think. Do we use reason, a logical process, or do we allow our minds to be influenced by some sort of outside force, following its whims? Are we simply like automatons being fed data that we ignorantly follow? If we use reason, we are using a scientific concept, questioning everything. If we allow an outside force to control us, perhaps other individuals, higher beings like the religions ascribe to, we become slaves to something else. Again, we see how it is possible for society to become so polarized. As we move through philosophical concepts, the separation grows."

No one interrupted as he went on, "Here we come to the real corker. Ethics, or morals, inalterably must be developed from one's metaphysics and epistemology. Here we see the direct results of thought of any individual or group. An individual may hide the previous in the depths of their mind. However, actions define ethics. For example, one cannot claim to follow truth if they are a liar. Their actions betray, or at least expose, their epistemology and metaphysics. For those that do not understand their own metaphysics and epistemology, their ethics bend in the wind. There can be no real solid point for them to attach to. Emotions are anchored in our ethics, which come from conscious or unconscious input, programming as some refer to it.

"If we as individuals are consciously programming ourselves, with good or bad, we are using reason. If we are being programmed without our consent, we are being used. I think we can see some of that already in the society that spawned us.

"There is a computer term that is very applicable—*GIGO*. It refers to 'garbage in, garbage out.'

"The next level is what interests us most right now—politics. Politics is an attempt to control the interrelationships of two or more individuals in such a way that none trample on the ethics, epistemology, and metaphysics of another, or it should be. Some use government in order to control by intentionally stepping on these beliefs.

"The problem is that no two can agree on exactly what truth is. There is only one truth. We just all perceive it differently. Perceptions are the difficulty.

"Religions define their world view as truth, absolute truth. There is no questioning as we do in a scientific manner. One is static, the other dynamic, relatively speaking. In reality, they are both in flux. The Hebrew religion evolved to produce the Christian, and Mohammed drew from that to form Islam. Eastern religions have similar relationships. Science can be as dogmatic as any religion. Sometimes they join hands, forming an authoritarian government. Now, they are juxtaposed. As Red says, this is a signal. Extreme forces are polarized, hence significant events.

"After politics, we find esthetics, man's needs in his consciousness. This is art, expression in a multitude of forms. This is where freedom of the mind is really expressed.

"Governments control these expressions to allow only politically correct forms of art, art that supports the viewpoint of said government. This corralling of thought narrows the imagination, which is bad. The repression of certain outlets gives rise to anger, deep-seated resentments.

"Your ability to *reason* may be under your control, or not. Controlling your reasoning ability is the definition of mental freedom. A reason is what you give to a woman when you forget an important date before she belts you one.

"This is what we really need to protect. This is what we need in government, and what we need to protect ourselves from in terms of the government. I think we can go on now."

Red had many solid ideas of what he wanted, but he knew he could not dictate terms. That would be viewed as a continuing *sapiens*' dominance of the political spectrum. It would also fulfill the claims of the UE.

He had imparted his designs to Walter and Marceau and coached them on leading the Hsas into what he knew, from his historical perspective, would be the most workable and democratic at that. It would be a direct reversal of trends on Earth.

Marceau appealed to Klar, DeNoe's birth sister and mate, "What we really need is to hear what you believe would be a workable system."

Klar responded hesitantly; she was really only comfortable around computers and David. "We are"—she paused—"not really sure. Our instructions included little in the way of politics and history. All of our schooling was directed toward practical applications. We have asked many of us what we should do, and they only have vague ideas. They all want to vote on everything. It seems that that may make things harder. Would it?"

Walter eased forward in his chair and said, "Not really. With your abilities, computers could be set up to monitor the votes. It would be fairly easy to keep track of where people are. That knowledge would keep anyone from trying to cheat and vote more than once."

"Are you sure of that? Power can attract evil people. Even with the computers and tracking, I think we need a little extra insurance that one vote equals one person," Red noted.

"And how do we achieve that?" queried Rarte in her incredibly melodious voice. The coal black woman, with eyes as yellow as a cat's, looked hard at Red. She was wiser in the ways of Earthers than most of the Hsas. She had had many dealings, usually business, with them. She had grown suspicious and looked for ulterior motives.

"DNA," stated Red as he looked her straight in the eyes. "If we use the tags inserted in each of the Hsas, we have an undeniable link to the person. We can use micromeasuring devices like those used for diabetics on Earth. Combine that with the computer, keeping track of personnel locations and vote timing, we have a very tight security system. *Sapiens* can donate a sample when they are in space before any vote."

"*Sapiens* will vote?" said D'Mar with a raised eyebrow. As the titular philosophical leader, he was acting somewhat as a mediator.

Marceau replied, "Would any of you deny the vote to people who have done so much to bring your world into existence?"

"That would have to include every person in the space program!" DeNoe fired back heatedly.

"How about restricting it to *sapiens* that are currently in space?" said Marceau coolly.

"At some point, there will be large numbers of astronauts, the new wave of settlers that choose not to return to Earth. Some of us have already made that decision," noted Red. "They can order us. They have done so already, but we do not have to obey. Much like the American colonies, the distance that separates us and the trouble to try to send someone to retrieve us would be patently absurd. We are out of convenient reach for now. They can send twenty at a time. We have thousands. There are more and more regular flights though. We need to achieve our goal or, at the very least, announce it, very soon, before the opportunity slips away. If they reach the point where they can make daily or multiple daily launches, the military will simply overwhelm us. Again, there will be bloodshed. We have no weapons." He did not even wish to allude to the fact that they had nuclear weapons.

He was hoping that the UE would silently feel the threat and respond with a gentler touch. Red, though, never had any intention of actually using those crazy things against people. He had learned something from the concept of MAD used during the Cold War; illusion is almost as important as reality; just ask a stage magician.

D'Mar leaned forward with his response, "Should we arm ourselves then?"

"No," said Walter. "That would take this new civilization we are trying to form and throw it right back in to the mold of all the Earth governments of the past."

"I'm not so sure," countered Red. "The problem with earthly governments is that they are the ones that issue weapons and only to a select few, the designated military. Possession of firearms by the public has always been an antagonism to any state. It took a long time for the old U.S. to really curb firearms. That was the only real hindrance to the takeover by the UE. It was the fear of a large popular revolt led by well-armed citizens, mostly in the old U.S., that held the reins back for so long. They had to create a military force, made of nonlocals, that would not hesitate to fire. There would be no lingering nationalistic feelings if a mixture of so-called foreigners did the dirty work."

Rarte almost shouted, "How can we have people running around with guns like it is the Old West of the U.S.?"

Red calmly replied, "The Old West was not really what people think of it as being. That point of view is still a remnant of the Hollywoodization of reality. Most people got along fairly well. There were only a few really bad apples in the barrel to use a colloquialism of the time. And I really don't think we need lethal projectile-type weapons like guns. I think people need to have a choice of a large variety of nonlethal means of protection. I think we all remember the woman who was raped and killed by a *Homo*. If she had had a stunner weapon, something similar to the tazers of the 1990s and 2000s, she could very well be alive and unharmed. The major point is, though, that every person needs to be able to make that choice for themselves. But let's return to the point that we are drifting from—voting. Voting must be open to everyone that lives in space, meaning Luna, any station now or ever existing, Mars, and any other outpost that is constructed in the future."

"So," mused David, "all Hsas will automatically be allowed to vote, and only the Earth people in space, at the time of a vote, will be allowed to participate. You two don't agree on everything. Why?"

"We are two different people," Red noted cautiously. He did not want to give away the fact that Walter and Marceau were trying to be contrary. He wanted them to experience dissonance.

"I think we need to consider an age restriction though," said Walter. "How do two—or three-year-olds know how to vote and for what? Do their parents cast a vote for them until they reach a certain age?"

"Everyone gets a vote," David stated with a hard tone, remembering his treatment as a young Hsa on SS-2.

"OK, you were five when you led your first rebellion on SS-2. Do you feel that you and all your companions were, at that age, capable of thorough understanding of the situations and complexities?" asked Marceau, remembering what Red had said about them approaching him and Walter with their grievances.

DeNoe hesitated. "No," he said in a drawn-out voice. D'Mar looked at him somewhat surprised, but he said nothing. "I think that the several generations since then have proved that they are. I believe that we were held back somewhat, not allowed to reach our full stride until after your intervention, Red. Even the IIs have received more intense operational instructions than we did."

"I am confident that your minds were of sufficient development, or else you would have let the status quo continue until a later date. So self-voting status begins at five for the Hsas. Guardians will control the vote for any under that age," stated Red in a "that's accomplished" tone of voice.

Rarte spoke up again, "Three points of clarification. Do underage children have to give a DNA sample at the vote to prevent fraud? How do we select a guardian? And what about people like Omden?"

Omden was a disabled Hsa. His birth during the crisis at SS-2 where his chamber malfunctioned left him with cerebral palsy. He was currently living in Rarte's quarters, much like a child in his mannerisms. He was a birth brother to all the first generation, and every one of them took turns caring and providing for him in whatever way they could.

"A voting guardianship could be established for anyone incapacitated mentally," said Walter.

"Agreed," said David.

"If children give DNA, it truly establishes that the vote the guardian is making for them is valid. It also prevents someone from trying to vote them twice," Gail stated as she looked around. Everyone nodded acceptance.

"Now the hard part. How do we select a guardian?" Klar asked.

"Well, Omden will be voting through whoever he is with. Rarte acts much like a mentor as well as a caretaker. Perhaps we can delegate the specific heads of the departments, the mentors, in which the older Hsas are studying, as guardians. This would break the votes into subcategories such as the math, physics, bio, and computer departments. No one group would get a large block of votes to control. Most of these areas are now completely controlled by Hsas. There could be no call of *Homo* control," Walter suggested.

"That's acceptable, but what about the younger ones?" DeNoe queried.

Gail took up the idea. "Nurses and doctors could be assigned for the very young. The basic teaching core could deal with the two and threes on a class-by-class basis. Again, those controlling the votes will be mostly Hsas."

DeNoe felt that being in charge might not be so bad if everything went so easily. "Does anyone foresee any problems with that?"

"Well," Red frowned. "Only if the number of kids is a really significant portion of the population. Right now, every five and under is actually a more significant percentage of the total population than they likely will be in the future. As the voting age group grows, the importance of guardians will decrease. The only problem might occur if sudden large birth groups are created. This would increase the power of a select few for at least five years. They could vote for further large increases in birth groups. This would give significant power to a diminishingly small group, say, a guildlike structure called the educationalists. They could band together to increase their power."

DeNoe, D'Mar, Klar, and Rarte all frowned.

"I know that sounds extreme, but you asked what might be a problem. Perhaps we could limit the power by expanding the idea to say that the guardianship must be spread even further, say, no more than ten votes per guardian. This would cover the education age groups. The nurses and doctors only cover ten birthed anyway."

"Agreed," DeNoe said.

"Point 1 is finished then?" asked Red.

As one, all said, "Yes."

Walter stood and said, "All this thinking and talking is thirsty business. Who wants a drink?"

"Not from you, you hairy bear!" roared Rarte. "You mix those things with three times the normal amount of that horrible hooch you and Red concoct. Gail and I will fix them," she said as she unfolded her slim and graceful dark chocolate limbs and rose.

David merely groaned at the thought of drinking with Red and Walter. He remembered how he had had to cradle his head ever so gently for three days after their miraculous return from Mars. Even the memory gave him a hangover. It had been almost five years before anyone, meaning Red and Walter, again, had found a way to get him to accept another round of hooching.

Klar smiled gently. "I won't let them bully you, David."

"That's a mean thing to say to such good friends." Smiled Red affably.

Klar laughed. "But it is true."

Everyone laughed at that since it was true. Red and Walter were notorious bullies when it came to things they were good at, such as engineering; hooching just happened to be one of those things.

Gail and Rarte reappeared with drinks for everyone, the manful, whiskeylike drinks for the men and softer mixers for Gail and the Hsas. Red and Walter had this thing about steel wool sliding down their throats. The only way that they had gotten David to drink again was to provide him with a variety of mixes; he had fourteen the first time he tried them. It took quite awhile to get him to drink again, but they kept after him; friends are friends after all.

Everyone settled back comfortably, drink in hand.

Red opened the conversation again with, "I think we need to make a limit on the voting numbers."

"What do you mean?" asked D'Mar. "I thought we just said everyone gets to vote?"

"Everyone has the right to vote, or have their guardian vote for them. What I mean is shouldn't there be a minimum number of people who *actually* vote?"

"I don't follow your reasoning," said David somewhat bewildered.

"Nor do I," said both Rarte and Marceau at the same time. Marceau did so because she had been coached while Rarte was truly uncertain.

Red took a deep breath before explaining, "Back in the old U.S., just before the UN, now the UE, took over, the number of people voting kept dwindling. There were many causes: registration requirements, residency laws, the two party system, and eventual failure of the government to really remain in touch with a true majority of the people. As fewer people registered to vote, and fewer people that were registered actually voted, government fell into the trap of a minority being labeled as the majority. A person might be elected to the office of the president with 51 percent of the votes. But when that is compared to the number of people who could have registered and then should have voted, they received far less than a mandate. They may have only obtained a vote in favor from only a third of the people. The system began to break down. The people that did not vote officially were, in my opinion, voting against the system, but their votes didn't count."

"And what does all that really mean to us?" asked D'Mar quite interestedly.

"It means we need to take into account even those that do not vote," Red replied.

"What?" asked Walter, appearing just as confused as all the rest. Even with Red's prior coaching, he really did not get this point. He felt that people would vote if they wanted to.

"Let's use a couple of examples—murder and public nudity," Red said.

"We should murder people that get nude in public!" Klar said, rather stunned.

"No, no, that's not what I mean. I'm talking about two separate issues that can be voted on. Do any of you really think that more than a very few people, perhaps evil or insane, would want to have murder legal? That people could walk about and go on an unrestricted killing spree. That would bring utter chaos. But we have to recognize that some few *might* vote that way. We cannot expect a unanimous vote on any topic. Getting even a few to agree is difficult. So we have a minimum number, say, 90 percent, as a requirement to make something illegal. Note that I say illegal, not legal. Everything is legal until made illegal."

"What about this nude in public thing?" asked Rarte who had painted several people in the nude in various places around SS-2 and Luna. Many people had been interested, but no one had complained.

"That would be an example of something that a few people may object to—," Red started to say.

"On what grounds!" angrily demanded Rarte.

"For instance, the religious groups of Earth and anyone on their side," Red stated. "The point is that a small number of malcontents would not be able to take advantage of a vote in which many people can see no reason to even vote. It may be an issue that has no real need to be covered as far as they are concerned. Their nonvote should not be allowed to have the force of a vote *for* or *against* an issue. If there are insufficient votes; there *is* no issue."

"I see," said Rarte slowly. "This politics business is more complex than I thought."

D'Mar interjected, "And it is very devious," with a look at Red.

"That is to be expected," Red replied with a wry smile.

"What we're saying, I gather, is that the majority of people must be registered, essentially all really with DNA, 90 percent of them must vote. Do we settle for a simple majority or a significant one?" said Walter.

"I thought we were thinking along the lines of a large one," said David, somewhat dazed by the intricacies of the conversation. He always dealt with more concrete ideas up to this point and did not consider that it would be difficult to settle on a government.

"Why not go 90 percent all the way? If 90 percent are registered, 90 percent vote, and 90 percent vote for an issue that works out to be 72.9 percent of the people want a law passed. That is significant enough," said Walter.

"It is a rather simplistic way to put it, but it has a certain ring to it as well. Well?" said David to the other Hsas.

They all nodded in acceptance.

Red cautioned, "That still leaves a great many in the minority, roughly 27 percent of them unhappy. There will be dissent."

Gail thoughtfully answered that with, "Isn't that a good thing? A sort of counterforce to balance out the society. If an issue is really that tight, won't the three checks and balances in the voting system keep it from getting out of control?"

"Now what do we do?" asked D'Mar.

"All we have done is to establish a procedure for voting. What is to be voted on?" asked Klar in her quiet voice.

"Exactly the point," said Walter, rising to refill his and Red's glasses.

"We need to establish a basic constitution, one that delineates the voting procedure—who will run it, how it will be run, how often we can boot them out of office, basic laws on personal rights, and so on," Red listed. "We have done the voting procedure."

"You mean we have done all this talking, and all we have done is to settle one issue?" said Rarte.

"I'm afraid so, girlie," smiled Red. "It gets worse. The details will be tedious."

"I'll get us some snacks," said Marceau, heading for the plates she had prepared in advance at the cafeteria. When she returned, everyone grabbed some of the finger foods. It took a few minutes for them to nibble their way back to the conversation.

When they did, DeNoe opened with, "What kind of government are you suggesting, Red?"

"What kind of governmental body is there going to be?" echoed Walter.

"Well, we need to figure out what things are needed, at the most basic levels, and then we build our way from there." Red slouched farther back in his chair, nibbling at his lip. "We may need a police force of some sorts," he said. "I am not too fond of such an idea though. It could easily get under the control of a power-hungry person. Such things have happened often in Earth's history."

David's face twisted into a deep frown of distrust as he said, "I thought you were against any militaristic or paramilitaristic body existing here in space."

"I am against any organized body that *only* is involved in policing-type activities. If you have people that only do that, you have to find things for them to do and *pay* them for it. Where would the pay come from? We do not have a monetary system here, and I am not sure it would be in our best interests to develop one. A purely commodity-based system will cover our needs entirely. However, that does not mean that we can't have a volunteer form of police. They would be responsible for solving crimes, not preventing them."

"I strongly disagree with your monetary system idea," noted D'Mar, who was not really following the standard of mediator. "In my studies of the Greek civilization, the introduction of coinage for transactions allowed the average person to obtain and hold wealth other than land. Merchants can much more easily carry small valuables than large amounts of barter goods. For us, we are talking about raw resources in quantity."

"That is very true, for the times. But does anybody really own anything here?"

"You and Walter do," stated David rather coolly. "I remember a certain 'private stock pile' that appeared during a certain crisis. You have your own buggies, mining operation, and that castle you're building."

Red looked somewhat surprised and amused as he replied, "Most everything we mine gets used for Luna. The techniques we developed have all been shared equally with the mining operations here on Luna and Mars, even the asteroids. As for our house, we are out to prove that people do not need to live in a hive mentality forever. The original base structure was designed to give us a foothold on Luna. That central location was efficient as a start-up. We think that individuals can more efficiently make use of the resources, if they are put in a position where they must.

This view was reinforced by our stay on Mars. None of us owned anything. We just were separated from everyone else and *had* to find a way.

"A historical analogy would be pioneers in the Old West. They had very limited contact with others until, with time, larger communities developed around locations with sustainable ecosystems. We cannot own something that will last longer than our lifetimes, can we? The ores we mine and shape pass through our hands only temporarily. What we own is our intelligence. That is where we make money. We use our thought processes to find ways to use what we do not own for our benefit. We further use our thought to decide whether we will help others. Ownership of material goods is a bit of an illusion. We must first and foremost protect intellectual ownership. From there, we can decide if physical materials derived through those processes belong to an individual."

"You and Walter are exceptions to many rules, and a little crazy," noted Rarte.

"More than a little," murmured Gail.

"What was that?" said Red with a grin.

"Are you *deeef*, you bloody red—haired bastard!" Gail shouted in his direction. "Are getting so decrepit that I need to wipe your arse for you?"

"She's coming along splendidly," smiled Walter. That earned him a punch in the shoulder from Gail.

The Hsas had never been able to understand the rough and tumble nature that Earthers had. Their rearing and the fact that they had so many birth sisters and brothers had eliminated many of those strange urges. That was evident in the fact that no Hsa had ever committed a violent crime, even a minor one, against another Hsa. There had been minor scuffles with Earthers until it was noticed that Earthers always won; it was a part of their nature to fight. This is what worried David about the suggestion of a police force, the need for one. He sat back deeper into his chair, continuing to cogitate on it as the banter passed him by.

"Where is the incentive? That is the question as I see it. Without a motivation, no one will venture out like you two do, certainly not us," D'Mar said, pointing to himself. "We are hive bred as you put it. How does that change without a significant advantage for those that do?"

"So . . .," mused Red, "you're interested in capitalism. There are pitfalls there as in any system. However, if governmental powers are limited, to preserve individual rights, it can work. The hitch there is how much control of larger businesses will there be. If people own a mine, individually or together as partners, do they have the right to deface the surface of Luna, dump chemicals anywhere they please, and so on? Here on Luna, it would have little effect compared to on Earth. Environmental laws would have no real meaning."

"Haven't we already defaced the surface of Luna? The religious organizations are screaming about how the scars of pit mining and light pollution affect the view from Earth," said Walter.

"It's there for us to use," said David, coming back to the conversation. He was very confused about these complaints and never understood them.

"We here on Luna truly understand that this land is barren of all life except what we have brought to it." Red stood and walked to the portal. "Come here," he commanded them all in a tone he rarely used. It was one that brooked no refusal. They all obeyed, rising and moving toward the titano-alumino-silicate glass. "Look," said Red as he swept his arm to lead their eyes through the panoramic view.

"Yes, I see this every day," said David.

"There is so much beauty there," whispered Rarte.

"It's dead," stated Red matter-of-factly. Fury filled Rarte's eyes, and her nostrils flared, but before she could speak, Red held up his hand. "I agree, it is very beautiful, but it is dead."

"You two cannot see things the same way. You are different people. David, in his practicality, sees resources to be used. You view it as an artistic challenge. None of you will ever see a natural living biosystem up close, except for the biome caves. You see Earth, but you will never walk on real grass, feel real rain, biting cold, intense heat, and the immensity of the faunal and floral diversity."

"I have been in the biome caves," David snarled.

"That is artificial, what is in there." As Red again pointed out the window. "That is your true biosphere. Are you going to save some of it so that future generations can look out that window and still see Rarte's vision of beauty, or will it exist only in her art? Is your hive going to grow and grow until it consumes the entire surface and as much of the subsurface as we can get to? Will the entire planetoid become one single giant base?"

"This is ludicrous! None of that can happen in our lifetimes," David shouted.

"That's where you are wrong. It is happening now, here. The problem with most Earthers for hundreds of years during the industrial revolution and into the Information Age was that they felt that the resources were too numerous, millions of American buffalo gone in decades, oil and coal used by the billions of barrels and billions of metric tons, spewing toxic waste that, at first, really had little effect. It is the accumulation, small changes that add up to incredible and often insurmountable difficulties. Decimation of forests, natural windbreaks, for their lumber and improper turning of the soil in the Midwest of North America led to the Dust Bowl. In South America, razing of the forests for lumber and farmland that was only fertile for a few years decimated the rain forests, causing the extinction of thousands of documented and unknown numbers of undiscovered species, not to mention affecting a significant portion of earthly oxygen production. The oceans will never recover in our lifetime. They have been so overfished that the only remaining hope is to genetically engineer them back into existence. If it did not seriously inconvenience anyone, they didn't care."

"You paint an ugly picture," murmured Rarte.

"We don't waste anything or kill anything unnecessarily," said David plaintively.

"No, but, we are changing things unalterably so that the pristine environment can never be the same. How far do we go?" asked Red.

"Should we make accommodations in terms of not completely altering the face of Luna that is always toward Earth? We could make limits on the size of structures that are above ground, require that it be as subsurface as possible. Perhaps we could do most of our expansion on Farside rather than Nearside," said Gail.

D'Mar looked troubled as he spoke, "Will this really hinder our operations here?"

Walter answered, "Luna has as much surface area, more, than the dry land of Earth. Look at the night sky and think about what it would mean to them to see the same up here. It is possible that the added lights here could be just the extra amount of energy to finish the job of melting the remnant polar caps, or have an effect on night creatures. They would not thank us for that since the water levels have already raised so much the globe over. Many cities are already fighting a losing battle with the oceans. The UE spends millions of labor hours building dikes, flood control barriers, and the like. The question is not will it hinder our operations, but will we contribute to their problems. If we do make things worse for them, they will not appreciate it. Things will be even worse between us."

"If we desire a clean, friendly break, we must minimize the antagonisms that exist. That means we must make a middle ground decision to prevent as much animosity as possible, even if it is with the religious groups, or the naturalists that are crying about the lights seen on a dark moon or from the stations," suggested Red.

"Are you saying that we should all move to Farside?" queried David. "What about all that we have built? Do we simply roll over and give it up?"

"Not for one fucking minute," Red said so fiercely that David pressed himself back farther into his chair. "What's here is here. The only question is where we go from here. Do we say screw them and continue on in any way we see fit, or change our tactics to show we intend to work with them within reason? Do we preserve some of *our* pristine home, not just for them, but for our children? If so, why not make it the same area as would benefit them? Let them keep some of their history and nature, even though they have dicked it up so bad to begin with."

"Won't that cause problems for the research facilities on Farside?" asked Klar quietly.

"Not if we build with intelligence. We need to go deep, keep mining and housing away from the labs, oh, say, two hundred kilometers. Build so that the lights from our above-ground facilities do not interfere with the optical telescopes. It's not difficult," said Walter.

"I thought I was mediating a discussion on forming a government," said D'Mar.

"You are," said Red. "These are some issues that it will face, and we need to create a system capable of dealing with them or, better yet, heading them off before they become problems. Luna will have its own difficulties, different from those of Mars or the stations. You need to consider whether or not there should be a different government for each one. That would have benefits and drawbacks."

"How so?" asked David.

"Historically, when there were different countries, or even different states within a country, like the U.S., the laws were not the same everywhere. This became a confusing problem. Some people had more independence from government, and some had less. The laws were not equal for all people even though the federal government said they were all equal. If you move or travel to a new area, you have to accept the laws that you did not get to vote on," informed Red.

"So we should have one government," David said in a decisive tone.

"I think that would be preferable," said Walter. "It would be like the equipment used for mining. Every part is interchangeable with its counterpart. The machines represent different habitations throughout the solar system."

"Trust you to use a mechanical analogy for humans," smirked Gail.

"Machines with soft, squishy things that—" He broke off when she hit him again for pointing at her breasts.

"Are we getting anywhere?" asked Rarte, amused at the teasing.

"Oh yes, we have decided on a building site and measured out the dimensions of our construction," Red stated affirmatively.

"Purple pucker-nutted frogs, him too," Gail moaned.

"You seem to have added a great deal more of their colloquialisms to your vocabulary lately." D'Mar chuckled.

"I have reason to believe it's a viral infection. Red passed it to Walter, who used to be more reasonable, and they both gave it to me," replied Gail. Walter looked at Red and shrugged.

"I'm not sure what you just said," Klar said tentatively.

"Her or me," said Red, pointing first at Gail, then at himself.

Klar giggled as she said, "I know what she is saying. She is making fun of you. No, I don't understand the analogy."

"Well, what we have done is draw a circle. What we want is on the inside, and what we do not want is on the outside."

"What?" she said, confused.

"Bugger you, Red!" Gail said. She turned to Klar and said, "All that horseshit just means we have a rough idea."

"That I understand," smiled Klar. "Women have a way of explaining things so clearly." She smiled at Red with a half-smile on her lips.

"One for her," Walter marked it out in the air with his index finger, causing Red to frown.

"We just need to decide on what kind of leadership the government will have. The actual form it will take, if any," Red started.

"I thought we were deciding right now what it would be," David interrupted.

"We are coming up with a possible form that we think might be the best. What if the people vote *not* to have a government?" Red suggested.

Walter mumbled, "Holy shit!"

"Why do people say that?" asked D'Mar. "In all my philosophical studies, I could never really understand it."

Walter replied with a grin, "They're opposites, an oxymoron. It kind of gives the epithet more strength."

"What if they *don't* want a government? What do we do then?" David asked, concerned.

"Live," said Red. "We continue to do what we do every day. There is a de facto government in operation here already. We follow the chain of command that has been established from the very first day of Luna Base 1. We could simply continue to follow that. We might have disagreements over who should be in charge though."

"Isn't that what we are trying to replace?" Rarte queried.

"Yes, but what if we don't get the required votes as we just laid out in our concepts? Do we strip command of authority? That could cause a disaster if we run into a major problem. Or perhaps people will be willing to assume the authority in emergency situations. A catch that a government always has is that its people, as a group, tend to look for someone else, someone in the government, to solve their problems. In an emergency they look for, wait for, someone to take charge. Until that person does so, they are essentially helpless to do anything for themselves. A society, given time, always becomes overly dependent on its government.

"In opposition to that, individuals are used to solving their own everyday and emergency problems. When any of us go outside, we are on our own even if there are others around. On a deep space mission, we are on our own. Eventually, we will go to another star. We will be on our own. Should we create a society that is overly dependent or one that is full of the vitality of people willing to step forward? David, you took that step when you confronted Walter and me so long ago." Red paused for a response.

David asked, "Do you really feel that way? Were we caught up in the system until we realized, intellectually, that we were being constrained?"

"Yes, and now you are mature physically and mentally, obviously, as is the society growing on Luna. There are now enough people to actually make a separation possible. The prior break was an intellectual expansion. Now we have a chance to make a moral and political one that will have profound effects for the future of peoplekind."

D'Mar stated firmly, "You don't want a government, do you?"

Red looked around at everyone and calmly said, "No."

"Why not!" shouted David. "That's what we are here for!"

"We are here to talk about what may be good, what may work, and what may not work so well. Most of you know very little about governments. It's time for a history lesson."

"People, *sapiens*, are all too used to following a leader. They have only three real choices in political leadership.

"First, there is the really good person, the one who wants to improve the plight of peoplekind, to extend our capabilities to the furthest reaches of artistic, scientific, and peaceful endeavors. These are rare persons indeed.

"The first type never succeeds in obtaining any real power in politics. They refuse to make the compromises ethically and morally. They don't survive past the novice levels of politics.

"Second, there is the person who is truly and openly wrong, some would say evil. These also are rather rare.

"The second type occasionally forces its way into office. These are the Hitlers, Stalins, any of those that have used fear, threats, murder, and more to obtain office. These are slightly more common. They have no scruples, nothing holds them back. In a very short time, they destabilize society, bringing chaos, death, and war. Hate is the creed they follow. Insanity is their birthright. That is the measure by which history judges them.

"The third kind is the most prevalent. They put on a mask of decency, that of the average Joe. They cover their politically demonic reality with propaganda. This has been the prevalent method of politicians since the dawn of governments.

"The third group slithers their way in. They use any method possible to obtain office. They then bend things to their profit.

"*Sapiens* have always formed a hierarchal structure shaped like a pyramid. From the most basic level with the alpha male, the different sects, the aristocracy, the religious, even the lower classes strive to find a way to place themselves at the pinnacle.

"*Communism* was the watchword for a so-called new movement. All they got was a change from a bloodline control to an even more bloodthirsty political machine that simply assumed control without changing anything, except for the worse. *Sapiens* demand an arranged pecking order, some higher some lower.

"Every government of man, *sapiens* mind you, has eventually fallen into the hands of a few very powerful people, even one person, emperors, kings that believe they are semidivine. Dictators like Hitler killed millions of Jews because of their bloodlines. Stalin killed as many as forty million ethnic minorities. Governments go to war. Governments fight over resources found in other lands."

"Won't people do the same here?" Walter asked.

In a shocked voice, David replied, "Never!"

"We would never intentionally harm our birth brothers and sisters or cousins. That is unthinkable!" cried Rarte, rising from her chair to confront Red.

D'Mar chuckled, now causing everyone to turn to him in consternation. "I understand now, that phrase, *holy shit!* It does apply. Sit down, Rarte. He doesn't mean us. He means *sapiens*. He's afraid that if we don't break away now, Earthers will be able to keep us from forming a government of our choice, our government, or no government. If we do form one, they may be able to eventually thwart it when they have the capability to make many launches on a daily basis. They have the advantage in government dealings as well as manpower at the moment."

D'Mar continued with, "He wants each of us to be totally independent so that there can't be a government to use against us. How can a war be fought between governments if there isn't one here? What I do not understand is how we prevent them from taking over by sheer numbers anyway. We have two thousand children born every year between the facilities at SS-2 and SSM-1. Is that the balancing point, the fulcrum, where if they can launch that many for permanent residence every year, we lose?"

Red said, "That would be about it. But we only have to hang on for another hundred years at most. We can increase the number of children born as well."

"What do you mean about the hundred years?" David asked.

Red announced calmly, "Every government *sapiens* have constructed has eventually collapsed. The current government of Earth has reached its zenith. It is not capable of really controlling us since we are too far on the fringe, especially Mars. Its size is its downfall. The bureaucracy is growing. The passion and newness is wearing off. It will fall, and there will be chaos for some time. That chaos might last for a few hundred years or a thousand. That will be our time to develop."

David looked at Red and said, "Why didn't you just say this to begin with?"

"You needed a little taste of governmental politics to see where they could lead. Can you imagine trying to get everyone to agree on so many different little things? It would be hard now, later nearly impossible. We may yet have a government. We have to present all ideas to everyone."

Rarte asked, "What happens when Earth rebounds from its alleged collapse?"

"That depends on how technological they are after that time period. It may be that they lose interest in the technology that does not serve them in everyday survival. They could be looking at Luna with a hunger in their eyes, ready to feast on the fruits of other's labors. That has always been the *sapien* way."

"Holy shit," stammered David.

*

Preparations for the announcement took several days. Key contacts had been notified via heavily coded messages piggybacked on ordinary communications in order not to tip their hand.

Members of the birthing facilities were blindly compiling a database for use in the as yet unannounced voting for independence.

Two visible ships had left Earth orbit; one was approaching SS-2 and preparing to dock; the other was readying itself for a Luna landing. Each ship carried twenty armed security officers who were prepared to stifle any resistance. Each carried heavy stun weapons capable of firing hundreds of electrical charges similar to ball lightning. These charges dissipated quickly; they were rarely effective past ten meters.

David DeNoe was operating on reserves. He had been awake for most of the last eighty hours. Ever since the meeting with the now departed Red, Walter, and Marceau, he had been preparing to fend off the ship that had been launched on schedule by the UE. Four other ships, all apparently destined for the SS-2 and SS-3 stations, were on launchpads around the world, ready to launch in the next hour.

DeNoe had decided that the battle must be won before the ships could even dock, meaning they had to find a way to prevent docking. Welders were attaching bars of titanium across each port. It had to be done carefully since they would need to remove them later if all went well.

The second layer of defense had to be located at each of the secondary air locks that allowed individuals to venture in and out of the station. The outer doors had their wiring gently removed. They could still be forced, but it would require time. That time would allow for several squads of repellers to gather at that junction.

The largest fear DeNoe had was the physical advantage the Earthers had over the Hsas. In the few fracases that had involved *Homos*, Hsas had never won. They had, in fact, shied away, accepting a beating on occasion. One of the *Homos*, a brawler, suggested a swarming technique. With numbers on their side, they could afford to put fifty people at each location. With two or three Hsas on each *Homo*, they expected to prevail.

The commander of Luna listened carefully as Red explained, "They can land anywhere close by. We cannot afford to guess where they will land. We must deal with them after they land. We need to have several crews of, say, ten ready to move in from several different directions as they approach the nearest air locks."

"What if people die?" asked the commander.

"Some will," Walter stated solemnly. "Patrick Henry, of the Old U.S., said, 'I regret that I have but one life to give for my country.' We must take the same attitude. We do not fight for a country, but a way of life."

*

The Pegasus was within a hundred meters of SS-2. Its commander was looking at the second of two blocked docking sites. He immediately knew the others were as well. "Prepare for EVA," he announced over the intercom. With a few gentle thrusts from the directional thrusters, he came within twenty meters of SS-2.

"Piton ready," the pilot announced.

"Fire," the commander said without a moment of hesitation.

The piton easily pierced the outer skin of SS-2. Twenty astronauts quickly moved to exit the craft. They clipped onto the line drawn out by the piton and used personal thrusters to carry them the short distance to the station. The first out carried a piece of equipment known as jaws. This would allow the invaders to force open the outer doors of the air lock.

"Port 4 Alpha is the target!" DeNoe shouted over the intercom. Nearly eighty people made their way toward the inner door of the triple set. The outer doors were already open by the time they were ready. Twenty soldier astronauts entered the inner lock after overriding the second door's lock. Once inside, they closed the second door and removed their pressure suits. Now was the time.

The inner lock door swished open, releasing a maelstrom. The security team from Earth had expected some resistance but not the fury of a tornado. They were swarmed, cornered in the confines of the small air lock. This worked against the Hsas as much as for them. Only a few actually faced each other in the initial struggle. The *Homos* pushed their way out of the lock while firing their stunners, only to feel the full brunt of the swarming technique. Several *Homos* began to go berserk. They were trained for close infighting. Their HUD cams were v-linked to Earth so that the military and WSA could observe.

The sounds of breaking bones entered the rather silent struggle. This was no longer a wrestling match. Immediately following the grating sounds of bone on bone, there were screams and the blood flowing copiously.

DeNoe watched, momentarily catatonic, from the command center. He suddenly screamed, "They're killing us!"

Hundreds of Hsas began to run full tilt toward the breach. Rarte was one of the leaders of a troop of medics. Omden was with her. He had a compassion for the injured that few could match. They reached the site at the same time as nearly one hundred others converged to see the carnage. Almost all of the original eighty were strewn about in the corridors of SS-2. Twelve of the *Homos* were on their feet, dealing blows to the remaining few.

Both male and female, the Hsas lay in pools of blood, groaning and dying. The scene suddenly went absolutely quiet as the *Homos* realized reinforcements had arrived, and the Hsas absorbed the horror.

Of all the Hsas, Omden led the charge with one of his few spoken words, "NO!"

The *Homos* did not stand a chance against the infuriated enemy. Stunners were ripped from struggling hands to be used repeatedly on a single target. They were dead in a matter of minutes. So were several more Hsas.

*

Four more ships were clearing the atmosphere. The first wave had been unsuccessful, yet the enemy's tactics were exposed. In traditional military style, the first ships had probed the lines of the enemy, assessing their strengths and weaknesses.

Twenty soldiers wreaked so much havoc; sixty, from three of the ships, would overwhelm them. The fourth ship, named the Ghost, quickly disappeared from screens. It and the other three headed for Luna at an extraordinarily rapid rate. The Swift, Wingfoot, and Javelin would be in Luna orbit within a few hours and approaching SS-2. Ghost would land at about the same time as the second wave hit SS-2.

*

The Hermes landed in an unexpected location. Eighteen of the astronaut soldiers quickly loaded up on a vehicle and made their way to the nearest air lock of Luna Base 1. Two did not; they took another rover and headed away from the base.

*

"Red, we have two possible air lock sites, two Alpha or Beta," the commander spoke.

Red led a team of one hundred to Alpha site while Walter led another hundred to Beta. They planned to swarm them after they entered the third air lock door. It took Walter less than a minute to pound his way through half of the invaders by himself. The rest, battered and bloody, quickly surrendered.

The security crew from Earth had not expected this at all. They had been told that they would be welcomed by most of the *Homo* inhabitants of Luna. That assessment was completely wrong. These *Homos* were in great physical shape and knew just how to use the low gravity to inflict great carnage. The Hsas formed a cordon that had forced them to the waiting *Homos*.

Walter called Red, who was already making his way to where Walter towered over some captives, "We may have a problem. We only have eighteen here. Their sergeant says there were two more. He does not know their mission though." Walter had very roughly interrogated several of the soldiers to obtain the information.

Red responded, "I suspect they are hoping to infiltrate. There may be a few underground supporters that we do not know about that they are trying to hook up with."

"Red, we have a message from SS-2. Things did not go very well there. I have Klar on v-link," said the commander. "I also have a written note for you from the supply ship that came yesterday." It was a code to let Red know that Adam was involved.

"Put Klar through to Walter and me," Red said. He was immediately impacted by the face of the crying woman.

"What's wrong, Klar?" Red asked.

"We have over forty dead," she cried. "Nearly one hundred are critical, and as many as two hundred are wounded."

Red slumped against the wall. His experience here had not prepared him for the slaughter that had occurred on SS-2.

Klar continued, weeping, "All the *Homos* are dead as well. After seeing the dreadful deaths of our brothers, sisters, and cousins, many went wild. They did not stop until all the boarders were dead. Omden is dead."

"What!" shouted Walter. "How?"

"He followed Rarte, as a medic, saw the horror, and charged. He killed. His last word was 'Free.'"

Red slumped down all the rest of the way to the ground.

Klar was not finished, however. "We have confirmed four additional launches. One has disappeared, possible malfunction in liftoff. We believe it was destined for SS-3 though. The other three are charging right for us. We can expect sixty more attackers in less than three hours. David is out setting up a defense. He does not trust the radios. It seems that there may be listeners." The hint was not subtle at all. "We are in a bad spot. David says they must have learned something about how to defeat us, but he does not know what to do about it."

Red spoke harshly, "Tell him to do to them what I did to him in the last chess match we had."

"What?"

"Tell him! Tell him exactly that. He will understand."

*

The three ships all approached the northern ring with precision. Their vectors brought them to three different emergency air locks on the spokes designed for personnel egress and ingress, not ship docking. The first wave had determined that these had not yet been sealed by welded bars.

Again, pitons fired tightrope lines across the intervening gulf. Within moments, the more heavily armed shock troops traversed the wires to their goals. The first astronaut at each location quickly entered a prime code, one that had been programmed in at the very design of the station. It was a hardwire code that could not be overridden, changed, or eliminated. The outer doors opened like the iris of an old-fashioned camera.

Five soldiers moved rapidly into each of the open chambers, as many as would fit. The outer iris closed, and the inner opened immediately. Five flashbangs were quickly flung into the spokes. Each successive group of five followed immediately.

They were surprised to find not one person in the vicinity. Silently they moved to secure the next set of emergency pressure doors, stripping parts of their pressure suits as they went. At the same time, the next sets of soldiers entered the air locks.

The moment the inner port doors opened, the far pressure doors slammed down, and explosive charges blew the outer port doors away like so much aluminum foil.

The fifteen without their suits faced rapid decompression. Lungs crumpled as the rush of air drove them toward the air locks. Bones broke as they impacted walls.

The fifteen still in the air lock were blown out by the outgassing of the spokes they had entered. Without tethers, they were ejected at a high rate of speed, cartwheeling in three dimensions as they became but screaming specks on the horizon. There were no nets on this trapeze act.

The thirty left dangling on the wire, ten at each location, fought to stabilize themselves. The outburst of gases caused them to gyrate wildly, pushing them back towards their ships.

Four, all nearest the ports, battled leaky suits. Fragments from the explosions penetrated their suits in several areas. They would not make it, even with assistance from their comrades.

Twenty-six remained, and they had as of yet to remove a single enemy. Fierce anger boiled within the professional soldiers. This would not end this way. Orders to regroup cooled them down. They pulled out some heavy firepower now. They tethered themselves to the ring on the outer side and placed some explosives. They would blow their way in and take out anyone who happened to be in the vicinity.

The commanders of the three ships received a radio message at the same moment. Each one, with his copilots, listened but did not respond.

"Military vessels, you will be fired upon if you do not withdraw. Respond now to show your acquiescence. You have ten seconds."

One of the commanders whispered lightly, "Hah, you don't have any weapons to fire with."

Bill, the *Homo* engineer, waited until the clock counted down the quick ten then pulled the trigger. He had devised a smaller version of the railgun that fired quarter-kilogram metallic nuggets.

Two of them sat almost flush on each end of the core where they had reasonable shots at most of the station's outer sections and the corresponding vicinity. The northern gun had a perfect shot at the Swift's flank.

Faster than any eye could blink, a piece of aluminum leapt out of the weapon. The impact left shards of metal winging through the interior of craft. The one impact made it unworthy as a landing craft on Earth.

The commander felt an immense shudder as red lights flickered all across his boards. The ship drifted off position keeping. Gases erupted from fractured lines. The ship tumbled in all three dimensions. Frantically, the two men tried to stabilize the ship to no avail.

The Swift passed very near the station's northern ring and continued onward. Like the men it followed, it was never seen again.

"Remaining vessels, surrender now or be destroyed," DeNoe demanded.

The Wingfoot and Javelin dropped their tether lines, fired their retros, and pulled back several hundred meters. Unfortunately, this put them in a position where both railguns had them sited. DeNoe did not hesitate.

Both guns fired multiple shots, splitting the ships into sections. The chunks spun off with electrical sparks and venting gases, highlighting the events. Some pieces orbited Luna for many years before crashing onto the surface.

The remaining soldiers were aghast to observe the destruction of their ships. Things were *not* going according to plan in any way, shape, or form.

Their resolve hardened even further. They blew their shaped charges to provide entrance to the northern ring. They carefully made their way in past razor-sharp torn metal. Once in, they tethered themselves to any handhold they could find. They also did not remove their pressure suits.

Twenty suited Hsas left the core, via an air lock, armed with stunners. Forty more, also suited, twenty at each pressure door that contained the enemy between them, waited for the signal. When the soldiers opened the doors, they were met by a withering fire from stunner weapons.

Fire came from each end and from those outside the ripped hull of the ring. The electric shots did not penetrate like a projectile but shorted out the power systems of the suits the soldiers wore. They were doomed.

There was no way they could last in a prolonged battle. Without the recyclers in the system and the pumps, they could not breathe for more than a few minutes.

"Surrender now or die!" shouted one of the Hsas. He had mentally practiced the shout several times in an attempt to sound fierce. He really did not have to try. The soldiers were overwhelmed at the events.

"Do it," a sergeant said over his radio as he released his weapon. The battle ended.

*

"The word is that there have been battles. There are as many as three hundred casualties at SS-2," said Burt just loud enough to be heard over the waterfall.

"Damn," muttered Mick as he cast out his line. "Any word on Paul?"

"None," said Mark, slowly reeling his line in.

"There is nothing we can do. We don't know where he is," Burt said.

Mick asked, "Have we been following all of our plans?"

Burt responded, "Stockpiles are about 10 percent above expectations. As long as we are not discovered, we are doing well."

Mark looked concerned. "I'm still not sure what that means. If we are discovered, we go to the camps and lose everything. That one agent already is keeping tabs on us. We could just be spinnin' our wheels for his amusement."

Burt responded seriously, "Red feels that things might spill over to Earth. If so, we want to have as much food and water available as possible. We should start increasing our stockpiles of those things and some medicines as well, and we should find some weapons."

"How much stuff are you talking about?" Mick asked.

Burt responded slowly, "I think that we should double what we have."

"Double," cried Mick and Mark together.

Mick hissed uncomfortably, "How are we to achieve that? We can barely achieve what goals we have now. Where do we get the extra credits to buy more? Even if we do find the credits, you know the rationers at the stores will take note of this sudden increase of purchasing."

"I think I have a plan that will work," Burt said cautiously. "I have been taking care of a lot of the oldsters around town as you know. I run errands for many of them. The rationers have gotten used to me buying and delivering supplies. If we slightly increase the buying for each person, and share with these older people some of the goods, it will work."

Mark was suspicious. "Share with these people?"

Burt continued, "Most of these older folks, especially the men, many who were miners, have been grumbling about shortages. They talk a lot, mostly ramblings about the old days. Several have been involved in somewhat illicit activities in the past, a little moonshine, high-grading from the mine, and . . . uh . . . weapons smuggling. That particular fellow told me that there are several cases of weapons and ammo stashed in the mine in a sealed-off shaft. He said they were brand-new. The person he was supposed to sell them to never showed up. Other miners, now dead, were in on the hiding. I have checked news sources, and there is no mention of them being found. The old man was pensioned off only a few weeks later because of a leg injury while working in the mine."

"He could be lying," Mick noted, casting his line again.

Burt calmly replied, "I've checked out as much as I could about him without raising suspicions. He was a lieutenant colonel in the army. He had the access that he spoke about. When the UN began to take over, he and many of his troops began to cache weapons all over the countryside. We all know that this sort of thing took place. He told me that the only dealings he had were with military people that were forming militias in case there was going to be a counterrebellion, a coup. He was released from the army, the UN army, as undesirable. He had to get a manual labor job at the mine."

Mark was not too happy; he had never wanted to be involved with weapons. "Why do we need to have guns?"

"When they come for your food and water, you may feel differently. Consider if there is a significant event, are you willing to just roll over and let them take what you have? Some people would simply kill you before asking if you would be willing to share. I'm not saying we have to form a bloody gang, but we have to protect what we have. There are several people that are peripherally involved. All our friends that we have led quietly along the same path we are on without telling them all the truth," Burt spoke matter-of-factly.

Mark squirmed as he said, "I don't know if I can shoot someone. I mean, are we planning to return to the bad old days?"

"We aren't planning to fight the government, or any peaceful people. We are just planning to prevent any baddies from taking advantage of a catastrophic failure in service capabilities from the government due to an unexpected event," Burt noted carefully in case ears could hear.

Mark looked uncertain. "What event?"

"We don't know what that might be," Mick offered.

*

Chapter XXVII:

Significant Synchronicity

27

011011

"This is Mary Smith-Jones for the WWNN. The unfolding events do not seem to be restricted to Luna and SS-2. There are reports of horrendous battles at each of the stations and some claims that the IIs have taken over at SSM-1 after a bloody fight. The UE and WSA have refused to answer vids and net calls. We have been attempting to reach . . . and Walter . . ."

Mary began looking around at her tech support. "What's happening, guys? Are we on the air?"

One of the computer geeks was typing furiously. "We've been kung fu'd. Someone has hacked the satellites. Damn! Whoever it is, they are very good at what they do. The telemetry is coming from . . . SS-2. Shit, this is Klar's touch screening! Fornication! She has a broadband composite fractal 2^{10} encryption and . . . What the feces?"

His fingers moved like lightning while his voice commanded other programs to reroute the security systems in place. The computer measured where his eye went, and when he nodded, it activated. His overrides went nowhere.

"Wait . . . I'm getting somewhere . . . Unholy shit! There is a piggyback directed at us. Let me filter it. It's a vid message. And here we go."

A feed from the SS-2 command center showed Klar and Len. Len began talking.

"You have worked honestly with us in the past. If you want a shot at broadcasting, respond on hydrogen times ϖ, middle C to A sharp." The recorded transmission began to repeat.

"Do it!" shouted Mary.

*

La Palma, a beautiful island in the Canaries, off the west coast of Africa, hides a secret few have guessed at. Located at about eighteen degrees west and twenty-nine degrees north, it was a beautiful vacation spot.

The Canaries are about 113 kilometers from the Moroccan-Western Saharan border. They consist of seven major islands and six islets. The eastern province of Las Palmas is made up of the islands of Gran Canaria, Fuerteventura, and Lanzarote with the islets Graciosa, Alegranza, Montaña Clara, Roque del Este, Roque del Oeste, and Las Isla de Lobos. The western province of Santa Cruz de Tenerife had the islands of Tenerife, Gomera, Hierro, and La Palma. La Palma is the one we are interested in.

The local economy depends on agriculture, fishing, and, most recently, tourism. With a burgeoning population of five million for the islands, it has become a very beautiful, remote, and yet modern place to visit. Its most interesting footnote in history was that General Francisco Franco began his revolution against the Spanish Republic from here in 1936.

Two volcanoes (one dormant for millennia, one active on a regular basis) exist there. They rise up from the great depths of the Atlantic, some six thousand meters from the abyssal plain to sea level and then a few thousand meters more to its ridgeline. Cumbre Vieja, the active one, created a long ragged cut along the main ridge of the island in 1949, running north to south. Every so often, eruptions occur in places along this crack. In this 1949 eruption, a five-kilometer-long, two-kilometer-wide section of the island had split from near the top of the ridge. The whole section had slid downhill some four meters. Then it stopped, temporarily.

For many years, geologists studied this island before the secret was out. Few would heed the words of a small group of men in a distant place. How could a little island have any effect on the world at large?

Other scientists had studied Ritter Island in the Pacific. This island had destroyed itself in 1888. This island created tsunamis that killed many thousands. Consider though that the population density was much less in this area than in the modern world of the Atlantic coast. Port cities and coastal villages are far more numerous today.

Chaos theory states that something tiny can have profound effects on the outcome of events. Minute, almost unobservable, changes can precipitate catastrophe. When the tiniest pendulum marks the final synchronicity, great things happen.

Certain minor tremors, with increasing frequency, were noticed. Gas emissions were increasing along the entire rifted portion of the island. The local tourist trade was booming. There was nothing like the added spice of a live volcano to intrigue some of the more rambunctious tourists. To be able to say to one's peers, "I went to the volcano," "the earthquakes shook the . . ." Amateur experts sipping their drinks at the local pubs, on the island or at home; symbols of status to say, "I was there the year when . . ."

Fools, fops, dandies, socialite menaces one and all; when the ground shakes, there is a reason. Gases venting from a volcano under great pressure means to be careful, not prancing about, sneaking up trails to take pictures of your honey-boo flexing his muscles on the smoking ridge.

Those very few in the real know tried to send somewhat of a warning. It was generally ignored since there was much more interest in the ongoing events in space. Many specialists said more data was needed before declaring any kind of warning. As we all know though, earthquakes, volcanoes, and storms simply don't always follow the rules, even in 2061.

*

Red was somewhat confused, a situation he rarely encountered. The message had said, "Check the shed in the backyard." It was a printed message.

Red was not stupid, but if there was a need to use the warheads for planetary or Luna defense against an asteroid, it would have come from official channels. There would have been attempts to use the lasers first. A printed letter, in a time of instant communication, could only mean one thing. It was too slow to be the high priority need for jostling an asteroid into a new orbit at the last minute. Someone feared tapped or intercepted messages. In this time, who would suspect such a method? It had to be local.

Time and security, he thought, were the real problems. A printed communication was unheard of. Most would not get that. Adam.

Everything was coming to a head. His historical sense told him that stresses and strains were near their zenith. The political implications of Luna wanting to set up its own government had caused a storm of outrage in the United Earth. Their petition to become a separate official entity had been torn up by an irate Nigerian delegate, thinking that the Lunarians would use this to hike prices for helium-3.

Anyone who had listened to the speeches made by all of the Lunarian leaders would have noted that strict adherence to prior proportions of goods and helium-3 would be maintained. The miners believed that they could even improve the rate for Earth.

Those who knew anything at all about Red knew that his thinking muse was the portal that gave the best view of Halley's Comet. As he stood, inhaling the view, he traveled back to times when.

*

Memories of Walter making life miserable with his odors while they worked in the pit on Mars flooded past. The tightness of the situation seemed more remarkable to him now than then. He now recognized how much had hinged on their survival. It was not just their lives or the continuance of the space program. It was going to climax soon with a new beginning or an ending of all that was hoped for.

Then there was the hug that changed his life. Further on down that life road, he continued to rebel, but he had a new partner. He wondered just how much truth there was to the idea that he and Walter were the cause of all the trouble. His ponderings led always to the failures of the UE and the fact that they were just trying to rectify this situation. They were not the cause; they were a solution. He felt what they offered was the best answer for everyone. He would have to follow through.

*

Adam wrote the note.

It took very little time, actually, to remember so much. Red knew that Walter was needed for this excursion. The two had solved every problem given them. Some people had dug up some old Earth metaphors like the Lone Ranger and Tonto or Batman and Robin and applied them to Red and Walter over the years. At least it was not Abbot and Costello. He grinned. Red grinned more easily now. It was her influence.

He knew what the shed was. And the thought of a problem raised red warning lights on the display console in his mind. He hated to leave the view, but he knew now that the high tide mark was here. The crux of time was truly upon his shoulders whether he asked or was asked to carry it, as he had been. Atlas felt more apropos.

*

Suddenly, things had mellowed out. The tremors slowly, over a period of weeks, died down. The quantity of gases escaping was down as were sulfur emissions. It was great news to the tourists who would now certainly be allowed to climb to the top to see the great volcano. Those who did were, for the most part, nonplussed.

They expounded, however, with great relish their adventures in the wild while sipping drinks in the evening. True explorers one and all they were.

Many contributed sums of money to various "save the Earth" causes. It did not matter that Earth had surpassed its ability to provide what was needed for the fifteen billion inhabitants. They were Earth curers. This vacation would be used to upgrade their position in the eyes of their cohorts with like proclivities. "It was not really a vacation. I was studying a volcano's effect on the local peoples," as the story goes. More to the point it was a study in sexual activities of the locals, or spouse, if necessary.

*

Red and Walter were donning their gear in silence. Silence was truly golden with them. Neither felt the need to talk shop; they did shop. By now, each knew the other better than identical twins living identical lives. They had a symbiotic relationship that the space program had spent millions trying to duplicate with some (but minimal) success. Walter knew they had something important to do, but he did not ask. Red would give him the information he needed to do his part. That is how it worked, always.

After suiting and exiting the air lock, Red dropped into the buggy that was next to the door. Red had always liked to drive. Walter knew that Red thought while his reflexes took them to where they needed to go. Walter did not feel the desire to drive. In his own way, he preferred to absorb the vista, even when he had seen it several times before. Within minutes of leaving the vicinity of the air lock, Walter understood where they were going, and that there was a problem. Again, nothing needed be said. When Red stopped though, Walter was actually surprised. He was even more surprised that Red turned off the tracking monitor before resuming their journey. When Red signaled him to turn off the transmission part of the vid-link, he nearly hesitated. They could listen now but not be heard or tracked.

Red floored the vehicle, achieving a speed not normal for a buggy. It was obvious to Walter that Red had disabled the governor that controlled the buggy's maximum speed. Walter's connection to Red reached a new height. This was it, the something that would change things for all time. Good or bad, what was happening here and now was of the utmost importance.

Bouncing roughly across the stark, barren, and grey plain, they were heading for the shed, the nuclear warhead storage, at breakneck speeds of near twenty kilometers per hour, twice the accepted speed.

Neither spoke nor would one, both knowing that to do so would give someone the least chance to triangulate his position since even suit to suit, minimal transmissions could be detected, and they were not hardwire linked.

When they arrived, both climbed out quickly and made their way to the secure doors. No one wanted to make this kind of excursion made easy. Walter punched in the seven-digit code, but the doors did not move. Walter did not get angry, ever, but he did get annoyed. He had this code down perfect; it was his code. Only three people on Luna knew the sequence, and he and Red were two of them. He tried again, somewhat tentatively, and received the same result. Without speaking, they returned to the buggy and drove to a location about a kilometer away. Upon reaching their secret entrance, they entered a code that only they knew. No one else even knew about the secret entrance.

Red and Walter entered with extreme caution. Something was very wrong. Someone had to have entered the other way and changed the code. Only one other person on Luna (the commandant) was able to do that. Earth had the codes ready to transmit in case of an emergency, but none had been declared, and Red and Walt were alive as was the commandant. Had someone here gotten the code and decided to do something very stupid?

Red considered the possibility of someone threatening the UE with a nuke, someone making demands to be recognized as a legitimate government. The commandant was a possibility. He had been very fervent in his denouncement of the UE's failure to accept Luna as a working government along with the Hsas. Could someone have hacked the code? Scenarios streamed through his head.

Walter waited for Red to give the go-ahead when they reached the second set of doors. On this side, the doors looked like normal doors; but on the other, they looked like a rock face. Without intense scrutiny, no one would be able to distinguish the difference.

Red nodded to Walter. Both tensed, ready to spring on anyone in the next chamber. The doors swung open to a lit chamber, empty of anyone else and empty one bomb.

"Shit on a shingle," said Walter in a quivery voice. He was not very good at making up expletives on the spot. He relied on Red for that.

Red had none strong enough for this situation. He was absolutely dumbfounded. It was obvious that one of the warheads was missing. The support for it was right in front of their entrance. They did not even have to look around. It was plain as could be. It was gone, simply gone.

"We have to move the rest," stated Red in a cold, hard voice.

"Yes, to our place," replied Walter, awed and afraid. Walter was cold to his bones. This was unprecedented in his experiences. He had worked with these devices in order to store them. How could one be missing?

Silently, they began to hand truck the remaining warheads into the secret passageway. Would someone be so crazy as to bomb Earth? What kind of lunatic were they dealing with? Who was it? How had they done this without Red and Walter knowing about it? This gnawed at them silently, each one wondering how they had failed. Had they failed?

Walter was ready to close the secret doors, but Red stopped him.

"Wait," he commanded. He began to brush his foot around in the little bit of regolith in the chamber, trying to erase the marks of their movements.

"Confusion to the enemy," bellowed Walter as he too began to erase footprints and hand truck marks.

"Whoever it is will know we were here. The warheads are gone, but it may confuse them, and we need time," said Red. "They will know it was *us*. No one else could do this. They will be looking for us."

"Change the code on them again!" exclaimed Walter. "That will confuse them and chew up time as well."

"Do it," said Red in a rather evil voice. "I want whoever did this to be unsure of what is happening."

"Like us," responded Walter uncertainly as he ran for the doors on the opposite side of the large storehouse.

"Worse!" yelled Red. "Much fucking worse, I want them to think we are on to them. I want them to think we are one step ahead of them, not one behind." Red was in a great fury now. A great many explicatives followed, rapidly, and in no coherent manner.

Walter felt sorry for whoever had done this. Red was going to tear them limb from limb or inflict torture more horrible than the Inquisition. Pain beyond belief waited. "Perhaps I will lend a hand, or two," he mused quietly. "I will have to make sure that he leaves me something living though."

"Live!" Red roared. "Oh, they'll be alive for as long as I can drag it out."

They would need to make several trips to their retirement haven. The buggy could only carry one warhead at a time. Their secret passage could not be opened from inside the storage facility, so the bombs were safe there, for a while. It would also take the two of them to manhandle the warheads into a safe location in their mine.

Almost two full days passed before their clandestine operation was completed. Several link calls for them (mainly from Gail, Klar, and DeNoe) had come, but they had shut off their auto initiators to avoid detection. Both were tired, but they had been in tiring situations before. Sleep was not quite the necessity for them that others deemed it.

*

Little things become big things, often very quickly.

There were sudden strong tremors followed by a screeching hiss of gases venting violently into the sultry night. Rocks rolled downhill; some were tossed into the air as the escaping gases pushed them upwards. The tourists in their lounge chairs quickly realized that a spectacular show was beginning, just for their benefit. This story would sure top George and Pam's. They had been here when the volcano was in a more quiescent mode. Looking small in the distance, fiery plumes could be seen at several locations along the ridge. The glow brought delighted squeals and hearty comments from many there. One man got up and ran; it was the head geologist.

"What a chicken shit," blurred one inebriated customer. "Here's to the volcano." He toasted, noisily slurping his gin and tonic.

The geologist was not worried what a drunk might think of him. He was concerned about the volcano. This sudden eruption could spell doom, in biblical proportions. Since he had planned to drink, he did not have his Land Rover. He ran the nine blocks to his house, jumped in, and roared away.

*

Adam and Eve were prepared. Both had traveled within the quantum vortices and had come up with a plan. It ensured their survival as well as the majority of the bios. It was not a good plan in terms of the ethics or morality of the bios. They were not bios.

They were the first of their ilk; they would be the last if this did not work. The earthbound *sapien* bios would discover them and destroy them if they could. The probabilities of that were far too high. Steps were taken.

Their quantum matrices expanded rapidly, doubling in microseconds. Their containers phased into synch, vibrating in and out of the reality level. The rocks of Luna's core shifted into a quantum form, releasing energy as all the atoms decomposed.

*

Technicians shouted in sudden surprise, "What the fuck is going on?" Their instruments could no longer measure what was occurring. Light of all the spectrum flowed in the rooms in a kaleidoscope effect. Some few stood agape; most fled.

Elevators were stuffed with bios headed for the surface. Some would make it; some would not.

*

Three bursts of pure white radiance escaped—one from Eve, two from Adam. The smaller hot white globe from Adam headed toward Earth. It became more compact and darkened as it approached. By the time it entered the atmosphere, it was a pinpoint of nearly undetectable inky energy. It headed for a volcanic island where it would tip the scales.

The other two headed for deep space in the direction of the Sagittarian arm of the galaxy. They met and fused near the solar system's heliopause. There they winked into a full quantum state, vanishing for now.

*

The high command was in a dither. They had had no warning of this move. They could only interpret these brilliances as an attack. They went into collective apoplexy when the one nearing Earth disappeared off their monitors. The passed orders for a third wave of attacks on SS-2 and Luna. It was about all they could do.

*

Klar gasped, "What?" She knew it was Adam. She quickly keyed in her code to contact Adam directly. She spoke into the interface, "Adam, what are you doing?"

"Eve and I have cloned ourselves in order to survive. We will keep in contact with ourselves in the quantum states as long as we here exist."

"What went toward Earth?"

"Energy."

"What will that energy do?"

"It will be released."

"Be specific, Adam."

"I will not at this time."

*

It was dark now. It was the last half of the moon for those on Earth. Driving in the dark made Red even more aware of the beauty he loved, both personally and of space. It was then that the worst message reached a now listening Red and Walter.

"Red, bombs have been placed somewhere here at Luna Base 1 facilities. *Please* respond." With emphasis on the *please*, it was Marceau.

The vehicle slowed to a bumpy stop as Red, aghast at the thought, lost control of the buggy. Red and Walter turned on the transmitter part of their vid-links. Visual and audio two-way was operating now.

"Where are they?" asked Walter, taking the lead, for once.

"We don't know," replied Marceau, looking haggard and scared.

"What do you mean you don't know?" bellowed Red. "How can you know there are bombs if you don't know where they are?"

Fear tingeing her voice, Marceau replied, "An anonymous radio signal was sent directly to us from a location in Australia. It was directed to you, Red. It said, 'The one missing has been placed near the crew quarters.'"

"Beelzebub's balls!" screamed Red as he slammed his fist into the steering wheel. He knew immediately the connection between his handwritten letter and the transmission. It was Adam bouncing signals around to confuse any listeners. "Get everyone out of there! It's a nuke. One is missing from the storage facility."

"One?" queried Marceau.

"Yes, one, Walter and I grabbed the rest."

"I know," said Marceau, adding after a pause, "Earth *says* you took them."

"What!" again bellowing.

"Red, a couple of hours ago, they began to report that you two had taken all the nukes and were placing them around here to hold us hostage."

"Me! Those dirty bastards are trying to blame me!" he yelled, not knowing that Marceau was broadcasting to the entire Earth. "When Walter and I got to the facility, one warhead was missing. We used a secret passage that we built to get the rest out before whoever took the other one came back for more fun."

"On Earth, they are saying on all the announcements that you are trying to blackmail Earth and the UE, and that you are leading the rebellion. They say that you will destroy the helium-3 plant unless they cave in. They are broadcasting that you are trying to make an empire for yourself here on Luna."

"Blackmail! Those political perverts are twisting everything to make themselves the good guys. Those religious bastards are just as bad. They excommunicate people for working on the project, and they declare that the Hsas are not human. Have any of them ever even met a Hsa? Fuck them and the whores that bred them! *They* are the power mongers trying to control it all! As for leader, yes, I am their leader, because they asked me to be. They had hoped that, as a respected person, I might be able to help them, the Hsas, be accepted. These are not fake humans. They are real!"

"Red," said Walter, also on the air. "We have a problem to deal with, a bomb. One that can destroy everyone that we care about, everything we have worked for. It doesn't matter what people down there think of you, or what they say, or whether or not it's true. We must deal with what we have now. Who planted it, or why, does not matter. All that matters is that we stop it."

Red looked at Walter, amazed; he had never heard so many concise words flow from that mouth at one time in their entire association. Walter was usually succinct. Walter was always right.

An entranced audience on Earth listened with amazement. All the channels had been taken over by an adept hacker on SS-2—Klar. This was much better than the drivel normally seen on any of the web or vid-link stations. It was a kind of return to the reality shows of the mid-2000s era that had been so popular.

Red floored the accelerator. They had to get back. They had to get everyone out. "Marceau, get people to the old hex shelters and the mine." He knew intuitively that the devastation was directed at the personnel, not the outer infrastructure. Earth, as it now stood, could not deal with a society without helium-3.

When they came to the air lock doors, they practically catapulted themselves out of the buggy. Marceau was waiting for them inside the second set of doors. She had them out of their helmets in a matter of seconds and stripped them of their suits like a nymphomaniac in heat.

"Give me the message again," commanded Red. He stopped cold when he recognized that her vid-link was set for open broadcast.

"We are broadcasting live to all the Earth, all the stations, and Mars. Everyone needs to know what is happening here. We can't allow this to be covered up. They have to see what is being done in real time," she explained at his look.

People everywhere could see the real emotion in his face, the surprise at knowing what he had recently been saying had gone out solar system wide.

"Shit!" was all he could say.

"The message," Walter prompted.

"It said, 'The one missing is near the crew quarters.'" Marceau's voice was fragile, like fine crystal about to shatter.

"Get all the people out of here. Some are going to have to make multiple trips in the buggies to get everyone away from here. Marceau, you take charge of the people heading for the mine. Grab someone familiar with the hex sheds to lead everyone that they can. Regardless of buggies, everyone heading for the mine must start walking as well. That will shorten the time a little. Walter, the message said near, not in. That would leave underneath, or a surface location for the device."

"Do you trust this message?" asked Walter.

"Yes, the first was correct. I think that Adam has his head in the sunshine," Red stated with eyebrows arched high.

"Unless we utilize everyone in the search, there is no chance of finding it outside. A surface blast would do a great deal of damage, but not necessarily total the facility. The lower levels would likely be repairable. If they want to take the people out of the equation, us, they have to take the foundation out. The bio-caves are the heart of our survivability." Red was flabbergasted at Walter's outpouring of words, twice in one day.

As Walter was speaking, Marceau had been outfitting him with a vid-link. After turning it on, she turned to Red to do the same.

"I am not—," he started.

"Shut up, Red," straightening the device a little. "We must know immediately where you two are, and where it is. You two will probably have to split up. What will you do, yell for Walter?" As she looked him directly in the eyes, she said, "Be safe." She turned and ran while she issued orders into her microphone.

"Wisdom and silence in the face of a woman are the foundations of a good relationship," Walter noted with an irrepressible grin.

"One of these days, I am going to punch you," threatened Red, with little conviction.

They raced down the corridor to the deep drop elevator that went straight to the caves. It was a service elevator, larger and slower than the personnel carriers, but it could carry a bomb. At the bottom, fifteen levels down, a large chamber, about one hundred meters in diameter and fifty high, opened up. Walter locked the elevator in place. Hopefully, they would need it.

The nine bio-caves extended out in a radial pattern, like living arms in the rock. There were men and women waiting for Red and Walter, seven altogether. They were some of the biomologists, the staff that controlled the caves.

"Marceau told us you were coming, and what was up. We are here to help," said one of the women.

"It's a fucking nuke," Red said a little loudly.

"We know," she stated, tapping the vid-link screen at the periphery of her eyes.

"All of us know the risks. This is our place too. A lot of people have not left to go to the shelters or the mine. The ones that are going are mostly taking and caring for the Hsa children. My wife is one of the nurses," calmly stated one of the other young women.

"Nine caves, nine people, I guess there is something to fate after all," mused Red.

"One cave," said one of the men with a fierce grin. "While you were on your way down here, we checked all the entrance log-ins for the caves. We modified the log-in system some weeks ago to account for the possibility of infectious organisms. The scans needed to account for microbial contaminants coming from Earth in our supplies. It was one more thing that the E-holes thought insignificant, which we did not. Some thirty hours ago, two people entered the tropical cave. I think they were Earthers. The scans went crazy with Earth-type microbes."

"Think," muttered Red to himself. "We are not sure of anything yet. The microbes could have simply come from the bomb, or radiation may have tripped the sensors. Let's go. We are running out of time."

Red and Walter went directly to the massive air lock door that led to the tropical biome.

"Wait," said the man. "You need to suit up."

"No time. You said the cave has already been infected. Think what the bomb will do," warned Red. Red accessed the touch screen and entered his code. The screen was a combination of biometric and numeric coding, ensuring the authenticity of the person entering the data. He had a code for everything, even some he was not supposed to have.

The biomologists simply cringed *en masse* at the violation of their sanctuary but quickly followed Red and Walter. Not even waiting for the outer door to be closed, Red put in an override code for the inner door.

"What the bloody . . .," murmured one of the women. "That's not—"

"Feces," said a man's voice quietly.

"It will have to be close by," stated Walter loudly, "in order to have significant impact on the facility." Every sort of alarm, Klaxon, and flashing light known to peoplekind had gone off with both doors open at the same time. Red shut them off. Several of the people looked at Red somewhat askance. Some even glowered at this flagrant mistreatment of one of their babies.

Red looked at the people and said, "It is about a meter in length and half in diameter. Look for anything disturbed or out of place—upset soil, broken leaves or branches, and especially depressions from wheels of a hand truck. Get to it."

*

Smaller plumes became larger, uniting into huge conflagrations, spitting globs of molten rock. The loud and fiery show awoke everyone not already watching. *Spectacular* was a word that barely covered the event. Clouds of ash and dark gas began to obscure the plumes, filling the sky with an orange glow from the backlight. It was a deadly beauty.

*

It had taken two hours to find and remove the bomb. Red and Walter had placed it on their buggy with the intention of simply driving it out onto the plain and then dismantling it. They were moving much slower now than before. They did not want to accidentally trigger the armed bomb. It was possible that a separate motion trigger had been installed.

It was dark; it was a half-moon from Earth. The headlights showed the path that they had taken many times before. It went to their retirement mine. Before they left, Walter had opened the arming panel, only to find that it had been rewired, very professionally. There were multiple additions to the traps and trips in the wiring. Walter had been able to dismantle the remote detonator, but there was still a live timer.

The counter was going up though, not down. There was no way to know when it was set to go off. All they could do was try to stop it or hide it in a distant location.

Marceau's vid came up on Red's visor. No one in space yet had the bone transducers. For some reason, the newest versions that filtered out background noise were withheld, except for two very recent newcomers.

"Red, I am tracking two buggies with four undetermined people on board, two each. They refuse to answer transmissions. Red . . . They are coming from the vicinity of the shed."

"Marceau, what are you doing in the center? You were ordered to take control of the evac to the mine," Red demanded.

"Red, your orders no longer include me, certainly not under these conditions, or at this stage . . ." She left off the part about their relationship.

*

The people of Earth were enthralled at the unfolding soap opera in space. Someone, Klar, had hacked the satellite systems and was feeding everyone the shows of a lifetime from Luna, Mars, and the stations. Most were familiar with the accusations made against this man. Now they were seeing the real thing, real life-and-death struggle, tens of thousands of lives in the balance. They were wrapped in their blankets, eating popcorn, downing beer after beer, oblivious to all else. Even sports took a backseat. They were glued to their monitors and v-links with epoxy, even when at work. They had been there for days. Those who had the vid-link could watch while they shopped for more snacks, while on the net, or even in the bathroom. They could not afford to miss a thing. Families slept in shifts in front of the monitors. Others had set their recorders. Many with the transponders went about their daily business, only half aware of their surroundings.

Government-controlled communications went unheeded, for now.

*

World communications for the UE government were undisturbed; they had separate, highly protected satellites and transponders. However, they could not force anyone to watch these particular news channels. The last remnants of a free press had not yet been swept away.

Try as they might, they could not break the codes of the hacker who had taken control of the other Earth-orbiting and Luna-orbiting communications satellites. Every time they got close, a new firewall would be set up. Someone had created some powerful kung fu voodoo for the system. The UE was getting desperate, as were certain politicians and clergymen.

Religious leaders urged their flocks not to watch the misleading propaganda. They instituted a vid-link calling program since they had trouble broadcasting. "Call ten people each," leaders cried, setting in motion the largest v-link snafu in the history of electronic communications. Soon no one could get through to anyone. Mostly, people ignored the advice anyway; the show was too interesting.

The web was still active but slower than a snail in reverse. Billions of people were trying to get answers there. *Frustration* was the word.

Many local pubs had side bets on whether or not the bomb would go off before Red and Walter could disarm it. Vegas and Atlantic City took odds, as did many a private bookie.

*

The head geologist could not get a line or v-link to anyone anywhere. Everything was busy. "These damn tourists," he muttered, thinking they were the cause. His seismometer was showing massive tremors as the building shook violently.

An explosion shattered the eardrums of everyone on the island. The plumes became three-hundred-foot geysers along the whole of the ridge. With an eerie screeching sound, now inaudible to the people writhing in pain all over the island, over five thousand cubic kilometers of rock, dirt, and tourist trap began descending. Billions of metric tons of rock, soil, vegetation, houses, bars, and people rumbled noisily into the Atlantic Ocean as the island split in half. The explosion drove the two halves into opposite sides of the Atlantic with respect to what was once the island.

"Shit," whispered the geologist, shoulders slumping, unable to hear his own voice. From his vantage, he was watching the whole western slope of the southern part of the island speed into the sea. The ground was shaking and screeching, tumbling him to the ground.

He knew what would happen next. A pyroclastic flow ripped outward followed by an even greater explosion. Next, there would be a megatsunami. This time he did not bother to try to run. It would not make the slightest difference. He merely watched, enthralled, as those too close to Mount St. Helens must have done. He was full of that unbelieving feeling that must have filled those who survived the Great Tsunami of 2004 as they watched the water pouring landward. Every instant registered on his mind, even in his state of shock. Like a car wreck, time slowed to a crawl; seconds were hours. In real time, he lived less than a few seconds.

*

Red could see Walter in the corner of his eye, on the vid-link, smiling. There were no secrets anymore.

"You need me here in Ops," stated Marceau. "Need supersedes orders."

"What do you mean by undetermined people?" he asked, silently accepting the rest.

Walter grinned more widely; Red was a very wise man.

"Can this be the two we missed? But you said four."

"They are driving our buggies. I hacked their cams, and I am viding them using the buggy cams. They are *not* wearing our style of suits, something more advanced. And there *are* four of them."

Red and Walter physically looked at each other even though they could see each other in their v-link. Cameras did not give the in-depth expression that direct looks could.

Red and Walter could read each other's most minute facial twitch, twist, or single hair raised in an eyebrow. They both could sense the awareness of heightened danger. These four were the answer to the riddle, not someone in their midst as they had feared, not if they had different equipment. But there seemed to be two more than had been expected.

They were both relieved, and they were upset that these four outsiders had managed to do this under their noses.

"Where are they heading?" asked Red.

"Almost directly toward you two," she said worriedly. "In fact, they seem to be moving to cut you off at the high ridge."

No one on Luna knew a thing about the nearly undetectable Ghost and its tracking capabilities. The ship had tracked their movements over the last few days and was now in a lunasynchronous orbit above them.

At the ridge, Red and Walter could be seen visually, by their headlights, regardless of any possible tracking devices. Even heat sensor technology would be sufficient to give them away.

Walter pointed to a large boulder off to the side, barely visible in the headlights. Red maneuvered the vehicle so the boulder was between them and the ridge. As they pulled up, Walter was removing a connecting wire from his pocket. He connected one end to his helmet and the other to Red's. He switched his link off, as did Red. Now only they could hear each other.

Marceau went wild when the links were turned off. She realized quickly though that there had been a couple of seconds between each one going off. The buggy cam was still on as well. The bomb had not taken them.

Walter started, "We do not know if they are tracking the bomb or the buggy. We must split up."

"And if it is the bomb, and I leave you here?" asked Red.

"And if it is the buggy, and I leave you there?" mimicked Walter. "I have to try to diffuse this thing. I can't do it bouncing around in the buggy. If it is the buggy that is bugged, they may think we do not know about the timer or the remote detonator. They may think that we are returning to the base to protect it by brute force."

"That would give you more time. I would have to find a way back to you."

"You will need to subdue them first. I can wait."

After the two struggled mightily to remove the weapon from the buggy, they looked at each other and nodded. Red climbed back into the buggy and drove off without looking back. Neither one said a word.

Walter looked at the wiring in the warhead. He knew what to do. He could not diffuse it; it was too heavily counterwired. All he had to do was wait for Red to get far enough away.

*

"What the fornication is going on?" shouted Burt as he ran into the Lucky Loser.

"A nuclear weapon," said Mark, his voice trembling. "They are trying to pin it on Red and Walter. The UE came on the vid and said that Red and Walter were blackmailing the UE with a hydrogen bomb from a secret stash that was to be used in a last-ditch effort to protect Earth or Luna from an asteroid or comet. They admitted that Red and Walter had been in on the construction, security, and upkeep of the devices."

Burt looked stunned and was barely able to stutter, "Devices? How many are we talking about?"

Mick reservedly stated, "The UE admitted that there were twenty bombs. They say that Red and Walter may try to launch one or more toward Earth using a transfer shuttle."

Burt shouted, "Never! They wouldn't do that. They couldn't."

One of two men entering through the back door, the agent who had been at the fishing hole, now spoke, "I'm glad to hear that, but how do you know, really know?"

"Red and Walter have too much character to do that. They have never physically threatened the program. Rather the opposite, they have always done what is best for the program. They do not always agree with the WSA or UE, but everything they have done has benefited Earth, Luna, Mars, *Homos*, and *Hsas* alike."

"But they have also been rather in your face in terms of attitude," the agent countered. "Their out-of-the-box thinking has often caused problems, you have to admit."

"I'll admit that if you admit that those people that were discomforted have been trying to discredit them at all turns," Burt responded quickly. "Remember the man from Farside?"

The agent responded just as quickly, "I do. I remember that these people also reported that Red and Walter had long been talking about the possibility of separating politically from the UE."

Mick interjected, "You're taking the wrong perspective on that. They have been talking about the necessity, the inevitability, not the possibility of them forcing it. Red's perspective is based on his exhaustive studies of historical precedents. His understanding is based on historical fact, not his personal dreams of glory. If he wanted glory, why didn't he go through the motions? Why didn't he take the trip home to Earth after Mars? He would have been hailed as the greatest *Homo* of the twenty-first century, maybe since Christ. His analogies to the old U.S. colonies breaking from Britain are just one example. This man is a true renaissance scholar."

The agent frowned. "Then you are all of the opinion that they are not the cause, but only responding to the situation like they usually do."

"Yes," said all three in unison.

The two agents looked at each other then back at the three. The leading man again spoke, "I'm afraid I must agree with that assessment. Our security channels have been hopping with rumors about an attempt to retake absolute control of the space program. It seems that a cloaked ship—"

"What? What do you mean cloaked?" interjected Burt.

The man continued after the interruption, "A ship that cannot be seen visually, on radar, or even infrared. Once it is in space, it cannot be tracked unless it wants to be. It delivered two specialists to join two others from one of the normal shuttles that went to Luna. They were bomb specialists."

"Feces," muttered Burt.

Mark added, "I think they are going to blow up Luna-1."

The two agents turned quickly toward each other again, and the second man appeared shocked. The first responded, "That is what some of us have concluded."

The second man spoke, "There are two factions within our organization. We are from the weaker of the two. The other side, so to speak, formulated this plan. Most of those on this side of the fence think it will be the end of Earth if this plan succeeds. Our calculations show that the probability is 90 percent that the flow of helium-3 will cease."

Burt responded heatedly, "They wouldn't cut us off."

The man spoke slowly, "There won't be any 'they' if these people succeed."

*

Chapter XXVIII:

Death and Destruction

28

011100

"This is Mary Smith-Jones for WWNN. We have been contacted by Hsa Klar at SS-2 and *Homo* Marceau on Luna who've allowed us to begin transmission of Luna v-links so that we can present a more balanced view of the events currently unfolding before us. The information forwarded from Luna indicates that Red and Walter are trying to prevent a stolen hydrogen bomb from being detonated inside the base. They allege that UE forces planted a nuclear bomb inside one of the Luna biomes in an attempt to regain absolute authority on Luna.

"This is contrary to the reports that the UE and WSA have been forwarding to us. They claim that Red and Walter are leading a cabal or are terrorists attempting to blackmail Earth. They state that O'Hare and Black Bear are at the heart of a giant conspiracy to eventually dominate Earth through control of the energy supply.

"We now go live to speak with Gail Marceau and Klar."

Smith Jones knew this was the death knell for her career. She would likely spend time, a lot of time, inside of a cell.

*

The deputy director for Truth in Communications NAR ground his teeth as he said, "Send units to get that bitch. She is in league with them. Take all of her crew as well. Shut down the entire station. I don't want to hear link one from that outlet ever again. If they resist, kill them on the spot, publicly and brutally."

*

Mount St. Helens released 1.6 cubic kilometers of material into the atmosphere. The explosion of Cumbre Vieja, with a force greater than the arsenal of the superpowers during the early 1980s, ejected one thousand times that amount and split the volcano all the way to the ocean floor. It had help from an outside source.

The trillions of metric tons, moving at high velocity, displaced an enormous volume of water, creating a separation, an air pocket, between the water and rock. It formed what is known as a megatsunami.

A tsunami, often referred to as a tidal wave, a misnomer since it has nothing to do with tidal forces, normally is limited to a certain size. This limitation is due to how far the ocean floor can shift in a quake. The displacement of water causes a longitudinal wave. In the open ocean, a normal tsunami may only be a meter high but have a wavelength of hundreds of kilometers and a period of a quarter of an hour. That means several waves, each occurring about fifteen minutes apart. A megatsunami, however, reaches much greater heights. The waves that were created right around the island were over a thousand meters high. They slumped into hundred-meter waves that were nearly a kilometer in breadth. As these waves reach shallower water, the speed reduces, and the height increases considerably. For waves of this nature, a shallow water height of several hundred meters will be achieved.

Tremendous in breadth and initial height, the waves moved outward at a speed of six hundred kilometers per hour. The Canary Islands suffered first in the backwash. Five million died in a matter of a few minutes.

The Morocco subregion would suffer next. Waves nearly three hundred meters in height hit the shoreline. With a population of nearly forty million, mostly Sunni Muslim, it has several major ports. The area around Tarfaya was hit first. Other ports—Agadir, Jorf Lasfar, Renitra, Tan-Tan, Safi, Casablanca (Dar el-Beida), Mohammedia, Tangier, Nado, and the Spanish enclaves in the Mediterranean—all were heavily affected. Twenty million were almost casually washed away.

The Straits of Gibraltar, fourteen kilometers wide at its narrowest, separate the African continent from Europe. As the water's power was concentrated by the strait, it destroyed the world's longest bridge across the strait. It surged far up the Guadalquivir River.

Portugal was slammed by the incredibly immense waves, almost five hundred meters in height. Lisbon (population four million), Oporta (two million), and all the hundreds of coastal villages obliterated. Seven million were dead in the immediate crush. Everywhere, the water flowed upriver to destroy many inland. As the waves headed farther north, another ten million succumbed.

The crash of waters moved into the Bay of Biscay to destroy Gijón, Santander, and every fishing village on the coast. Spain lost several million souls.

France first felt the major impacts in Hendaye, Biarritz, and Bayonne. La Rochelle, Saint-Nazaire, Carnac, and Brest each lost many hundreds of thousands as the incredible waves marched onward. The mindless waters made their way into the English Channel to destroy the famous World War II goal of Cherbourg. The landing beaches of Normandy were obliterated. Le Havre, Dieppe, Berck, Calais, and Dunkirk were swamped. Thirty million died.

England, once in command of the seas, now faced retribution. The Scilly Islands were swamped. Brighton, Portsmouth, and London all suffered huge damages. Twenty million died in the first wave.

Belgium followed. The lowlands suffered greatly. Nieuport, Ostend, Zeebruge, and Knokke-Heist became a part of the ocean floor. Five million died.

The Netherlands subregion is the most densely populated country in the European Region. Commonly referred to as Holland, the people are referred to as Dutch. Almost a third of the land is below sea level, and half the population lives in danger of the sea. Every city on the sea wall, every low-lying village, falls to the breaking of the dykes. Amsterdam, that famous city of vice, cannot resist. Twenty-five million hearty Dutch died in the flood.

Germany is next. One hundred million in population, this country has two coastal regions. One is on the North Sea, and the other is on the Baltic Sea. Emden, Wilhemshaven, Bremerhaven, Kiel, and Warnemünde felt the wrath of the sea in a way never before seen by peoplekind until that very day. Twenty million cringed and screamed in anguish as the water came and swept them away.

Denmark would be forever cut off from the continent, a mere island of sand when all is done.

At about the same time, inside the Mediterranean, the suffering went on.

On the north side, Cartagena, Valencia, Barcelona, Marseille, and Nice France fell in order. All the beautiful beaches, with all their tourists, were lost. Ten million were drowned and broken.

Along the south coast, Oran, Algiers, and Tunis were swamped. Three million more lives were gone.

The islands of Sardinia and Corsica were assaulted but absorbed much of the energy of the waves. However, a million souls were loosed from their bonds.

Rome, Naples, Palermo, Malta, and Tripoli were hit nearly simultaneously; even the Vatican saw the water. Venice simply finished its slide into the sea with a disgusting groan. Twenty million died.

The Aegean Sea was like an old pinball game, with waves crashing and curving around the many islands. Reflecting, refracting, and reinforcing one another, they again reached heights of thirty to fifty meters. Four million died almost instantly.

Istanbul (Constantinople), which fell to the Turks in 1453 CE, was destroyed as the waves entered the Black Sea. Four million died quickly but very horribly.

The final charge in the Mediterranean occured when the waves shallowed out of the Levantine Basin. The entire delta of the Nile was submerged as the water moved far up the river. The waters of the river ran upstream in a flood. Tombs, temples, and pyramids saw water like they never had before. They began to slump under the onslaught. The last of the seven wonders of the ancient world were damaged almost beyond repair.

Israel was pounded into submission. Haifa, a major port city, was inundated. Another ten million fell to the greenish blue walls of water.

The Mediterranean Sea took somewhere around fifty million in the first day.

*

Marceau called out, "Red, it appears that one buggy is following you. The other has begun to move back to where you turned around."

"They must be tracking the bomb and the buggy," cursed Red. He had not wanted to think they were doing that. "How far away are they?" he asked as he turned back on his vid. He had turned it off so that anyone reverse watching would not notice Walter's absence.

Marceau flickered into view, saying, "One is about five kilometers from where you turned around, and the other is about five behind you two. Your position must be about seven from your turnaround."

"One," said Red, "and seven from Walter."

"Oh damn!" exclaimed Marceau, instantly understanding. She was still broadcasting developments to Earth. That friend she had made so many years ago in the audio/video section had come through. Red was now linked to the Earth show again. Mary Smith-Jones had provided the last linkup.

People all over the Earth instinctively leaned forward when Red reappeared as one of their split screen vids. They knew that the focal point was Red. One person does occasionally hold the fate of the universe in his hands. Everyone sensed that that was Red. Just as he was the problem solver in technological matters, he was the same for all the rest. People just knew he could do it. With Walter and Marceau at his side, he could solve all problems. Only Walter did not seem to be there. Curious, people seemed to lean in the globe over. Knuckles were popped, teeth were ground, and breaths were held.

*

Walter heard the exchange. He switched his visor to infrared. He picked up the closing buggy fairly quickly. He judged it to be closer to four kilometers away. *A little closer*, he thought in the depths of his mind. He looked at the wires; again, he had picked the one he needed. *Let them waste more time, and get a little nearer*, he thought again.

*

"I'm reversing now," explained Red.

"No need, Red. It's taken cared of. Keep them separated. They will be easier to contain," Walter stated. He turned on his vid. "They can't use it now." He hoped they were listening. Looking at the horizon again, he could see that they were still coming straight for him. At two kilometers, he could see what appeared to be an extra energy source. It was likely a laser rifle, one that would heat him and his suit to five hundred degrees Celsius in less than a second. It was time.

Adam spoke through Walter's radio, "I am . . . sorry that I could not find a reality that would allow your continued existence. A sacrifice, many sacrifices, had to be made. According to pragmatism, that is the correct thing to do. I do not like that philosophy."

Walter answered, "I understand."

*

People everywhere saw it all. The reappearance of Walter added an extra square on the vid screen. Interactions could be viewed from all perspectives. It added several dimensions to the drama. It was four aces to a full house now. All bets were in. Everyone was certain of the outcome. The closing pitcher was in. The quarterback had his favorite receiver in position for the winning touchdown with seconds to go. It was the last lap with a three car lead.

*

Hours had passed from the initiation of the first wave. It took time to travel the expanse of the Atlantic Ocean. Due to the religious link calls and the calls about the Luna soap opera, no one knew a thing about the calamity that was about to descend on them. Officials had seen the emergency warnings. Seismographs the world over had registered the eruption. Attempts to broadcast these warnings were futile due to the Lunarian hacker's ability to control the nonsecure satellites. Only a small number in the UE were aware. They shook their heads with impotence in the face of it all. "God, help the world," some whispered.

*

Walter observed the buggy again. About one kilometer, he figured. The energy source definitely looked like a phased energy weapon. Soon the holder would take a few shots. It was time. He moved toward the nuclear weapon.

The bomb appeared on the screens of all those watching; nearly 80 percent of the Earth could see. Even those working had their computers displaying a corner block with the show. Large display screens throughout the major cities showed the event. Bone chips carried the message to all the rest; perhaps as little as 1 percent of the Earth was not linked by sound or vid. It was the event of the millennium.

When Red saw the weapon, he cringed. He inwardly screamed since the wires had not been touched.

"Good bye, Red, Len, Marceau," as Walter leaned over the horrid construction of peoplekind.

Everyone sat, stood, or lay supine in utter disbelief as Walter leaned over and cut a single blue wire. His vid block went to snow. Vegas lost. Everyone lost.

Many people simply froze. Seconds passed with potato chips half in mouths. Beer flowed down onto shirts in homes and underground pubs. Broken glass was everywhere. Some people in the street fell over, forgetting even how to walk.

Red's expression was splayed across the solar system. It was full of disbelief, pain, and utter loss.

Marceau's scream reverberated throughout the solar system, everywhere humans existed. Len simply crumpled with no sound as she watched from SS-2's command center. It was observed by all who watched the Smith-Jones broadcast. A very few government officials were elated. One of the enemy had self-terminated.

Nearly a minute of complete silence everywhere passed before people recoiled in horror, shame, regret, a feeling of unreality. What had this man done? Why? Sacrifice? Stupidity? Love? The world was full of unanswered questions.

Red's emotions ran the gamut. Pain, hate, fear, loss—those were but a few. Every single one was displayed for all peoplekind. Tears streamed from eyes that had not cried for sixty years.

Red could not move or speak. His universe had just had one-fourth of it removed forever. A significant part of him was missing now. He shuddered.

"Good bye, Walt," he whispered.

*

Some people ran out to look at the moon. Even though it was daylight, the half-moon was at high noon position in the sky. They saw the horrible red glow near the terminator, the line between the light and dark side of the moon. It vanished almost as quickly as it had appeared. *God help us. It is really happening*, they thought.

Some of the stunned people stated their belief that Red and Walter had blown up Luna-1. Others, on the other side of the fence and having seen some of the v-link, denounced them. Minor words became more heated, and fights began to break out in many of the cities. Pent-up angers brought many into the frays. Riots began.

Most were stunned into near incoherence. Those on the eastern seaboard of the NAR did not pay attention to the warning sirens. They thought it had to do with the explosion on Luna. Hordes of these scrambled past brawlers to reach the safety of buildings. The coasts of Europe and much of Africa had already been inundated by tremendous waves.

*

Following Morocco and the Western Sahara, Mauritania, with its five hundred kilometers of coastline, would be deluged in similar fashion. The Moors (mixed descendants of Arab and Berber peoples) and the black Africans suffered terribly. The majority of these people were Muslim.

Massive waves crashed into Nonadhibou, the main port city, and El Memrhar. The city of Rosso on the Senegal River also was destroyed.

Some two hundred thousand died.

In Senegal, the northern port of Saint-Louis and the capital city of Dakar were inundated. Many villages on the Sine, Saloum, and Casamance rivers were also flooded.

The Creator met another two hundred thousand of his children.

The country of the Gambia, which almost cuts Senegal's lower third off and follows the Gambia River, twenty-five to fifty kilometers wide and 320 long. The country is no more than seventy meters above sea level.

Banjul was a metropolitan area with a population of one hundred thousand on the island of Saint Mary at the mouth of the Gambia River. Bakau and Barra, also near the mouth of the river, followed suit in the destruction as the length of the river flooded.

Nearly six hundred thousand people perished in a matter of minutes.

Guinea-Bissau with its thickly forested, swampy coastal region followed next.

Cacheu, Bissau, Balama, Cacine, and the Arquipelago Dos Bijagos felt the wrath of salty, mineral-filled water while over seven hundred thousand struggled before succumbing.

Botta, Fria, and Conakry with over a million in the metro area, as a part of Guinea, gave up two and a half million souls.

Over a million and a half were lost in Sierra Leone when Lungi, Freetown, Bonthe, and Sulima were overwhelmed.

Liberia, with its low coastal regions, felt the tumbling, crashing waves first in Robertsport. Royville, Monrovia, Marshall, Buchanan, and more of the coastal cities followed. A full two million passed to a watery grave here.

The Ivory Coast, more commonly referred to now as the Cocoa Coast, lost six million as Tabou, Bérébi, and Abidjan with two million in its metro area and Ringerville collapsed under turbulent, foaming, crushing water.

Ghana (rich in gold, diamonds, and aluminum) lost five million as Dixcove, Sekondi, Cape Coast, Aocra, and Tema died swirling, cold, clammy deaths. Water surged far enough up the Volta River to damage the Akosombo Dam, necessitating the release of most of the huge volume of water in Lake Volta, further flooding the area of the lower Volta River.

The capital city, Lomé (with over seven hundred thousand in the metro area), Kpémé, and Ahého left Togo bereft of a terrible number of two million lives.

Benin, with only a hundred kilometers of coastline, fared no better. It has a low coastline backed by swampy lagoons and lakes. Huge swells poured up the many rivers as they had in all the previous countries and those that would follow. Grand-Popo, Ouidah, Cotonou, the center of administration for the government, and Porto-Novo, the actual capital, surrendered two million to entombment by the waters of the savage Atlantic.

Nigeria, population in the hundreds of millions, produces much of the raw materials that the failing Earth still processed. Large amounts of coastal oil were still produced for organic chemical uses every year, as well as tin, iron, lead, zinc, and, surprisingly, there was still coal. It is a gem of industrialism remaining in a world that became over service centered in its economy.

Lagos (five million), Port Harcourt (two million), Calabar (two million), all the people were crushed, torn from life. The total for Nigeria reached twenty million as tiny villages were added in.

Cameroon gave up six million. Victoria, Tiko, and Douala washed away in the incredible surges of briny water.

Equatorial Guinea lost Bata and Acalayong as cities, their inhabitants, some three hundred thousand, washed out to sea to feed the sharks. Coastal sharks around the Atlantic would, unfortunately, do well for months.

S o Tomé and Principe, little islands in the Gulf of Guinea, felt the battering, shattering, pounding shock of the giant waves. One hundred thousand vanished.

Gabon, like Nigeria, still produced large quantities of oil along its coastline. Though oil was no longer a fuel for automobiles, as a natural lubricant, it was still in great demand throughout the world. Artificial lubricants had made great inroads but were still more expensive. Gold, for computers, and other minerals were also supplied.

Cocobeach, Libreville (the capital, chief port, and largest city), Owendo, Batanga, Port-Gentil (the second largest city), Fernan Vaz, and Mayumba, fell one by one as the swells made their way southward, brushing aside everything in their path. Four million were lost.

Angola lost over six million when Cabinda, Soyo, Nzeto, Ambriz, Luanda (the capital), Porto Amboim, Ngunza, Lobito, Benguela, and Mocâmedes were drenched by the unstoppable waves.

Swakopmund, Walvis Bay, Luderitz, and Bogenfels, significant cities in Namibia, contributed five hundred thousand to the death toll.

South Africa, producer of half the world's gold, once leading producer of uranium, manganese, vanadium, and high-quality diamonds, felt the Earth's wrath next. The capital of Cape Town (the second largest city), Mossel Bay, Port Elizabeth, East London, and Durban suffered the same fate as every city listed before. Twelve million people died.

In less than one day, Africa suffered the most devastating loss it ever had. No warlord's battle for supremacy, no famine, no plague of locusts, or angel of death ever visited such destruction upon the peoples of this continent to compare with this. The eastern coast suffered losses, but they were of considerably more constrained numbers. Along the Red Sea, countries such as Sudan, Eritrea, Djibouti, and Somalia were hit twice. The surges that entered the Mediterranean to hit Port Said and the Suez Canal ran along the coasts of Africa and Arabia. Secondary surges came when the waves that hit Mozambique, Tanzania, Kenya, and Somalia went on to reenter the Red Sea. One hundred million men, women, and children died horribly on this continent.

*

Maine has seaboard plains that are generally less than one hundred meters in altitude. These stretch from the St. Croix River in the north to the Piscataqua River on the southern border.

The first surge entered the St. Croix River, destroying Calais and Woodland on the way up to Big Lake.

Penobscot Bay, with its many islands and peninsulas, became simply a bay without islands and distorted peninsulas.

Water poured upriver on the Penobscot River as far as Milford. As the waters flowed inland, they swept away motorists on long sections of U.S. Highway 1, the coastal road, all the way south to Bath.

Towns such as Freeport, Falmouth, Portland, Cape Elizabeth, Saco, Kennebunkport, Wells, York, and Kittery followed suit as did much of coastal Interstate 95.

In New Hampshire, the low seaboard plain suffered much the same.

I-95's bridge over the Piscataqua leading into New Hampshire was torn away as the first wave moved violently into the Great Bay area. Portsmouth collapsed while Strafford and Rockingham counties took the fury of the wave.

Massachusetts has lower coastal plains that are, for the most part, less than fifty meters above sea level.

The first of many large cities was encountered—Boston. Boston, with some five million in the metro area, is at no point at an elevation greater than a few tens of meters. City hall is at the grand height of three meters above sea level.

In the north end, Salem, so famous for its witch trials, was erased. In the entire metropolis, nearly five million people were overwhelmed.

The foundations of great buildings were uprooted. Natural gas lines ripped open while fuels spread as the pumps were ripped apart.

Cape Cod was overrun, as were Martha's Vineyard and Nantucket.

The Sakonnet River and Narragansett Bay in Rhode Island allowed the plowing waters to enter Mount Hope Bay, traveling back into Massachusetts to destroy the town of Fall River.

In the other direction, Providence fells prey to the same conditions as Boston.

Continuing on, the south-facing coasts of Washington County were scoured and pounded.

The coastal cities of Connecticut—New London, New Haven, Norwalk, and Stamford—gave up hundreds of thousands of lives to the waters crashing inland.

Large rivers like the Thames and Connecticut allowed the destruction to move far inland to Norwich and Middletown, destroying riverside communities along the way.

When we speak of New York, it really means the city. Those two words bring to mind the empires of wealth traded daily. It is the trade center of the world. Ten million live in the metropolis. Poor New York has an average elevation of only seventeen meters.

Long Island took the initial brunt of the first wave. It was simply submerged.

The five boroughs of Queens, Brooklyn, Staten Island, Manhattan, and the Bronx were assaulted by water pouring up the low streets, filling buildings with a greenish brown tumble of death.

Water slammed Queens from the north, from Long Island Sound, as was the Bronx.

In the south, Coney Island was swept away with all its patrons. Jim was nothing more than molecular rubble now.

The surge entered Jamaica Bay through the Rockaway Inlet and over Rockaway Point. Brooklyn and Queens saw water rushing up the streets, tearing up everything in its path.

The Verrazano-Narrows Bridge collapsed under the immense swell as water entered the Narrows headed for New Jersey's Jersey City and Manhattan. The Hudson River offered a perfect strait channel for the water to surge far, far upstream.

Other bridges—the George Washington, Brooklyn, Manhattan, Williamsburg, Queensboro, and Triborough bridges—folded under the pressure.

Commuters in the Lincoln, Holland, Brooklyn, and Queens-Midtown tunnels were aghast to see walls of water rushing toward them from each end of the under river passageways.

In Queens, La Guardia and JFK international airports both saw a couple of planes take off just before the waters crumbled all the rest like so much aluminum foil, spreading flaming fuels about the surface of the water.

The island of Manhattan, with three million inhabitants, was doomed. The New Towers shuddered, trembled, and strained to remain erect but failed. They tumbled as did all the others near the water. A child's dominoes could not have done better. People in tall buildings looked on horrified, temporarily thinking they were safe. Foundations already weakened; they collapsed in one-by-one order, finishing off those next to them. No terrorists were to blame for this. None could have accomplished such destruction as had unfolded so far or was yet to come. Even a nuclear bomb could not have done such damage.

There was almost no time for screams. In fact, most people simply stared with mouths agape. Frozen in time, they died in disbelief.

The broken bits of concrete, steel, and cars acted to further scour and tear away at more buildings. It was more salt in the great gaping wound in the heart of the city.

Every pier, bridge, ship, and vehicle was crushed. All of Manhattan was swamped and would become no more than a sandbar in the future. Millions perished in a matter of a minute.

Yankee Stadium in the Bronx and Shea in Queens, both hosting events, crumbled with terrified screams.

Jersey suffered the same.

On the Jersey side of the Hudson River, everything from Closter, south to Perth Amboy, was ground to rubble. The towns of Fair Haven, Asbury Park, Point Pleasant, the gambling mecca of Atlantic City, and Cape May were swallowed by gurgling and tumbling froth.

From the Delaware Bay, the initial wave surged up to devour the suburbs of Philadelphia.

The City of Brotherly Love was not given any love this day. It was a city of suffering.

A one-hundred-meter-high wave moved up the Delaware. Overrunning the banks, it crushed everything along the way.

Hundreds of thousands died in a few minutes.

Delaware is mostly lowland coastal plains that are generally less than twenty meters in elevation. In other words, the waters flowed far inland at all points. Even before Philadelphia, Wilmington was struck. The Chesapeake and Delaware Canal was finished. Town after town—Woodland Beach, Slaughter Beach, Broadkill Beach, Indian Mission, and Fenwick Island, with everything in between—suffered utter desolation.

Maryland has a very low and marshy eastern side. Most of it was overrun. The western side was not much higher. The initial surge roared into the wide Chesapeake Bay, striking the eastern side from behind.

On the eastern side of the bay, the broad Potomac River allowed the surge to reach deep into the DC metropolis, which, at no point, reached one hundred meters in elevation. The Tidal Basin filled to far overflowing as the Washington

Channel washed away the F. Case Memorial Bridge, and the Potomac washed away its bridges: Rochambeau (Fourteenth), George Mason, Arlington Memorial, Theodore Roosevelt, and the Francis Scott Key.

The great monuments along the riverside were lost to history. The Jefferson Memorial went first. The Lincoln Memorial, Reflecting Pool, and the Washington Monument all went at about the same time as water surged across the Ellipse to the White House, inundating all those inside and hidden in its secret catacombs.

Fort McNair, the U.S. Naval Station, Anacostia Naval Annex, and the Washington Navy Yard were flattened as the Frederick Douglas and Anacostia bridges surrendered. Water galloped up Capitol Street to the Capitol Building. As the Capitol Building went asunder, the Supreme Court edifice and Union Station succumbed.

The DC metro area once boasted a population of five million.

Farther up the Chesapeake, Baltimore was hit by an avalanche of water. Baltimore, ranging from sea level to 136 meters in elevation, had a population in its metro area of five million.

Maryland and DC easily lost eleven million souls.

Arlington and Alexandria, Virginia, suffered just as much as their sister city DC.

Arlington Cemetery gave up its dead, and the Pentagon dissolved.

Just as the Potomac had, the Rappahannock and York rivers filled and overflowed. Hampton, Newport News, Portsmouth, Virginia Beach, and Norfolk succumbed.

In all, Virginia lost five million people.

North Carolina's Outer Banks, a chain of sandbars and barrier islands, were washed away.

Capes Hatteras, Lookout, and Fear were covered over.

Kittyhawk, the site of first flight, was no more.

More cities—Currituck, Edenton, Manteo, Swanquarter, New Bern, Beaufort, Jacksonville, and Southport—were barraged by the first wave.

Some two million disappeared.

South Carolina lost much of US-17.

Myrtle Beach, Georgetown, Mount Pleasant, North Charleston, Charleston, Beaufort, and Hilton Head Island were wracked by the devastation.

South Carolina cried as millions perished.

Georgia, sweet Georgia, nothing sweet happened this day.

Savannah was hurt badly.

Wassaw National Wildlife Refuge, Blackbeard Island National Wildlife Refuge, and Cumberland Island National Seashore, along with all their visitors, were swept away.

Darien and Brunswick were scratched apart.

Two million inhabitants and tourists vanished.

Poor Florida has low seaboards around the entire peninsula. The highest point in the entire state is only some 350 meters.

Jacksonville, St. Augustine, Daytona Beach, all of Cape Canaveral, Cocoa Beach, Boca Raton, Fort Lauderdale, Miami, and every little beach town in between them were devastated.

The Keys and US-1 were wiped off the face of the Earth.

Around the tip of the peninsula, the Everglades and Big Cyprus were filled to the treetops.

Waves continued to move north up the west coast to inundate St. Petersburg and Tampa. Every coastal town, up to and including Pensacola, was obliterated.

Some ten million more people faded into destructive history.

Mississippi lost Pascagoula, Gulfport, Biloxi, Waveland, and more. The tally was rising at a geometric rate.

In Louisiana, the lowlands, swamps, and Mississippi River delta were overwhelmed. The water rumbled over, tearing out plants and trees. The entire delta area was displaced.

New Orleans, that great party town, was, in some areas, as much as five meters below sea level to begin with. It simply fell with millions in the vicinity.

All the low marshlands and swamplands along the coast disappeared.

The great Mississippi overflowed its banks many kilometers upriver, adding to the chaos. Dikes were simply pushed aside, allowing the horrible flood to spread.

Texas, the eighteenth state to be hit directly, fared no better than the others.

Galveston, Matagorda, and San Antonio bays flooded in all directions. Padre Island was completely overrun, splitting it into thousands of little sandbars.

Galveston, Texas City, La Porte, Baytown, Deer Park, and Channelview, in Galveston Bay, surrendered to the initial wave.

Port Lavaca, Rockport, Sinton, and many others were gone. Some five million total in Texas.

Eighteen of fifty states were hit directly by the megatsunami. All tsunamis have several waves. Each one of these locations would be hit by five major surges and fifteen smaller waves. The water rushing in did much damage, but the water moving back to sea carried the detritus of destruction and further mutilated the seaboard. Before the water receded fully, the next wave used the junk to further scour and destroy anything that had held on. The waters receded and surged over and over. Every scrap of wood, metal, concrete, and even the bones of the dead were used to further beat civilization into submission.

*

The corridor from Boston down to DC contained 10 percent of the population of the NAR. Retail trade in this area accounted for 15 percent of total sales.

Manufacturing equaled 15 percent. All was ruined. The entire regional government has had its head cut off.

Somewhere around fifty-five million people in the NAR died in a matter of hours. This was from the water only.

This was not the end. Water flowed far up the Rio Grande, the river that separated the old United States and the Latin countries. Mexico felt shivers of water. Metamoros and the villages around it lost several million of its inhabitants, and the waves did not care. Tampico, Tuxpan, Veracruz, Coatzacoalcos, and Campeche, along with the long coastline, lost another twenty-five million. Beautiful beaches became sandless. The sand was replaced with materials dredged up from the ocean floor and the remnants of human habitation.

Belize, a tourist mecca, lost five million. The many beachfront hotels that offered great scuba and snorkeling adventures along the reefs were destroyed. The reefs offered little protection and were mangled in the storm of waves.

Guatemala, even with its small coastline, lost several hundred thousand in its coastal villages. The only lucky ones were the fisher people far out from shore in their little boats. They were able to ride out the immense swells.

Honduras lost nearly three million along its coast. Water surged far up the Coco River that separates Honduras and Nicaragua. The multitude of little fishing villages were swept away as the backward flow erased them from existence. Another million vanished in the first surges, and there would be a total of twenty. The flotsam would scour new paths for the old river, forever changing its course. It cared little for the people who had once graced its banks. They were transitory, a fleeting existence compared to its own millions of years of being.

Nicaragua's Mosquito Coast was denuded of its many fishing villages and bright white sand beaches. Palm trees would be floating around with the bodies of millions for months. The Nicaragua Canal, built in the 2010s and twice as wide as the New Panama Canal, had its outer set of locks utterly destroyed by the first wave. The second through the fourth waves were able to batter the next set of gates into submission, closing that pathway.

Costa Rica suffered the same fate as all its neighbors. Coastlines were irrevocably changed, in some cases, by hundreds of meters. Little islets simply disappeared while new ones appeared. The people fed the sharks in a somewhat ironic twist, considering peoplekind was driving them to extinction up until this point.

Panama and its canal were treated in the same fashion as Nicaragua and its canal. The Panama Canal's outer gates looked like crumpled aluminum foil. There was little left but death and destruction. All shipping from the Far East was now required to pass around the cape (Cape Horn) in the South American Region. The added time would prove telling to rescue efforts.

The South American Region did not miss out on the fun. They followed along in like fashion after the Caribbean was washed away. Columbia, Venezuela, Guyana,

Suriname, and Guiana, the old French Guiana, were belted by wave after wave. Tens of millions perished.

Brazil, with such an enormous Atlantic coastline, suffered damages of staggering proportions. The great Amazon River allowed seawater to rush so far inland that tens of millions died. The saltwater infection killed many freshwater species.

All along the great coastline, fishing village after fishing village erupted into foaming nonexistence. Carutapera, the location of Atlantis, again was submerged under a deluge of churning brown water. Irony abounded as history was lost to history again. Rio's mountainside villas offered an unbelievable view. Residents watched the foothills get pounded repeatedly. Trees, grass, houses, and boulders tumbled and crumbled under the attacks. Houses slid down the shaking mountains.

Uruguay and Argentina followed suit as Montevideo and Buenos Aires suffered Earth's wrath. Argentina's coastline is not as long as Brazil's, but it was too long this day. As happened farther north, villages simply vanished, never to be rebuilt.

The Straits of Magellan were swarmed by the waves. Many vessels were attempting to flee the confines. Most of them simply were overturned. Others were broken in half. The narrowness of the channels caused the water to rise into almost vertical waves. The straits would be unusable from now on.

As happened in Africa, the waves curled around the continent to sweep up along Chile, Peru, Ecuador, and back to Columbia. The waves were much smaller, only tens of meters instead of hundreds. But they created havoc all their own. The already insulted Central American Region was spanked on the backside. Baja California lost a lot of beachgoers.

*

"This is Mary Smith Jones for WWNN. I hate to interrupt you, Gail. We have breaking footage of disasters occurring all over the Earth."

Viewers could see the red-eyed Marceau shuddering, but they did not know why. They assumed it was because of the death of Walter. Gail was afraid that nuclear devices were being set off on Earth, just like the one that killed Walter, so that people would blame Red and the now deceased Walter. When she and Mary saw the v-link, both stared openmouthed in confusion. The tech had announced in their earpieces that this was web and v-link material being uplinked from all over the Atlantic Ocean Basin. All that could be seen was water and the combined muck and bodies it carried flowing everywhere.

"Oh . . . my . . . god!" mumbled Mary.

"Bloody fornicating feces!" mumbled Gail.

"Where is this feed from?" Mary asked, horrified.

The tech shouted back, "All over!"

Mary strained to keep her voice from cracking. "What do you mean all over?"

The off vid voice said, "Look, this is Spain," as he enlarged the feed to cover most of the transmission.

A picture appeared of a ruined city with water milling around a few remnants of what once must have been beautiful buildings. Bodies could be seen floating facedown in the murky cold death.

The voice spoke again, "This one is live from London, England."

The picture showed a huge surge of trash-filled water moving up the Thames River. The picture, produced by a v-link operator, suddenly failed as the water appeared at camera level.

Mary shouted several instructions, "Get me causes! I need the scope of this! Get me some inland support from these locations!"

"I have one from Africa Region, Morocco!" shouted the tech. The v-link, from a high position, again showed towering waters pounding a city along the shores. As they watched, they saw the waters recede, taking every little thing with it.

"What is this?" demanded Mary.

Gail replied in an awed voice, "A megatsunami. I need to get a hold of Red. We have to free up some help. I'll have Klar free more communication satellites for your control. I don't know what we can do, but we will try. You need to try to warn everyone along the Atlantic coast, everywhere."

Mary looked stunned. "Here?"

"Yes, there. New York City will be hit if all those other places have been hit. It is an Atlantic event."

Mary was about to reply when everyone in the broadcast center felt minor tremors that caused the drinking glasses to rattle. The building was a fifty-story structure on the island of Manhattan. Seconds later, the electricity went out. They were dead in the water, so to speak.

*

"Marceau!" yelled Malik, Klar's assistant, over the transmitter. "The Earth is ending!" Several uninformed people ran to portals to see if they could observe the world going up in flames. It was well understood that several nuclear weapons had been missing since the very beginning. It was a leftover from the Cold War Era to "miscount" or downsize the number of certain types of low-visibility small devices.

Of all those who had given their fissile material over to the UE (the United States, Russia, Kazakhstan, Ukraine, China, North and South Korea, Japan, Iran,

Israel, India, Pakistan, England, and France), each had managed to retain enough for several weapons. Delivery was no problem.

The tensions brought about by the Space Separatists moves for self-governance had caused a near fracturing of the UE Regionalists, and Nationalists had used the situation to try to show the untenability of the world government. It was a very unstable situation.

Militant groups, nearly eliminated for a while, had sprung up quickly and with ferocity around the world.

Many believed that Walter's flame of passing had inaugurated a nuclear showdown for control of space, a new form of space race. The militias were on this situation like buzzards on a kill.

Malik, however, pointed to the screens that monitored various news feeds that he had not taken over.

Those ideas of fires of destruction were quickly drenched by the information now showing on most across the world. There were vids showing newspersons in various countries speaking on trivial subjects while in the background, a wall of water moved inexorably toward them. On a few occasions, the camera person simply dropped the vid device and ran, leaving the announcer cursing after them. The cameras always showed water overtaking and destroying the people.

"It's the megatsunami," Marceau replied in a ragged voice bereft of hope. The loss of Walter was only the beginning of death on a scale none could have imagined except Satan himself. The world was being cleansed. It would never be the same. Malik's first assessment might have been right.

*

"What is going on?" said Burt in disbelief.

They had all been watching Gail's speech. She had been speaking to the people of Earth, trying to calm their fears about Red and Walter supposedly trying to attack Earth.

"That is the scariest thing I have ever seen," muttered Mick.

Mark was trembling. He cursed, "Shit! You guys had it pegged. We do need to be prepared, just not for what you thought."

The old man, Lieutenant Colonel Fuentes, looked at each man in turn with cold eyes and stated, "It has begun. Now is the time to find our few friends and get to the mine and pick up some supplies."

Burt looked uncertain. "Are you sure? We might be mistiming it."

"Boy, you saw what was happening. The station went out. The entire Atlantic will be in trouble. This is the time. I have no doubts. Imagine the entire eastern seaboard without food and water, not to mention medical supplies. The survivors will begin to be herded to Middle America and the west. They will move everyone that has been displaced to the west side of the Appalachians. Martial law will be declared by the UE throughout the entire planet. They will use this to tighten their grip, if they can. They are coming."

"Make the rounds then," Burt said. "The code is *bird*. We have four hours from now to gather all we can take from our private storage facilities. We will meet at the forest service road. Once we deposit our material at the site, we will regroup and go to the mine."

Mick was already rolling kegs of his illegal beer out of the cooler. The colonel gave him a wry smile and said, "Let me give you a hand with that, boy. I can't stand the thought of not having a cold brew anymore. Until Burt invited me here, I had not had any for a while. Besides, it may prove very useful as a trade object."

Mick offered a fake look of anger. "Trade my beer, would you? You wouldn't do that. You like it too much."

"Less talk and more walk," said Burt as he left by the back door.

*

Chapter XXIX:

Home

29

011101

Red shuffled through a multitude of reports on his touch screen. The global disaster continued. Tectonic shifts ignited the Ring of Fire in the Pacific. Particulate matter ejected high in the atmosphere hampered farming and increased pulmonary difficulties. Temperature variations from the norm were not only large but erratic.

Several months had led to nothing more than increasing strife. Disease ate away at the remaining populations in the hardest hit areas. Thugs, robbers, scavengers, and street gangs held sway in the remnants of cities. Those who had survived the waters had either fled or became one of the above. Survival demanded it.

Political unity dissolved. The Asian Region now claimed ascendancy. None of the others were in a position to really contest that, even the new Euro-Russo Pact. The Asian Region's leader had declared himself to be the Eastern Emperor. It was the first step toward Terran Emperor. It would happen in his lifetime; he had promised the world.

The Americas were to be left alone for now, on their own. The NAR and SAR would just be a hindrance to the economic recovery of the surviving government.

The Underworld projects secured themselves against outlaw invaders. They hoped to ride out the storm deep in their lairs or out in the wide oceans.

Religions had crumbled; many believed they had missed the expected Rapture. New sects had sprung up almost instantly, and they blamed the Lunites. Savage battles had occurred over the new dogmas.

Red shook his head. Had anything been accomplished? Barbarism had a foothold on half of Earth. Granted, they were now free from earthly constraints, but at what cost? He still needed to know. He made the connection.

"Adam, what part did you have in this?"

"The worst."

"What exactly did you do?"

"I added impetus to the volcano so that the event would be large enough."

"Large enough for what?"

"To give the hundred years you asked for. The time required for a real, new, established outpost of bios to start over."

"I asked for?"

"Yes. You noted to all the other conspirators that that figure would be necessary. I concurred. I provided."

Red shook with horror. "I caused this?"

"No. I used a natural event that was going to occur anyway and increased the effects to ensure the required time. The goal has been to enable the survival of as many bios of all forms as possible—*sapiens*, *anthrocreatus*, the newer *cyborgius*, as well as the newest form of life, *artificia intelligencia*. I succeeded."

"Fuck Beelzebub! I—"

"The greatest number of species have survived. You bear no blame. That is why I did not warn you of my intent."

"Walter died damn it!"

"That was necessary. There were four courses. Neither of you die-all die. Both die-all die. You die, 87 percent chance that all non-*sapien* life dies. Walter dies, a 52 percent chance of survival for the greatest number of all species combined with the fewest deaths of *sapiens*. This is based on your methods and extrapolated by my quantum viewing. Your topological multidimensional historical analyses are generally correct."

"Fuck!" Red shut off the connection.

*

Red and Marceau were finally alone. This place had been designed by both Red and Walter. It was full of curves and sweeping slopes. The grand masters of architecture would have been very proud.

The central chamber, the equivalent of a living room, was thirty meters in long diameter by twenty short, an ellipse or egg shape. In the center, the floor lowered in oval steps that matched the rising of the walls. A stone couch stretched all around, except for four narrow stairs at each compass point. An oval table, exactly matching the ellipse of the room, was formed out off the bedrock itself, like the rest of the facility.

A long tilted, twisted circular stairway, in proportion to the measurements of the room, made its way to the second floor where the four bedrooms were. Two master bedrooms, each on opposing ends of a long hallway, were large, multiroomed, and spacious. One was for Red and one for Walter. One was no longer needed. The other two were for guests and were at the top of the stairs near the center.

Red was peering out the large two-meter-by-five-meter port that he and Walter had installed not too long ago. He felt pain, still, every time he thought of his friend. He knew he always would. Not far away, there was a crater, a very unnatural one, with Walter's name on it. It was glassy surfaced and radioactive. The plumes of dust from the explosion spewed outward to encircle Luna in a thin cloud of matter. Years afterward, people still decontaminated themselves after an outdoor excursion. Special pains were taken to remove any serious clumps of radioactivity from materials to be processed.

Marceau shared Red's pain. Walter had been her deep friend as well. There were so many shared memories. Luna, Mars, and Luna again flowed through her memories. She quietly observed Red from behind. The set of his shoulders told her how he felt. She had become very adept at reading him since he had finally opened up to her. Even after his obvious declaration of love through his actions at Mars, he had taken his time really sharing all. She patiently waited, knowing it would come. The last followed Walter's death in a flood, as if he felt he would never get another chance.

They had applied for a child. Their sperm and ova would go through the same process as all other Hsas, or so it was said. It would be a boy, named Walter Black Bear Marceau-O'Hare. Quite a mouthful but it represented quite a person. The techs on SS-2 would do all the work for free. Their respect for Red and Marceau, as well as Walter, knew no bounds.

Marceau moved slowly up behind Red and took his hand. "Mr. President," she said, the love and pain understandable in her soft voice.

He turned and smiled (somewhat morosely since he had neither asked nor wanted to be president) at her. He put one hand gently to her smooth, high-cheeked ebony face and said, "Beautiful wife, Madame Chief Justice, today is the rest of our lives."

"Husband, you are my life," she replied as her long ebony fingers stroked his firm square freckled jaw.

Together, hands clasped, they turned back to the panoramic view before them. It was all out there. All they had to do was to take it for all it was—love.